Differential Equations: Theory and Applications

Second Edition

David Betounes

Differential Equations: Theory and Applications

Second Edition

 Springer

David Betounes
Department of Physics, Astronomy
  and Geosciences
Valdosta State University
1500 N. Patterson Street
Valdosta, GA 31698
USA

Additional material to this book can be downloaded from http://extra.springer.com

ISBN 978-1-4899-8265-0   ISBN 978-1-4419-1163-6 (eBook)
DOI 10.1007/978-1-4419-1163-6
Springer New York Dordrecht Heidelberg London

Mathematics Subject Classification (2000): 34A34, 34A30, 34A12, 34D05, 34D20, 70E15, 70H05, 70H06, 70H15

Springer is part of Springer Science+Business Media (www.springer.com)

Preface

This book was written as a comprehensive introduction to the *theory* of ordinary differential equations with a focus on mechanics and dynamical systems as time-honored and important applications of this theory. Historically, these were the applications that spurred the development of the mathematical theory and in hindsight they are still the best applications for illustrating the concepts, ideas, and impact of the theory.

While the book is intended for traditional graduate students in mathematics, the material is organized so that the book can also be used in a wider setting within today's modern university and society (see "Ways to Use the Book" below). In particular, it is hoped that interdisciplinary programs with courses that combine students in mathematics, physics, engineering, and other sciences can benefit from using this text. Working professionals in any of these fields should be able to profit too by study of this text.

An important, but optional component of the book (based on the instructor's or reader's preferences) is its computer material. The book is one of the few *graduate* differential equations texts that use the computer to enhance the concepts and theory normally taught to first- and second-year graduate students in mathematics. I have made every attempt to blend together the traditional theoretical material on differential equations and the new, exciting techniques afforded by computer algebra systems (CAS), like Maple, Mathematica, or Matlab. The electronic material for mastering and enjoying the computer component of this book is on Springer's website.

Ways to Use the Book

The book is designed for use in a one- or two-semester course (preferably two). The core material, which can be covered in a single course, consists of Chapters 1, 2, 3, 4, and 5 (maybe Chapter 6, depending on what you think is basic). The other chapters consist of applications of the core material to integrable systems (Chapter 7), Newtonian mechanics (Chapters 8), and Hamiltonian systems (Chapter 9). These applications can be covered in a sequel to a first course on the core material, or, depending on the depth with which the core material is treated, parts of these applications (like Chapter 9) can be squeezed into the first course. There is also much additional material in the appendices, ranging from background material on analysis and linear algebra (Appendices A and C) to additional theory on Lipschitz maps and a proof of the Hartman-Grobman Linearization Theorem (Appendix B). The electronic material serves as an extensive supplement the text.

The material is structured so that the book can be used in a number of different ways based on the instructor's preferences, the type of course, and the types of students. Basically the book can be used in one of three ways:

- theoretical emphasis

- applied emphasis

- combination of theoretical and applied emphasis

Here the designation *applied* means, for the students, not being required to prove theorems in the text or results in the exercises. Besides using the first emphasis, I have also had success using the third emphasis in classes that have had math and physics students. For such classes, I would require that students understand major theorems and definitions, be able to prove some of the easier results, and work most of the non-theoretical exercises (especially the ones requiring a computer).

Guide to the Chapters

Chapters 1 and 2 are intended to develop the students' backgrounds and give them plenty of examples and experiences with particular systems of differential equations (DEs) before they begin studying the theory in Chapter 3. I have found this works well, because it gives students concrete exercises to study and work on while I am covering the existence and uniqueness results in Chapter 3.

Chapter 3 is devoted both to proving existence and uniqueness results for systems of differential equations and to introducing the important concept of the flow generated by the vector field associated with such systems. Additional material on Lipshitz maps and Gronwall's inequality is presented in Appendix B.

Chapter 4 presents the basic theory for linear systems of differential equations, and this material, given its heavy dependence on concepts from linear algebra, can take awhile to cover. Some might argue that this material ought to be in Chapter 1, because the theory for linear systems is the simplest and most detailed. However, I have found, over the years developing this book, that starting with this material can put too much emphasis on the linear theory and can tend to consume half of the semester in doing so.

Chapter 5 describes the linearization technique for analyzing the behavior of a nonlinear system in a neighborhood of a hyperbolic fixed point. The proof of the validity of the technique (the Hartman-Grobman Theorem) is contained in Appendix B. Another, perhaps more important, purpose of the chapter is the introduction of the concept of transforming vector fields, and thus of transforming systems of differential equations. Indeed, this concept is the basis for classifying equivalent systems—topologically, diferentiably, linearly equivalent—and helps clarify the basis of the Linearization Theorem.

Chapter 6 covers a number of results connected with the stability of systems of differential equations. The standard result for the stability of fixed points for linear systems in terms of the eigenvalues of their coefficient matrices leads, via the Linearization Theorem, to the corresponding result for nonlinear systems. The basic theorem on Liapunov stability, determined by a Liapunov function, is discussed and shown to be most useful in the chapters on mechanics and Hamiltonian systems. A brief introduction to the stability of periodic solutions, characteristic multipliers, and the Poincaré map is also provided as an illustration of the analogies and differences between the stability techniques for fixed points and closed integral curves (cycles).

Chapter 7 is a brief introduction to the topic of integrable systems, which is a special case of the more general theory for integrable systems of partial differential equations (in particular, Pffafian systems). The ideas are very simple, geometrically oriented, and are particularly suited to study with computer graphics.

Chapter 8 begins the application of the theory to the topic of Newtonian mechanics and, together with Chapter 9, can serve as a short course on mechanics and dynamical systems. A large part of Chapter 8 deals with rigid-body motion, which serves to illustrate a number of the concepts studied for linear systems and integrable systems.

Chapter 9 comprises the elementary theory of Hamiltonian systems and includes proofs of Arnold's Theorem, the Transport Theorem, Liouville's Theorem, and the Poincaré Recurrence Theorem. This chapter is independent of Chapter 8, but certainly can serve as an important complement to that chapter. Because of the independence there is a certain amount of repetition of ideas (conservation laws, first integrals, sketching integral curves for 1-D systems). However, if your students studied the prior chapters, this can help reinforce their learning process.

Additional Features

There are several features of the book that were specifically included to enhance the study and comprehension of the theory, ideas, and concepts. They include the following:

- **Proofs:** Complete proofs for almost every major result are provided either in the text or in the appendices (with the exception of the Inverse Function Theorem, Taylor's Theorem, and the change of variables formula). Minor results are often proved or the proofs are assigned as exercises. Even if the book is used in an applied way, without an emphasis on proofs, students may at some later point in their careers become more interested in, and in fact need, the additional understanding of the theory that the proofs provide.

- **Blackboard Drawings:** An extensive number of figures is provided to either illustrate and enhance the concepts or to exhibit the often complex nature of the solutions of differential equations. Most of these have been done with Corel Draw and Maple. However, the text has a number of hand-drawn figures, which are reproduced so as to appear as blackboard sketches. These are meant to convey the belief that many aspects of visualization are still easiest and best done by hand.

- **Electronic Component:** The electronic material on Springer's website is provided to complement and supplement the material in the text. It is a major resource for the book. Many of the pertinent examples in the text that use Maple are on the website, in the form of Maple worksheets, along with extensions and additional commentary on these examples. These can be beneficial to students in working related computer exercises and can greatly reduce the amount of time spent on these exercises.

  An important part of the electronic component of the book is the supplementary material it contains on *discrete dynamical systems*, or the theory of iterated maps, which has many analogies, similarities, and direct relations to systems of differential equations. However, to eliminate confusion, to add flexibility of use, and to save space, all the theory, applications, and examples of this subject have been relegated to the electronic component of the book. This can serve as material for a short course in itself.

The electronic component also contains many worksheets that are tutorial in nature (like how to plot phase portraits in Maple). There is also some special-purpose Maple code for performing tasks such as (1) plotting the curves of intersection of a family of surfaces with a given surface, (2) plotting integral curves dynamically as they are traced out in time (an extension of Maple's **DEplot** command), (3) implementating the Euler numerical scheme for the planar N-body problem, (4) animating rigid-body motions, (5) animating the motion of a body constrained to a given curve or surface, and (6) animating discrete dynamical systems in dimensions 1 and 2.

The electronic material is organized by chapters, corresponding to the chapters in the text. You can access all the Maple worksheets constituting a given chapter by opening the table of contents worksheet, **cdtoc.mws**, and using the hyperlinks there. Appendix D has the table of contents for the electronic material.

Preface to the 2nd Edition

In this the 2nd Edition of the book, all the chapters have been revised and enhanced, and in particular, extensive additional examples, exercises, and commentary have been added to Chapter 4 (Linear Systems), Chapter 7 (Stability Theory), and Chapter 9 (Hamiltonian Systems). The electronic material (obtained from Springer's website) has been revised and extended too, with all files now compatible with any version of Maple from Maple 5 to Maple 12.

Contents

Chapter 1

Introduction

This book is about the study of systems of differential equations (DEs), or more precisely, systems of 1st-order ordinary differential equations. As we shall see, any higher-order system of differential equations can be reduced to a 1st-order system and thus the study of first-order systems suffices for the general theory. Many systems of differential equations model the motion of something, and for this reason systems of DEs are often referred to as dynamical systems.

The purpose of this introduction is not only to present a number of examples which serve to illustrate the main concepts and ideas involved in the study of dynamical systems, but also to give some of the initial mathematical definitions for these concepts. Do not be concerned with the complexity of some of these examples or the physics behind them. For now, just read and get an overview of what dynamical systems are (if you do not already have experience with them from elsewhere).

1.1 Examples of Dynamical Systems

Most of the examples presented here arise in either fluid mechanics or Newtonian mechanics, two areas that provide a rich supply of dynamical systems. The examples are further discussed and elaborated in the accompanying electronic component.

Example 1.1 (Circular Flow) Our first example, although very simple, is important and connected with many topics. The system is

$$x' = y \tag{1.1}$$
$$y' = -x. \tag{1.2}$$

D. Betounes, *Differential Equations: Theory and Applications*, DOI 10.1007/978-1-4419-1163-6_1,
© Springer Science + Business Media, LLC 2010

In this system there are two unknown functions $x = x(t)$ and $y = y(t)$, each being a function of t (the time), and the prime $'$ stands for the time derivative: $x' = dx/dt$, $y' = dy/dt$. A solution of the system (1.1)-(1.2) consists of a pair of functions x, y which satisfy the equations simultaneously. Simply put: the derivative of x is y and the derivative of y is $-x$. One well-known pair of such functions is $x = \sin t$, $y = \cos t$, which is easily verified:

$$(\sin t)' = \cos t \qquad \text{and} \qquad (\cos t)' = -\sin t.$$

Any such solution can be thought of as a (parametrized) *curve*

$$\alpha(t) = (x(t), y(t)) = (\sin t, \cos t)$$

in the x-y plane (i.e., in the 2-dimensional space $\mathbb{R}^2$). The electronic component provides some review material on curves in the plane (as α is here), curves in space, i.e., in $\mathbb{R}^3$, and more generally curves in $\mathbb{R}^n$. In this case α parametrizes the unit circle $x^2 + y^2 = 1$.

There are many other solutions of the system and, in fact, we will see later how to get the "general solution" of this system. It is

$$\begin{aligned} x(t) &= x_0 \cos t + y_0 \sin t \\ y(t) &= -x_0 \sin t + y_0 \cos t, \end{aligned}$$

or in terms of a curve, the general solution is

$$\alpha(t) = (x_0 \cos t + y_0 \sin t, -x_0 \sin t + y_0 \cos t),$$

where x_0, y_0 are constants. It is easy to check that (x, y) is a solution of the system. Also note that in this general solution, $\alpha(0) = (x_0, y_0)$ and so x_0, y_0 are the coordinates of the point that α passes through at time $t = 0$. An easy computation also shows that $x(t)^2 + y(t)^2 = x_0^2 + y_0^2$, for all t, and so α lies on the circle $x^2 + y^2 = r_0^2$, where $r_0^2 = x_0^2 + y_0^2$. Thus, all solutions of this system parametrize circles of various radii and centers at the origin. See Figure 1.1.

Example 1.2 (A Row of Stagnation Points) This example comes from fluid mechanics and, as in the last example, consists of a system of two equations

$$\begin{aligned} x' &= \sinh y &&\text{(1.3)} \\ y' &= -\sin x, &&\text{(1.4)} \end{aligned}$$

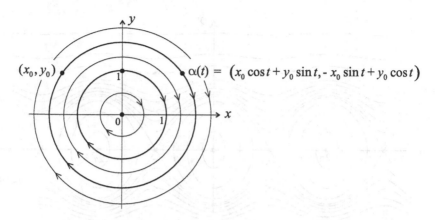

$$\alpha(t) = \left(x_0 \cos t + y_0 \sin t, - x_0 \sin t + y_0 \cos t\right)$$

Figure 1.1: *Solution curves of the system $x' = y$, $y' = -x$ which represents circular flow in the plane.*

with two unknown functions $x = x(t)$, $y = y(t)$, and any solution x, y of the system (1.3)-(1.4) can be thought of as a curve: $\alpha(t) = (x(t), y(t))$ in the $\mathbb{R}^2$.

What's different here, however, is that we do not have any analytical expressions for particular solutions x, y of this system and certainly cannot write down the general solution. Nevertheless, we can still study this system by computer and other means. For instance, the system in the first example (which is known as a linear system) is similar in some respects to this system (which is a nonlinear system). More specifically, since $\sinh y \approx y$ for y near zero and $-\sin x \approx -x$ for x near zero, we might expect that the nonlinear system here have solutions near $(0,0)$ which are similar to the circular solutions in the linear system. We will see that this is indeed the case.

Knowing something about what the system represents can also be helpful in studying it. The particular system of DEs in this example provides a model of a planar fluid flow, with any particular solution α representing a *streamline* in the flow, i.e., the path a particular particle of the fluid would take as it is carried along by the overall motion of the fluid. Thus, $\alpha'(t) = (x'(t), y'(t))$ represents the velocity of the fluid particle at time t and the system (1.3)-(1.4) is interpreted as saying that the velocity at any point and any time should match the given expressions on the right hand sides of these equations. Figure 1.2 shows the plots of 38 different streamlines (solutions α of the system).

It is important to note the system (1.3)-(1.4) has some solutions of a very special and simple type. Such a solution is one where the position of

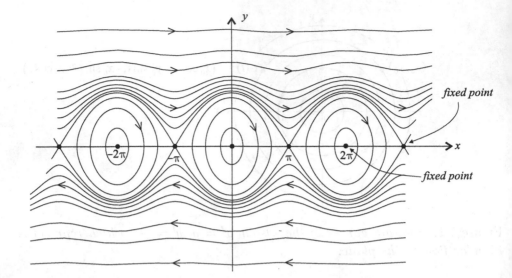

Figure 1.2: *Solution curves of the system* $x' = \sinh y$, $y' = -\sin x$ *which represents a fluid flow past a row of stagnation points.*

the fluid particle is constant in time and therefore the solution "curve" is actually just a point. This is known as a *fixed point* for a dynamical system in general and, for this example, is known as a *stagnation point*. Here the system (1.3)-(1.4) has infinitely many fixed points (seven of which are shown in Figure 1.2). These points occur where the velocity is zero, i.e., at points (x, y) which satisfy the algebraic system of equations

$$\sinh y = 0 \quad \text{and} \quad \sin x = 0.$$

This system is easily solved to give $\{(k\pi, 0) | k \in \mathbb{Z}\}$ as the set of fixed points (stagnation points).

The interpretation of Figure 1.2 is that the fluid flows from left-to-right above the stagnation points and from right-to-left below them. The further the streamline is from the x-axis the greater the speed of flow. This comes from the calculation of the speed

$$v \equiv (x'(t)^2 + y'(t)^2)^{1/2} = (\sinh^2 y + \sin^2 x)^{1/2}.$$

Now $\sin x$ is bounded, taking values between -1 and 1, while $\sinh y$ approaches $\pm\infty$ as $y \to \pm\infty$. Thus, the speed $v \to \infty$ as $y \to \pm\infty$.

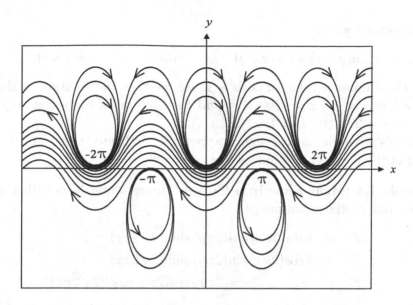

Figure 1.3: *Solution curves of the system (1.5)-(1.6) which represents a fluid flow past two rows of vortices with opposite circulations.*

Example 1.3 (Karmen Vortex Sheet) A more complicated system than the last one, but one with solution curves which are more interesting is

$$x' = \frac{\cos x \sinh y - \sinh c}{M(x,y)} \tag{1.5}$$

$$y' = \frac{-\sin x \cosh y}{M(x,y)}. \tag{1.6}$$

Here $c = 0.8828$ and the function in the denominators is

$$M(x,y) = \frac{[\cosh(y+c) + \cos x][\cosh(y-c) - \cos x]}{(2\cosh c)} \tag{1.7}$$

The solution curves of this system again represent the streamlines in a planar fluid flow that contains two rows of vortices (infinitely many in each row) with opposite circulation in each row. Figure 1.3 shows 24 of these curves.

A vortex is the complete opposite of a stagnation point in a fluid flow. The circulation of the flow around a vortex becomes faster and faster as the center is approached and at the vortex center the speed of the flow is infinite. In this example, according to equations (1.5)-(1.6), the velocity $\alpha'(t) = (x'(t), y'(t))$ becomes unbounded where the denominator $M(x,y) =$

0. This occurs where

$$\cosh(y + c) + \cos x = 0 \quad \text{or} \quad \cosh(y - c) - \cos x = 0.$$

Since the minimum value of cosh is 1, we see that the solution of the first equation is $y = -c$, $x = (2k + 1)\pi$ and the solution of the second equation is $y = c$, $x = 2k\pi$ (for $k = 0, \pm 1, \pm 2, \ldots$).

In this example, as the figure appears to indicate, there are no fixed points of the flow. This is indeed the case.

Example 1.4 (Heat Flow in a Cube) This example deals with a system of three differential equations,

$$\begin{align}
x' &= b\sin(2\pi x)\cos(2\pi y)\sinh(2\sqrt{2}\pi z) \\
y' &= b\cos(2\pi x)\sin(2\pi y)\sinh(2\sqrt{2}\pi z) \qquad\qquad (1.8) \\
z' &= -1 - \sqrt{2}b\cos(2\pi x)\cos(2\pi y)\cosh(2\sqrt{2}\pi z)
\end{align}$$

where $b = 2\pi/\sinh(2\sqrt{2}\pi)$. This system describes the flow of heat in a unit cube $U = [0, 1] \times [0, 1] \times [0, 1]$, and for this reason we restrict attention to solution curves $\alpha(t) = (x(t), y(t), z(t))$ which lie in U, even though the system (1.8) is defined on all of $\mathbb{R}^3$. Visualizing a curve in three dimensional space can be difficult, especially if the curve is complicated, and visualizing a number of them simultaneously can be even more confusing. Thus, it helps to know something about what to expect.

Without going into all the details of the heat problem here, we just mention that the system arises from a certain distribution S of temperatures in the cube. Thus, $S(x, y, z)$ gives the temperature at the point (x, y, z). The heat flux vector field is $-\nabla S$, and $-\nabla S(x, y, z)$ is the direction of greatest decrease in S at the point (x, y, z). The theory is that the heat flows in this direction at each point and thus the heat flow lines are curves $\alpha(t) = (x(t), y(t), z(t))$ that satisfy $\alpha'(t) = -\nabla S(\alpha(t))$. In terms of components this is the system of DEs:

$$\begin{align}
x' &= -S_x(x, y, z) \qquad\qquad\qquad (1.9) \\
y' &= -S_y(x, y, z) \qquad\qquad\qquad (1.10) \\
z' &= -S_z(x, y, z). \qquad\qquad\qquad (1.11)
\end{align}$$

This system of DEs results in the particular system (1.8) when the temperature function is

$$S(x, y, z) = z + \Big(\cos(2\pi x)\cos(2\pi y)\sinh(2\sqrt{2}\pi z) \Big) / \sinh(2\sqrt{2}\pi).$$

The function S is the solution of the partial differential equation $S_{xx} + S_{yy} + S_{zz} = 0$ which also satisfies certain conditions on the boundary of the cube. In this example, these conditions are the following: the vertical sides are insulated, the bottom is held at 0 degrees, and the top is held at a temperature that is $S(x, y, 1) = 1 + \cos(2\pi x)\cos(2\pi y)$ degrees at the point $(1, x, y)$ on the top. Note that the distribution of temperatures across the top of the cube varies from 0 to 2 degrees (in a rather complicated way) and the distribution on the bottom is a constant 0 degrees. Using the principle that *heat flows from hot to cold*, we expect that some of the heat flow lines that start on the top of the cube will end up on the bottom of the cube. By judiciously choosing some starting points on the top and plotting the solution curves of the system (1.8) that start at these points, we get the picture shown in Figure 1.4. Note that the system in this example has no

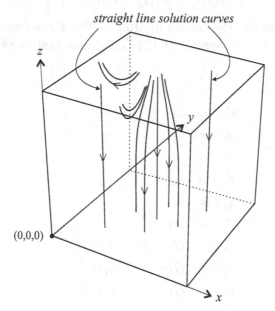

Figure 1.4: *Heat flow lines in a unit cube which arise as solutions of the system* (1.8).

fixed points and that it has three solutions which are straight lines (exercise).

Example 1.5 (The Two-Body Problem) One of the oldest examples of a dynamical system is the *two-body system*: two bodies with masses m_1, m_2, attract each other mutually with the force of attraction along the line joining the bodies and of magnitude reciprocally as the square of the distance. In terms of Newton's second law of dynamics (mass $\times$ acceleration = forces), the model for this system is

$$m_1\mathbf{r}_1'' = Gm_1m_2(\mathbf{r}_2 - \mathbf{r}_1)/r_{12}^3 \tag{1.12}$$
$$m_2\mathbf{r}_2'' = Gm_1m_2(\mathbf{r}_1 - \mathbf{r}_2)/r_{12}^3 \tag{1.13}$$

This is actually a second-order system of DEs written in vector form. Here

$$\mathbf{r}_1(t) = (x_1(t), y_1(t), z_1(t))$$
$$\mathbf{r}_2(t) = (x_2(t), y_2(t), z_2(t)),$$

are the position vectors for the two bodies, and for convenience we've used the notation

$$r_{12}(t) \equiv |\mathbf{r}_1(t) - \mathbf{r}_2(t)|$$

for the distance between the bodies at time t. For the sake of comparison, we can write the two-body system (1.12)-(1.13) as a system of 12 first-order DEs:

$$
\begin{aligned}
x_1' &= u_1 \\
y_1' &= v_1 \\
z_1' &= w_1 \\
x_2' &= u_2 \\
y_2' &= v_2 \\
z_2' &= w_2 \\
u_1' &= Gm_2(x_2 - x_1)/r_{12}^3 \\
v_1' &= Gm_2(y_2 - y_1)/r_{12}^3 \\
w_1' &= Gm_2(z_2 - z_1)/r_{12}^3 \\
u_2' &= Gm_1(x_1 - x_2)/r_{12}^3 \\
v_2' &= Gm_1(y_1 - y_2)/r_{12}^3 \\
w_2' &= Gm_1(z_1 - z_2)/r_{12}^3
\end{aligned}
\tag{1.14}
$$

In addition to the six unknown functions $x_i, y_i, z_i, i = 1, 2$, the above system involves the six functions $u_i, v_i, w_i, i = 1, 2$, which you recognize (via the first six DEs in the system) as the components of the velocity vectors for the two bodies:

$$\mathbf{v}_i \equiv (u_i, v_i, w_i) = \mathbf{r}_i'.$$

This system (1.14) of 1st-order scalar DEs is not as convenient as the 2nd-order vector form (1.12)-(1.13), but exhibits a general technique for reducing higher order systems to 1st-order systems (introduce extra unknown functions!). A solution of the first-order system (1.14) is a curve in the 12-dimensional space: $\mathbb{R}^{12} \cong \mathbb{R}^6 \times \mathbb{R}^6$, of positions and velocities. Thus, α has the form

$$\alpha(t) = \Big(\mathbf{r}_1(t), \mathbf{r}_2(t), \mathbf{v}_1(t), \mathbf{v}_2(t) \Big),$$

for t in some interval I. More precisely, since r_{12} cannot be zero on the right side of the two-body system (1.14), each solution curve lies in the 12-dimensional submanifold $\mathcal{O}$ of $\mathbb{R}^{12}$ defined by:

$$\mathcal{O} \equiv U \times \mathbb{R}^6,$$

where $U \equiv \{\, (\mathbf{r}_1, \mathbf{r}_2) \,|\, \mathbf{r}_1 \neq \mathbf{r}_2 \,\}$. This space $\mathcal{O}$ is known as the *phase space* for the two-body dynamical system. While we cannot visualize (except possibly mentally) the graph of solution curve in this 12-dimensional phase space $\mathcal{O}$, it is nevertheless a useful theoretical notion. By using projections, or in this special case, center of mass coordinates, we can get around this visualization limitation and plot the orbits of the two bodies relative to the center of mass. We will look at the details for this later.

By analogy with the two-body system, the space $\mathcal{O} = \mathbb{R}^2 \setminus C$ for the row of vortices example and the space $\mathcal{O} = U$ for the heat flow in a cube example are also called the phase spaces for those dynamical systems, even though historically the term phase space referred to spaces of positions and velocities (or positions and momenta).

Example 1.6 (Pendulum/Ball in a Hoop) An example from mechanics that is considerably simpler than the two-body problem (and which you might have studied as an undergraduate) is the motion of a pendulum. A weight is suspended from a point with a string and the motion of the weight, as it swings back and forth, is modeled by the second-order DE:

$$\theta'' = -k \sin(\theta),$$

where $\theta = \theta(t)$ is the angle, at time t, that the string makes with the vertical. See Figure 1.5. As in the two body problem, if we introduce the velocity $v = \theta'$ (which is actually the angular velocity), then it's easy to see that the above 2nd-order DE can be written as the following 1st-order system

$$\theta' = v \tag{1.15}$$
$$v' = -k\sin(\theta). \tag{1.16}$$

The phase space here is the whole plane $\mathcal{O} = \mathbb{R}^2$, and while this makes sense mathematically, several comments are necessary for the physical interpretation. First, the angle $\theta \in \mathbb{R}$ describing the position of the weight must be interpreted in terms of its related angle in the interval $[0, 2\pi]$. Thus, for example, $\theta = 0, 2\pi, 4\pi, \ldots$ all refer to the position where the weight hangs vertically (straight down). Second, the physical apparatus of a weight suspended by a string does not correspond to the full phase space. For example, if the weight is displaced to position $\theta = 3\pi/4$ and released from rest, it will temporarily fall straight down before the slack in the string is taken up. Then the above DE fails to model the actual motion. Thus, it is best to revise the physical apparatus to one consisting of a hollow, circular tube, in a vertical plane, with a ball rolling around inside, i.e., a ball in a hoop (see Figure 1.5). The ball in the hoop (or pendulum) is one example of what is known as *constrained motion*.

By plotting a sufficient number of solution curves, $\alpha(t) = (\theta(t), v(t))$, of the system (1.15)-(1.16), we obtain an overall picture, or portrait, of the behavior of the ball in the hoop under various initial conditions. This is known as the *phase portrait* for the system. The phase portrait for this example is shown in Figure 1.6. As you can see, this phase portrait appears similar to the one for the row of vortices in Figure 1.2, but has important differences. Here the closed curves (orbits) correspond to motions of the ball

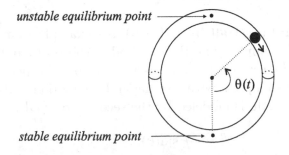

Figure 1.5: *Motion of a ball enclosed in a hollow, circular tube (a hoop). Under the force of gravity only, the hoop constrains the motion to being along the prescribed circular path.*

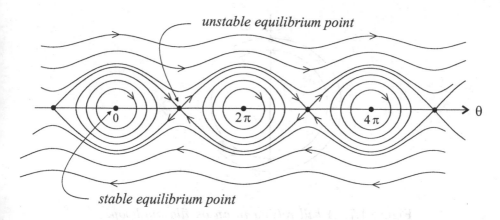

Figure 1.6: *Phase portrait for the system* (1.15)-(1.16), *which models the constrained motion of a ball in a circular hoop (pendulum model)*

where it oscillates periodically, back and forth, about the low point in the hoop, which is the stable, equilibrium point. This physical point corresponds to a set $C = \{(2k\pi, 0)|k \in \mathbb{Z}\}$ in the phase space and is a set of fixed points for the system. Each point in C corresponds to an initial condition where the ball is placed at the low point in the hoop and released with no angular velocity. Thus, obviously the ball remains stationary (fixed). The points in C are also the centers for the various sets of closed orbits. By contrast, in the example for the row of vortices, the centers for the closed orbits are not fixed points and indeed the system is not even defined at these points. The set $\{((2k+1)\pi, 0)|k \in \mathbb{Z}\}$ corresponds to the high point on the hoop, which is an unstable, fixed point, since small displacements of the ball from this point result in motions that take it far away from the summit. The wavy, non closed, solution curves (shown above and below the orbital solution curves) correspond to motions where the ball has enough initial energy (potential and kinetic) to cycle perpetually around the hoop. The solution curves that divide, or separate, the orbital and wavy solution curves are known as *separatrices* for the system.

Example 1.7 (Perturbed Pendulum/Ball in a Hoop) One more example, whose many interesting properties we will explore later, is the model obtained by perturbing the motion of the ball in the hoop from the last example. One way to do this is to have the hoop oscillate periodically about its vertical axis as shown in Figure 1.7. Suppose the hoop has radius 1 and is initially in the vertical x-z plane and oscillates between angular dis-

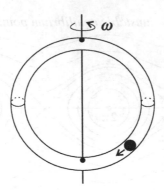

Figure 1.7: *A ball rolling in an oscillating hoop.*

placements of $\omega = \pm a$, with frequency b. The position of the ball on the
hoop is again determined by the angular position function $\theta = \theta(t)$, with
$\theta = 0$ corresponding to the position $(0, 0, -1)$ on the axis of oscillation. The
differential equation for θ is

$$\theta'' = -g \sin(\theta) + \frac{a^2 b^2}{2} \sin^2(bt) \sin(2\theta). \tag{1.17}$$

Here g is the acceleration of gravity near the earth's surface. If we again
introduce $v = \theta'$ for the angular velocity, then the second-order DE (1.17),
can be written as the following 1st-order system for unknowns θ, v

$$\theta' = v \tag{1.18}$$

$$v' = -g \sin(\theta) + \frac{a^2 b^2}{2} \sin^2(bt) \sin(2\theta). \tag{1.19}$$

One aspect of this system that is different from the systems considered
above is that the right side of the system (1.18)-(1.19) explicitly involves
the time t. For this reason this system is called *nonautonomous* or *time-
dependent*. The five previous systems we have considered are *autonomous*
or *time-independent* systems. Figure 1.8 shows the plot of just *one* integral
curve $\alpha(t) = (\theta(t), v(t))$ of this system in the θ-v plane from times $t = 0$
to $t = 40.5$. The parameter values $g = 1, a = 1, b = 2$ were used when
producing this figure. Thus, the hoop oscillates with period π between angular displacements $\omega = \pm 1$ (radians) from the x-z plane. The integral
curve shown in Figure 1.8 describes how the angular position and velocity
change over time for a ball initially placed at the low point in the hoop and

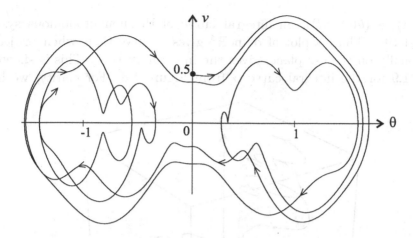

Figure 1.8: *Plot of a single integral curve for the perturbed ball on the hoop model with initial condition* $(\theta(0), \theta'(0)) = (0, 0.5)$.

given and initial angular velocity of 0.5. Note that the curve crosses itself several times (this happens for nonautonomous systems, but cannot occur for autonomous ones) and does not appear to be periodic (so the ball never returns to the same position with exactly the same velocity).

The erratic behavior of the ball in the hoop is more easily analyzed by converting the nonautonomous system to an autonomous one. This requires introducing an additional dimension and is a technique that works in general. The conversion amounts to introducing an additional function, $\tau(t) = t$, to hide the time dependence and also adding an equation for τ to the system. In this example the resulting autonomous system is

$$
\begin{aligned}
\tau' &= 1 \\
\theta' &= v \\
v' &= -g\sin(\theta) + \frac{a^2 b^2}{2}\sin^2(b\tau)\sin(2\theta).
\end{aligned}
\tag{1.20}
$$

Of course, this additional equation just says that $\tau(t) = t + c$ for some constant c. So in a certain sense (which can be made precise) the autonomous system (1.20) is equivalent to the nonautonomous system (1.18)-(1.19) (exercise). If the initial time is $c = 0$, then the integral curves of the autonomous system (1.20) have the form

$$
\tilde{\alpha}(t) = (t, \theta(t), v(t)),
$$

with $\alpha(t) \equiv (\theta(t), v(t))$, an integral curve of the nonautonomous system
(1.18)-(1.19). Thus, a plot of $\tilde{\alpha}$ in $\mathbb{R}^3$ gives a curve that, when projected
orthogonally on the θ-v plane, gives the integral curve α. This is shown in
Figure 1.9 for the integral curve α from Figure 1.8. For clarity, we have

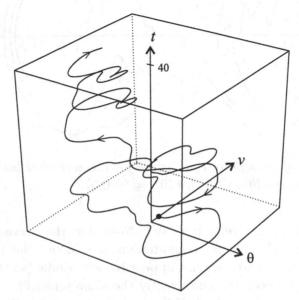

Figure 1.9: *Plot of the integral curve* $\tilde{\alpha}(t) = (t, \theta(t), v(t))$ *with initial condi-*
tions $t = 0, \theta(0) = 0,$ *and* $\theta'(0) = 0.5$ *for the perturbed ball in the hoop.*

rotated the view in Figure 1.9 so that the time axis (axis for τ) is vertical.
This makes it easier to visualize how the curve projects onto the curve shown
in Figure 1.8.

Having examined some concrete examples of dynamical systems, we need
to introduce some precise terminology and definitions in an abstract setting
for the notions and ideas contained in these examples (systems of DEs, solution curves, phase space, etc.). This is done in the next section.

Exercises 1.1

1. If you need some review on curves in the plane and curves in space, read
   the Maple worksheets `plcurves.mws` and `spcurves.mws` on the electronic
   component and work the suggested exercises there.

2. Learn how to plot phase portraits for systems in the plane using some computer software package. If you wish to use Maple for this, study the material

on the electronic component in the Maple worksheet: `deguide1a.mws`. Apply what you learn to do the following problems. *Note*: If you are using Maple, you might want to cut and paste some of the code from `deguide1a.mws` directly into your worksheet, which you create for solving the problems.

(a) Plot the phase portrait for the system in Example 1.3 in the text (the Karmen vortex sheet). Make your figure look like, or better than, Figure 1.3 in the text. Mark the direction of flow on several integral curves.

(b) Plot the phase portrait for the system in Example 1.6 in the text (the ball in the hoop). Make your figure look like, or better than, Figure 1.6 in the text. Mark the direction of flow on several integral curves. Describe the motion of the ball corresponding to each type of integral curve in the phase portrait. There are four types in this example. *Note*: θ is measured in radians.

(c) For the system in Example 1.7 (perturbed pendulum), use a computer to draw the single integral curve shown in Figures 1.8 and 1.9. Describe the motion of the ball which corresponds to this integral curve. *Note*: θ is measured in radians.

3. Consider the unit cube $U = [0, 1] \times [0, 1] \times [0, 1]$ with temperature function

$$S(x, y, z) = z + k \cos(\pi x)\, \cos(2\pi y)\, \sinh(\sqrt{5}\pi z),$$

where $k = 1/\sinh(\sqrt{5}\pi)$. As in Example 1.4, this distribution of temperatures in U corresponds to boundary conditions where the temperature is held at 0 degrees on the bottom of U, varies as $S(x, y, 1)$ on the top, and no heat is allowed to escape through the (insulated) vertical sides of U. Use a computer to draw the heat flow lines that start at appropriately selected points on the top face. *Suggestion*: Among others, use the following groups of points: $G_x = \{(x, .55, 1), (x, .6, 1), (x, .7, 1), (x, .8, 1), (x, .9, 1)\}$ for $x = .4, .5, .6$. Draw numerous pictures, some with projections of the flow lines on the coordinate planes, in order to adequately describe the phase portrait. Choose your pictures judiciously. Do not waste paper by handing in too many ill-conceived pictures.

Determine if there are any flow lines that are straight lines. Show that the heat flow lines that start at points in the plane $y = .5$, remain in this plane.

If you are using Maple, read the worksheet `deguide1b.mws` first and make this assignment easier by using portions of that worksheet in yours.

4. **(First Integrals/Conservation Laws)** In Example 1.1 the solution curves of the system were shown to lie on circles $y^2 + y^2 = r^2$ of various radii r. This was determined by using the formula for the general solution curve. We can, however, determine this by using the system of DEs directly. To see this, take the system

$$x' = y, \qquad y' = -x,$$

and multiply the first equation by y' and the second equation by x' to get

$$x'y' = yy', \qquad x'y' = -xx'.$$

Hence

$$yy' = -xx' \quad \text{or} \quad xx' + yy' = 0.$$

Consequently

$$\frac{d}{dt}(x^2 + y^2) = 0,$$

and so $x^2 + y^2$ is a constant: $x^2 + y^2 = k$, for some constant k. This equation is known as a *conservation law*. The function $F(x, y) = x^2 + y^2$ is called a *first integral* for the system of DEs. The above argument indicates that each solution (x, y) of the system of DEs lies on a level curve $F(x, y) = k$ of the function F. Thus, a plot of the level curves of F gives a picture of the solution curves.

(a) Find a first integral F for the system $x' = \sinh y$, $y' = \sin x$ in Example 1.2. Plot the graph of F and a number of its level curves $F(x, y) = k$.

(b) **(A Row of Vortices)** Consider the following system, which is related to the system in Example 1.2,

$$x' = \frac{\sinh y}{M(x, y)}$$
$$y' = \frac{-\sin x}{M(x, y)},$$

where $M(x, y) = \cosh y - \cos x$. Show that this system has the same first integral as the system in Example 1.2. What can you conclude from this about the solution curves of this system? Show that this system has vortices (places where the speed is infinite) as well as stagnation points. Locate where these occur.

(c) Can you find a first integral for the system in Example 1.3?

1.2 Vector Fields and Dynamical Systems

In this section we define the important concept of a vector field on an open set in $\mathbb{R}^n$, which allows us to think of a 1st-order system of differential equations in a geometric way.

There is a discussion of certain aspects of multivariable calculus and analysis in Appendix A. You should review it if necessary, since some of the concepts and notation from it will be used here and elsewhere throughout the book. However, a brief word about terminology is appropriate here.

It is standard to denote by $\mathbb{R}^n$ the set of all n-tuples,

$$x = (x_1, \ldots, x_n),$$

of real numbers $x_i \in \mathbb{R}$, $i = 1, \ldots, n$. We will view $\mathbb{R}^n$ either as the canonical n-dimensional Euclidean space, whose elements $x \in \mathbb{R}^n$ are points in this space, or alternatively as an n-dimensional vector space, whose elements x are vectors (position vectors). Of course, when $n = 2$ or $n = 3$, the subscripting notation is sometimes not used (as in the previous section) and instead of (x_1, x_2) we often use (x, y) or (θ, v) for a point or vector in $\mathbb{R}^2$. Similarly, (x, y, z) or other variants are often used instead of (x_1, x_2, x_3) for a point or vector in $\mathbb{R}^3$.

Generally we will *not* denote vectors or elements $x \in \mathbb{R}^n$ by bold face, like **x**, or with arrows drawn over them, like $\vec{x}$. It seems easier to teach a class and write on the blackboard without having to embellish vectors with these extra notations. Thus, the distinction between vectors and scalars (i.e., numbers) will have to, and usually can be, understood from the context. The one exception to this rule, as you have noticed in the example of the 2-body problem, is in the places where we discuss mechanics (Chapters 8 and 9). There we will use boldface to denote vectors.

Definition 1.1 (Vector Fields) Suppose $\mathcal{O} \subseteq \mathbb{R}^n$ is an open set. A *vector field* on $\mathcal{O}$ is just a function, or map:

$$X : \mathcal{O} \to \mathbb{R}^n.$$

In component form, the map X is given by:

$$X(x) = \left(X^1(x), \ldots, X^n(x) \right),$$

where $x \in \mathcal{O}$ and $X^i : \mathcal{O} \to \mathbb{R}$, $i = 1, \ldots, n$, are the component functions of X. The terminology of differentiability, $C^1, \ldots, C^\infty$, applies to vector fields (Appendix A), so we speak of C^1 vector fields, $\ldots$, C^∞ vector fields, and so on. Usually in the sequel we will assume some degree of differentiability of our vector fields, and not mention this explicity except where it is necessary to be more precise.

Vector fields have an important geometric interpretation: for a point $x \in \mathcal{O}$, one interprets $X(x)$ as a vector attached to the point x. Namely, instead of plotting the vector $X(x)$ with its initial point at the origin, take

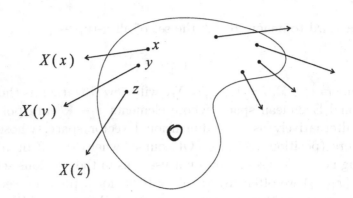

Figure 1.10: *A plot of X at a number of points in O.*

its initial point to be x. See Figure 1.10. Doing plots like this at a number of different points $x, y, z, \ldots$ in $\mathcal{O}$, as shown, gives a geometric picture of a *field of vectors* (in some respects similar to a field of Kansas wheat). This geometric picture also explains the origin of the name *vector field*.

In our study of systems of DEs, the notion of a vector field is more or less synonymous with a system of (autonomous) DEs. Thus, if X is a vector field on $\mathcal{O}$, the corresponding system of DEs is

$$x' = X(x). \tag{1.21}$$

Written out fully in component form the system is

$$\begin{aligned}
x_1' &= X^1(x_1, \ldots, x_n) \\
x_2' &= X^2(x_1, \ldots, x_n) \\
&\vdots \\
x_n' &= X^n(x_1, \ldots, x_n).
\end{aligned}$$

Example 1.8 Suppose $X : \mathbb{R}^2 \to \mathbb{R}^2$ is the vector field given by

$$X(x_1, x_2) = (x_1 x_2, x_1 + x_1^3 x_2^4).$$

The two component functions of X are $X^1(x_1, x_2) = x_1 x_2$ and $X^2(x_1, x_2) = x_1 + x_1^3 x_2^4$. The system of DEs associated with X is

$$\begin{aligned}
x_1' &= x_1 x_2 \\
x_2' &= x_1 + x_1^3 x_2^4.
\end{aligned}$$

The precise description of what is meant by a solution of a system of DEs is contained in the following definition.

Definition 1.2 (Solutions of Autonomous Systems)

(1) A *curve* in $\mathbb{R}^n$ is just a map: $\alpha : I \to \mathbb{R}^n$ from some interval I into $\mathbb{R}^n$. If α is differentiable, then it is called a *differentiable curve*.

(2) If X is a vector field on $\mathcal{O}$, then a *solution* of the system

$$x' = X(x)$$

is a differentiable curve $\alpha : I \to \mathbb{R}^n$ with the properties:

(a) $\alpha(t) \in \mathcal{O}$, for all $t \in I$;

(b) $\alpha'(t) = X(\alpha(t))$, for all $t \in I$.

Such a curve α is also called a *solution curve* of the system, or an *integral curve* of the vector field X (or an integral curve of the system of DEs). Geometrically, property (a) says that the curve α lies in the open set $\mathcal{O}$, which of course is necessary in order for the expression $X(\alpha(t))$ in property (b) to make sense ($X(x)$ is only defined at points x in $\mathcal{O}$). Property (b) is the important requirement on α. It just says that α satisfies the system of DEs. In component form property (b) is:

$$
\begin{aligned}
\alpha_1'(t) &= X^1(\alpha_1(t), \ldots, \alpha_n(t)) \\
\alpha_2'(t) &= X^2(\alpha_1(t), \ldots, \alpha_n(t)) \\
&\ \vdots \\
\alpha_n'(t) &= X^n(\alpha_1(t), \ldots, \alpha_n(t)),
\end{aligned}
$$

for all $t \in I$. From a geometrical point of view, this property: $\alpha'(t) = X(\alpha(t))$, means that α is a curve whose tangent vector to the curve at the point $\alpha(t)$ coincides with vector $X(\alpha(t))$ at the same point. See Figure 1.11.

There are many concrete realizations of this in physics. For example if X is an electrostatic force field on a region $\mathcal{O} \subseteq \mathbb{R}^3$, then the integral curves of X are the force field lines. If $X = -\nabla\phi$ is derived from a potential, then a positively charged particle in the force field will move along a force field line towards regions of lower potential. Another physical interpretation, as we have seen in Examples 1 and 2 above, is where $\mathcal{O}$ is a vessel or tank

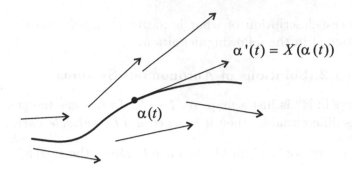

Figure 1.11: *Geometric view of an integral curve α of X*

in which a fluid is circulating with a steady flow, and $X(x)$ represents the velocity of the fluid flowing through the point x at any time. If we select any one particle in the fluid and follow its motion over time, then it will describe a trajectory that is an integral curve of the vector field X. The heat flow in Example 3 above is yet another situation where the vector field and its integral curves have physical meaning. There $\mathcal{O}$ is the unit cube, X is the heat flux vector and the integral curves of X are the heat flow lines, or lines along which the heat must flow in order to maintain the distribution of temperatures.

The idea of viewing solutions of systems of DEs as integral curves of a vector field, with the geometrical interpretation this implies, is of great importance to the theory of differential equations. In fact, an accurate plot of the vector field at a large number of points in $\mathcal{O}$ will almost delineate the picture of what all the integral curves look like.

Example 1.9 The system of DEs

$$
\begin{aligned}
x_1' &= \tfrac{1}{2}(x_1 + x_2) \\
x_2' &= -\tfrac{1}{2}x_2,
\end{aligned}
$$

is associated with the vector field $X : \mathbb{R}^2 \to \mathbb{R}^2$ given by

$$
X(x_1, x_2) = \left(\tfrac{1}{2}(x_1 + x_2), \ -\tfrac{1}{2}x_2 \right).
$$

For this example, it's relatively easy to plot the vector $X(x)$ *by hand* at a number of different points x in $\mathbb{R}^2$. The plot of this field of vectors is shown

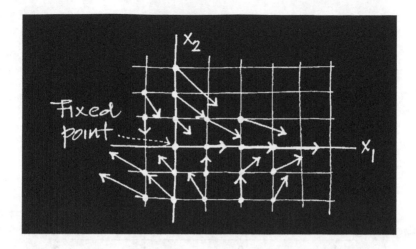

Figure 1.12: *Hand-drawn plot of* $X(x_1, x_2) = (\frac{1}{2}(x_1 + x_2), -\frac{1}{2}x_2)$.

in Figure 1.12 and gives a rough picture of what X looks like. Even as simple as this vector field is, the plot shown in the figure can take 15 minutes or so to construct by hand. For more complicated vector fields, the process of plotting by hand is far too tedious to be practical. Prior to the advent of the computer, and especially the PC, all plots of vector fields were limited to the simplest examples. Now we can take advantage of plotting software to exhibit vector field plots for even the most complicated vector fields. Most software actually plots what is known as the *direction field* for X. This is the plot of $kX(x)/|X(x)|$ at all the points x on a specified grid (Here $|X(x)|$ denotes the length of the vector $X(x)$). All these vectors have the same length, namely k, so the resulting plot only indicates the direction of X at each point in the grid. The scale factor k depends on the software package and on the grid size. For this example the plot of the direction field on the rectangle $[-1, 2] \times [-3, 3]$, divided into a grid of 20×20 points, is shown in Figure 1.13.

Since each integral curve of X traces out a path in $\mathbb{R}^2$ that is tangent to the direction field element at each point, its not hard to roughly discern from the figure what some of the integral curves look like (and that some are straight lines).

The simplicity of this example also allows us to analytically solve the system and explicitly exhibit formulas for the solutions. Thus, solving the

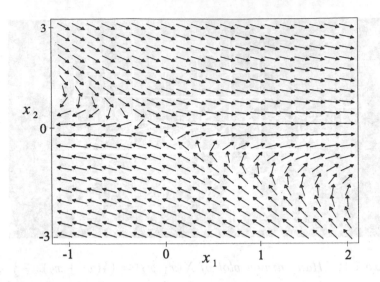

Figure 1.13: *A computer-generated direction field for the vector field* $X(x_1, x_2) = (\frac{1}{2}(x_1 + x_2), -\frac{1}{2}x_2)$.

second equation in the system: $x_2' = -x_2/2$, gives

$$x_2 = b_2 e^{-t/2} \tag{1.22}$$

as the general solution involving an arbitrary constant b_2. Substituting this in the first equation and solving the resulting DE for x_1 gives

$$x_1 = b_1 e^{t/2} - \tfrac{1}{2} b_2 e^{-t/2} \tag{1.23}$$

as the general solution, involving yet another arbitrary constant b_1. The resulting integral curve is then

$$\alpha(t) = \left(b_1 e^{t/2} - \tfrac{1}{2} b_2 e^{-t/2}, \; b_2 e^{-t/2} \right). \tag{1.24}$$

This can be thought of as a formula for the general integral curve since the arbitrary constants b_1, b_2 can be chosen so that $\alpha(0) = c$, where $c = (c_1, c_2)$ is any specified point. This gives the system: $b_1 - b_2/2 = c_1, b_2 = c_2$, which is easily solved for b_1, b_2. Then using these values for b_1, b_2 in the formula (1.24) for the general integral curve, we get the specific integral curve that starts at c at time $t = 0$. For example if $c = (1, 2)$, then one easily finds that $b_1 = 2, b_2 = 2$ and thus the curve

$$\alpha(t) = \left(2e^{t/2} - e^{-t/2}, \; 2e^{-t/2} \right),$$

is the integral curve that satisfies $\alpha(0) = (1, 2)$.

Even with this explicit formula for α, the plot of the curve (by hand) might be tedious, unless more information is used. Generally for curves in the plane that are given parametrically (which are easier for computers to graph), one can, in theory, eliminate the parameter to get a Cartesian equation for the curve. Often this is impossible to do by hand, and when it is, the resulting equation is not one for a well-known type of curve. In this example, however, we observe that each integral curve must lie on a branch of a hyperbola or on a straight line. To see this eliminate the parameter t in the parametric equations for the curve and arrive at the single equation

$$2x_1 x_2 + x_2^2 = 2b_1 b_2. \tag{1.25}$$

To get this, it is necessary to make the assumption that $b_2 \neq 0$. Thus, any integral curve with $b_2 \neq 0$ will lie on the curve given by equation(1.25). Since this is a second-degree equation, the corresponding curve is a conic section (possibly a degenerate one). From the plot of the direction field in Figure 1.13, it is easy to guess that this curve is a hyperbola or a straight line. The straight line case occurs for $b_1 = 0$ and then equation(1.25) gives a pair of lines: $2x_1 + x_2 = 0$ and $x_2 = 0$. The integral curves that lie on these lines are also discernible from the plot of the direction field as well. Note that in the case when $b_2 = 0$, then equation(1.24) for the general integral curve reduces to $\alpha(t) = (b_1 e^{t/2}, 0)$ and this (when $b_1 \neq 0$) lies on the straight line $x_2 = 0$.

Example 1.10 For the system

$$\begin{aligned} x' &= \sinh y \\ y' &= -\sin x, \end{aligned}$$

which models the infinite row of stagnation points, the corresponding vector field is

$$X(x, y) = (\sinh y, -\sin x).$$

Doing a plot of this vector field, or its direction field, by hand is not so easy. However, a computer-generated plot is shown in Figure 1.14. Solving this system, by hand or computer, explicitly in terms of elementary functions is not possible. Numerical solutions are always possible and, since explicitly solvable systems are in some sense rare, the numerical method will be our main tool in studying systems. This is discussed in Chapter 2 and is the

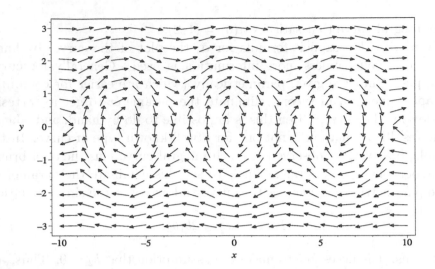

Figure 1.14: *The direction field for the system* (1.3)-(1.4) *which represents a fluid flow past a row of stagnation points.*

method used in drawing the phase portraits for the figures in this chapter. Even when drawing the integral curves with a computer, the plot of the direction field for the system is an essential aid and should perhaps always be done before plotting any of the curves. A plot of the direction field gives an overall view of what to expect, gives the direction of flow for the system, and helps locate fixed points, if any.

Exercises 1.2

1. This exercise is related to the activities described in Example 1.9. The system there and those here are very simple, so these activities are practical and of pedagogical value. For more general systems these activities are not practical. For each of the following vector fields X on $\mathbb{R}^2$:

   (i) Plot (by hand) the vector $X(x)$ at each of the points $x = (x_1, x_2)$ in the grid of points (x_1, x_2), with integer components in the specified rectangle $R = [a, b] \times [c, d]$. Do this in pencil or black ink. Then plot, in red ink, the vectors $X(x)/(2|X(x)|)$ at the same points.

   (ii) Solve (by hand) the system $x' = X(x)$ exactly, using two arbitrary constants b_1, b_2 in your answer. Eliminate the parameter t in the solution and find the x_1-x_2 equations for the integral curves.

   (iii) Find the straight line integral curves, if any. These are straight lines in the plane such that any integral curve starting at a point on the line, remains on the line for all time.

(a) $X(x_1, x_2) = (-x_1 + 2x_2, x_1)$. $R = [-2, 4] \times [-2, -2]$. *Hint for part* (ii): Differentiate the second equation in the system $x' = X(x)$ and substitute the result in the first equation to get a 2nd-order DE involving only x_2.

(b) $X(x_1, x_2) = (x_1, -x_1 + x_2)$. $R = [-2, 3] \times [-1, 2]$.

(c) $X(x_1, x_2) = (x_1, -x_1 - x_2)$. $R = [-2, 3] \times [-1, 2]$.

2. Suppose $\alpha : I \to \mathbb{R}^n$ is a solution of the autonomous system: $x' = X(x)$. Show that for any number r, the curve β defined by

$$\beta(t) = \alpha(t + r),$$

for $t \in I - r$, is also a solution of the autonomous system. This is a basic property of autonomous systems. *Note:* By definition, $I - r = \{s - r | s \in I\}$. Thus, if the interval $I = (a, b)$, then $I - r = (a - r, b - r)$.

3. Suppose $X : \mathcal{O} \to \mathbb{R}^n$ is a vector field on an open set $\mathcal{O} \subseteq \mathbb{R}^n$ and that $\mathcal{O}$ is symmetric about the origin, i.e., if $x \in \mathcal{O}$ then $-x \in \mathcal{O}$. Suppose X has the property:

$$X(-x) = -X(x),$$

for every $x \in \mathcal{O}$. Show that for each integral curve $\alpha : I \to \mathbb{R}^n$ of X, the curve β defined by

$$\beta(t) = -\alpha(t),$$

$t \in I$, is also an integral curve of X. Interpret, geometrically, what this means for the phase portrait of $x' = X(x)$. An example of such a vector field is $X(x_1, x_2) = (x_1^2 x_2, x_1 + x_2)$.

4. With the same supposition as in the last problem, but now with the assumption that

$$X(-x) = X(x),$$

for every $x \in \mathcal{O}$, show that for every integral curve $\alpha : I \to \mathbb{R}^n$ of X, the curve:

$$\beta(t) = -\alpha(-t),$$

for $t \in -I$, is also an integral curve of X. Interpret this geometrically. An example of such a vector field is $X(x_1, x_2) = (x_1 x_2, x_1^2 + x_2^2)$.

5. As a generalization of Exercise 3, suppose $X : \mathcal{O} \to \mathbb{R}^n$ is a vector field on an open set $\mathcal{O} \subseteq \mathbb{R}^n$ and that A is an $n \times n$ matrix. Assume that $\mathcal{O}$ is invariant under A, i.e., if $x \in \mathcal{O}$, then $Ax \in \mathcal{O}$. Suppose X has the property

$$X(Ax) = AX(x),$$

for every $x \in \mathcal{O}$. Show that for each integral curve $\alpha : I \to \mathbb{R}^n$ of X, the curve β defined by:

$$\beta(t) = A\alpha(t),$$

$t \in I$, is also an integral curve of X.

1.3 Nonautonomous Systems

So far we have only developed the abstract setting for autonomous dynamical systems. They are modeled by $x' = X(x)$, where X is a vector field on some open subset $\mathcal{O} \subseteq \mathbb{R}^n$. The example of the externally driven pendulum we discussed above requires a setting where we use time-dependent vector fields, with the corresponding system denoted by

$$x' = X(t, x).$$

The following definition give the specifics of this.

Definition 1.3 (Nonautonomous Systems)

(1) A *time-dependent vector field* is a map

$$X : B \subseteq \mathbb{R}^{n+1} \to \mathbb{R}^n,$$

defined on an open subset B of $\mathbb{R}^{n+1} \cong \mathbb{R} \times \mathbb{R}^n$. Denoting the points in B by (t, x), with $x \in \mathbb{R}^n, t \in \mathbb{R}$, gives a component form for X:

$$X(t, x) = \Big(X^1(t, x), \ldots, X^n(t, x) \Big).$$

For each $t \in \mathbb{R}$ we let

$$B_t = \{\, x \in \mathbb{R}^n \,|\, (t, x) \in B \,\}.$$

This is a *slice* through B at time t (and may be empty). At this instant t in time we get a vector field on B_t in the previous sense (when B_t is open). It is the map: $X_t(x) \equiv X(t, x)$. Plotting the direction field for X_t gives us a snapshot of X at time t. As t varies so do the plots of the direction fields. See Figure 1.15. Often B is a product: $B = J \times \mathcal{O}$, of an open interval J and an open set $\mathcal{O}$ in $\mathbb{R}^n$, which greatly simplifies things.

(2) Suppose $X : B \to \mathbb{R}^n$ is a time-dependent vector field. A *solution*, or *integral curve*, of the corresponding nonautonomous system $x' = X(t, x)$ is a curve: $\alpha : I \to \mathbb{R}^n$, in $\mathbb{R}^n$, with the following properties:

    (a) $(t, \alpha(t)) \in B$, for every $t \in I$;

    (b) $\alpha'(t) = X(t, \alpha(t))$, for every $t \in I$.

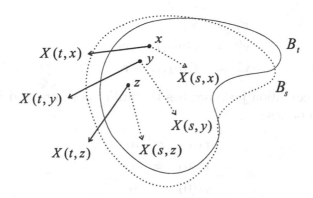

Figure 1.15: *Plots of the direction fields for X_t and X_s.*

(3) An *initial value problem* (IVP) consists of finding a solution of a system of DEs which passes through a given point at a given time. More specifically, suppose $(t_0, x_0) \in B$ is a given. The corresponding initial value problem is written symbolically as

$$x' = X(t, x) \qquad (1.26)$$
$$x(t_0) = x_0. \qquad (1.27)$$

A *solution of the initial value problem* is a curve $\alpha : I \to \mathbb{R}^n$ that is a solution of the system (1.26) and which also satisfies $\alpha(t_0) = x_0$ (note this implies that the interval I contains t_0. The condition (1.27) is known as an *initial condition*. Often x_0 is called an *initial point* for the integral curve in the IVP.

Later we will look at some theorems and propositions concerning the solvability of the initial value problem in general. These results come under the heading of *existence and uniqueness theorems*. Basically we are interested in knowing whether a given initial value problem (IVP) has a solution at all. This is the existence part. When this is the case, we also wish to know if there is only one solution (uniqueness), or if possibly there are several solutions. For dynamical systems that arise in physics and the other sciences, existence and uniqueness of solutions to IVPs is often taken for granted.

Example 1.11 In the two-body system (1.14) discussed above, the system is (in vector form)

$$\mathbf{r}_1' = \mathbf{v}_1$$

$$\begin{aligned}
\mathbf{r}_2' &= \mathbf{v}_2 \\
\mathbf{v}_1' &= Gm_2(\mathbf{r}_2 - \mathbf{r}_1)/r_{12}^3 \\
\mathbf{v}_2' &= Gm_1(\mathbf{r}_1 - \mathbf{r}_2)/r_{12}^3,
\end{aligned} \qquad (1.28)$$

and the initial condition just specifies the initial positions and initial velocities of the two bodies:

$$\begin{aligned}
\mathbf{r}_1(0) &= \mathbf{a}_1 \\
\mathbf{r}_2(0) &= \mathbf{a}_2 \\
\mathbf{v}_1(0) &= \mathbf{b}_1 \\
\mathbf{v}_1(0) &= \mathbf{b}_2.
\end{aligned}$$

Here $\mathbf{a}_1 \neq \mathbf{a}_2, \mathbf{b}_1, \mathbf{b}_2$ are given points (vectors) in $\mathbb{R}^3$. With some effort we can actually exhibit a solution of this IVP (see the next chapter) with explicit dependence on the initial condition data that gives uniqueness of the solution. For a larger number of bodies, 3-bodies (like the earth, moon, sun), 4-bodies, . . . , or in general N-bodies, the possibility of exhibiting an exact solution of the IVP (except in highly special cases) is too much to hope for. However, proving existence and uniqueness by other means is more tractable. The existence and uniqueness question here can be interpreted as saying that Newtonian mechanics is deterministic, i.e., knowing the initial positions and velocities of the all the bodies determines uniquely their evolution in time thereafter.

Exercises 1.3

1. In the following parts, a solution, involving an arbitrary constant k, of a nonautonomous DE $x' = X(t,x)$ on the real line $\mathbb{R}$, is given. Verify that the stated solution actually satisfies the DE and then find the two particular solutions that satisfy the two given initial conditions $x(t_0) = x_0$. Determine the largest domain $B \subseteq \mathbb{R}^2$ for the time-dependent vector field $X : B \to \mathbb{R}$, associated with the system and describe the sets B_t. Determine the largest intervals I, on which the solutions, $\alpha : I \to \mathbb{R}^2$, of the initial value problems are defined. Plot, in the same figure, the graphs of the two particular solutions and mark, on the figure, the points (t_0, x_0) for the initial data.

   (a) $x = t \pm \sqrt{t^2 - t^3 + k}$ is a solution of

   $$x' = \frac{2x - 3t^2}{2(x - t)}$$

   for any value of k (and each choice of the $\pm$ sign). Initial conditions $x(1) = 0$ and $x(1) = 2$.

(b) $x = t\tan(\ln|t| - k)$ is a solution of

$$x' = \frac{t^2 + xt + x^2}{t^2},$$

for any value of k. Initial conditions $x(1) = 1$ and $x(-1) = 1$.

1.4 Fixed Points

Many autonomous systems have very special solutions, called fixed points, which are important not only because they are very simple, but also because they often determine the behavior of the other integral curves around them.

Definition 1.4 (Fixed Points) A point $c \in \mathcal{O}$ is called a *fixed point* of the vector field $X : \mathcal{O} \to \mathbb{R}^n$, if $X(c) = 0$.

Any fixed point gives rise to a very simple integral curve (also called a fixed point) of the autonomous system: $x' = X(x)$. Namely, define $\alpha(t) = c$, for every $t \in \mathbb{R}$. Then since $\alpha'(t) = 0$ and $X(\alpha(t)) = X(c) = 0$ for every t, it is clear that α is an integral curve. Fixed points are also called *equilibrium points* because, for example, in the N-body problem, a fixed point of the system gives positions at which the N bodies would remain forever if placed there initially with no initial velocity. Another name for fixed points that is often used in the literature is *critical points*.

Not all systems have fixed points, but the existence of a fixed point is often helpful in analyzing the other solutions of the system, as you will see in the later chapters. Some of the previous examples discussed the physical significance of fixed points, which further indicates the importance of fixed points in the study of systems of DEs.

You should realize that the determination of fixed points amounts to solving the system of algebraic equations $X(x) = 0$, i.e., $X^1(x_1, \ldots, x_n) = 0, \ldots, X^n(x_1, \ldots, x_n) = 0$, for the n unknowns $x_1, \ldots, x_n$. In general this can be difficult to do, both theoretically and numerically, but the following exercises give examples where the fixed points can easily be calculated by hand.

Exercises 1.4

1. For each of the following systems, write down the formula for the associated vector field X and find all the fixed points of X.

(a)

$$x' = y^2 + x - 2$$
$$y' = x^2 - y^2$$

(b)

$$x' = (x - 5)(y - 1)$$
$$y' = (x - 3)(y + 2)$$

(c)

$$x' = (e - ax - by)x$$
$$y' = (f - cx - dy)y$$

(d)

$$x' = y(z - 1)$$
$$y' = x(z + 1)$$
$$z' = -2xy$$

1.5   Reduction to 1st-Order, Autonomous

Not all systems of differential equations are first-order systems, but, from a theoretical viewpoint, a system of any order can be replaced by a corresponding first-order system that is equivalent in some sense to the original system. This is called *reduction to first-order* and because of this it suffices to just study the theory for first-order systems.

The technique for reducing a system of nth-order DEs to first-order is quite natural and easy to do in practice. After reducing to a 1st-order system, we remove the time dependence to obtain an autonomous system.

Many higher order systems are commonly 2nd-order, having their origin in physics, chemistry, etc., where the laws of motion of complex systems seem to dictate 2nd-order equations. For 2nd-order systems, reduction to 1st-order amounts to considering the velocities, i.e., the 1st derivatives, as extra unknown functions and adding extra equations to the system.

The general 2nd-order system has the form

$$f(t, x, x', x'') = 0,$$

involving n unknown functions of t, i.e., the component functions of $x = (x_1, \ldots, x_n)$. In the above system $f : E \to \mathbb{R}^n$ is a vector-valued function defined on some open subset E of $\mathbb{R} \times \mathbb{R}^n \times \mathbb{R}^n \times \mathbb{R}^n$, and we will assume that system can be manipulated algebraically so as to solve for x'' in terms of t, x, and x'. Namely so as to put it in *normal form:*

$$x'' = F(t, x, x'). \tag{1.29}$$

By resorting to the Implicit Function Theorem (with some restrictions on f), we are guaranteed that this is possible (in theory). The relation between the solutions of the original system and (one) of its normal forms can be delineated, but in order to circumvent all these details we will just consider systems in normal form (1.29). The function F is a map: $F : U \to \mathbb{R}^n$, with domain some open subset U of $\mathbb{R} \times \mathbb{R}^n \times \mathbb{R}^n$.

The 1st-order system to which (1.29) reduces is

$$x' = v \tag{1.30}$$
$$v' = F(t, x, v). \tag{1.31}$$

This is a system of $2n$ equations for the $2n$ component functions of $x = (x_1, \ldots, x_n)$ and $v = (v_1, \ldots, v_n)$. For the sake of clarity, we explicitly write the system (1.30)-(1.31) in component form:

$$x_1' = v_1$$
$$\vdots$$
$$x_n' = v_n$$
$$v_1' = F^1(t, x_1, \ldots, x_n, v_1, \ldots, v_n)$$
$$\vdots$$
$$v_n' = F^n(t, x_1, \ldots, x_n, v_1, \ldots, v_n).$$

We can be a little more formal than this by introducing a time-dependent vector field $X : U \to \mathbb{R}^n \times \mathbb{R}^n$, defined by

$$X(t, x, v) = (v, F(t, x, v)),$$

and then exhibiting the explicit relationship between the solution curves of the 1st-order system determined by X and the solution curves of the original system. This level of formality is necessary to be precise about the reduction to first-order technique, but is rather simple to carry out, so the details are left to the reader.

The reduction of an kth-order system (in normal form):

$$x^{(k)} = F(t, x, x', \ldots, x^{(k-1)}), \tag{1.32}$$

to a 1st-order system is entirely similar to what we just did for a 2nd-order system: just introduce (vector-valued) functions $z_1 = x, z_2 = x', \ldots, z_k = x^{(k-1)}$, for x and all its derivatives up to order $k - 1$. Then the kth order system reduces to the 1st-order system:

$$
\begin{aligned}
z_1' &= z_2 \\
z_2' &= z_3 \\
&\vdots \\
z_{k-1}' &= z_k \\
z_k' &= F(t, z_1, z_2, \ldots, z_k).
\end{aligned}
$$

The unknowns here are curves: $z_j : I \to \mathbb{R}^n$, $j = 1, \ldots, k$, in $\mathbb{R}^n$, and so this system in vector form is

$$z' = Z(t, z),$$

where $Z : \mathcal{O} \to \mathbb{R}^{kn}$ is an appropriate time-dependent vector field on an open set $\mathcal{O}$ of $\mathbb{R} \times \mathbb{R}^{kn}$. Again, we could be more formal about this, but the main idea is clear. The exercises will explore some of the formalities.

The technique for reducing a 1st-order, nonautonomous system

$$x' = X(t, x),$$

to an autonomous system, amounts to introducing another equation for the time, considered as a new unknown function. Thus, one considers:

$$
\begin{aligned}
x_0' &= 1 \\
x' &= X(x_0, x).
\end{aligned}
$$

The exact relation of the integral curves of this autonomous system to those of the system we began with will be studied in the exercises.

Exercises 1.5

1. Consider a second-order system in normal form:

$$x'' = F(t, x, x'), \tag{1.33}$$

where F is a map: $F : U \to \mathbb{R}^n$, with domain some open subset U of $\mathbb{R} \times \mathbb{R}^n \times \mathbb{R}^n$. The corresponding first-order system is:

$$(x', v') = X(t, x, v),$$

where X is the time-dependent vector field: $X : U \to \mathbb{R}^n \times \mathbb{R}^n$, defined by

$$X(t, x, v) = (v, F(t, x, v)).$$

The purpose of this exercise is to describe the precise relationship between the solutions of the 2nd-order system and its corresponding 1st-order system.

(a) Give a definition of what is meant by a solution: $\alpha : I \to \mathbb{R}^n$, of the second-order system.

(b) Show that for each solution $\alpha : I \to \mathbb{R}^n$ of the second order system, the curve $\tilde{\alpha} : I \to \mathbb{R}^n \times \mathbb{R}^n$ defined by:

$$\tilde{\alpha}(t) = (\alpha(t), \alpha'(t)),$$

is a solution of the corresponding first-order system.

(c) Suppose $\tilde{\alpha} : I \to \mathbb{R}^n \times \mathbb{R}^n$ is a solution of the corresponding 1st-order system. Being a curve in $\mathbb{R}^n \times \mathbb{R}^n$ it can be written as

$$\tilde{\alpha}(t) = (\alpha(t), \beta(t)),$$

where α and β are curves in $\mathbb{R}^n$. Show that α is a solution of the original 2nd-order system.

2. Consider the 1st-order, nonautonomous system:

$$x' = X(t, x),$$

where $X : B \to \mathbb{R}^n$ is a time-dependent vector field on an open set $B \subseteq \mathbb{R}^n \times \mathbb{R}$. The corresponding autonomous system is

$$
\begin{aligned}
x_0' &= 1 \\
x' &= X(x_0, x).
\end{aligned}
$$

More formally, introduce a vector field: $\tilde{X} : B \to \mathbb{R} \times \mathbb{R}^n$ on B, defined by

$$\tilde{X}(x_0, x) = (1, X(x_0, x)).$$

This exercise exhibits the relation between the solutions of the original nonautonomous system and the autonomous system determined by $\tilde{X}$.

(a) Suppose $\alpha : I \to \mathbb{R}^n$ is a solution of the nonautonomous system and define a curve: $\tilde{\alpha} : I \to \mathbb{R} \times \mathbb{R}^n$ by

$$\tilde{\alpha}(t) = (t, \alpha(t)).$$

Show that $\tilde{\alpha}$ is a solution of the corresponding autonomous system: $z' = \tilde{X}(z)$.

(b) Suppose that $\tilde{\alpha} : I \to \mathbb{R} \times \mathbb{R}^n$ is a solution of the autonomous system: $z' = \tilde{X}(z)$. Since $\tilde{\alpha}$ is curve in $\mathbb{R} \times \mathbb{R}^n$, it can be written in the form

$$\tilde{\alpha}(t) = (\tau(t), \alpha(t)),$$

where τ and α are curves in $\mathbb{R}$ and $\mathbb{R}^n$, respectively. Show that there exists a constant t_0 such that the curve β defined by:

$$\beta(t) = \alpha(t - t_0),$$

for $t \in I + t_0$, is a solution of the original nonautonomous system.

3. Reduce each of the following systems to a 1st-order, autonomous system.

(a) $x''' = x'x + x^2 + x''$.

(b) $x''' = x'x + x^2t + x''$.

(c) $x''' = 5x'' + 2x' - x + \sin 3t$.

(d)

$$
\begin{aligned}
x'' &= x'y + x^2y' + y \\
y'' &= xy' + y^2 - x'.
\end{aligned}
$$

4. For each of the DEs you were assigned in Exercise 1, Section 3, reduce the DE to a 1st-order, autonomous system and plot, in the same figure, the direction field for the autonomous system along with the graphs of the two particular solutions you found of the nonautonomous system.

1.6 Summary

In this introduction we have attempted to give an overview of the subject of study—dynamical systems/systems of DEs. This we have done by means of specific examples, general discussions of the important features, and precise definitions of *some* of the objects necessary to formulate the theory in mathematical terms. The rest of the book will add to this, not only by digging deeper into the details of the topics which arose here in the discussion, but also by introducing new topics that suggest themselves naturally.

The next chapter will introduce some further concepts and techniques that arise in the theory of differential equations, and, for the most part, the discussion is intuitive. The mathematical tools, theorems, and results necessary to make these concepts rigorous will, beginning in Chapter 3, eventually be covered.

While your previous experience in the study of DEs has probably dealt primarily with writing down exact, closed-form solutions of DEs and IVPs, or with deriving series solutions where possible, you will now find very little of that activity stressed here (although from time to time we will explicity solve some DEs). In essence the emphasis here is on the *qualitative theory* of systems of DEs, since in practice most nonlinear systems do not have solutions that can be written down explicitly in terms of elementary functions (Being able to do this is the quantitative aspect of the study of systems of DEs.)

Chapter 2

Techniques, Concepts and Examples

In this chapter we look at a number of examples of dynamical systems (systems of DEs) in detail and use this as an opportunity to introduce many concepts, such as gradient vector fields, stable/unstable fixed points, separatrices, limit cycles, transformations of DEs, and conservation laws, which will be studied more formally later. Our goal is to give the reader some experience with looking at, working with, studying, and analyzing some typical examples of systems that can occur. The computer exercises here, in the previous chapter, and on the electronic component are intended to aid the student in obtaining this experience and to help establish an intuitive feel for the concepts well before we study the underlying theory for these concepts. Waiting until after the development of the theory to begin computer analyses of dynamical systems is too long to wait.

A basic understanding of the concepts is easily obtained with a good computer and a computer algebra system (CAS), like Maple, which we will use here and throughout the text. Maple will be used to *numerically* plot the integral curves for the systems we study and not as a tool for finding closed-form formulas for these integral curves (which rarely is possible). In essence the vector field X defining the system

$$x' = X(t, x)$$

contains all the information we need to numerically study the integral curves of the system via their plots. It also contains much geometric information about the phase portrait of the system as well.

The mathematics behind the various numerical schemes for solving differential equations is discussed in many numerical analysis courses and books,

D. Betounes, *Differential Equations: Theory and Applications*, DOI 10.1007/978-1-4419-1163-6_2,

and is not part of the scope of this book. However, it is very important, both for building a geometric intuition about systems and their flows and for using Maple's numerical software properly, to have some some understanding of how these numerical schemes work. With this in mind we briefly discuss the *Euler method* for numerically solving systems.

2.1   Euler's Numerical Method

For simplicity here, we limit the discussion to autonomous systems $x' = X(x)$, with $X : \mathcal{O} \to \mathbb{R}^n$ a vector field on $\mathcal{O}$.

One can motivate Euler's numerical method in several ways. We begin with the geometric way.

2.1.1   The Geometric View

Suppose we want to determine an approximation to the integral curve γ which starts the point $c \in \mathcal{O}$ at time $t = 0$, i.e., to the curve γ which satisfies

$$\gamma'(t) = X(\gamma(t))$$
$$\gamma(0) = c$$

Plotting the vector $X(c)$ at the point c gives us the direction the integral curve will go in flowing away from c (since $X(c) = X(\gamma(0)) = \gamma'(0)$ is the tangent to γ at $c = \gamma(0)$). Visualize constructing (actually drawing) a polygonal approximation as follows. Fix a small positive number h and move from c along the tangent line a distance of $h|X(c)|$, arriving at the point

$$c_1 \equiv c + hX(c).$$

See Figure 2.1. *Note*: $|X(c)|$ stands for the length of the vector $X(c)$. If the tangent vector $X(c)$ is not too large in magnitude, then the point c_1 will be approximately on the integral curve γ. Thus, joining c and c_1 gives the first side in a polygonal approximation to γ. The next side starts at the point c_1, where the flow has tangent vector $X(c_1)$ As in the first step, now move from c_1 along the tangent line a distance of $h|X(c_1)|$, arriving at the point

$$c_2 \equiv c_1 + hX(c_1).$$

Joining c_1 to c_2 with a straight line gives the second side in the polygonal approximation. By repeating the process for a total of N steps, we get

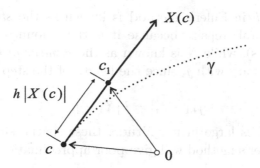

Figure 2.1: *The geometry behind Euler's numerical method for approximating integral curves of the vector field X.*

a polygonal approximation with vertices $c_0 = c, c_1, c_2, \ldots, c_N$, which are successively computed by

Euler's Method:

$$c_{j+1} = c_j + hX(c_j), \qquad (2.1)$$

for $j = 0, 1, \ldots, N - 1$.

Quite simply, Euler's method amounts to computing the points $c_1, c_2, \ldots, c_N$ from equation (2.1), with $c_0 = c$ as the initially given point, and then joining these points successively with line segments to get the polygonal approximation to the desired integral curve. See Figure 2.2.

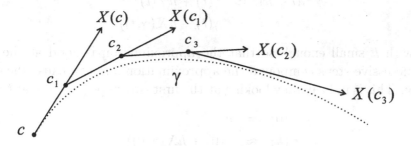

Figure 2.2: *The geometry behind Euler's numerical method for approximating an integral curve γ of the vector field X. Starting at the initial point c the points $c_1, \ldots, c_N$ are generated by Formula (2.1) and the resulting polygonal approximation to γ is obtained.*

The polygonal approximation to γ is on the time interval $[0, T]$, where

$T = Nh$, and the h in Euler's method is known as the *stepsize* (or more precisely the temporal stepsize because it is the amount of time between the successive points), while N is known as the *number of time steps*. The spatial stepsize can vary with j, since the length of the step from c_j to c_{j+1} is

$$|c_{j+1} - c_j| = h|X(c_j)|.$$

In regions where X is large in magnitude, these spatial steps can become quite large and Euler's method will give poor approximations when h is not sufficiently small.

2.1.2   The Analytical View

Euler's method, and it's accuracy, is based on the Taylor series expansion of γ

$$\gamma(t + h) = \sum_{k=0}^{\infty} \frac{\gamma^{(k)}(t)}{k!} h^k$$

$$= \gamma(t) + h\gamma'(t) + \frac{h^2}{2}\gamma''(t) + \cdots$$

Note: In the second equation, we have written the scalars $h^k/k!$ on the left of the vectors $\gamma^{(k)}(t)$, as is the custom. Using just the first two terms of the Taylor series and the fact that $\gamma'(t) = X(\gamma(t))$, we get the approximation by the first two terms in its Taylor series expansion:

$$\gamma(t + h) \approx \gamma(t) + h\gamma'(t)$$

$$= \gamma(t) + hX(\gamma(t)).$$

Even with h small enough so that the approximation is good at the first step, successive steps compound the approximation and can cause the error to grow. This is indicated by looking at the first two steps (starting at $t = 0$):

$$\gamma(0) = c$$

$$\gamma(h) \approx \gamma(0) + hX(\gamma(0))$$

$$= c + hX(c) = c_1$$

$$\gamma(2h) \approx \gamma(h) + hX(\gamma(h))$$

$$\approx c_1 + hX(c_1) = c_2.$$

Here, as you can see, c_1 involves one approximation while c_2 involves two approximations. Similarly, the approximations to $\gamma(3h), \ldots, \gamma(Nh)$ involve the Taylor series approximation *and* the approximations from the previous steps. You can well imagine (see Figure 2.2) how the overall error could accumulate over a large time interval. Thus, controlling the stepsize h is crucial to obtaining accurate plots.

Most computer algebra systems like Maple use more refined numerical methods than the Euler method, but the basic ideas on approximating solutions, stepsize, accumulation of errors, and so on, are similar.

Example 2.1 (Row of Vortices) Exercise 4(b) from Section 1.1 will serve well to illustrate the need to set appropriate stepsizes to achieve accurate plots of integral curves. The system of DEs is

$$x' = \frac{\sinh y}{M(x,y)}$$
$$y' = \frac{-\sin x}{M(x,y)},$$

where $M(x,y) = \cosh y - \cos x$. So the system is closely related to Example 1.2, except now the function M occurs in the denominators. The vector field for the system is

$$X(x,y) = \left(\frac{\sinh y}{M(x,y)}, \frac{-\sin x}{M(x,y)} \right),$$

and this becomes infinite where $M(x,y) = 0$ (i.e., at the vortex centers). Let's consider plotting an integral curve near the vortex center $(0,0)$.

If we choose a point c close to $(0,0)$, say $c = (0,0.1)$, then the integral curve γ that starts there at time zero, $\gamma(0) = c$, is a closed, circular-like, curve. Since the speed of flow along a streamline (integral curve) is very great near the center of a vortex, the numerical approximations to γ will be quite poor unless the stepsize is small.

This is illustrated in Figure 2.3, which shows four approximations to γ, only one of which is reasonably accurate. The plot in the upper left of the figure is for the time interval $t = 0 \ldots 1$ and stepsize $h = 0.05$. This means that there are $N = T/h = 1/0.05 = 20$ spatial steps in the numerical method and the resulting approximation is a polygon with 20 sides (which can clearly be counted in the figure). However, due to the high velocity, the actual integral curve γ wraps around the vortex center three times in the 1 second time interval. Thus, 20 steps are not enough to give a good approximation and this is readily apparent in the figure. Taking a stepsize of

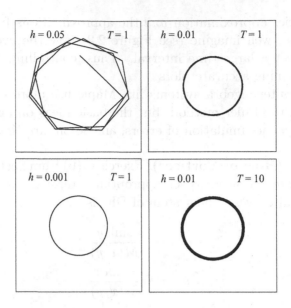

Figure 2.3: *Plots of the approximations to a single integral curve near a vortex center in Example* 2.1. **Upper Left:** *stepsize* = 0.05 *and* $t = 0 \ldots 1$. **Upper Right:** *stepsize* = 0.01 *and* $t = 0 \ldots 1$. **Lower Left:** *stepsize* = 0.001 *and* $t = 0 \ldots 1$. **Lower Right:** *stepsize* = 0.01 *and* $t = 0 \ldots 10$.

$h = 0.01$ gives $N = 1/0.01 = 100$ steps and results in a better approximation as shown in the upper right of Figure 2.3. This is still inaccurate because the heavy line thickness indicates that the approximation is "wandering" somewhat as it wraps around the vortex center. This is verified by taking $h = 0.001$ (one thousand steps) to get the plot in the lower left of the figure. The "wandering" problem in approximations is exhibited more vividly in the plot at the lower right of the figure. This is for $t = 0 \ldots 10$, so that γ winds about the vortex center a greater number of times, and $h = 0.01$. Even though the approximation has the right shape, its thickness indicates the inaccuracy.

Having said this, we should also say that there are systems, like the perturbed ball in the hoop example from Chapter 1, where an integral curve *will* wander around the phase space and appear to fill out whole regions. This is known to occur theoretically and not to be a result of inaccuracies in the approximations. This points out the need for having some theory to guide the experimental studies done on a computer.

Exercises 2.1

1. Use Euler's method to construct, by hand, polygonal approximations to the specified integral curves of the following systems. In each case the integral curve starts at c at time $t = 0$. Two stepsizes h are specified for the given time interval $[0, T]$ and both of the corresponding approximating polygons should be drawn in the same figure for comparison.

   (a) $c = (1,1), T = 2, h = 0.5$, and $h = 0.2$. The system is

   $$\begin{aligned} x_1' &= x_1 \\ x_2' &= -x_1 + x_2. \end{aligned}$$

   (b) $c = (3, -1), T = 1, h = 0.2$, and $h = 0.1$. The system is

   $$\begin{aligned} x_1' &= -x_1 + 2x_2 \\ x_2' &= x_1. \end{aligned}$$

2. For the system in 1(a), but now with h and T unspecified, compute the points $c_1, \ldots, c_5$ in Euler's method in terms of h.

3. As in Example 2.1, but now for the Karmen vortices, use a computer to draw the polygonal approximations to the integral curve that starts at $c = (0, 0.25)$ at time zero and stepsizes and time intervals: (a) $h = 0.05, T = 1$, (b) $h = 0.01, T = 1$, (c) $h = 0.001, T = 1$, and (d) $h = 0.01, T = 10$. Compare and contrast these results with those shown in Figure 2.3.

2.2 Gradient Vector Fields

Many systems $x' = X(x)$ arise from vector fields X which are gradients of scalar fields $X = \nabla F$. All heat flow systems and some fluid flow systems are of this type. These systems are special and the extra information about the system that we get from F is often helpful.

Definition 2.1 A vector field $X : \mathcal{O} \to \mathbb{R}^n$ on an open set $\mathcal{O} \subseteq \mathbb{R}^n$ is called a *gradient vector field* if there is a (scalar) function $F : \mathcal{O} \to \mathbb{R}$, such that

$$\nabla F(x) = X(x),$$

for every $x \in \mathcal{O}$. In terms of components of X, the condition is

$$\frac{\partial F}{\partial x_i} = X^i(x),$$

for $i = 1, \ldots, n$ and all $x \in \mathcal{O}$. The function F is called a *potential* (or *potential function*) for X. If X represents a force field (and is a gradient vector field), then it is called a *conservative* force field.

Clearly not all vector fields are gradient vector fields, i.e., have potentials. Indeed if X has a potential (and is continuously differentiable on $\mathcal{O}$), then necessarily the components of X must satisfy the equations:

Integrability Conditions:

$$\frac{\partial X^j}{\partial x_i} = \frac{\partial X^i}{\partial x_j}, \tag{2.2}$$

on $\mathcal{O}$ for all $i, j = 1, \ldots, n$ (exercise). *Note*: For $n = 3$, i.e., for a vector field X on an open set $\mathcal{O}$ in $\mathbb{R}^3$, the integrability conditions are equivalent to the condition that $\text{curl}(X) = 0$ on $\mathcal{O}$. In general equations (2.2) are *necessary*, but not *sufficient*, for X to be the gradient of some function. If the domain $\mathcal{O}$ is simply connected, in particular if $\mathcal{O} = \mathbb{R}^n$, then there are integral formulas for constructing potentials for any X that satisfies equations (2.2). Maple and other CASs have built-in procedures for calculating potentials for vector fields (exercises). When X has a potential F, then for any constant c, the function $F + c$ is also a potential for X. Thus, potentials are not unique.

If X is a gradient vector field with potential F, then since ∇F is perpendicular to each hypersurface

$$S_F^k = \{\, x \in \mathbb{R}^n \mid F(x) = k \,\},$$

(called a level set of F), it follows that $X = \nabla F$ is also perpendicular to each of these surfaces. Consequently, each integral curve of X intersects each level set of F orthogonally. In dimensions $n = 2$ and $n = 3$ this is often useful information, as the next example shows.

Example 2.2 (Heat Flow in a Square Plate) All heat flows arise from gradient vector fields, because the heat flow lines are the integral curves of the heat flux vector field X. The heat flux vector field, by definition, is the negative of the gradient of the temperature function F, i.e., $X \equiv -\nabla F$. The minus sign in this definition is there because ∇F points in the direction of greatest increase in temperature and physically heat flows from hot to cold. Thus, $-\nabla F$ is the direction of heat flow. As a specific example consider the temperature function

$$F(x, y) = b \cos(2\pi x) \sinh(2\pi y),$$

where $b = 1/\sinh(2\pi)$. This temperature distribution F arises from boundary conditions where the bottom of the square is held at temperature 0, the

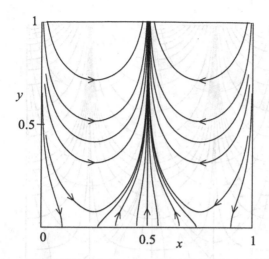

Figure 2.4: *Plot of the heat flow lines for the temperature function* $F(x,y) = b\cos(2\pi x)\sinh(2\pi y)$.

top is held at a temperature that varies as $F(x,1) = \cos(2\pi x)$, and the sides are insulated (i.e., the heat flux vectors $X(0,y)$ and $X(1,y)$ are tangent to these sides, respectively). The heat flux vector field for F is

$$X(x,y) = \left(2\pi b\sin(2\pi x)\sinh(2\pi y), -2\pi b\cos(2\pi x)\cosh(2\pi y)\right).$$

The plots of some of the integral curves of X, i.e., some of the heat flow lines for the temperature distribution F, are shown in Figure 2.4.

By what was said above, each heat flow line must intersect each level curve of F orthogonally. A level curve of a temperature function is known as a *isotherm*, since the temperature is the same at all points along such a curve. Figure 2.5 shows the plots of some isotherms for this example. With enough isotherms drawn in, it is possible to construct a relatively good plot of the heat flow lines by starting at a point on the top boundary of the square and drawing a curve, by hand, that intersects each isotherm at a right angle. While Figure 2.5 does not indicate which temperature corresponds to which isotherm, you can get a relative idea of this correspondence by plotting the graph of the temperature function F and using a style for the rendering that has the isotherms drawn in on the surface. This is shown in Figure 2.6.

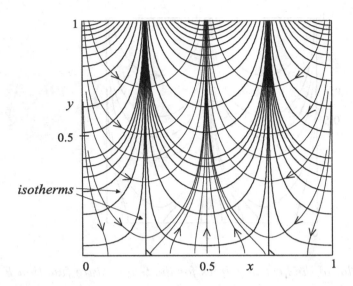

Figure 2.5: *Plot of some level curves (isotherms) for the temperature function* $F(x,y) = b\cos(2\pi x)\sinh(2\pi y)$ *and some heat flow lines.*

Example 2.3 (Flow Past a Cylinder) In fluid mechanics there is a class of planar fluid flows where the velocity vector field X for the fluid is a gradient vector field. These are often introduced using complex function theory and, indeed, the construction of many of these is best understood using complex variables. We will not go into the details here (see [Ma 73], [Be 98]), but just mention that is where the vortex examples in the Introduction came from. This example comes about in a similar way.

Thus, consider the vector field on the plane given by

$$X(x,y) = \left(a + \frac{b(y^2 - x^2)}{(x^2 + y^2)^2}, \frac{-2bxy}{(x^2 + y^2)^2} \right),$$

where a and b are positive constants. This vector field is defined on the whole plane minus the origin, $\mathcal{O} = \mathbb{R}^2 \setminus \{0\}$.

From the form of X, it is not too hard to recognize that it is the gradient of the function

$$F(x,y) = ax + \frac{bx}{x^2 + y^2}.$$

(Verify!) If you wish, you can use Maple to find this function, even though in this case this is easy to do by hand. The role of the potential function F is not

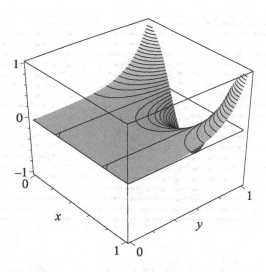

Figure 2.6: *Plot of temperature function* $F(x, y) = b \cos(2\pi x) \sinh(2\pi y)$ *with some of the isotherms drawn in on the surface.*

very physically meaningful here, although it could be used to visualize the integral curves as suggested in the last example. Namely, a plot of sufficiently many level curves of F will enable you to draw the integral curves, by hand (see the exercises).

The plot of X (or rather, its direction field) is shown in Figure 2.7. The figure indicates the presence of a circle of radius $r = 1$ with center at the origin. The points (x, y) on this circle are special since they have the property that $X(x, y)$ is tangent to the circle at this point (exercise). This is important in fluid mechanics because the theory requires that the velocity vector field for a certain fluids should be tangent to the boundary of an obstacle placed in the fluid. Thus, if we disregard the integral curves on the interior of the circle, then the corresponding picture represents an ideal fluid flow past a cylinder. This is shown in Figure 2.8.

You can see from the figure that far from the cylinder, the flow is essentially uniform. Check also that $X(x, y) \approx (a, 0)$ for $x^2 + y^2$ large, so that the parameter a represents the speed of the uniform flow.

A related vector field, which contains the one above as a special case, is

$$X(x, y) = \left(a + \frac{b(y^2 - x^2) + cy(x^2 + y^2)}{(x^2 + y^2)^2}, \frac{-(2bxy + cx(x^2 + y^2))}{(x^2 + y^2)^2} \right),$$

The choice of $c = 0$ gives the vector field for the flow past a cylinder . For

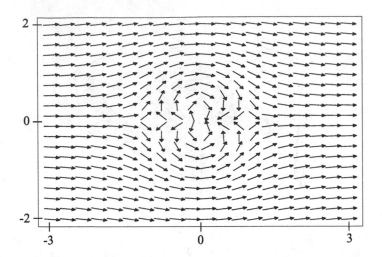

Figure 2.7: *Plot of the direction field for the vector field X in Example 2.3. The parameter values are $a = 1, b = 1$.*

$c \neq 0$, this vector field gives a model for a spinning (or rotating) cylinder in a uniform flow. Thus, the extra terms change things slightly and give a very interesting fluid flow (which you can study more in the exercises). A potential for this flow is

$$F(x, y) = ax + \frac{bx}{x^2 + y^2} - c \tan^{-1}\left(\frac{y}{x}\right).$$

(Verify this!) Besides having a different form, this new potential has a different domain, in fact a smaller one:

$$U = \{\, (x, y) \,|\, x \neq 0 \,\} = \mathbb{R}^2 \setminus \{\, y\text{-axis} \,\}.$$

Thus, unless the domain for X, which is still $\mathcal{O}$, is changed, we cannot claim that X is gradient vector field according to the above definition. This is in the nature of things. One can show that there does *not* exist a function G defined on all of $\mathcal{O}$, such that $\nabla G = X$ on $\mathcal{O}$ (exercise).

Exercises 2.2

1. **(Flow past a cylinder)** Consider the vector field from Example 2.3:

$$X(x, y) = \left(a + \frac{b(y^2 - x^2)}{(x^2 + y^2)^2}, \frac{-2bxy}{(x^2 + y^2)^2}\right),$$

where a and b are positive constants. Do the following

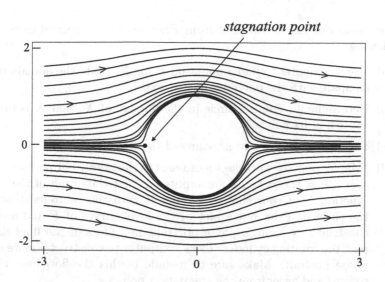

Figure 2.8: *Ideal fluid flow past a cylinder in a uniform flow. The parameter values are $a = 1$, $b = 1$, the radius of the cylinder is $r = \sqrt{b/a} = 1$, and the speed of the flow far from the cylinder is approximately $a = 1$.*

(a) Verify that the function

$$F(x, y) = ax + \frac{bx}{x^2 + y^2}$$

is a potential function for X.

(b) Show that for a point (x, y) on the circle of radius $r = \sqrt{b/a}$ (i.e., $x^2 + y^2 = b/a$), the vector $X(x, y)$ is tangent to the circle at the point. *Hint:* Recall that a tangent vector to a circle at a point is perpendicular to the radius vector from the center to the point.

(c) Show that $(\pm\sqrt{b/a}, 0)$ are the only fixed points (stagnation points) of the flow.

(d) (Maple users see worksheet `gradvecfields.mws`.) For parameter values $a = 1$, $b = 1$, use a computer to plot some level curves of the function F. Print this out and then, by hand, use the level curve plots to construct integral curves of X (i.e., flow lines past a cylinder). Make sure the original computer plot contains enough level curves to achieve an accurate plot of the flow lines (Cf. Figure 2.8.)

2. **(Flow past a rotating cylinder)** Consider the vector field

$$X(x, y) = \left(a + \frac{b(y^2 - x^2) + cy(x^2 + y^2)}{(x^2 + y^2)^2}, \frac{-(2bxy + cx(x^2 + y^2))}{(x^2 + y^2)^2} \right),$$

which contains the vector field from Exercise 1 as a special case. Do the
following

(a) Use a computer to find a potential F for X. State the domain of F and
    compare with the potential in the text.

(b) Determine if there is a circle in the domain of X, that X is tangent to
    at each of its points.

(c) Determine the stagnation points of the fluid flow.

(d) (Maple users see worksheet `gradvecfields.mws`.) For parameter values
    $a = 1$, $b = 1$, $c = 1$, use a computer to plot the graph of the potential
    function F of your choice. Render it with the contours drawn in on
    the surface. Print out a plot of the level curves of F and use this to
    construct, by hand, a phase portrait, i.e., plots of the fluid flow lines
    past the rotating cylinder. Use a computer to construct a more accurate
    phase portrait. Make sure to include in this the flow lines that flow
    toward and away from the stagnation points.

2.3   Fixed Points and Stability

We consider a few examples to illustrate the technique of finding fixed points
both by hand and by using Maple.

Example 2.4 We revisit the last example and consider the vector field

$$X(x, y) = \left(a + \frac{b(y^2 - x^2)}{(x^2 + y^2)^2}, \frac{-2bxy}{(x^2 + y^2)^2} \right).$$

The fixed points of X (places where it vanishes) are the solutions (x, y) of
the equation $X(x, y) = 0$, i.e., of the algebraic system of equations

$$a + \frac{b(y^2 - x^2)}{(x^2 + y^2)^2} = 0$$

$$\frac{-2bxy}{(x^2 + y^2)^2} = 0.$$

This is easy to solve by hand. Thus, the second equation gives that either
$x = 0$ or $y = 0$. If $x = 0$ then the first equation reduces to $a + b/y^2 = 0$, which has no (real) solutions. On the other hand if $y = 0$, then the
first equation reduces to $a - b/x^2 = 0$, which has solutions $x = \pm\sqrt{b/a} = \pm r$. Thus, there are two fixed points $(\pm r, 0)$, which in this case are called
stagnation points. These points are clearly shown in Figure 2.8, and each is

an *unstable* fixed point. Later we will give a precise definition of stability for systems of DEs and develop some tools for determining stability. However, the instability of the two stagnation points here should be easy to understand intuitively. If a particle is placed at either stagnation point, it will remain there forever. However, if a particle is placed near, but ever so slightly away from, a stagnation point, then the fluid flow will carry it away downstream to infinity. That is to say, integral curves of X that start near a fixed point do not remain near it as time evolves.

Example 2.5 An interesting abstract system that does not represent any particular physical situation is

$$x' = (x^2 - 1)y \tag{2.3}$$
$$y' = (x+2)(y-1)(y+2). \tag{2.4}$$

The fixed points of this system are the solutions of the algebraic system

$$(x^2 - 1)y = 0$$
$$(x+2)(y-1)(y+2) = 0.$$

Because of the factored form, this is easy to solve by hand. The reasoning is as follows. For the first equation to hold, either $x = \pm 1$ or $y = 0$. We examine what each of these implies in the second equation of the system. (A) If $x = \pm 1$, then the second equation is $(2 \pm 1)(y-1)(y+2) = 0$ and so either $y = 1$ or $y = -2$. From this we get four fixed points $(1,1), (-1,1), (1,-2)$, and $(-1,-2)$. (B) If $y = 0$, then the second equation is $-2(x+2) = 0$ and so $x = -2$. This gives the fixed point $(0,-2)$. Thus, altogether this system of DEs has five fixed points.

The plot of the direction field for the system is shown in Figure 2.9. Here, as is often the case, the number of fixed points can be discerned from the figure and even their approximate locations can be found by clicking on the plot window at approximately where the fixed points appear to occur.

The system of DEs has some easily found integral curves which lie on horizontal or vertical straight lines. Thus, if we take $x = 1$ in the first equation of the system, i.e., look for an integral curve of the form $\alpha(t) = (1, \alpha_2(t))$, we see that such integral curves are possible because the first equation is satisfied automatically, while the second equation becomes $y' = (y-1)(y+2)$, which is easily solved for y (i.e., for $\alpha_2(t)$). Thus, the vertical line through $x = 1$ contains straight-line integral curves and in fact three

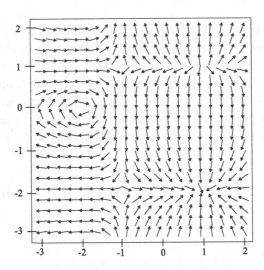

Figure 2.9: *The direction field for the system* (2.3)-(2.4).

distinct ones, since this line is divided into three parts by the two fixed points
(1,1) and (1, −2) which lie on it.

Similar reasoning will lead you to discover the other vertical and horizontal integral curves in the system. These, along with numerous other (curved)
integral curves are shown in Figure 2.10.

You can see that some of the fixed points have nearby integral curves
which are qualitatively different from the others. The fixed point (1, −2)
is an *asymptotically stable* fixed point, since all integral curves that start
near enough to it will tend toward it over time. On the other hand the
fixed point (1, 1) is *unstable*, since all integral curves that start in a suitable
small neighborhood of it will leave this neighborhood in a finite amount
of time. The fixed points (−1, −2) and (−1, 1) are likewise unstable fixed
points since, in a small enough neighborhood of either point, most integral
curves starting in the neighborhood will eventually leave the neighborhood.
The last fixed point (−2, 0) is *stable* (but not asymptotically stable), since
(roughly speaking) integral curves that start near it, will stay near it, but
not approach it in the limit.

Example 2.6 This example illustrates the need to use numerical methods
to find the fixed points of a system. The system of DEs is

$$x' = x^5 y^3 + x^2 + y^2 - 4$$
$$y' = x^3 y + x^2 - y^2 + 1.$$

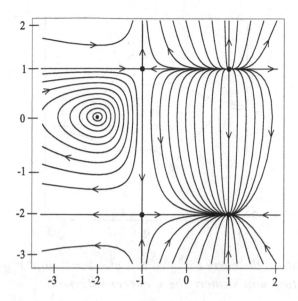

Figure 2.10: *The phase portrait for the system* (2.3)-(2.4).

The corresponding system of algebraic equations for the fixed points of the system is

$$x^5 y^3 + x^2 + y^2 - 4 \; = \; 0 \qquad (2.5)$$
$$x^3 y + x^2 - y^2 + 1 \; = \; 0, \qquad (2.6)$$

and since the individual equations do not factor completely, the system is not readily solvable by hand. You can find the solutions of this system numerically by using Maple's `fsolve` command (see the Maple worksheets for details on this). This command generally tries to find all real solutions if the equations are polynomial equations (as they are here), but often will return only one solution. This is the case for this example. Maple returns the single solution

$$(x, y) = (-.7817431847, -1.530451384).$$

Determinig the number of solutions of such a system can be theoretically and practically difficult. It is often useful, for systems with two equations and two unknowns, to try to determine the number of solutions and their approximate values by graphical methods. This is based on the following observation. Each equation in the algebraic system (2.5)-(2.6) represents a

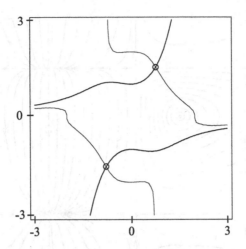

Figure 2.11: *Plots of the curves* $x^5y^3+x^2+y^2-4=0$ *and* $x^3y+x^2-y^2+1=0$, *showing the two points where these curves intersect.*

curve in the plane, and solving the system amounts to finding the points of intersection of the two curves in the system. Hence, plotting both curves in the same picture (with an appropriate window size) can help find the number of solutions and their approximate values. Figure 2.11 shows the two curves in the system here. Using this information, you can specify a rectangle for Maple to use in searching for the other solution. As it turns out, the other solution is

$$(x, y) = (.7817431847, 1.530451384),$$

which is the negative of the one found above. One could have exploited the symmetry in the system to predict this (exercise), and thus the plot in Figure 2.11 only serves to verify that there are only two solution to the system. *Note*: This is not absolute proof that there are only two solutions.

Thus, the system of DEs has just two fixed points. The plots of the direction field and a collection of integral curves of the system is left as an exercise.

Exercises 2.3

1. Plot the curves of the algebraic system

$$
\begin{aligned}
(x^2 - 1)y &= 0 \\
(x + 2)(y - 1)(y + 2) &= 0,
\end{aligned}
$$

using different colors for each curve. This system comes from Example 2.5 and its solutions are the fixed points of the corresponding system of DEs. Because of the algebraic factors, these fixed points are easy to determine. Verify that the fixed points are the points of intersection of the curves you plotted. Explain why the equations for the curves, because they are factored, influences the nature of these "curves."

2. Plot the phase portrait for the system in Example 2.6, i.e., the system:

$$\begin{aligned} x' &= x^5 y^3 + x^2 + y^2 - 4 \\ y' &= x^3 y + x^2 - y^2 + 1, \end{aligned}$$

Determine the stability of the fixed points. See the material on the worksheet `fixedpts.mws`.

3. Consider the system

$$\begin{aligned} x' &= x^2 y + xy^2 - x - 1 \\ y' &= x^4 + xy^2 + y - 4. \end{aligned}$$

Do a complete study of this system, that is, do the following.

   (a) Find all the fixed points. Plot, using different colors, the curves in the algebraic system that determines these fixed points. Justify that you have found all the fixed points.

   (b) Draw a good, complete phase portrait for the system. Choose the viewing rectangle of a suitable size so that all the features of the system are showing.

2.4 Limit Cycles

A limit cycle for a system of DEs is a *closed* integral curve which has the property that it attracts or repels nearby integral curves. Limit cycles will be classified as stable or unstable, much like fixed points, and indeed you can consider a limit cycle as sort of a one dimensional analog of a fixed point (which is zero-dimensional). As with fixed points, a given system of DEs may *not* have any limit cycles.

Example 2.7 For convenience of notation let $r = (x^2 + y^2)^{1/2}$ and consider the following system in the plane:

$$\begin{aligned} x' &= (1 - r^2)x - (a + r^2)y & (2.7) \\ y' &= (a + r^2)x + (1 - r^2)y, & (2.8) \end{aligned}$$

where a is a constant which we will take to be $a = -4$. (The exercises will cover some other interesting choice for a.) This gives a system with a limit cycle which is a circle of radius 1 centered at the origin. (In general, limit cycles need not be circular.) Figure 2.12, shows this limit cycle and some of the other integral curves of the system. As you can see, this limit cycle

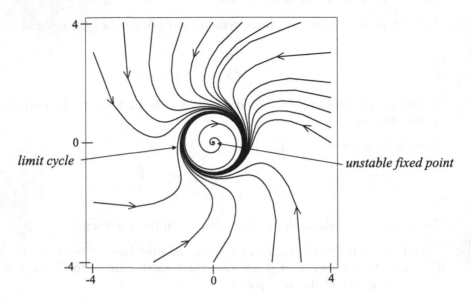

Figure 2.12: *Plots of flow lines (integral curves) for the vector field X in Example 2.7. A limit cycle, i.e. a closed integral curve which is approached in the limit by other integral curves, is clearly visible in the picture.*

is *stable*, since all the integral curves of the system which start at points either inside or outside the circle approach it asymptotically. Chapter 6 gives a general technique for transforming a system of DEs and shows how, in particular, the system here can be transformed to polar coordinates. By doing so, we can prove that the circle of radius 1 here is actually a limit cycle and not just a manifestation of poor numerical approximations. Rather than using the general transformation theory, here we motivate the idea by showing how it works for *transforming to polar coordinates*.

Thus, suppose we look for solutions of the system (2.7)-(2.8) that have the special form

$$x = r\cos\theta$$
$$y = r\sin\theta,$$

where $r = r(t)$ and $\theta = \theta(t)$ are two new unknown functions of t. The above relation between the two sets of unknowns x, y and r, θ allows us to calculate the relation between their derivatives. It is easy to see that this relation is

$$
\begin{aligned}
x' &= r' \cos \theta - r\theta' \sin \theta \\
y' &= r' \sin \theta + r\theta' \cos \theta.
\end{aligned}
$$

Using all of these relations, we can rewrite the system (2.7)-(2.8) entirely in terms of the new unknown functions r, θ. This gives a new system:

$$
\begin{aligned}
r' \cos \theta - r\theta' \sin \theta &= (1 - r^2)r \cos \theta - (a + r^2)r \sin \theta & (2.9) \\
r' \sin \theta + r\theta' \cos \theta &= (a + r^2)r \cos \theta + (1 - r^2)r \sin \theta. & (2.10)
\end{aligned}
$$

Multiplying the first equation by $\cos \theta$, the second equation by $\sin \theta$, and adding the two resulting equation will yield an equation involving only r'. A similar combination will give an equation involving only θ'. These two new equations are quite simply

$$
\begin{aligned}
r' &= (1 - r^2)r & (2.11) \\
\theta' &= r^2 - 4. & (2.12)
\end{aligned}
$$

(exercise). This new system is known as the *polar coordinate version* of the original or as the *transformed system* under the polar coordinate map. You can see now why transforming the original system is helpful. Indeed the new system (2.11)-(2.12) is completely solvable by elementary methods (exercise). The first equation does not involve θ and is a separable DE. Solving this for r and using this in the second equation, allows θ to be found by integration. This will give the general solution of the system and then using $x = r \cos \theta$ and $y = r \sin \theta$ will give the general solution of the original system.

Here, however, we wish to look at only three particular solutions. In the system (2.11)-(2.12), the first equation is just a 1st-order, one-dimensional system and is easily seen to have three fixed points $r = 0, 1, -1$. The use of these values in the second equation of the system gives $\theta' = k$, where $k = -4, -3, -3$, respectively, is a negative constant. Thus, $\theta(t) = kt + \theta_0$. These three solutions of the r-θ system give us two solutions of the x-y system which are, respectively, (1) a fixed point at the origin $(x, y) = (0, 0)$ corresponding to the zero radius $r = 0$, and (2) the limit cycle, or circle of radius one, corresponding to $r = \pm 1$. The fact that this is a limit cycle and is stable can be proven by using the explicit general solution of the r-θ

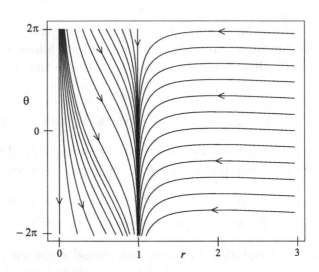

Figure 2.13: *Plots of flow lines for the polar system* (2.11)-(2.12) *Example*
2.7.

system (exercise). Figure 2.13 shows the phase portrait for the polar system
(2.11)-(2.12). The two vertical, straight-line integral curves correspond to
the fixed point at the origin ($r = 0$) and the limit cycle ($r = 1$). The plots
of the other integral curves lend experimental evidence to the assertion that
the $r = 1$ integral curve corresponds to an asymptotically stable limit cycle
in the original system.

There is a method for constructing, by hand, a rough sketch of the phase
portrait for the original system just by interpreting the polar system (2.11)-
(2.12) geometrically. The circle $r = 1$ divides the plane into two regions
and we can examine the behavior of the integral curves $\alpha(t) = (x(t), y(t))$
which start in one of these two regions. For an integral curve α having
corresponding polar version $\beta(t) = (r(t), \theta(t))$ with $r(0) = r_0 > 1$, the
quantity $(1 - r^2)r$ is negative in a neighborhood of r_0. Thus, $r' = (1 -
r^2)r < 0$ and so r will be a decreasing function near time zero (and in
fact for all time) and we would expect α to approach the limit cycle $r = 1$
in the long run. In addition, the quantity $\theta' = r^2 - 4$ will change from
positive to negative as r goes from being greater than to less than 2. This
indicates that the integral curves that are outside the limit cycle will have
a turning point in their approaches to the limit cycle and that this point
occurs as they cross the circle $r = 2$. This is exhibited in Figure 2.12.
It is predicted from the above comment, since where $\theta' > 0$, the angle θ

is increasing (moving counterclockwise) and where $\theta' < 0$, the angle θ is decreasing (moving clockwise). In a similar fashion, for an integral curve α that starts inside the limit cycle (and not at the fixed point), $0 < r_0 < 1$, we can predict from $r' = (1 - r^2)r > 0$ that r is increasing and from $\theta' = r^2 - 4 < 0$ that the angle θ is always a clockwise rotation (no turning points). This analysis enables us to draw a reasonably accurate phase portrait.

Exercises 2.4

1. In Example 2.7, the system

$$x' = (1 - r^2)x - (a + r^2)y \tag{2.13}$$
$$y' = (a + r^2)x + (1 - r^2)y, \tag{2.14}$$

in the x-y plane was transformed into the polar system

$$r' = (1 - r^2)r \tag{2.15}$$
$$\theta' = r^2 - 4, \tag{2.16}$$

in the r-θ plane, and it was shown how the polar system can be used to easily sketch, by hand even, the phase portrait of the original system in the x-y plane. While this is the main use of the polar system, it has other uses as well.

(a) The DE (2.15) in the polar system is a separable DE. Solve this explicitly and write out the particular solution which satisfies $r(0) = r_0$. Use this to show that the limit cycle in the x-y plane corresponding to $r = 1$ is stable, and indeed is asymptotically stable. This means that if α is an integral curve, distinct from the limit cycle and the fixed point at the origin, then

$$\lim_{t \to \infty} |\alpha(t)| = 1.$$

This is apparent intuitively from the plot of the phase portrait in Figure 2.12, but needs some rigorous proof nonetheless. You may use the fact that any integral curve α of the system (2.13)-(2.14) has the form:

$$\alpha(t) = (r(t) \cos \theta(t), \, r(t) \sin \theta(t)),$$

where r, θ is a solution of the polar system (2.15)-(2.16).

(b) Figure 2.13 shows the phase portrait of the polar system for a viewing rectangle $[0, 3] \times [-6, 6]$, i.e., $r = 0..3$, $\theta = -6..6$ (radians). Look at the phase portrait on the larger viewing rectangle $[-3, 3] \times [-6, 6]$ shown on the worksheet limitcycles.mws. There is an apparent symmetry of the phase portrait about the θ-axis (the vertical axis). Show that this is indeed the case by proving that if $\beta(t) = (r(t), \theta(t))$, for $t \in J$, is a solution of the polar system (2.15)-(2.16), then the curves defined by

$$
\begin{aligned}
\gamma(t) &\equiv (-r(t), \theta(t)) \\
\mu(t) &\equiv (r(t), \theta(t) + \pi),
\end{aligned}
$$

for $t \in J$, are also solutions to the polar system. Show also that β and μ correspond to the same integral curve in the x-y plane. Discuss how these assertions explain the symmetry in the polar phase portrait and why.

2. The system

$$
\begin{aligned}
x' &= (1 - r^2)x - (a + r^2)y & (2.17) \\
y' &= (a + r^2)x + (1 - r^2)y + b, & (2.18)
\end{aligned}
$$

with parameters a, b has dramatically different features for different values of a and b. In Exercise 1 above, $a = -4, b = 0$. In this exercise study the system for the two cases (i) $a = -1, b = 0$ and (ii) $a = -1, b = 1/2$. Specifically, find all the fixed points and limit cycles (if any) and plot a phase portrait (for the x-y system only) which displays all the prominent features.

2.5   The Two-Body Problem

We return to the two-body problem from the introduction and use it to illustrate several concepts and techniques—conservations laws, Jacobi coordinates, and transformations of systems of DEs, which will help us explicitly solve this system. The second-order version of this system

$$
m_1 \mathbf{r}_1'' = \frac{Gm_1 m_2}{r_{12}^3}(\mathbf{r}_2 - \mathbf{r}_1) \tag{2.19}
$$

$$
m_2 \mathbf{r}_2'' = \frac{Gm_1 m_2}{r_{12}^3}(\mathbf{r}_1 - \mathbf{r}_2), \tag{2.20}
$$

will be the most convenient form of the system to use in the theoretical discussion.

We first derive a *conservation law* associated with this system, which comes simply from the observation that if $\mathbf{r}_1 = \mathbf{r}_1(t)$, $\mathbf{r}_2 = \mathbf{r}_2(t)$, for $t \in I$, is a solution of the system (2.19)-(2.20), then these functions also satisfy the equation obtained by adding equations (2.19)-(2.20) together, i.e.,

$$
m_1 \mathbf{r}_1'' + m_2 \mathbf{r}_2'' = 0,
$$

or equivalently:

$$\frac{d}{dt}\left(m_1\mathbf{r}_1' + m_2\mathbf{r}_2'\right) = 0.$$

This implies there exists a constant (vector) $\mathbf{P}$ such that $\mathbf{r}_1, \mathbf{r}_2$ satisfy the following:

Conservation of Linear Momentum:

$$m_1\mathbf{r}_1'(t) + m_2\mathbf{r}_2'(t) = \mathbf{P}, \qquad (2.21)$$

for all $t \in I$. The constant vector $\mathbf{P} = (P_1, P_2, P_3)$ is known as the *total linear momentum* of the system, and the conservation law just says that no matter how the individual momenta $m_1\mathbf{r}_1'(t), m_2\mathbf{r}_2'(t)$ change over time, their sum remains constant. Thus, if the initial velocities $\mathbf{r}_1'(0), \mathbf{r}_2'(0)$ are known, then the total linear momentum is

$$\mathbf{P} = m_1\mathbf{r}_1'(0) + m_2\mathbf{r}_2'(0).$$

Another way to write the conservation of linear momentum law (2.21) is

$$\frac{d}{dt}\left(m_1\mathbf{r}_1(t) + m_2\mathbf{r}_2(t)\right) = \mathbf{P},$$

and thus we see (by integration) that there exists a constant vector $\mathbf{B}$ such that

$$m_1\mathbf{r}_1(t) + m_2\mathbf{r}_2(t) = \mathbf{P}t + \mathbf{B},$$

for all $t \in I$. If we let $M = m_1 + m_2$ denote the *total mass* of the system and divide the last equation by M, we get the following:

Uniform Motion of the Center of Mass:

$$\frac{m_1}{M}\mathbf{r}_1(t) + \frac{m_2}{M}\mathbf{r}_2(t) = \mathbf{V}t + \mathbf{C}, \qquad (2.22)$$

for all $t \in I$. Here $m_1\mathbf{r}_1(t)/M + m_2\mathbf{r}_2(t)/M$ is (by definition) the position of the *center of mass* of the system at time t and the above equation is just the law that the center of mass moves with uniform (i.e., constant) velocity $\mathbf{V} = \mathbf{P}/M$ and has position $\mathbf{C} = \mathbf{B}/M$ at time $t = 0$. This is illustrated in Figure 2.14.

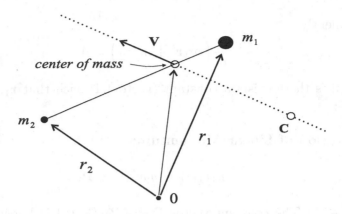

Figure 2.14: *Line of motion for the center of mass in the two-body system.*

2.5.1 Jacobi Coordinates

Having seen above that the motion of the center of mass is as simple as possible, i.e., rectilinear, we separate this motion from the more complex motion of the bodies relative to one another. Technically, this amounts to transforming to new coordinates.

Definition 2.2 (Jacobi Coordinates) If $\mathbf{r}_1 = \mathbf{r}_1(t)$, $\mathbf{r}_2 = \mathbf{r}_2(t)$, for $t \in I$, is a solution of the two-body system (2.19)-(2.20), let

$$\mathbf{R} = \frac{m_1}{M}\mathbf{r}_1 + \frac{m_2}{M}\mathbf{r}_2 \qquad (2.23)$$

$$\mathbf{r} = \mathbf{r}_2 - \mathbf{r}_1. \qquad (2.24)$$

Then $\mathbf{R}, \mathbf{r}$ are known as the *Jacobi coordinates* for the system. The first is a vector-valued function pointing, at each time t, from the origin to the center of mass, and the second is a vector-valued function pointing, at each time t, from the first body to the second (see Figure 2.15).

Note that the above definition expresses $\mathbf{R}$ and $\mathbf{r}$ as linear combinations of $\mathbf{r}_1, \mathbf{r}_2$. In matrix form this is

$$\begin{bmatrix} \mathbf{r} \\ \mathbf{R} \end{bmatrix} = A \begin{bmatrix} \mathbf{r}_1 \\ \mathbf{r}_2 \end{bmatrix},$$

where A is a certain 2×2 matrix and the other matrices shown are 2×3 matrices with $\mathbf{r}, \mathbf{R}$ and $\mathbf{r}_1, \mathbf{r}_2$ as their rows, respectively. Thus, transforming to Jacobi coordinates is a *linear* transformation of the system of DEs.

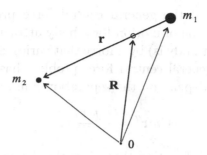

Figure 2.15: *The Jacobi coordinates* $\mathbf{R}, \mathbf{r}$ *for the two-body system.*

The transformation to polar coordinates in the last example was a *nonlinear* transformation. Since A is invertible we can express $\mathbf{r}_1, \mathbf{r}_2$ as a linear combination of $\mathbf{R}, \mathbf{r}$ as well (exercise). Thus, knowing $\mathbf{R}, \mathbf{r}$ will give us $\mathbf{r}_1, \mathbf{r}_2$.

To get the transformed system of DEs for $\mathbf{R}, \mathbf{r}$, we use the equations of motion (2.19)-(2.20) and the conservation of linear momentum. We use the notation

$$r \equiv |\mathbf{r}| = |\mathbf{r}_1 - \mathbf{r}_2| \equiv r_{12}.$$

From the equations of motion, we easily find that $\mathbf{R}, \mathbf{r}$ satisfy the equations

$$
\begin{aligned}
\mathbf{R}'' &= 0 \\
\mathbf{r}'' &= \mathbf{r}_2'' - \mathbf{r}_1'' \\
&= -\frac{Gm_1}{r^3}\mathbf{r} - \frac{Gm_2}{r^3}\mathbf{r} \\
&= -\frac{GM}{r^3}\mathbf{r}.
\end{aligned}
$$

These equations are uncoupled and the general solution of the first equation, as we have seen above, is $\mathbf{R}(t) = \mathbf{V}t + \mathbf{C}$. Thus, all we need to do is solve the second equation for $\mathbf{r}$ and that will complete the solution.

2.5.2 The Central Force Problem

It is traditional to introduce the *reduced mass* for the two-body system:

$$m \equiv \frac{m_1 m_2}{m_1 + m_2}.$$

Then the second equation we found above can be written as

$$mr'' = -\frac{Gm_1 m_2}{r^3}\mathbf{r}.$$

This is a special case of the general central force problem. It is simply interpreted as the equation of motion for a body attracted toward the origin (which is where the first body is) by a force that varies inversely as the square of the distance. The general central force problem has the same form, but allows for the force to depend on the separation r in a different manner:

$$m\mathbf{r}'' = -\frac{f(r)}{r}\mathbf{r}.$$

Here $f : (0, \infty) \to \mathbb{R}$ is a function that gives the dependence of the force on the distance r. For the inverse square law, $f(r) = Gm_1m_2r^{-2}$.

There is a further conservation law that comes immediately from the above DE. To get it, note that

$$\mathbf{r} \times m\mathbf{r}'' = -\mathbf{r} \times \frac{f(r)}{r}\mathbf{r} = \mathbf{0}.$$

Then recall that there is a product rule for derivatives of cross products (see Appendix C), which when applied here gives

$$\frac{d}{dt}\left(\mathbf{r} \times m\mathbf{r}'\right) = \mathbf{r}' \times m\mathbf{r}' + \mathbf{r} \times m\mathbf{r}'' = \mathbf{r} \times m\mathbf{r}'' = 0.$$

Since this hold for all $t \in I$, there is a constant vector $\mathbf{L}$, such that the following law holds.

Conservation of Angular Momentum:

$$\mathbf{r}(t) \times m\mathbf{r}'(t) = \mathbf{L}, \tag{2.25}$$

for all $t \in I$. The conservation law says that this (vector) quantity does not change during the motion.

The angular momentum conservation law has the important geometric consequence that the motion of the body lies in one plane, namely the plane through the origin with normal vector $\mathbf{L}$ (see Figure 2.16). We will take advantage of this fact and assume, without loss of generality, that the motion is actually in the x-y plane. If not, we could make a linear transformation of the system into one whose motion is in the x-y plane. If we were tracking satellites, the details of this transformation would be important, but we do not need this extra information here.

Thus, we assume the DE for equation of motion has the following form

$$\mathbf{r}'' = -\frac{kf(r)}{r}\mathbf{r},$$

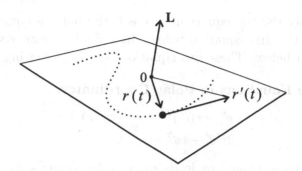

Figure 2.16: *In a central force problem the motion is in the plane through the origin with the angular momentum vector* **L** *as normal.*

where $\mathbf{r} = (x, y)$, $r = (x^2 + y^2)^{1/2}$, and $k = m^{-1}$.

Writing out the above DE in terms of components, we get two 2nd-order DEs:

$$x'' = -\frac{kf(r)}{r}\, x \qquad\qquad (2.26)$$

$$y'' = -\frac{kf(r)}{r}\, y, \qquad\qquad (2.27)$$

and the form of this suggests transforming to polar coordinates $x = r\cos\theta$, $y = r\sin\theta$, with $r = r(t), \theta = \theta(t)$ as the new unknown functions. Doing this just as we did for the example with the limit cycle (except now we need the second derivatives), we obtain first

$$\begin{aligned} x' &= r'\cos\theta - r\theta'\sin\theta \\ y' &= r'\sin\theta + r\theta'\cos\theta. \end{aligned}$$

Then differentiating these, we get

$$\begin{aligned} x'' &= r''\cos\theta - 2r'\theta'\sin\theta - r\theta''\sin\theta - r(\theta')^2\cos\theta \\ y'' &= r''\sin\theta + 2r'\theta'\cos\theta + r\theta''\cos\theta - r(\theta')^2\sin\theta. \end{aligned}$$

Substituting these in the x-y equations above and collecting terms, gives the system in polar coordinates:

$$\begin{aligned} ((r'' - r(\theta')^2)\cos\theta - (2r'\theta' + r\theta'')\sin\theta &= -kf(r)\cos\theta \\ (r'' - r(\theta')^2)\sin\theta + (2r'\theta' + r\theta'')\cos\theta &= -kf(r)\sin\theta. \end{aligned}$$

Now we multiply the top equation by $\cos\theta$, the bottom equation by $\sin\theta$, and add to get the first equation below. In a similar manner we obtain the second equation below. These two equations are the following:

Central Force Equations in Polar Coordinates:

$$r'' - r(\theta')^2 \;=\; -kf(r) \tag{2.28}$$
$$2r'\theta' + r\theta'' \;=\; 0. \tag{2.29}$$

The second of these equations leads directly to Kepler's 2nd law. To see this, multiply both sides of it by r to get

$$2rr'\theta' + r^2\theta'' = 0,$$

or, equivalently

$$\frac{d}{dt}\left(r^2\theta'\right) = 0.$$

Thus, we get that, if $r = r(t), \theta = \theta(t)$ is a solution of the system (2.28)-(2.29), then there exists a constant such that the following law holds.

Kepler's Second Law:

$$r(t)^2\theta'(t) = c, \tag{2.30}$$

for all $t \in I$. Kepler worked prior to the invention of calculus, so he did not state his law like this, but rather phrased it as: *equal areas are swept out in equal times as a planet orbits the sun.* To see why equation (2.30) says the same thing, multiply both sides by $1/2$ and then integrate both sides from t_1 to t_2 (two arbitrary times) to get

$$\tfrac{1}{2}\int_{t_1}^{t_2} r^2(t)\theta'(t)dt = \tfrac{1}{2}c(t_2 - t_1).$$

The integral on the left side is precisely the area swept out by the curve in polar coordinates between the two times $t_1 < t_2$. The equation says the area depends only on the duration of time $t_2 - t_1$. (See Figure 2.17.) Of course, the above derivation gives a much more general result than Kepler's. It says that *Kepler's law on equal areas holds in all central force problems.*

It is also important to note that the constant c is related to the magnitude of the angular momentum. In the coordinates we have chosen $\mathbf{r} = (r\cos\theta, r\sin\theta, 0)$ and so it is easy to see that

$$\begin{aligned}
\mathbf{L} \;&=\; \mathbf{r} \times m\mathbf{r}' \\
&=\; (0, 0, mr^2\theta') = (0, 0, mc) = (0, 0, k^{-1}c).
\end{aligned}$$

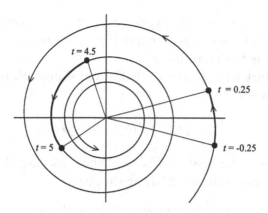

Figure 2.17: *Illustration of Kepler's 2nd law: equal areas are swept out in equal times in any central force problem.*

The next step in the explicit solution of the central force system is to transform it once again, this time with a nonlinear transformation. The transformation arises from the assumption that r and θ are functionally related. More specifically, assume there is a real-valued function ρ, of a real variable, such that for each solution $r = r(t), \theta = \theta(t)$, with $t \in I$, we have:

$$r(t) = \rho(\theta(t))^{-1} = \frac{1}{\rho(\theta(t))}, \qquad (2.31)$$

for all $t \in I$. For simplicity in the calculations below, we will suppress the explicit dependence of the functions on t in equation (2.31) and the resulting equations. Thus, we write the functional relation that ρ establishes as

$$r = \rho(\theta)^{-1}. \qquad (2.32)$$

The strategy is now to use the system (2.28)-(2.29) to deduce something about ρ. However, first note that knowing ρ will enable us to find solutions r, θ of the system. To see this, observe that Kepler's law $r^2\theta' = c$ can be written as

$$\theta' = c\rho(\theta)^2. \qquad (2.33)$$

So if we know ρ then this becomes a separable DE for θ and has its solution given implicitly by

$$\int \rho(\theta)^{-2} d\theta = ct + b.$$

Theoretically we can solve this explicity for θ as a function of t and then from equation (2.31), we can get r explicitly as a function of t as well. Thus, all we need to do is to determine ρ.

To do this, start with equation (2.32) and differentiate both sides with respect to t. This will give

$$r' = -\rho(\theta)^{-2}\rho'(\theta)\theta' = -c\rho'(\theta),$$

where we have used the relation (2.33). Differentiating the above equation once again and using relation (2.33) again gives

$$r'' = -c\rho''(\theta)\theta' = -c^2\rho(\theta)^2\rho''(\theta).$$

For brevity, write this last equation as $r'' = -c^2\rho^2\rho''$ and write relation (2.33) as $\theta' = c\rho^2$ (thus we are suppressing the explicit dependence on θ as well as t). Substituting these into equation (2.28), i.e., into the equation

$$r'' - r(\theta')^2 = -kf(r),$$

gives, after minor simplification,

$$\rho'' + \rho = a\rho^{-2}f(\rho^{-1}). \tag{2.34}$$

Here we have, in the simplification process, let $a = k/c^2$. This assumes that $c \neq 0$. The case when $c = 0$ is important because the motion then is along a straight line through the center of force. The study of this case is left to the exercises.

Equation (2.34) is viewed as a 2nd-order DE for ρ as a function of θ. This DE is simplest for an inverse square law $f(z) = 1/z^2$, since then $f(\rho^{-1}) = \rho^2$ and the DE reduces to

$$\rho'' + \rho = a.$$

This elementary, 2nd-order, linear DE has general solution

$$\rho = a + A\cos\theta + B\sin\theta,$$

where A, B are constants determined by the initial conditions. Since $r = 1/\rho$, we get the following functional relation between r and θ:

$$r = \frac{1}{a + A\cos\theta + B\sin\theta}, \tag{2.35}$$

Thus, the arguments have led us to the conclusion that any solution $r = r(t), \theta = \theta(t)$ of the polar system for the central force problem with an inverse square law will satisfy

$$r(t) = \frac{1}{a + A\cos(\theta(t)) + B\sin(\theta(t))}, \qquad (2.36)$$

The constants a, A, B can be determined from the initial conditions and it can be shown that equation (2.35) is an equation for a conic section in polar coordinates with pole located at a focus of the conic section (see the exercises). Thus, the body moves on an elliptical, parabolic, or hyperbolic trajectory about the center of force. The exception to this is when the trajectory is a straight line through the origin. This is the case $c = 0$ which we mentioned above.

Figure 2.18 shows three trajectories for an inverse square law of attraction with $k = 1$, initial positions $(x_0, y_0) = (1,0), (1,0), (1,0)$, and initial velocities $(x_0', y_0') = (-0.5, 0.5), (-1, 1), (-1, 1.25)$, respectively.

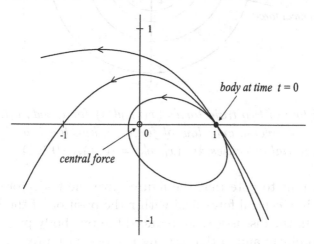

Figure 2.18: *Plots of three trajectories $(x(t), y(t))$ for a body attracted toward the origin with a inverse square law force. The initial positions are all the same $(x_0, y_0) = (1,0)$, while the initial velocities are $(x_0', y_0') = (-0.5, 0.5), (-1, 1), (-1, 1.25)$, respectively.*

The initial velocity $(-0.5, 0.5)$ gives the elliptical orbit shown, while the initial velocities $(-1, 1)$ and $(-1, -1.25)$ give trajectories that appear to be a parabola and a hyperbola, respectively. Of course, the plots of the latter two do not have enough information to rule out the possibility that these

trajectories are elliptical or even to decide definitively if they are parabolic
or hyperbolic. However, this can be determined theoretically (exercise).

Figure 2.19 shows two trajectories of a body under an inverse cube law
of attraction $f(r) = 1/r^3$. The initial position in each case is the same,
$(x_0, y_0) = (1, 0)$, while the initial velocities are $(x_0', y_0') = (0, 1), (-0.1, 1)$,
respectively. With the first initial velocity, the body travels in a closed orbit
around the origin, while the latter initial velocity gives a trajectory that
spirals in toward the origin.

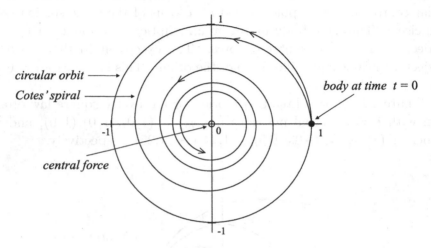

Figure 2.19: *Plots of two trajectories $(x(t), y(t))$ for a body attracted toward
the origin with a inverse cube law of force. While the initial positions are
the same, the initial velocities are $(x_0', y_0') = (0, 1), (-0.1, 1)$.*

It is important to note that the figures give the trajectories of a hypothetical body in a central force field with **r** the position of the body relative
to the origin. In the discussion, we reduced the two-body problem to such a
central force problem and in that setting $\mathbf{r} = \mathbf{r}_2 - \mathbf{r}_1$ represents the position
of the second body relative to the first. The actual motion of the two bodies
comes from writing $\mathbf{r}_1, \mathbf{r}_2$ in terms of $\mathbf{R}, \mathbf{r}$ and then plotting each trajectory,
i.e., curves $\mathbf{r}_1, \mathbf{r}_2$. Each of these trajectories can look quite different than
those in Figures (2.17)-(2.19). See Figure 9.4 in Chapter 9 for an example
of what we mean. Generally each body will trace a path about the center of
mass, which may be stationary or in motion (exercise). In the cases where,
say m_1 is very large compared with m_2 (like the sun's mass as compared
with the earth's), the center of mass essentially corresponds to the position
of the first body, given by $\mathbf{r}_1$. Thus, for all practical purposes the solution **r**

of the central force problem gives the trajectory of the second body orbiting the mass first body. In this case figures (2.17)-(2.19) give pictures of the "actual" motion of the second body.

Exercises 2.5

1. The transformation to Jacobi coordinates, in matrix form, is

$$\begin{bmatrix} \mathbf{r} \\ \mathbf{R} \end{bmatrix} = A \begin{bmatrix} \mathbf{r}_1 \\ \mathbf{r}_2 \end{bmatrix}.$$

Find the matrix A and it's inverse A^{-1}. Use this to give the formulas that explicitly express $\mathbf{r}_1, \mathbf{r}_2$ as linear combinations of $\mathbf{r}, \mathbf{R}$.

2. Suppose $f : (0, \infty) \to \mathbb{R}$ is any function. Show that the system of DEs

$$\begin{aligned} m_1 \mathbf{r}_1'' &= G m_1 m_2 f(r_{12})(\mathbf{r}_2 - \mathbf{r}_1)/r_{12} \\ m_2 \mathbf{r}_2'' &= G m_1 m_2 f(r_{12})(\mathbf{r}_1 - \mathbf{r}_2)/r_{12}, \end{aligned}$$

transforms into the system

$$\begin{aligned} \mathbf{r}'' &= -\frac{kf(r)}{r} \mathbf{r} \\ \mathbf{R}'' &= 0, \end{aligned}$$

in Jacobi coordinates. That is, any solution $\mathbf{r}_1, \mathbf{r}_2$ of the first system gives a solution $\mathbf{r}, \mathbf{R}$ of the second system, and conversely. Thus, the two-body problem with a general force of interaction f is reduced to the general central force problem. This generality is useful, for example, in electromagnetism where some particles repel one another and where some molecular interactions are modeled by something other than an inverse square law.

3. Use Kepler's 2nd law to show that the central force equations (2.28)-(2.29) are equivalent to the system

$$r'' - \frac{c^2}{r^3} = -kf(r) \tag{2.37}$$

$$r^2 \theta' = c. \tag{2.38}$$

For some choices of f, one can explicitly do the integrals necessary to solve (2.37) for r. Then θ is determined directly from equation (2.38).

4. In the central force problem, a particle's trajectory is in a straight line when its angular momentum is zero, i.e., when $c = 0$. This is to be expected physically, but mathematically it follows from the fact that the system of DEs (2.37)-(2.38) in Exercise 1 reduces to

$$r'' = -kf(r) \tag{2.39}$$

$$\theta = \theta_0. \tag{2.40}$$

This says that the particle moves along the line through the origin which makes an angle θ_0 with the x-axis and its position $r = r(t)$ on this line is determined from the differential equation for r. The motion is toward or away from the origin depending on the initial radial velocity r_0'. Even if $r_0' > 0$, the velocity may not be great enough for the particle to "escape" the central force, i.e., at some time the particle may change direction and move back toward the origin. This exercise studies the precise type of motions possible for the inverse square law: $f(r) = r^{-2}$. Specifically:

(a) Show that for any solution $r : I \to \mathbb{R}$ of the DE (2.39), there is a constant E such that

$$\tfrac{1}{2} r'(t)^2 = \frac{k}{r(t)} + E, \tag{2.41}$$

for all $t \in I$. Assuming $0 \in I$ and $r(0) = r_0, r'(0) = r_0'$ are given initial values, the value of E is

$$E = \tfrac{1}{2}(r_0')^2 - \frac{k}{r_0}.$$

Show that equation (2.41), after some algebraic manipulation, becomes a separable DE with solution formally given by

$$t = \pm \int \frac{\sqrt{r}}{\sqrt{2}\sqrt{Er + k}}\, dr. \tag{2.42}$$

Here the $\pm$ sign is chosen as $+$ if $r_0' \geq 0$ and as $-$ if $r_0' < 0$.

(b) In the case $E > 0$, show that the relation between r and t described by equation (2.42) is

$$t = \pm \frac{E^{-3/2}}{\sqrt{2}} \left(\sqrt{E}\sqrt{r}\sqrt{Er + k} - k \ln \left(\frac{\sqrt{Er + k} + \sqrt{Er}}{\sqrt{k}} \right) \right) + b. \tag{2.43}$$

This requires computing the indefinite integral by using a rationalizing substitution, a trig substitution, and integration by parts. Do *not* use a computer for this, but rather show your work by hand.

(c) In the case $E < 0$, show that the relation between r and t described by equation (2.42) is

$$t = \pm \frac{E^{-3/2}}{\sqrt{2}} \left(\sqrt{E}\sqrt{r}\sqrt{Er + k} + k \sin^{-1} \left(\frac{\sqrt{Er + k}}{\sqrt{k}} \right) \right) + b. \tag{2.44}$$

(d) In the case $E = 0$, show that the relation between r and t described by equation (2.42) is

$$t = \pm \frac{\sqrt{2}}{3\sqrt{k}} r^{3/2} + b. \tag{2.45}$$

(e) For a given initial position r_0 (always assumed to be positive) analyze the nature of the motion of the particle with initial radial velocity r_0'. For this, sketch (by hand) graphs of r as a function of t, by using the information in the DE: $r'' = -kr^{-2}$ and the equation (2.41). Do graphs illustrating the differences that occur for choices of r_0' and E that are positive, zero, and negative. *Note*: Equations (2.43)-(2.45) give t precisely as a function of r (the constant of integration b is determined from $r(0) = r_0$). However, you should be able to do the sketches without this information.

(f) Let $r : I \to \mathbb{R}$ be a solution of the DE (2.39) with initial conditions $r(0) = r_0, r'(0) = r_0'$. For the cases when $r_0' > 0$, show that there is a radial velocity $\varepsilon > 0$, the *escape velocity*, such that (i) if $r_0' \geq \varepsilon$, then $r'(t) > 0$, for all t and so the particle escapes the central force, and (ii) if $r_0' < \varepsilon$, then $r'(t_*) = 0$, at some time t_* and $r'(t) < 0$ for all $t > t_*$. Thus, the particle stops and reverses its direction of travel. In case (ii), find the time t_*.

(g) Let τ be the function defined by

$$\tau(r_1) = \pm \int_{r_0}^{r_1} \frac{\sqrt{r}}{\sqrt{2}\sqrt{Er+k}} \, dr. \qquad (2.46)$$

This depends on E as well as the choice of the $\pm$ sign. Use equations (2.43)-(2.45) to find explicit formulas for τ in the three cases $E > 0, E < 0, E = 0$. Express your answers, in the first two cases, with the escape velocity ε in them. For particles with initial radial velocity that is negative, or positive but less than the escape velocity, show that the particle collides with the origin in a finite amount of time and find the exact time at which this occurs.

5. **(Conic sections in polar coordinates)** Ellipses, parabolas, and hyperbolas are particularly simple to describe in polar coordinates and indeed all three are described by the same equation. This is so *provided* the origin is one of the foci. Geometrically the description is one and the same as well. Thus, let F be a given point (a focus), D a given line (the directrix) not passing through F, and $e > 0$ a given number (the eccentricity). Consider the collection of all points P in the plane such that the distance $\overline{PF}$ to the focus in ratio to the distance to the $\overline{PD}$ to the directrix is equal to e. Specifically,

$$\frac{\overline{PF}}{\overline{PD}} = e.$$

See Figure 2.20. Take F as the origin and the x-axis perpendicular to D as shown. Let b be the distance from the focus to the directrix, $r = \overline{PF}$, $u = \overline{PD}$, and θ the angle as shown in the figure.

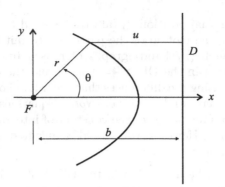

Figure 2.20: *Geometrical description of a conic section in polar coordinates.*

(a) Show that the polar equation for the curve with the above geometrical description is

$$r = \frac{eb}{1 + e \cos \theta}. \tag{2.47}$$

(b) Find the Cartesian equation of the curve with polar equation (2.47) and use this to show that the curve is a parabola if $e = 1$, an ellipse if $e < 1$, and a hyperbola if $e > 1$. In terms of e and b, compute all the standard quantities for these conics (coordinates of the center, major and minor axis lengths, etc.). Sketch the curves in the three cases.

6. **(Inverse square law)** For the inverse square law of attraction, we found that each integral curve, in polar coordinates, of the central force problem, lies on a curve with polar equation

$$r = \frac{1}{a + A \cos \theta + B \sin \theta}. \tag{2.48}$$

This exercise is to identify this latter curve as a conic section and to relate a, A, B with the constants from the initial conditions.

(a) Show that the curve (2.48) is a conic section with focus at the origin. *Hint:* Convert equation (2.48) into one of the form (2.47) as follows. Let $R = \sqrt{A^2 + B^2}$ and let δ be the unique angle in $[0, 2\pi]$ such that

$$\cos \delta = \frac{A}{R}, \qquad \sin \delta = \frac{B}{R}.$$

Then show that

$$A \cos \theta + B \sin \theta = R \cos(\theta - \delta),$$

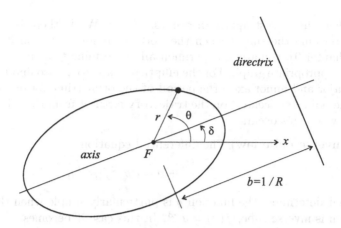

Figure 2.21: *The curve* $r = (a + A \cos \theta + B \sin \theta)^{-1}$ *is a conic section with focus, axis, and directrix as shown.*

and that equation (2.48) can be written in form (2.47) with

$$e = \frac{R}{a}, \qquad b = \frac{1}{R}.$$

Explain how this gives the picture shown in Figure 2.21. In particular describe how to locate the axis of the conic section and determine the point on the curve closest to the origin. This is known as the *pericenter* for the conic section. For the ellipse also determine the point with greatest distance from the origin (know as the *apocenter*).

(b) Suppose $t \mapsto (r(t), \theta(t))$ is an integral curve of the central force equations (2.28)-(2.29) in polar coordinates, with initial conditions $(r(0), \theta(0)) = (r_0, \theta_0)$ and $(r'(0), \theta'(0)) = (r_0', \theta_0')$. For simplicity assume that $\theta_0 = 0$. Let c and k be the constants introduced in the text, and, assuming the angular momentum is not zero ($c \neq 0$), let

$$a = \frac{k}{c^2}.$$

Use equation (2.36) to show that

$$A = \frac{1}{r_0} - a, \qquad B = -\frac{r_0^4 r_0'}{c}. \tag{2.49}$$

7. Use the results in Exercise 6 to prove that the three trajectories in Figure 2.18 are an ellipse, a parabola, and a hyperbola. For each trajectory, find the angle δ that the axis of the conic section makes with the x-axis, the point (pericenter) of closest approach to the origin, and the approximate time t_1

when the closest approach occurs. *Note*: While there is a way to exactly
determine the time t when the body is at any point on its trajectory (see
Chapter 9), you are to experimentally determine t_1 by using a computer and
an appropriate graph. For the elliptical trajectory, also find the lengths of the
major and minor axes, the period of the orbit (time for one complete cycle),
the point (apocenter) on the trajectory farthest from the origin, and the time
t_2 when this occurs.

8. **(Inverse cube law)** The differential equation

$$\rho'' + \rho = a\rho^{-2} f(\rho^{-1}),$$

that determines the function ρ is particularly simple when the law of attraction is inverse cube: $f(u) = u^{-3}$. In this case it becomes

$$\rho'' + (1 - a)\rho, \tag{2.50}$$

which has well-known solutions depending on the value of $1 - a$. Recall that
$a = k/c^2$.

(a) Suppose $a \equiv k/c^2 < 1$ and let $\omega = \sqrt{1 - a}$. Then the general solution
of (2.50) is

$$\rho = A\cos(\omega\theta) + B\sin(\omega\theta).$$

Use the technique from Exercise 6 to rewrite this with only a cosine
term and thus show that when the angular momentum $c > \sqrt{k}$ for a
solution $t \mapsto (r(t), \theta(t))$ of the equations of motion lies on the curve
with polar equation of the form.

$$r = \frac{R^{-1}}{\cos\omega(\theta - \delta)}. \tag{2.51}$$

Analyze the nature of this curve, explaining the significance of R, δ and
what happens when the denominator is zero. Is there an axis for the
curve, apocenter, or pericenter (or does the curve tend to infinity)?
Determine the relationship of A, B, R, δ to the initial data r_0, r_0' for the
DE (you may assume that $\theta_0 = 0$).

(b) Suppose $a \equiv k/c^2 > 1$ and let $\omega = \sqrt{a - 1}$. Then the general solution
of (2.50) is

$$\rho = A\cosh(\omega\theta) + B\sinh(\omega\theta). \tag{2.52}$$

Since the hyperbolic functions have entirely similar properties to the
corresponding trig functions, we can use a technique like that in Exercise
4 to rewrite the above as a single hyperbolic cosine function. However,
we do need to divide into cases depending on where the point (A, B)
lies in the x-y plane. Note that every point in the plane lies on a unique
hyperbola $x^2 - y^2 = \pm R^2$, or on the pair of lines $x^2 - y^2 = 0$.

(i) If $B = \pm A$ show that equation (2.52) reduces to $\rho = A e^{\pm \omega \theta}$ and thus the solutions of the polar equations of motion lie on the curve with polar equation

$$r = A^{-1} e^{\mp \omega \theta}.$$

Analyze the nature of the motion in each case (choice of $\pm$) and relate the constants to the initial data r_0, r_0'.

(ii) If $R^2 \equiv A^2 - B^2 > 0$, show that the solutions of the polar equations of motion lie on the curve with polar equation having one of the following forms:

$$r = \frac{R}{\cosh \omega (\theta + \delta)} \quad \text{or} \quad r = \frac{-R}{\cosh \omega (\theta - \delta)}.$$

Analyze the nature of the motion in each case (choice of $\pm$) and relate the constants to the initial data r_0, r_0'.

(iii) If $R^2 \equiv A^2 - B^2 < 0$, do an analysis like that in (ii).

Note: The curves in these cases (i), (ii), and (iii) are known as *Cotes' spirals*.

9. Give a precise meaning and explanation for the statement that, in the two-body problem, each body executes an orbit about the center of mass.

2.6 Summary

In using a computer to numerically approximate solutions and plot phase portraits for systems of DEs, it is essential to have some understanding of the algorithm used by the computer to obtain these approximations. This is especially so in effectively controlling and interpreting the computer plots. This chapter discussed the Euler numerical scheme as the most basic and easily understood numerical method for solving systems of differential equations. In practice, most software packages, like Maple, use more refined numerical techniques, but the Euler method is best for pedagogical purposes and also helps in understanding the more refined methods.

The chapter discussed gradient systems $x' = \nabla F(x)$, where $F : \mathcal{O} \subseteq \mathbb{R}^n \to \mathbb{R}$ is a given differentiable function, as special examples of systems that arise in heat flow and potential fluid flow. The form of gradient systems leads to the result that each integral curve of the system intersects each level set of F orthogonally. In dimension $n = 2$ this allows one to use the time-honored technique of sketching the integral curves from plots of a sequence of level curves.

The determination of fixed points and the classification of their stability was studied via a number of representative examples of systems in $\mathbb{R}^2$. While stability of fixed points for systems in the plane can usually be discerned from plots of the phase portrait, a rigorous definition of stability (see Chapter 6) is needed, especially in higher dimensions where visualization is difficult.

Limit cycles were introduced and their stability discussed by means of the technique of transforming a planar system into a system in polar coordinates. Again Chapter 6 will give precise definitions for these concepts, but here the discussion was directed toward showing how, like fixed points, these 1-dimensional geometrical objects influence the nature of the phase portrait and how the technique of transforming systems of DEs (studied in Chapter 5) is an important analytical tool.

The discussion of the two-body problem provides a prelude to the study of the N-body problem in Chapter 8. Even if you do not intend to study the material in Chapter 8, the discussion here is valuable as a motivation for why conservation laws are important in the analysis of systems of DEs. Conservation laws are synonymous with the existence of 1st integrals, and this topic is discussed in more detail in the chapter on integrable systems (Chapter 7) and, of course, in the chapters related to mechanics (Chapters 8 and 9).

Chapter 3

Existence and Uniqueness: The Flow Map

In this chapter we describe in detail several general results concerning the initial value problem (IVP):

$$x' = X(t, x)$$
$$x(t_0) = c.$$

The main result is the Existence and Uniqueness Theorem, from which many additional results can be derived. Throughout, $X : B \to \mathbb{R}^n$ is a time-dependent vector field on an open set $B \subseteq \mathbb{R}^{n+1}$. Various continuity and differentiability conditions will be imposed on X in order to get the results, but at the start we assume, at the bare minimum, that X is continuous on B. For the sake of reference we include the proofs of most of the results, although understanding some of the details requires being comfortable with several basic ideas from functional analysis.

The most fundamental and important construct to arise from the results presented here is the *flow map*, or simply the *flow*, ϕ generated by the vector field X. It is an indispensable notion and tool in many fields of study ranging from differential geometry to continuum mechanics.

A key ingredient in the proof of existence and uniqueness theorems is the fact that any initial value problem (IVP) can be reformulated as an integral equation. To see how to do this, we need a definition.

Definition 3.1 Suppose $\alpha : I \to \mathbb{R}^n$ is a continuous curve:

$$\alpha(s) = (\alpha_1(s), \ldots, \alpha_n(s)),$$

D. Betounes, *Differential Equations: Theory and Applications*, DOI 10.1007/978-1-4419-1163-6_3,

for $s \in I$. Assume I contains the given initial time t_0. For $t \in I$, the integral of α from t_0 to t is, by definition:

$$\int_{t_0}^{t} \alpha(s)\, ds = \left(\int_{t_0}^{t} \alpha_1(s)\, ds, \ldots, \int_{t_0}^{t} \alpha_n(s)\, ds \right).$$

Thus, the value of the integral is a point (or vector) in $\mathbb{R}^n$.

It is easy to see that the Fundamental Theorem of Calculus extends to such integrals of vector-valued functions. Specifically:

For a curve α as above, we get a curve $\beta : I \to \mathbb{R}^n$, given by:

$$\beta(t) \equiv \int_{t_0}^{t} \alpha(s)\, ds.$$

Here we use the convention that $\int_{t_0}^{t} = - \int_{t}^{t_0}$, if $t < t_0$. The curve β is differentiable and

$$\beta'(t) = \frac{d}{dt} \int_{t_0}^{t} \alpha(s)\, ds = \alpha(t),$$

for every $t \in I$. This is the first part of the Fundamental Theorem of Calculus. The second part is:

$$\int_{t_0}^{t} \gamma'(s)\, ds = \gamma(t) - \gamma(t_0),$$

for any differentiable curve: $\gamma : I \to \mathbb{R}^n$.

The reformulation of the IVP: $x' = X(t,x)$, $x(t_0) = c$, in terms of an equivalent integral equation, involves two observations:

Observation 1: Suppose that $\alpha : I \to \mathbb{R}^n$ is a solution of the IVP, i.e., $t_0 \in I$, $(s, \alpha(s)) \in B$, for all $s \in I$,

$$\alpha'(s) = X(s, \alpha(s)), \tag{3.1}$$

for all $s \in I$, and

$$\alpha(t_0) = c.$$

Then it's legitimate to integrate both sides of equation (3.1) from t_0 to t. Doing so, using the Fundamental Theorem of Calculus and the initial condition, shows that α satisfies the integral equation:

Integral Version of the IVP:

$$\alpha(t) = c + \int_{t_0}^{t} X(s, \alpha(s))\, ds \qquad (3.2)$$

for all $t \in I$.

In order to justify the name *integral version of the IVP*, we must also show than any solution α of equation (3.2), does indeed satisfy the original IVP. This comes from the following:

Observation 2: Suppose that $\alpha : I \to \mathbb{R}^n$ is a continuous curve that satisfies the integral equation (3.2). This means that $t_0 \in I$, $(s, \alpha(s)) \in B$, for all $s \in I$, and that (3.2) holds. Then clearly $\alpha(t_0) = c$. By the first part of the Fundamental Theorem of Calculus, the curve defined by the right side of equation (3.2) is differentiable, and thus so is α, and

$$\alpha'(t) = \frac{d}{dt}\left[c + \int_{t_0}^{t} X(s, \alpha(s))ds \right] = X(t, \alpha(t)),$$

for all $t \in I$.

This shows the equivalence of the differential and integral versions of the IVP.

Example 3.1 The initial value problem:

$$
\begin{aligned}
x' &= (x+1)y \\
y' &= x - y,
\end{aligned}
$$

$x(0) = 3$, $y(0) = 5$, is equivalent to the system of integral equations:

$$
\begin{aligned}
\alpha_1(t) &= 3 + \int_0^t (\alpha_1(s) + 1)\, \alpha_2(s)\, ds \\
\alpha_2(t) &= 5 + \int_0^t (\alpha_1(s) - \alpha_2(s))\, ds.
\end{aligned}
$$

Of course reformulating, an IVP as an integral equation does not make it any easier to solve by hand. However, from a theoretical point of view integral equations are easier to deal with than differential equations.

Inherent in the integral equation version of the IVP is a certain operator, or transformation, T acting on curves. This is defined as follows.

Definition 3.2 Suppose $\beta : I \to \mathbb{R}^n$ is any continuous curve with $(s, \beta(s)) \in B$, for all $s \in I$. Let $T(\beta)$ denote a new curve on I (a transformation of β) defined by:

$$T(\beta)(t) = c + \int_{t_0}^{t} X(s, \beta(s))ds, \qquad (3.3)$$

for $t \in I$.

Using the notation from the definition, it is clear that the integral version (3.2) of the IVP is

$$\alpha = T(\alpha).$$

This equation just says that α is a *fixed point* of the map T. This is not to be confused with the notion of a fixed point of an autonomous vector field (although we shall see later that a fixed point of a vector field is indeed a fixed point of the flow it generates). This viewpoint of considering a solution of the IVP as nothing other than a fixed point of the map T has led to generalizations and broad abstractions of the material present here. In addition, the key ingredient in the proof of the existence and uniqueness theorem below is the idea that a fixed point of T can be found by iterating the action of the map T. This gives the modern version of Picard's iteration scheme.

3.1   Picard Iteration

The Picard iteration scheme is a method used not only to prove the existence of a solution α of an IVP (i.e., a fixed point of T), but also to give us explicit approximations to α. The method is actually quite simple and is described as follows.

   Again assume that I is an interval containing the initial time t_0 for the IVP. Let $\mathcal{C}$ be the set of all continuous curves $\beta : I \to \mathbb{R}^n$, and for simplicity in this section on Picard iteration, assume that $B = I \times \mathbb{R}^n$. Then the map T, defined by equation (3.3), has the set $\mathcal{C}$ as its domain as well as its codomain, i.e., $T : \mathcal{C} \to \mathcal{C}$. Note that for a general open set B in $\mathbb{R}^{n+1}$, there is no guarantee that $T(\beta) \in \mathcal{C}$ for each $\beta \in \mathcal{C}$. In the Existence and Uniqueness Theorem in the next section, where there is no restriction on B, we will have to choose a somewhat smaller set of curves $\mathcal{C}$, which T, in the general case, will map into itself.

Definition 3.3 For a given curve $\beta \in \mathcal{C}$, the *Picard iterates* of β are the elements in the sequence $\{T^k(\beta)\}_{k=1}^{\infty}$,

$$
\begin{aligned}
T^0(\beta) &= \beta \\
T^1(\beta) &= T(\beta) \\
T^2(\beta) &= T(T(\beta)) \\
T^3(\beta) &= T(T(T(\beta))) \\
&\vdots \\
T^k(\beta) &= T(T^{k-1}(\beta)).
\end{aligned}
$$

It will be shown in the Existence and Uniqueness Theorem that, under appropriate assumptions, a sequence of Picard iterates always converges to a fixed point α of T, *regardless of the choice for the initial curve* β. This is a rather remarkable fact. The assumptions we use also guarantee that α is unique. Thus, we can start with *any* curve β and take the limit of its iterates to find α. Because of this, a standard choice for the initial curve is the constant curve: $\beta(t) = c$, for all $t \in I$.

Example 3.2 If the system of DEs is not too complicated, the first few Picard iterates are easy to compute by hand. For example, consider the system:

$$
\begin{aligned}
x' &= x(y+1) \\
y' &= x^2 - y^2,
\end{aligned}
$$

with initial conditions $x(0) = 1, y(0) = 1$. Then $c = (1,1)$ and the vector field X is

$$
X(x, y) = (x(y+1), x^2 - y^2).
$$

If we take the constant map

$$
\beta(t) = (1, 1),
$$

for all $t \in \mathbb{R}$, as the initial curve, then since $X(1, 1) = (2, 0)$, the first Picard iterate is

$$
T(\beta)(t) = (1, 1) + \int_0^t (2, 0)ds = (1 + 2t, 1).
$$

Since $X(1 + 2s, 1) = (2 + 4s, 4s + 4s^2)$, the 2nd Picard iterate is

$$T^2(\beta)(t) = (1, 1) + \int_0^t (2 + 4s, 4s + 4s^2)ds$$
$$= (1 + 2t + 2t^2, 1 + 2t^2 + \tfrac{4}{3}t^3).$$

The higher-order Picard iterates get rather complicated algebraically and while the calculations at each step involve only polynomial integration, the work quickly becomes tedious if done by hand. For example, it takes several minutes to calculate the 3rd iterate:

$$T^3(\beta)(t) =$$
$$\left(1 + 2t + 2t^2 + 2t^3 + \tfrac{4}{3}t^4 + \tfrac{4}{3}t^5 + \tfrac{4}{9}t^6, 1 + 2t^2 + \tfrac{4}{3}t^3 + \tfrac{4}{3}t^4 - \tfrac{8}{9}t^6 - \tfrac{16}{63}t^7\right)$$

by hand. For $T^4(\beta)$, which has polynomial components of the 14th and 15th degrees, and the rest of the iterates, it is best to use a computer algebra system to calculate the results. One can also use a computer to see how well the iterates approximate the actual solution α near c. Figure 3.1 illustrates this for the example here. As the figure indicates, the 1st Picard iterate is

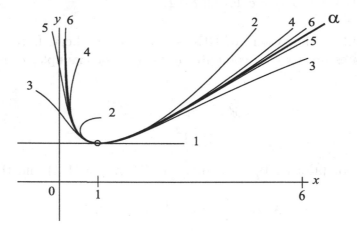

Figure 3.1: *Plots of the Picard iterates $T^k(\beta)$, $k = 1, \ldots, 6$ that approximate the solution α if the IVP in Example 3.2.*

the tangent line to α. It's not hard to show, in general, that whenever X is an autonomous vector field and for the initial curve we take $\beta(t) = c$, for all t, then the first Picard iterate $T(\beta)$ is the tangent line to the actual solution α at $\alpha(0) = c$.

Exercises 3.1

1. Let $\beta^k = T^k(\beta)$, $k = 1, 2, 3, \ldots$, be the sequence of Picard iterates for an initial value problem $x' = X(x)$, $x(0) = c$. For each of the following IVPs compute, *by hand*, the first few Picard iterates: β_1, β_2, and β_3, starting with $\beta(t) = \beta_0 \equiv c$. This will be informative and good practice for future tests. Then use a computer to compute the iterates β_k, $k = 1, \ldots, 6$ and graph these, in the same figure, along with the actual solution α of the IVP. Choose suitable ranges for the plots and annotate the resulting figure, identifying the iterates and the exact solution α. Also, do the additional studies asked for in each specific problem. You may find it helpful to read and use the Maple worksheet picard.mws on the electronic component.

   (a)
   $$x' = 1 + x^2$$
   $$x(0) = 0.$$

   Also, for this problem, find the general solution of the DE.

   (b)
   $$x' = -y$$
   $$y' = x(y + 1)$$
   $$(x(0), y(0)) = (1, 1).$$

   Also, do the following additional study:

   (i) The actual solution: $\alpha(t) = (x(t), y(t))$, of the initial value problem is a closed curve. Plot, in the same figure, the graphs of the component functions x, y on a sufficient time interval to determine the approximate period p of α.

   (ii) Study how well the Picard iterates $\beta(t) = (u(t), v(t))$, $k = 1, \ldots, 6$, approximate α on the whole time interval $[-p, p]$. Include plots of the graphs of x and u_4, u_5, u_6 in the same figure and plots of y and v_4, v_5, v_6 in the same figure (properly annotated). If your computer is capable, extend the study to the iterates $\beta_7, \beta_8, \beta_9$, but be careful of memory requirements. The theory is that the Picard iterate β_k, for large k, will approximate α well only on a neighborhood, i.e., interval I, of the initial time $t = 0$. From your figures determine, for each iterate, the largest interval on which the approximation appears good. Why, for the particular approximations here, would you expect them only to be good for t close to zero ?

   (c)
   $$x' = y - x$$
   $$y' = x(y - 2)$$
   $$(x(0), y(0)) = (-1, 1).$$

Also do the following additional study. The actual solution: $\alpha(t) = (x(t), y(t))$, of the initial value problem is a spiral that tends to the origin as $t \to \infty$. Study how well the Picard iterates approximate α by following the instructions in part (b)(ii) of the last problem, but now use the time interval $[0, 20]$ instead of $[-p, p]$.

2. Let $\beta^k = T^k(\beta)$, $k = 1, 2, 3, \ldots$, be the sequence of Picard iterates for an initial value problem $x' = X(x)$, $x(0) = c$, and assume that the initial curve β is the constant curve $\beta(t) = c$, for every $t \in \mathbb{R}$. Let $\alpha : I \to \mathbb{R}^n$ be an (the) actual solution of the initial value problem. Show that the 1st Picard iterate β_1, is the tangent line to the curve α at c. Show that the 2nd Picard iterate β_2 is a curve that passes through c and has the same velocity and acceleration as α at time 0, i.e., $\beta_2'(0) = \alpha'(0)$ and $\beta_2''(0) = \alpha''(0)$. Can you generalize?

3.2   Existence and Uniqueness Theorems

There are numerous existence and uniqueness results in the literature and this section discusses just one from the many available. By assuming less than we do here, more general results on the existence of solutions to systems of DEs can be proven. In addition, existence *and* uniqueness can be proven by using the more general idea of a Lipschitz condition. These extensions will be discussed in the exercises and Appendix B.

Theorem 3.1 (Existence and Uniqueness Theorem) *Suppose $X : B \to \mathbb{R}^n$ is a time-dependent vector field on $B \subseteq \mathbb{R} \times \mathbb{R}^n$. Assume that all the partials $\partial X^i / \partial x_j, i, j = 1, \ldots, n$, exist and are continuous on B. Then for each point $(t_0, c) \in B$, there exists a curve $\alpha : I \to \mathbb{R}^n$, with $t_0 \in I$, which satisfies the initial value problem*

$$x' = X(t, x)$$
$$x(t_0) = c.$$

Furthermore, if $\gamma : J \to \mathbb{R}^n$ is any other solution for the initial value problem, then there is an interval $Q \subseteq I \cap J$, with $t_0 \in Q$ such that:

$$\alpha(t) = \gamma(t) \qquad\qquad \text{for every } t \in Q. \qquad (3.4)$$

Hence any two solutions of the initial value problem agree on a neighborhood of t_0.

Note: Corollary 3.1 below improves on the uniqueness result by showing that equation (3.4) holds for all $t \in I \cap J$. For now it's more expedient to just prove the restricted result stated in the theorem.

Proof: We will need to make some estimates that require a norm on the elements $x = (x_1, \ldots, x_n)$ in $\mathbb{R}^n$. The usual norm $|x| = $ the length of x, while being geometrically preferred ($|x - y|$ gives the distance between the points $x, y \in \mathbb{R}^n$), is not as easy to use here as is the norm $\| \cdot \|$ defined by

$$\|x\| = \sum_{i=1}^{n} |x_i|.$$

This is called the ℓ_1 norm, and $\|x - y\|$ is abstractly still considered as measuring the "distance" between x and y. With respect to this norm, we use the notation:

$$\overline{B}(c, r) \equiv \{ x \in \mathbb{R}^n \mid \|x - c\| \leq r \},$$

for the closed ball in $\mathbb{R}^n$, centered at c, and having radius r.

Since B is an open set and $(t_0, c) \in B$, we can choose $r > 0, b > 0$ so that $[t_0 - b, t_0 + b] \times \overline{B}(c, r) \subset B$. Since the functions $X^i, \partial X^i / \partial x_j, i, j = 1, \ldots, n$ are continuous on B, and therefore also on $[t_0 - b, t_0 + b] \times \overline{B}(c, r)$, and the latter set is compact, there exists a constant $K > 0$ such that:

$$|X^i(t, x)| \leq K \tag{3.5}$$

$$\left| \frac{\partial X^i}{\partial x_j}(t, x) \right| \leq K, \tag{3.6}$$

for all $(t, x) \in [t_0 - b, t_0 + b] \times \overline{B}(c, r)$, and all $i, j = 1, \ldots, n$. Inequality (3.5) immediately leads to the inequality:

$$\|X(t, x)\| = \sum_{i=1}^{n} |X^i(t, x)| \leq nK, \tag{3.7}$$

for all $(t, x) \in [t_0 - b, t_0 + b] \times \overline{B}(c, r)$. We also claim that inequality (3.6) leads to the inequality:

$$\|X(t, x) - X(t, y)\| \leq nK \|x - y\|, \tag{3.8}$$

for all $x, y \in \overline{B}(c, r)$ and $t \in [t_0 - b, t_0 + b]$. To see this, fix $x, y \in \overline{B}(c, r)$, $t \in [t_0 - b, t_0 + b]$, and for each i define a function h_i by:

$$h_i(\lambda) = X^i(t, \lambda y + (1 - \lambda)x),$$

for $\lambda \in [0, 1]$. Note that the point $\lambda y + (1 - \lambda)x$ lies on the line segment joining x and y, and so lies in the set $\overline{B}(c, r)$, since this set is convex. Thus, the above definition of h_i makes sense. Now apply the mean value theorem to h_i to get that there exists a $\lambda_0 \in (0,1)$ such that

$$h_i(1) - h_i(0) = h_i'(\lambda_0).$$

Using this, the chain rule to differentiate h_i, and inequality (3.6) gives us

$$
\begin{aligned}
|X^i(t,x) - X^i(t,y)| &= |h_i(1) - h_i(0)| = |h_i'(\lambda_0)| \\
&= \left| \sum_{j=1}^{n} \frac{\partial X^i}{\partial x_j} \left(t, \lambda_0 y + (1 - \lambda_0)x \right) [y_j - x_j] \right| \\
&\leq K \sum_{j=1}^{n} |y_j - x_j| = K\|y - x\|.
\end{aligned}
$$

Thus, summing both sides of this last inequality as i goes from 1 to n gives the desired inequality (3.8).

Now choose a number $a > 0$ such that

$$a < r/(nK), \qquad a < b \qquad a < 1/(nK),$$

and let $\mathcal{C}$ denote the following set of curves:

$$\mathcal{C} = \{ \beta : [t_0 - a, t_0 + a] \to \overline{B}(c, r) \,|\, \beta \text{ is continuous } \}.$$

Restricting the transformation T defined by

$$T(\beta)(t) = c + \int_{t_0}^{t} X(s, \beta(s)) \, ds$$

to the curves in the set $\mathcal{C}$, we claim that $T : \mathcal{C} \to \mathcal{C}$. That is, if β is in $\mathcal{C}$, then also $T(\beta)$ is in $\mathcal{C}$. Note that $T(\beta)$ is automatically continuous, since, by the Fundamental Theorem of Calculus, it is differentiable. Thus, all we have to show is that $T(\beta)(t) \in \overline{B}(c, r)$ for all $t \in [t_0 - a, t_0 + a]$. To see this, suppose first that $t > t_0$. Then

$$
\begin{aligned}
\|T(\beta)(t) - c\| &= \left\| \int_{t_0}^{t} X(s, \beta(s)) \, ds \right\| \\
&\leq \int_{t_0}^{t} \|X(s, \beta(s))\| \, ds \\
&\leq nK(t - t_0) \\
&< nKa < r.
\end{aligned}
$$

The first inequality in the above is a general result which follows from the definition of the integral of a curve from t_0 to t and the special choice of the norm $\|\cdot\|$. The other inequalities follow from the above work. The argument in the case that $t < t_0$ is similar.

We next show that $T : C \to C$ is a *contraction map*, i.e., there exists a constant $0 < q < 1$ such that

$$\|T(\beta) - T(\gamma)\| \le q\|\beta - \gamma\|, \qquad (3.9)$$

for all $\beta, \gamma \in C$. This inequality says that the distance between $T(\beta)$ and $T(\gamma)$ is strictly less than the distance between β and γ (since $q < 1$). Thus, T "contracts" the distance between points. This is the crucial property needed to ensure convergence of the Picard iterates.

The contraction property of T in (3.9) comes from an estimate involving yet another norm. This is a norm *not* on $\mathbb{R}^n$, but rather on the vector space:

$$C \equiv \{\beta : [t_0 - a, t_0 + a] \to \mathbb{R}^n \mid \beta \text{ is continuous }\},$$

of *all* continuous curves on the interval $[t_0 - a, t_0 + a]$ (not just the ones that lie in $\overline{B}(c, r)$). This vector space C is infinite-dimensional and contains the set $\mathcal{C}$ as a subset. The norm on C, which also applies to elements in $\mathcal{C}$, is a natural one, called the *sup* or *supremum norm*. It is defined as follows. If $\beta \in C$, then the map $t \mapsto \|\beta(t)\|$ is continuous on the closed interval $[t_0 - a, t_0 + a]$, and so by the Extreme Value Theorem from advanced calculus, attains its largest (or supreme) value on this interval. The value is the norm of β. Symbolically we write

$$\|\beta\| = \sup\{\|\beta(t)\| \mid t \in [t_0 - a, t_0 + a]\}.$$

With respect to this norm, we will show that T is a contraction on $\mathcal{C}$. Note: By the above definition, $\|\beta(t)\| \le \|\beta\|$, for all t.

Now suppose that $\beta, \gamma \in \mathcal{C}$ and $t \in [t_0 - a, t_0 + a]$. We assume that $t_0 < t$. The other case is similar. Then we get

$$\begin{aligned}
\|T(\beta)(t) - T(\gamma)(t)\| &= \left\| \int_{t_0}^{t} [X(s, \beta(s)) - X(s, \gamma(s))] \, ds \right\| \\
&\le \int_{t_0}^{t} \|X(s, \beta(s)) - X(s, \gamma(s))\| \, ds \\
&\le nK \int_{t_0}^{t} \|\beta(s) - \gamma(s)\| \, ds \\
&\le nK|t - t_0|\|\beta - \gamma\| \le nKa\|\beta - \gamma\|.
\end{aligned}$$

We get the same result when $t_0 > t$. Hence by definition of the supremum, we get

$$\|T(\beta) - T(\gamma)\| \le nKa\|\beta - \gamma\|. \tag{3.10}$$

Thus, we can take

$$q = nKa < 1$$

to get that T is a contraction mapping on $\mathcal{C}$. We now use this to show that the sequence of Picard iterates converges to a unique solution of the IVP, for *any* choice of $\beta \in \mathcal{C}$.

We have already seen that T maps $\mathcal{C}$ into itself: $T : \mathcal{C} \to \mathcal{C}$. So suppose we select any curve $\beta \in \mathcal{C}$ and apply T to it successively, resulting in the sequence of Picard iterates:

$$\{T^k(\beta)\}_{k=1}^{\infty},$$

which is a sequence of curves (or abstract points) in $\mathcal{C}$. We first get an estimate on how far apart $T^k(\beta)$ and $T^{k+p}(\beta)$ are with respect to the sup norm. Using the contraction property (3.9) repeatedly, we derive the inequality:

$$
\begin{aligned}
\|T^k(\beta) - T^{k+p}(\beta)\| &= \|T(T^{k-1}(\beta)) - T(T^{k+p-1})(\beta))\| \\
&< q\|T^{k-1}(\beta) - T^{k+p-1}(\beta)\| \\
&\vdots \\
&< q^k\|\beta - T^p(\beta)\|, \tag{3.11}
\end{aligned}
$$

for any k and p. A particular case of this inequality is for $p = 1$, from which we get

$$\|T^k(\beta) - T^{k+1}(\beta)\| \le q^k\|\beta - T(\beta)\|. \tag{3.12}$$

To analyze inequality (3.11) further, we ask how far apart can β and $T^p(\beta)$ be with respect to the sup norm? To answer this, note that the sup norm satisfies the triangle inequality: $\|\gamma + \mu\| \le \|\gamma\| + \|\mu\|$, for any two curves $\gamma, \mu \in C$. The triangle inequality extends to the sum of any number of curves. Using this and the contraction property (3.9), and inequality (3.12), we get

$$
\begin{aligned}
\|\beta - T^p(\beta)\| &= \|\beta - T(\beta) + T(\beta) - T^2(\beta) + \cdots + T^{p-1}(\beta) - T^p(\beta)\| \\
&\le \|\beta - T(\beta)\| + \|T(\beta) - T^2(\beta)\| + \cdots + \|T^{p-1}(\beta) - T^p(\beta)\| \\
&\le \|\beta - T(\beta)\|(1 + q + \cdots + q^{p-1}) \\
&\le \|\beta - T(\beta)\|/(1 - q) \tag{3.13}
\end{aligned}
$$

If we use this on the last part of inequality (3.11), we get:

$$\|T^k(\beta) - T^{k+p}(\beta)\| < q^k \|\beta - T(\beta)\|/(1-q), \qquad (3.14)$$

for all k and p. This is the result we have been working toward. It says (since $\lim_{k\to\infty} q^k = 0$) that the sequence $\{T^k(\beta)\}_{k=0}^{\infty}$ is a Cauchy sequence in $C \subseteq \mathcal{C}$. It is a standard result in functional analysis that C is *complete*, i.e., every Cauchy sequence in C converges to some element in C. Thus, there exists an $\alpha \in C$ such that

$$\lim_{k\to\infty} \|T^k(\beta) - \alpha\| = 0.$$

(This is equivalent to saying the sequence of curves: $\{T^k(\beta)\}_{k=0}^{\infty}$, converges uniformly on $[t_0 - a, t_0 + a]$ to the curve α.)

We now want to show that in fact $\alpha \in \mathcal{C}$ and is a fixed point of T. To see the first assertion, note that for every $t \in [t_0 - a, t_0 + a]$, and for all k

$$
\begin{aligned}
\|\alpha(t) - c\| &\leq \|\alpha - c\| \\
&\leq \|\alpha - T^k(\beta)\| + \|T^k(\beta) - c\| \\
&\leq \|\alpha - T^k(\beta)\| + r.
\end{aligned}
$$

Here we are abusing notation and writing c, when we really mean the constant curve: $\gamma(t) \equiv c$, for every t. Also in the above we used the fact that $T^k(\beta) \in \overline{B}(c, r)$. Now taking the limit as $k \to \infty$ in the above gives:

$$\|\alpha(t) - c\| \leq r,$$

for every $t \in [t_0 - a, t_0 + a]$, which says that the curve α lies in $\overline{B}(c, r)$.

As mentioned above the curve to which the sequence of iterates converges is an element of C, which means in the case at hand that α is continuous. Thus, we have verified that $\alpha \in \mathcal{C}$.

To see that α is a fixed point, it suffices to observe that since T is a continuous map:

$$
\begin{aligned}
T(\alpha) &= T\left(\lim_{k\to\infty} T^k(\beta)\right) \\
&= \lim_{k\to\infty} T(T^k(\beta)) \\
&= \lim_{k\to\infty} T^{k+1}(\beta) = \alpha.
\end{aligned}
$$

All of the above shows the existence of a solution α defined on the interval $I = [t_0 - a, t_0 + a]$, and we finally wish to show that α is unique in the sense that if $\gamma : J \to \mathbb{R}^n$ is any other solution of the IVP, then $\gamma = \alpha$ on an interval $Q = [t_0 - m, t_0 + m] \subseteq I \cap J$. We choose $m > 0$ as follows. By continuity of γ, there is a $\delta > 0$ such that $|\gamma(t) - c| \leq r$ for all t such that $|t - t_0| < \delta$. Now take $m = \min\{\delta, a\}$. By construction $\alpha(t)$ and $\beta(t)$ lie in $\overline{B}(c,r)$ for all $t \in Q$. Next let

$$M = \sup\{ \, \|\alpha(t) - \gamma(t)\| \, | \, t \in Q \, \}.$$

Then all we need to show is that $M = 0$. The argument for this is entirely similar to that used in deriving inequality (3.10). Thus, suppose $t \in Q$. Then since α and γ satisfy the IVP, we have

$$\alpha(t) \;=\; c + \int_{t_0}^{t} X(s, \alpha(s))ds$$

$$\gamma(t) \;=\; c + \int_{t_0}^{t} X(s, \gamma(s))ds,$$

and so we get (assuming that $t > t_0$)

$$
\begin{aligned}
\|\alpha(t) - \gamma(t)\| &= \left\| \int_{t_0}^{t} [X(\alpha(s)) - X(\gamma(s))] \, ds \right\| \\
&\leq \int_{t_0}^{t} \|X(\alpha(s)) - X(\gamma(s))\| \, ds \\
&\leq nK \int_{t_0}^{t} \|\alpha(s) - \gamma(s)\| \, ds \\
&\leq nK|t - t_0|M \;\leq\; nKaM.
\end{aligned}
$$

Note that the second inequality above comes from inequality (3.8) and requires that $\alpha(s), \gamma(s) \in \overline{B}(c,r)$ for all $s \in Q$. That is why we chose Q the way we did and is also why we must settle for a weaker version of uniqueness (which we strengthen in the next corollary). Many differential equations texts incorrectly handle the argument on uniqueness at this point, so beware.

Similar reasoning holds if $t < t_0$. From this last inequality and the definition of the sup, and with $q = nKa$ as before, we get:

$$M \leq qM.$$

Since $q < 1$, the only way for this inequality to hold is for $M = 0$, which is what we wanted to show. This completes the proof. $\square$

The proof just given, while quite lengthy, is certainly worth your study. It exhibits several ideas—contraction mapping, fixed points, and completeness of normed linear spaces—that have become essential to much of modern analysis. There are several results connected with the Existence and Uniqueness Theorem. These are directed to the all-important definition of the flow generated by the vector field X.

Corollary 3.1 (Global Uniqueness) *If $\alpha : I \to \mathbb{R}^n$ and $\beta : J \to \mathbb{R}^n$ are two solutions of the system $x' = X(t, x)$ and if $\alpha(t_0) = \beta(t_0)$ for some $t_0 \in I \cap J$, then*

$$\alpha(t) = \gamma(t) \qquad\qquad \textit{for every } t \in I \cap J.$$

Proof: Let a_0, b_0 be the left- and right-hand end points, respectively, of the interval $I \cap J$ (possibly $a_0 = -\infty$ or $b_0 = \infty$). Define two sets of numbers:

$$A = \{\, a \in I \cap J \,|\, \alpha = \beta \text{ on } (a, t_0] \,\}$$
$$B = \{\, b \in I \cap J \,|\, \alpha = \beta \text{ on } [t_0, b) \,\}.$$

If A is not bounded below, then $\alpha = \beta$ on $(-\infty, t_0]$ and if B is not bounded above, then $\alpha = \beta$ on $[t_0, \infty)$. Thus, $\alpha = \beta$ on $\mathbb{R}$, and so the corollary is certainly true.

Suppose B is bounded above and let $b_1 = \sup B \leq b_0$. Note that if $t \in [t_0, b_1)$, then t is not an upper bound for B and so there is a $b \in B$ such that $t < b$. Then $\alpha(t) = \beta(t)$. This shows that $\alpha = \beta$ on $[t_0, b_1)$.

Claim: $b_1 = b_0$.

If not, then $b_1 < b_0$, and we get a contradiction as follows. Since b_1 is not an end point of $I \cap J$, there is an $\varepsilon > 0$ such that $(b_1 - \varepsilon, b_1 + \varepsilon) \subset I \cap J$. See Figure 3.2. Now $\alpha = \beta$ on $(b_1 - \varepsilon, b_1)$. For if $t \in (b_1 - \varepsilon, b_1)$, then t is not an upper bound for B and so there is a $b \in B$ such that $t < b$. Hence

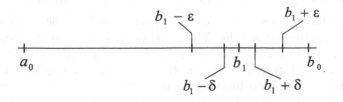

Figure 3.2: *The intervals $(b_1 - \varepsilon, b_1 + \varepsilon)$ and $(b_1 - \delta, b_1 + \delta)$ that occur in the proof.*

$\alpha(t) = \beta(t)$. Having thus shown that $\alpha = \beta$ on $(b_1 - \varepsilon, b_1)$, continuity of α and β gives in addition that $\alpha(b_1) = \beta(b_1)$. By the local uniqueness of solutions of initial value problems, we have that $\alpha = \beta$ on a neighborhood of b_1, say on $(b_1 - \delta, b_1 + \delta)$, for some δ. We have, in total, that $\alpha = \beta$ on the intervals $[t_0, b_1 - \varepsilon]$, $(b_1 - \varepsilon, b_1]$, and $[b_1, b_1 + \delta)$. Thus, if $b \in (b_1, b_1 + \delta)$, then $\alpha = \beta$ on $[t_0, b)$. Hence $b \in B$ and $b > b_1$. This contradicts the fact that b_1 is the least upper bound for the set B.

With the claim verified, we have that $\alpha = \beta$ on $[t_0, b_0)$. A similar argument can be constructed for the case when A is bounded below and gives that $\alpha = \beta$ on $(a_0, t_0]$. In total, we have $\alpha = \beta$ on (a_0, b_0), i.e., on $I \cap J$, except possibly at the endpoints a_0, b_0, if one or both of these belong to $I \cap J$. If $a_0 \in I \cap J$, then continuity gives that $\alpha(a_0) = \beta(a_0)$. Similarly if $b_0 \in I \cap J$. This completes the proof. $\square$

Exercises 3.2

1. **(Continuous Dependence on Parameters)** This exercise is a good one for reinforcing your understanding of the proof of Theorem 3.1. It also gives an important generalization of that theorem. The generalization involves a time-dependent vector field which also depends on additional parameters $u_1, \ldots, u_m$. For example, the perturbed pendulum model in Example 1.6 has a time-dependent vector field $X : \mathbb{R} \times \mathbb{R}^2 \to \mathbb{R}^2$ given by

$$X(t, \theta, v) = \left(v, \; -g\sin(\theta) + \frac{a^2 b^2}{2} \sin^2(bt) \sin(2\theta) \right),$$

where the parameters g, a, b are the acceleration of gravity, the string length (or hoop radius), and the magnitude of the oscillation about the vertical axis. Thus, the solutions of the corresponding system of DEs implicitly depend on the parameters g, a, b, and we would like to know how small variations in the parameters affect the corresponding solutions. For example, if b is very small, but nonzero, is the phase portrait for the system very "close" to the phase portrait shown in Figure 1.5 for the unperturbed pendulum?

To formulate a precise result, we explicitly include the parameters in the domain for the vector field. Thus, assume that G is an open subset of $\mathbb{R} \times \mathbb{R}^n \times \mathbb{R}^m$ and that $X : G \to \mathbb{R}^n$ is a time- and parameter-dependent vector field. Then $X(t, x, u) \in \mathbb{R}^n$ for $(t, x, u) \in G$. A solution of the system $x' = X(t, x, u)$ is, by definition, a map $\alpha : I \times U \to \mathbb{R}^n$, where I is an interval and U is an open set in $\mathbb{R}^m$, such that $(t, \alpha(t, u), u) \in G$ for all $(t, u) \in I \times U$, and

$$\frac{\partial \alpha}{\partial t}(t, u) = X(t, \alpha(t, u), u),$$

for all $(t, u) \in I \times U$. This assumes implicitly that α is differentiable with respect to t. With these definitions, prove the following generalization of Theorem 3.1

Theorem 3.2 *Suppose $X : G \to \mathbb{R}^n$ is a time- and parameter-dependent vector field on $G \subseteq \mathbb{R} \times \mathbb{R}^n \times \mathbb{R}^m$. Assume that all the partials $\partial X^i / \partial x_j, i, j = 1, \ldots, n$, exist and are continuous on G. Then for each point $(t_0, c, u_0) \in G$, there exists a differentiable map $\alpha : I \times U \to \mathbb{R}^n$, with $t_0 \in I$ and $u_0 \in U$, which satisfies the initial value problem:*

$$x' = X(t, x, u)$$
$$x(t_0, u_0) = c,$$

for all $u \in U$. Furthermore, if $\gamma : J \times V \to \mathbb{R}^n$ is any other solution for the initial value problem, then there is an interval $Q \subseteq I \cap J$, with $t_0 \in Q$, and an open set $W \subseteq U \cap V$, such that

$$\alpha(t, u) = \gamma(t, u) \qquad \text{for every } t \in Q \text{ and } u \in W. \tag{3.15}$$

Hence any two solutions of the initial value problem agree on a neighborhood of t_0.

Hints and Suggestions: Modify the proof of Theorem 3.1 by including a u in the notation, using the same constants a, r, b, K, and use the following set of maps for $\mathcal{C}$:

$$\mathcal{C} = \{\beta : [t_0 - a, t_0 + a] \times \overline{B}(u_0, r') \to \overline{B}(c, r) \,|\, \beta \text{ is continuous}\},$$

where $r' > 0$ is an appropriately chosen number.

2. **(Contraction Mapping Principle)** In proving Theorem 3.1, we constructed a set of curves $\mathcal{C}$ which the Picard iteration map T mapped into itself: $T : \mathcal{C} \to \mathcal{C}$ and on which T was a contraction. This technique, generalizes and abstracts to other situations and has been elevated to a basic principle in analysis and functional analysis. This exercise studies such a generalization.

Definition 3.4 (Metric Spaces) A map $d : M \times M \to \mathbb{R}$ on a set M is called a *metric* if

(1) $d(x, y) \geq 0$, for all $x, y \in M$, and $d(x, y) = 0$ if and only if $x = y$.
(2) $d(x, y) = d(y, x)$, for all $x, y \in M$. (Symmetry)
(3) $d(x, y) \leq d(x, z) + d(z, y)$, for all $x, y, z \in M$. (Triangle Inequality)

The pair (M, d), consisting of a set together with a metric on this set, is called a *metric space*. A map $T : M \to M$ is called a *contraction* if there is a positive constant q such that

$$d(T(x), T(y)) \leq q \, d(x, y), \qquad \text{for all } x, y \in M. \tag{3.16}$$

The metric d is considered as measuring the distance between points in M, with $d(x, y)$ being the distance between x and y. Thus, equation (3.16) says that a contraction map decreases the distances between points. As with the Picard map, we let T^k denote the composition of T with itself k times, i.e., $T^2 \equiv T \circ T$, $T^3 \equiv T \circ T \circ T$, and so on.

Relative to these concepts, suppose $T : M \to M$ is a contraction and do the following.

(a) Suppose x is any point in M. Imitate the pertinent steps in the proof of Theorem 3.1 to show that

$$d(T^k(x), T^{k+p}(x)) \le \frac{q^k(1 - q^p)}{1 - q}\, d(x, T(x)), \qquad (3.17)$$

for all positive integers k, p.

(b) A sequence $\{x_k\}_{k=1}^{\infty}$ in M is called a *Cauchy sequence* if for every $\varepsilon > 0$ there is a positive integer K such that $d(x_k, x_m) < \varepsilon$, for all $k, m \ge K$. Use the result in part (a) to show that $\{T^k(x)\}_{k=1}^{\infty}$ is a Cauchy sequence in M for any choice of $x \in M$.

(c) A map $f : M \to M$ is called *continuous* if for every $x \in M$ and every $\varepsilon > 0$, there is a $\delta > 0$ such that $d(f(x), f(y)) < \varepsilon$, for every x, y with $d(x, y) < \delta$. Show that f is continuous.

(d) A sequence $\{x_k\}_{k=1}^{\infty}$ in M is said to *converge* to $c \in M$, if for every $\varepsilon > 0$ there is a K such that $d(x_k, c) < \varepsilon$, for every $k \ge K$. In this case we write $\lim_{k \to \infty} x_k = c$. Show that if $f : M \to M$ is a continuous map and $\lim_{k \to \infty} x_k = c$, then $\lim_{k \to \infty} f(x_k) = f(c)$.

(e) A metric space (M, d) is called *complete* if every Cauchy sequence in M converges to a point in M. Use parts (b)-(d) to prove the following:

Theorem 3.3 (Contraction Mapping Principle) *If $T : M \to M$ is a contraction on a complete metric space (M, d), then there is a unique point $c \in M$, such that $T(c) = c$. Moreover, for each $x \in M$, one has $\lim_{k \to \infty} T^k(x) = c$.*

The point c in the theorem is called a *fixed point* of T.

3.3 Maximum Interval of Existence

The Existence and Uniqueness Theorem above guarantees a solution defined on *some* interval of times. Here we construct a solution that is defined on the largest possible interval of times.

Theorem 3.4 (Maximum Interval of Existence) *Suppose $X : B \to \mathbb{R}^n$ is a time-dependent vector field satisfying the hypotheses of the Existence and*

Uniqueness Theorem 3.1. If $(s, c) \in B$, then there exists an interval $I_{(s,c)}$ containing s, and a curve: $\alpha_{(s,c)} : I_{(s,c)} \to \mathbb{R}^n$, which satisfies the IVP:

$$x' = X(t, x)$$
$$x(s) = c,$$

and which has the further property that if $\beta : J \to \mathbb{R}^n$ also satisfies the IVP, then

$$J \subseteq I_{(s,c)}.$$

The interval $I_{(s,c)}$ is called the maximum interval of existence, and the integral curve $\alpha_{(s,c)}$ is the maximal integral curve passing through c at time s.

Proof: Let $\mathcal{I}$ denote the collection of all open intervals $J = (a, b)$ on which there exists a curve: $\beta : J \to \mathbb{R}^n$, which satisfies the IVP. Note that $s \in J$ for all $J \in \mathcal{I}$. Further, let $\mathcal{L}$ denote the set of numbers that are left-end points of some interval in $\mathcal{I}$, and let $\mathcal{R}$ denote the set of all numbers that are right-end points of some interval in $\mathcal{I}$. Finally, let:

$$\ell = \inf\{a | a \in \mathcal{L}\}$$
$$r = \sup\{b | b \in \mathcal{R}\},$$

be the greatest lower bound of $\mathcal{L}$ and least upper bound of $\mathcal{R}$, respectively. Note that it is possible that $\ell = -\infty$ and/or $r = \infty$. The maximum interval that we are looking for is then

$$I_{(s,c)} = (\ell, r).$$

We define a curve $\alpha_{(s,c)}$ on this interval as follows. Suppose $t \in (\ell, r)$. We consider two cases:

(1) $(t < s)$ Since $\ell < t$ and ℓ is the infimum of $\mathcal{L}$, there exists an $a \in \mathcal{L}$ with $a < t < s$. By definition of $\mathcal{L}$ there is an interval of the form (a, b) on which there is a solution $\beta : (a, b) \to \mathcal{O}$ of the IVP. Note that since $a < t < s$ and $s \in (a, b)$, it follows that $t \in (a, b)$. So we define

$$\alpha_{(s,c)}(t) = \beta(t).$$

(2) $(s < t)$ Since $t < r$ and r is the supremum of $\mathcal{R}$, there exists an $b \in \mathcal{R}$ with $s < t < b$. By definition of $\mathcal{R}$ there is an interval of the form

(a, b) on which there is a solution $\beta : (a, b) \to \mathcal{O}$ of the IVP. Note that since $s < t < b$ and $s \in (a, b)$, it follows that $t \in (a, b)$. So we define

$$\alpha_{(s,c)}(t) = \beta(t).$$

This defines $\alpha_{(s,c)}$. Note that in the definition there is no ambiguity in the choice of a (or b) and β for the given t. This is so because a different choice $\gamma : (c, d) \to \mathcal{O}$, being a solution of the same IVP, must, by the uniqueness of solutions, coincide with β on an interval about t, and therefore agree with β at t. A little thought shows that $\alpha_{(s,c)}$ satisfies the IVP and that $I_{(s,c)}$ contains every interval on which a solution is defined. $\Box$

A somewhat shorter proof of the last theorem is as follows. Let $\mathcal{I}$ be the collection of all open intervals I on which there is defined a solution of the IVP. Then let

$$I_{(s,c)} = \cup \{I | I \in \mathcal{I}\}.$$

Since $I_{(s,c)}$ is the union of open, connected sets with a point in common (namely s), it is also open and connected. A fact from basic topology says that the only connected subsets of $\mathbb{R}$ are the intervals. Thus, $I_{(s,c)}$ is an open interval. Define a curve $\alpha_{(s,c)}$ on this interval as follows. If $t \in I_{(s,c)}$, then there is an interval $I \in \mathcal{I}$, such that $t \in I$. By definition of $\mathcal{I}$, there is a solution β of the IVP defined on I. We take

$$\alpha_{(s,c)}(t) = \beta(t).$$

By the uniqueness part of the Existence and Uniqueness Theorem the choice of β (as opposed to some other solution γ of the IVP defined on I) does not matter, and so $\alpha_{(s,c)}$ is well defined. It is also easy to see that $\alpha_{(s,c)}$ satisfies the IVP. It is clear that $I_{(s,c)}$ is the maximal interval on which there is defined a solution.

Some vector fields are particularly important since the maximum interval of existence for each integral curve is as large as possible, viz., $\mathbb{R}$. Such vector fields are called *complete*, as the following definition explicitly records.

Definition 3.5 Suppose $X : B \subseteq \mathbb{R} \times \mathbb{R}^n \to \mathbb{R}^n$ is a (time-dependent) vector field. X is called a *complete* vector field if

$$I_{(s,c)} = \mathbb{R},$$

for every $(s, c) \in B$. A point $(s, c) \in B$ is called (+) *complete* if

$$[0, \infty) \subseteq I_{(s,c)}.$$

Similarly c is called (−) *complete* if

$$(-\infty, 0] \subseteq I_{(s,c)}.$$

The point (s, c) is called *complete* if $I_{(s,c)} = \mathbb{R}$.

Note that the definition includes the case when X does not depend on the time (i.e., is autonomous). Generally it is quite difficult, even impossible, to determine maximum intervals of existence and thus determine completeness of a vector field. The exercises here give some examples of how to compute $I_{(s,c)}$ for some simple DEs that can be solved explicitly. The next chapter shows how, for 1 dimensional systems, to determine maximal intervals and the domain for the flow. Theorem 3.8 below and Theorem B.1 in Appendix B give some general results about completeness.

Exercises 3.3

1. Determine the maximal interval existence $I_{(s,c)}$ for each of the following DEs and each of the two given initial conditions $x(s) = c$. Do this by solving the DE explicitly (show your work), choosing constants so the initial condition is satisfied and then determining the largest interval on which the particular solution is defined. if the maximum interval is not $\mathbb{R}$, describe how the solution behaves as t approaches an endpoint of the interval

   (a) Initial conditions $x(0) = 2$, $x(0) = 1/4$, and differential equation

   $$x' = x(1 - x),$$

   which is a *separable* DE.

   (b) Initial conditions $x(1) = 2$, $x(1/2) = 0$, and differential equation

   $$x' = \frac{1 - 2t}{\cos(x)},$$

   which is a *separable* DE.

   (c) Initial conditions $x(1) = 1/15$, $x(3) = -3/17$, and differential equation

   $$x' = t^{-1}x + 4t^2x^2,$$

   which is a *Bernoulli* DE.

   (d) Initial conditions $(x(0), y(0)) = (1, 1)$, $(x(0), y(0)) = (-1, 1)$, and system of differential equations

   $$\begin{aligned} x' &= x \\ y' &= xy^{-1}. \end{aligned}$$

3.4 The Flow Generated by a Time-Dependent Vector Field

We are now in a position to define the flow ϕ generated by time-dependent vector field $X : B \subseteq \mathbb{R} \times \mathbb{R}^n \to \mathbb{R}^n$ that satisfies the hypotheses of Theorem 3.1. *From now on these hypotheses will be implicitly assumed to hold for all vector fields.* Some texts define the flow only for autonomous vector fields and mention that the general case can be reduced to this, since the nonautonomous field X can always be extended to an autonomous one: $\widetilde{X} : B \to \mathbb{R} \times \mathbb{R}^n$, by:

$$\widetilde{X}(t, x) = (1, X(t, x)).$$

While this is one way to handle the general case (see the exercises), we do not take this approach here for two reasons. One is that the flow $\widetilde{\phi}$ for $\widetilde{X}$ is not the correct geometric object to use for the flow ϕ for X. A second reason is that geometric concepts connected with $\widetilde{X}$ on $\mathbb{R} \times \mathbb{R}^n$ can be confusing when related back to those for X in $\mathbb{R}^n$. In particular, it is not directly apparent how the semigroup property for autonomous vector fields should generalize to the nonautonomous case. Indeed, the general semigroup property presented below is not discussed in most textbooks on differential equations.

For $(s, c) \in B$, we let $\alpha_{(s,c)} : I_{(s,c)} \to \mathcal{O}$ denote the maximal integral curve of X that passes through c at time s. Thus, $I_{(s,c)}$ is the maximal interval of existence, and $\alpha_{(s,c)}$ satisfies

$$
\begin{aligned}
\alpha'_{(s,c)}(t) &= X(t, \alpha_{(s,c)}(t)), & \forall t \in I_{(s,c)} && (3.18) \\
\alpha_{(s,c)}(s) &= c.
\end{aligned}
$$

We now bundle all the integral curves together, as (s, c) varies over B, to get the flow map.

Definition 3.6 Let $\widetilde{\mathcal{D}}$ be the following subset of $\mathbb{R} \times \mathbb{R} \times \mathbb{R}^n$:

$$\widetilde{\mathcal{D}} = \{\, (t, s, x) \in \mathbb{R} \times \mathbb{R} \times \mathbb{R}^n \mid (s, x) \in B \text{ and } t \in I_{(s,x)} \,\}.$$

Note that if X is a complete vector field, then $I_{(s,x)} = \mathbb{R}, \forall x$, and so $\widetilde{\mathcal{D}} = \mathbb{R} \times B$. The *flow generated by* the vector field X is the map $\phi : \widetilde{\mathcal{D}} \to \mathbb{R}^n$ defined by

$$\phi(t, s, x) = \alpha_{(s,x)}(t). \tag{3.19}$$

An alternative notation, which will be convenient, is

$$\phi_t^s(x) = \phi(t, s, x).$$ (3.20)

Here $(s, x) \in B$ and $t \in I_{(s,x)}$, by definition of $\widetilde{\mathcal{D}}$. Note that since $\alpha_{(s,x)}$ is a solution of the system, we have

$$(t, \phi_t^s(x)) \in B,$$

for all $t \in I_{(s,x)}$.

If we fix $(s, x) \in B$, then $t \mapsto \phi_t^s(x)$ is the maximal integral curve of X which is at x at time s. In the new notation this last condition is

$$\phi_s^s(x) = x,$$ (3.21)

for all $(s, x) \in B$. This gives the first important property of the flow map.

There is more to the flow ϕ than just the convenient relabeling implied by its definition in (3.19). In fact, we would now like to fix s and t and consider the map: $x \mapsto \phi_t^s(x)$. This requires a little preciseness about the domain for the x variable. For this we first introduce the following:

Theorem 3.5 *The domain $\widetilde{\mathcal{D}}$ of the flow map ϕ is an open set.*

We do not prove this theorem here (cf. [Di 74, p. 5]), but rather just use it for our purposes.

Now take a point $(t_0, c) \in B$. Then since $(t_0, t_0, c) \in \widetilde{\mathcal{D}}$ and $\widetilde{\mathcal{D}}$ is open, there exists an open set $U \subseteq \mathbb{R}^n$ with $c \in U$, and an open interval I containing t_0, such that

$$I \times I \times U \subseteq \widetilde{\mathcal{D}}.$$

Restricting the flow map ϕ to $I \times I \times U$, we get the following construction.

For $s, t \in I$, we can consider the notation in (3.20) as the notation for a map $\phi_t^s : U \to \mathbb{R}^n$. Again this is just more notation, but the distinction is important. If you trace back through the above notation, you will find that $\alpha_{(s,x)} : I_{(s,x)} \to \mathbb{R}^n$ is the map (or curve) you get by fixing (s, x) in the flow map $\phi(t, s, x)$, and

$$\alpha_{(s,x)}(t) = \phi(t, s, x) = \phi_t^s(x),$$

for every $x \in U$ and $s, t \in I$. Thus, we get a two-parameter family,

$$\{\phi_t^s\}_{s,t \in I},$$

of maps $\phi_t^s : U \to \mathbb{R}^n$, with the property: $\phi_s^s = 1$ (the identity map on U), for all $s \in I$. The fact that two of these maps can sometimes be combined, via composition of functions, to give another map in the family, is known as the semigroup property. This comes from the following general result.

Theorem 3.6 (Semigroup Property for Nonautonomous Flows)
Suppose ϕ is the flow generated by an time-dependent vector field $X : B \subseteq \mathbb{R} \times \mathbb{R}^n \to \mathbb{R}^n$. Suppose $(s, x) \in B$ and $t \in I_{(s,x)}$. Then $I_{(t,\phi_t^s(x))} = I_{(s,x)}$ and

$$\phi_u^t(\phi_t^s(x)) = \phi_u^s(x), \tag{3.22}$$

for every $u \in I_{(s,x)}$.

Proof: Let $J = I_{(t,\phi_t^s(x))} \cap I_{(s,x)}$. Note that J is not empty since $t \in J$. The two maximal integral curves $\alpha_{(t,\phi_t^s(x))}$ and $\alpha_{(s,x)}$ have domains $I_{(t,\phi_t^s(x))}$ and $I_{(s,x)}$, respectively, and these curves have the same initial value at time t:

$$\alpha_{(t,\phi_t^s(x))}(t) = \phi_t^s(x) = \alpha_{(s,x)}(t).$$

Hence by definition of the maximal intervals (see Theorem 3.4), we have $I_{(t,\phi_t^s(x))} \subseteq I_{(s,x)}$ and $I_{(s,x)} \subseteq I_{(t,\phi_t^s(x))}$. Hence these intervals are the same and thus also the respective maximal integral curves are the same:

$$\alpha_{(t,\phi_t^s(x))} = \alpha_{(s,x)}. \tag{3.23}$$

But this identity is the same as the semigroup property (3.23) when properly interpreted with the established notation:

$$\phi_u^t(\phi_t^s(x)) = \alpha_{(t,\phi_t^s(x))}(u) = \alpha_{(s,x)}(u) = \phi_u^s(x),$$

for every $u \in I_{(s,x)}$. $\square$

To interpret the semigroup property (3.22) in terms of algebraic structures, we return to the situation described prior to the theorem. With $I \times I \times U \subseteq \tilde{\mathcal{D}}$, we are guaranteed, for every $x \in U$ and every $s, t \in I$, that $(s, x) \in B$ and $t \in I_{(s,x)}$. Further, because $\phi_{t_0}^{t_0}(x) = x$ for every $x \in U$ and because the flow $\phi : \mathcal{D} \to \mathbb{R}^n$ is continuous, we can assume that the interval I is small enough so that $\phi_t^s(U) \subseteq U$, for all $s, t \in I$. Thus, each of the maps in the two-parameter family $S = \{\phi_t^s\}_{s,t \in I}$ maps U into itself. Hence any two maps in the family can be combined via composition of functions and in particular

$$\phi_u^t \circ \phi_t^s = \phi_u^s, \tag{3.24}$$

for every $s, t, u \in I$. Further, since $\phi_s^s = 1$ is the identity map on U, a particular case of identity (3.24) is

$$\phi_s^t \circ \phi_t^s = 1, \tag{3.25}$$

for all $s, t \in I$. Thus, each map ϕ_t^s in the family S is invertible with inverse $(\phi_t^s)^{-1} = \phi_s^t$. Consequently, S would be a group (mathematically) if the composition operation $\circ$ were a closed operation on the elements of S. While this is not true for any pair of elements of S, the semigroup property (3.24) says that certain pairs of elements of S do combine to give another element of S.

The semigroup property also has important interpretations in continuum mechanics, a special case of which is fluid mechanics. For such a situation, $n = 3$ and $X(t, x)$ is interpreted as the velocity of the fluid flowing through the point x in a tank U at time t. For steady fluid flows this velocity does not change over time, but in the case where X varies with time, the flow through a given point varies continually. The motion of a particular portion the fluid therefore depends upon the time at which the observation begins. Thus, suppose at time s a portion W of the fluid located in a region $W \subseteq U$ of the tank is observed. Following the motion until time $t > s$, we find that W has moved and deformed into the region $\phi_t^s(W)$. From time t until time $u > t$, the fluid in the region $\phi_t^s(W)$ moves and deforms into the region $\phi_u^t(\phi_t^s(W))$. By the semigroup property, this region is the same as $\phi_u^s(W)$, and thus the time u shape and position does not depend on the intermediate shapes and positions, but rather only on the initial time s and the position and shape W. See Figure 3.3. Each of the maps $\phi_t^s : U \to U$ is considered as a *deformation* of the parts W of the continuum U (or, in the fluid case, as a *motion* of the fluid in U). For a starting time s, ending time $u > s$, and intermediate times $s < t_1 < t_2 < \cdots < t_k < u$, compounding the corresponding deformations $\phi_{t_{i+1}}^{t_i}$ gives the overall resultant deformation

$$\phi_u^{t_k} \circ \phi_{t_k}^{t_{k-1}} \circ \cdots \circ \phi_{t_2}^{t_1} \circ \phi_{t_1}^s = \phi_u^s. \tag{3.26}$$

Of course, this identity holds whether the times are ordered as mentioned or not, but the ordering helps focus on the concept involved.

A further interpretation of the semigroup property is that it provides a rationale for studying discrete dynamical systems as being modeled by a

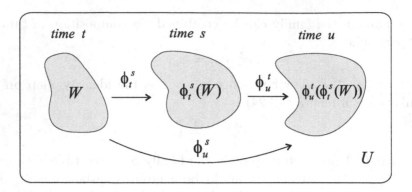

Figure 3.3: *Deformation of a set $W \subseteq U$ by the flow.*

sequence of iterates of a family of maps. Starting with the point $c \in U$, a first iterate $\phi_t^s(c)$ gives another point in U, a second iterate $\phi_u^t(\phi_t^s(c))$ gives yet another. The semigroup property says that all three of these points lie on the same integral curve. Thus, for a sequence of closely spaced times, the discrete iterates approximate the continuous motion along an integral curve. Additional details of this are discussed in Exercises 2 and 3 in the next section.

In the next example, we consider a planar, nonautonomous system, which is simple enough so that we can actually compute the formula for its flow and also analyze the flow lines without too much difficulty. You should realize that the flow is mainly a theoretical tool, since for almost all meaningful examples, such as the fluid flow examples in Chapters 1 and 2, the formulas for the flows are impossible to compute. One exception to this is for 1-dimensional dynamical systems, which are treated extensively in the next chapter.

Example 3.3 Consider the following nonautonomous system in the plane:

$$x' = \frac{x}{t}$$
$$y' = x.$$

This is a rather contrived system, not representing any real physics, but for the reasons stated above it can serve to illustrate the theory. Thus, we view $X(t, x, y) \equiv (xt^{-1}, x)$, with domain $B = \{(t, x, y) \in \mathbb{R}^3 \,|\, t \neq 0\}$, as representing a velocity vector field for a nonsteady, planar fluid.

Note that the y-axis, $L = \{(0, y) | y \in \mathbb{R}\}$, is a line of fixed points (stagnation points) for the system. We can find the other solutions of the system by solving the first equation in the system (it is a separable DE) and substituting the result in the second equation. Thus, the first equation in separated form is $x^{-1}dx = t^{-1}dt$ and when integrated gives $\ln x = \ln t + k$. Choosing k so that the initial condition $x(s) = c_1$ holds, with s and c_1 given, yields $k = \ln c_1 - \ln s$. Substituting this for k and solving for x leads to

$$x = \frac{t}{s}c_1. \tag{3.27}$$

The argument assumes implicitly that all the quantities are positive, but it is easy to see that the resulting formula actually is the solution of the initial value problem for any choice of c_1 and any $s \neq 0$.

Using this result gives $y' = \frac{c_1}{s}t$, for the second DE in the system. Integration of this results in $y = \frac{c_1}{2s}t^2 + k$. The constant k is again determined from the generic initial condition: $y(s) = c_2$. After some minor rearrangement, we find that

$$y = \left(\frac{t^2 - s^2}{2s}\right)c_1 + c_2. \tag{3.28}$$

Consequently, the formula for the flow is

$$\phi_t^s(c_1, c_2) = \left(\frac{t}{s}c_1, \left(\frac{t^2 - s^2}{2s}\right)c_1 + c_2\right), \tag{3.29}$$

where $s \neq 0$ and $t \in I_{(s,c)}$. It is easy to see that $I_{(s,c)}$ is $(-\infty, 0)$ or $(0, \infty)$ if $s < 0$ or $s > 0$ and c is not a fixed point. We will assume from now on that $s > 0$, and thus just analyze the flow for positive times.

We have expressed the formula for the flow with the constants c_1 to the right of t, which may seem to be a nonstandard order, but this makes verification of the semigroup property easier. Thus, direct calculation gives

$$
\begin{aligned}
\phi_u^t(\phi_t^s(c)) &= \left(\frac{u}{t}\frac{t}{s}c_1, \left(\frac{u^2 - t^2}{2t}\right)\frac{t}{s}c_1 + \left(\frac{t^2 - s^2}{2s}\right)c_1 + c_2\right) \\
&= \left(\frac{u}{s}c_1, \left(\frac{u^2 - s^2}{2s}\right)c_1 + c_2\right) \\
&= \phi_u^s(c).
\end{aligned}
$$

We can use the explicit formula for the flow to describe the behavior of points and subsets of $\mathbb{R}^2$ under deformation by the flow. As simple as this example is, this description is still somewhat complicated. Consider the flow as that of a fluid, and imagine dropping a small cork into the fluid at some point $c = (c_1, c_2)$ and observing its motion. What happens? Well,

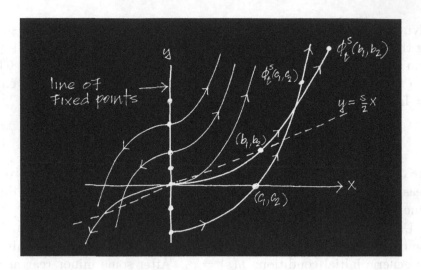

Figure 3.4: *The possible trajectories of a cork for various choices of initial positions c and choice of initial time s = 1.*

that depends on the initial time s when your observation starts. Wait a few minutes and the path followed by c will be completely different. But with s fixed, the cork follows the path given by the curve: $t \mapsto \phi_t^s(c)$.

Assume that c is not a stagnation point, and for definiteness that $s > 0$. By eliminating the parameter t in equations (3.27)-(3.28), we see that the path of the cork lies on the parabola with equation

$$y = \frac{s}{2c_1}x^2 + c_2 - \frac{s}{2}c_1. \tag{3.30}$$

It is easy enough to plot by hand a number of these parabolas for various initial positions c of the cork. Figure 3.4 shows these curves for the case $s = 1$. Note that the paths, or flow lines, intersect each other, but this should not seem contradictory, since particles of the fluid traveling along different paths will reach the point of intersection at different times. This together with the parabolic shape of the flow lines can also be better understood by considering the behavior of the velocity vector field $X = (xt^{-1}, x)$, that is generating the flow.

At any instant $t > 0$ in time, X has the same direction, namely $(t^{-1}, 1)$, at every point, but this direction changes continually and tends to a vertical direction $(0, 1)$, as $t \to \infty$. Thus, the longer we wait before dropping the cork in the tank, the more vertical will be its path of flow.

It is also instructive to examine how the flow deforms subsets W of $\mathbb{R}^2$, that is, how $\phi_t^s(W)$ changes over time. This again depends on the initial time s when the observation starts. Figure 3.5 shows this for a particular square W and two initial observation times: $s_1 = 0.2$ and $s_2 = 1$.

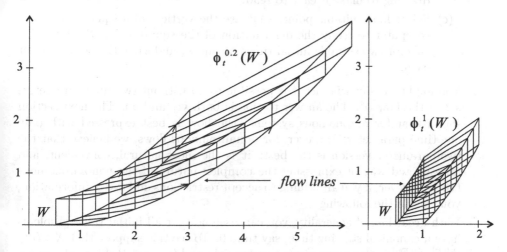

Figure 3.5: *Deformation of a square W under the flow. The deformation depends on the time s at which the observation begins. The deformation on the left begins at time $s = 0.2$, while the one on the right begins at $s = 1$.*

Exercises 3.4

1. For the nonautonomous, planar systems

(a)

$$x' = \left(\frac{2t}{1+t^2}\right) x$$
$$y' = 3x,$$

(b)

$$x' = e^{-t}x^2$$
$$y' = x,$$

do the following.

(a) Find an explicit formula for the flow map and use it to directly verify the semigroup property of the flow.

(b) For a selected initial time s, use a computer to draw the flow lines $t \mapsto \phi_t^s(c)$ for number of different initial points c. Use your judgment here since you will be graded on how well your drawing represents the nature of the flow. Mark directions on the flow lines and annotate the drawing to make it easy to read.

(c) Select four suitable points that are the vertices of a square and use a computer to study the deformation of the square under the flow. Do this for two suitable initial times s_1 and s_2 and same time duration for each.

2. You might wonder whether it is possible to do without two time parameters s, t in the flow ϕ_t^s. The answer to this is both yes and no. The next section shows that for autonomous systems, the flow is best expressed with only one time parameter. However, for nonautonmous flows, we believe that the two-parameter version is the best: it is the most natural, convenient, and unrestricted way of expressing the complex nature of the non-autonomous flow. However, if you are willing to accept restrictions and loss of information, you can do the following.

With great loss of generality, you can assume that all initial value problems have a canonical starting time, say time 0. To see this suppose that $X : B \subseteq \mathbb{R}^{n+1} \to \mathbb{R}^n$ is a nonautonomous vector field and $(t_*, c) \in B$. Since B is open, there is an open interval $L = (r_1, r_2)$ containing t_*, and an open set $\mathcal{O} \subseteq \mathbb{R}^n$, containing c, such that $L \times \mathcal{O} \subseteq B$. Restricting X to $L \times \mathcal{O}$, consider an IVP with arbitrary starting time $t_0 \in L$:

$$x' = X(t, x) \tag{3.31}$$
$$x(t_0) = c. \tag{3.32}$$

Let $M = L - t_0 = (r_1 - t_0, r_2 - t_0)$ and define $Y : M \times \mathcal{O} \to \mathbb{R}^n$ by $Y(s, y) = X(s - t_0, y)$ for $(s, y) \in M \times \mathcal{O}$. Show that each solution $\beta : J \to \mathbb{R}^n$ of the IVP

$$y' = Y(s, y)$$
$$y(0) = c,$$

determines a solution α of the IVP (3.31)-(3.32), and conversely each solution α of the first IVP gives rise to a solution β of the second IVP.

This result allows one to use 0 as the canonical starting time in the two-parameter flow ϕ_t^0. To see how restrictive this is, you might want to look at Example 4.5 in the next chapter.

3.5   The Flow for Autonomous Systems

Suppose now that $X : \mathcal{O} \to \mathbb{R}^n$ is autonomous and let $\phi : \tilde{\mathcal{D}} \to \mathbb{R}^n$ be the flow for X as defined in the previous section. Because of the time independence of X, the flow has additional properties and we can specialize it to a more useful map. These additional properties arise from the following elementary observation.

Proposition 3.1 (Time Translational Symmetry) *If $\alpha : J \to \mathbb{R}^n$ is an integral curve for an autonomous vector field X, then for any $t_0 \in J$, the curve $\beta : J - t_0 \to \mathbb{R}^n$, defined by*

$$\beta(t) = \alpha(t + t_0),$$

for $t \in J - t_0$, is also an integral curve of X.

Proof: This is easy to prove, and indeed is an exercise in Chapter 1.

Corollary 3.2 *For an autonomous vector field $X : \mathcal{O} \to \mathbb{R}^n$, the maximal intervals $I_{(s,c)}$ and integral curves $\alpha_{(s,c)} : I_{(s,c)} \to \mathbb{R}^n$, for $s \in \mathbb{R}$ and $c \in \mathcal{O}$, satisfy the relations*

$$
\begin{aligned}
I_{(s,c)} &= I_{(0,c)} + s, & &(3.33)\\
\alpha_{(s,c)}(t) &= \alpha_{(0,c)}(t - s), & \forall t \in I_{(s,c)}. \quad &(3.34)
\end{aligned}
$$

Consequently, the flow ϕ for X satisfies

$$\phi_t^s(c) = \phi_{t-s}^0(c), \qquad (3.35)$$

for all $t \in I_{(s,c)}$.

Proof: By the proposition, the curve β defined by

$$\beta(t) = \alpha_{(s,c)}(t + s),$$

for $t \in I_{(s,c)} - s$, is an integral curve of X. Further, $\beta(0) = \alpha_{(s,c)}(s) = c$, and consequently β satisfies the same initial value problem as $\alpha_{(0,c)}$. Hence, by maximality, $I_{(s,c)} - s \subseteq I_{(0,c)}$ and $\beta = \alpha_{(0,c)}$ on $I_{(s,c)} - s$.

In a similar fashion, the curve γ defined by

$$\gamma(t) = \alpha_{(0,c)}(t - s),$$

for $t \in I_{(0,c)} + s$, is an integral curve satisfying the same initial value problem as $\alpha_{(s,c)}$. Hence, by maximality, $I_{(0,c)} + s \subseteq I_{(s,c)}$ and $\gamma = \alpha_{(s,c)}$ on $I_{(0,c)} + s$.

Putting both parts together gives the results of the corollary. $\square$

Property (3.35) says that the flow map ϕ_t^s only depends on the difference $t-s$ of the two times s and t, i.e., on the amount of time elapsed between these two times. For this reason, in the autonomous case, the flow reduces from a dependence on two parameters to a dependence on only one parameter. We can thus revise the notation and redefine the domain of the flow map as follows.

Definition 3.7 (The Flow for Autonomous Systems) Let $X : \mathcal{O} \to \mathbb{R}^n$ be a vector field on $\mathcal{O} \subseteq \mathbb{R}^n$. We choose $0 \in \mathbb{R}$ as the standard initial time and, for $x \in \mathcal{O}$ let $I_x = I_{(0,x)}$ and $\alpha_x = \alpha_{(0,x)}$ denote the maximal interval and integral curve that passes through x at time 0. Let $\mathcal{D}$ be the following subset of $\mathbb{R} \times \mathbb{R}^n$:

$$\mathcal{D} = \{\, (t,x) \in \mathbb{R} \times \mathbb{R}^n \,|\, x \in \mathcal{O} \text{ and } t \in I_x \,\}.$$

Note that if X is a complete vector field, then: $I_x = \mathbb{R}, \forall x$, and so $\mathcal{D} = \mathbb{R} \times \mathcal{O}$.

The flow $\phi : \tilde{\mathcal{D}} \to \mathbb{R}^n$ defined previously in general can now be reduced to map $\phi : \mathcal{D} \to \mathbb{R}^n$. We use the same notation ϕ for this map (hopefully with no confusion). The *flow generated by* an autonomous vector field X is the map defined by

$$\phi_t(x) = \phi_t^0(x). \tag{3.36}$$

We also use the notation $\phi(t, x) = \phi_t(x)$. Viewing the integral curves of X as tracing out paths in $\mathcal{O}$ as time moves in the positive direction, we interpret (3.36) as saying that the integral curve that starts at x at time 0, is at $\phi(t, x)$ at time t.

The semigroup property discussed above now specializes to the following.

Theorem 3.7 (Semigroup Property for Autonomous Systems)
Suppose ϕ is the flow generated by an autonomous vector field on $\mathcal{O}$ and $x \in \mathcal{O}$. If $t \in I_x$ and $s \in I_{\phi_t(x)}$, then $s + t \in I_x$ and

$$\phi_s(\phi_t(x)) = \phi_{s+t}(x). \tag{3.37}$$

The proof of the theorem is left as an exercise.

To interpret what the theorem says, we restrict ϕ to a product neighborhood $I \times U \subseteq \mathcal{D}$, with $I = (t_0 - a, t_0 + a)$. Note that $I \subseteq I_x$ for all $x \in U$.

By continuity of ϕ, we can choose a small enough so that $\phi_t(U) \subseteq U$, for every $t \in I$. Then

$$\{\phi_t\}_{t \in I}$$

constitutes a *one parameter family* of maps: $\phi_t : U \to U$. The parameter is the time t. The semigroup property (3.37) now reads more simply as: If $x \in U$ and $s, t, s + t \in I$, then $\phi_s(\phi_t(x)) = \phi_{s+t}(x)$. A better way to put this is in terms of composition of maps:

$$\phi_s \circ \phi_t = \phi_{s+t}, \tag{3.38}$$

for every $s, t \in I$ *for which* $s + t \in I$. This last proviso about $s + t$ being in I is what restricts the family (or set) $\{\phi_t\}_{t \in I}$ from being a group. The semigroup operation is composition $\circ$ of maps, and property (3.38) is just the closure property of $\circ$. Note that

$$\phi_0 = I,$$

the identity transformation on U is the identity element of the semigroup. Also if $t \in I$, then

$$\phi_t \circ \phi_{-t} = \phi_{t-t} = \phi_0 = I,$$

and so ϕ_{-t} is the inverse of the transformation ϕ_t. In symbols:

$$\phi_{-t} = \phi_t^{-1}.$$

In the case when X is a complete vector field, we can take $U = \mathcal{O}$, $I = \mathbb{R}$, and get that $\{\phi_t\}_{t \in \mathbb{R}}$ is actually a group: a *one-parameter group* of invertible transformations.

Besides having a special semigroup property, autonomous vector fields X have many other special features. The following theorem gives some information on completeness of X and is useful in other regards too.

Theorem 3.8 *Suppose $X : \mathcal{O} \to \mathbb{R}^n$ is a C^1 vector field on $\mathcal{O}$ and that $c \in \mathcal{O}$. Let the $I_c = (a_c, b_c)$. If the forward flow through c remains in a compact, convex set $M \subseteq \mathcal{O}$, i.e., if*

$$\phi_t(c) \in M, \qquad \forall t \in [0, b_c),$$

then $b_c = \infty$. Likewise, if the backward flow through c remains in a compact, convex set $M \subseteq \mathcal{O}$, i.e., if

$$\phi_t(c) \in M, \qquad \forall t \in (a_c, 0],$$

then $a_c = -\infty$. Hence, if $\phi_t(c) \in M$ for all $t \in I_c$, then $I_c = \mathbb{R}$, i.e., c is a complete point if the entire forward and backward flow remains in a compact convex set.

Proof: As in the proof of Theorem 3.1, we use the ℓ_1 norm $\|x\|$ for $x \in \mathbb{R}^n$ and we choose a constant $K > 0$ such that

$$\|X(x)\| \leq K \tag{3.39}$$

$$\|X(x) - X(y)\| \leq K\|x - y\|, \tag{3.40}$$

for all $x, y \in M$. This relies on the compactness and convexity of M. We just prove the first part of the theorem which assumes that $\phi_t(c) \in M$ for all $t \in [0, b_c)$. We suppose $b_c < \infty$ and get a contradiction as follows.

Recall the notation: $\alpha_c(t) = \phi_t(c)$, which will be convenient to use here. If $t_1 < t_2 \in I_c$, then since α_c is an integral curve, it satisfies

$$\alpha_c(t_i) = c + \int_0^{t_i} X(\alpha_c(s)) \, ds,$$

for $i = 1, 2$. Thus, from inequality (3.39) we get

$$\|\alpha_c(t_2) - \alpha_c(t_1)\| = \left\| \int_{t_1}^{t_2} X(\alpha_c(s)) \right\| \leq K(t_2 - t_1).$$

Now let $\{t_k\}_{k=1}^{\infty}$ be a sequence in I_c that converges to b_c. Applying the last inequality above gives

$$\|\alpha_c(t_k) - \alpha_c(t_m)\| \leq K|t_k - t_m|,$$

for all k, m. Thus, $\{\alpha_c(t_k)\}_{k=1}^{\infty}$ is a Cauchy sequence in the compact set M and so converges to a point $p \in M$.

Let $\beta : J \to \mathcal{O}$ be a solution of the initial-value problem $x' = X(x)$, $x(b_c) = p$, with, say, $J = (b_c - r, b_c + r)$. Then we have

$$\alpha_c(t) = c + \int_0^t X(\alpha_c(s)) \, ds, \qquad \forall t \in (a_c, b_c)$$

$$\beta(t) = p + \int_{b_c}^t X(\beta(s)) \, ds, \qquad \forall t \in (b_c - r, b_c + r).$$

Note also that $p = c + \int_0^{b_c} X(\alpha_c(s)) \, ds$. Thus, if we define $\gamma : (a_c, b_c + r) \to \mathcal{O}$ by

$$\gamma(t) = \begin{cases} \alpha_c(t) & \text{if } t \in (a_c, b_c) \\ \beta(t) & \text{if } t \in [b_c, b_c + r), \end{cases}$$

then γ is continuous. Furthermore, we claim that γ satisfies the initial value problem: $x' = X(x)$, $x(0) = c$. To see this we check the integral version of the IVP.

First suppose $a_c < t < b_c$. Then

$$c + \int_0^t X(\gamma(s))\, ds = c + \int_0^t X(\alpha_c(s))\, ds = \alpha_c(t) = \gamma(t).$$

On the other hand, if $b_c \le t < b_c + r$ then

$$
\begin{aligned}
c + \int_0^t X(\gamma(s))\, ds &= c + \int_0^{b_c} X(\gamma(s))\, ds + \int_{b_c}^t X(\gamma(s))\, ds \\
&= c + \int_0^{b_c} X(\alpha_c(s))\, ds + \int_{b_c}^t X(\beta(s))\, ds \\
&= p + \int_{b_c}^t X(\beta(s))\, ds = \beta(t) = \gamma(t).
\end{aligned}
$$

This shows that γ is a solution of the IVP: $x' = X(x)$, $x(0) = c$, and so its interval of definition $(a_c, b_c + r)$ is contained in the maximal interval of existence $I_c = (a_c, b_c)$. This is a contradiction! $\square$

Exercises 3.5

1. Prove Theorem 3.7. *Hint*: Use Theorem 3.6 for the general nonautonomous flow and be careful with the notation. For example, the first part of Theorem 3.6 says that for any time t_1, if $t_2 \in I_{(t_1, x)}$ (i.e., if $t_2 - t_1 \in I_{(0,x)}$, because (3.33) holds in the present setting), then

$$I_{(t_2, \phi_{t_2}^{t_1}(x))} = I_{(t_1, x)}.$$

   Hence (in the present setting) it follows that if t_3 is a time such that $t_3 - t_2 \in I_{(0, \phi_{t_2 - t_1}(x))}$, then $t_3 - t_1 \in I_{(0,x)}$.

2. **(Discrete Dynamical Systems, Part I)** There is a area of mathematical study, called discrete dynamical systems, which is closely related to the study of systems of differential equations and which, even though the idea originated with Poincaré a century ago, has only within the last two decades received widespread attention and popularity, primarily because of the advent of the personal computer. This exercise and the next give a brief introduction to this topic. The electronic component has much additional material, theory, and exercises (see CDChapter 3), but this can only give you a glimpse extensive number of results, computer studies, and theorems arising from the study of discrete dynamical systems. See the texts [Rob 95], [Dev 86], [CE 83], [Mar 92] for more details.

The motivation behind discrete dynamical systems comes from the flow map and its semigroup property. Consider the autonomous case where ϕ is the flow for a vector field $X : \mathcal{O} \subseteq \mathbb{R}^n \to \mathbb{R}^n$. To make the discussion simple, suppose that X is complete, so that the domain for the flow map is $\mathcal{D} = \mathbb{R} \times \mathcal{O}$, and hence the flow $\phi : \mathbb{R} \times \mathcal{O} \to \mathcal{O}$ gives a 1-parameter group $\{\phi_t\}_{t \in \mathbb{R}}$ of maps $\phi_t : \mathcal{O} \to \mathcal{O}$. The dynamics of the system of differential equations is controlled by this group in the sense that $t \mapsto \phi_t(c)$ gives the continuous motion of each $c \in \mathcal{O}$. Instead of the continuous motion of c under the flow, we can look at its positions:

$$c, \ \phi_\tau(c), \ \phi_{2\tau}(c), \ \phi_{3\tau}(c), \ \dots \ ,$$

at a discrete set of times $0, \tau, 2\tau, 3\tau, \dots$, where τ is fixed (small) positive number. By the semigroup property of the flow, the position of c at time $t = k\tau$ is given by a composition of maps:

$$\phi_{k\tau}(c) = \phi_\tau \circ \phi_\tau \circ \cdots \circ \phi_\tau(c) = (\phi_\tau)^k(c),$$

i.e., repeated application of the map ϕ_τ gives the position of c at time t. Here, as in the discussion of Picard iterates, we use the customary notation for the repeated composition of a map with itself: If $f : S \to S$ is any map, let

$$f^k = f \circ f \circ \cdots \circ f,$$

denote the composition of f with itself k times. For example, $f^2 = f \circ f$ and $f^3 = f \circ f \circ f$.

Thus, for a small increment τ of time, we can replace the continuous dynamics of the flow with the discrete dynamics of the map

$$f \equiv \phi_\tau.$$

We would expect that studying how points $c \in \mathcal{O}$ behave under repeated applications of f would in some sense be similar to studying the true dynamics of the flow.

Abstracting from the above motivational discussion, this exercise studies several particular maps f of a set into itself. There are a few initial concepts comprising this study that are most natural and defined as follows.

Definition 3.8 A *discrete dynamical system* is a map $f : S \to S$, of a set S into itself. The *forward orbit* of a point c under f is the set of points:

$$O^+(c) = \{ f^k(c) \mid k = 0, 1, 2, \dots \}.$$

If f is 1-1 and onto, then we let $f^{-k} \equiv f^{-1} \circ f^{-1} \circ \cdots \circ f^{-1}$, denote the composition of the inverse f^{-1} with itself k times. Then the *backward orbit* of c under f makes sense and is defined as

$$O^-(c) = \{ f^{-k}(c) \mid k = 0, 1, 2, \dots \}.$$

A *fixed point* of f is a point $c \in S$ such that $f(c) = c$. A *periodic point* of f is a point $c \in S$ such that $f^k(c) = c$ for some positive integer k. The *period* of a periodic point is the least positive integer p such that $f^p(c) = c$. A point $c \in S$ is called *eventually periodic* if there is a positive integer k such that $f^k(c)$ is a periodic point of f.

It is clear that a periodic point c has orbit $O^+(c)$ consisting of p points, where p is the period of c, and this orbit corresponds to the notion of a closed integral curve for a system of DEs. An eventually periodic point c has orbit $O^+(c)$ consisting of a finite number of points and corresponds to an integral curve of a DE that asymptotically approaches a limit cycle of the DE.

If we let $x_k \equiv f^k(c)$, $k = 0, 1, 2, \ldots$, denote the sequence of iterates of c under the map f, then this sequence is a solution of the basic and most simplistic of all iteration schemes

$$x_{k+1} = f(x_k), \qquad (k = 1, 2, , 3, \ldots),$$

and satisfies the initial condition $x_0 = c$. This gives us another interpretation discrete dynamical systems and their analogy with systems of differential equations.

For the particular study in this exercise, consider the following maps:

(a) The logistic map $f : \mathbb{R} \to \mathbb{R}$, given by

$$f(x) = rx(1 - x),$$

where $r > 0$ is a parameter.

(b) A cubic map $f : \mathbb{R} \to \mathbb{R}$, given by

$$f(x) = rx(1 - x)(2 - x),$$

where $r > 0$ is a parameter.

(c) The Hénon map $f : \mathbb{R}^2 \to \mathbb{R}^2$, given by

$$f(x, y) = (\, r - qy - x^2, \, x \,),$$

where r and q are parameters.

For each of the maps assigned for study, do the following:

1. Find all the fixed points.
2. Read the material on the Maple worksheets, referenced by CDChapter 3 on the electronic component and work the exercises shown there. The electronic component contains some special Maple code for visualizing discrete dynamics in one and two dimensions. Using this and the discussion there, you will be able to experimentally discover periodic and eventually periodic points for f and look at how the dynamics change when the parameters r and q change.

3. **(Discrete Dynamical Systems, Part II)** In this exercise, we generalize the discussion of discrete dynamical systems in the last exercise to obtain the discrete analog of a nonautonomous system of differential equations $x' = X(t, x)$. Since the vector field X depends on the time, it is natural to obtain the discrete analog by allowing the map f in the iteration scheme to depend on the time step k. Thus, the scheme has the form

$$x_{k+1} = f_k(x_k), \qquad (k = 1, 2,, 3, \ldots).$$

So formally we define the concept by

Definition 3.9 A *nonautonomous, discrete dynamical system* is a sequence $\{f_k\}_{k=1}^{\infty}$ of maps $f_k : S \to S$ of a set S into itself.

This can also be motivated by the analogy with the flow ϕ generated by a time-dependent vector field. With the appropriate restrictions, the flow gives a two parameter family $\{\phi_t^s\}$ of maps $\phi_t^s : \mathcal{O} \to \mathcal{O}$, which has the semigroup property

$$\phi_u^t \circ \phi_t^s = \phi_u^s.$$

Then, relative to a starting time s, the continuous dynamics of each point $c \in \mathcal{O}$ is given by its flow line: $t \mapsto \phi_t^s(c)$. To make the motion discrete, suppose τ is a (small) positive number. Then by the semigroup property, we see that the position of c at time $t = k\tau$ is

$$\phi_{k\tau}^0(c) = \phi_{k\tau}^{(k-1)\tau} \circ \phi_{(k-1)\tau}^{(k-2)\tau} \circ \cdots \circ \phi_{\tau}^0(c).$$

Thus, the discrete dynamics is expressed by iterations:

$$f_k \circ f_{k-1} \circ \cdots \circ f_1,$$

of the sequence $\{f_k\}_{k=1}^{\infty}$ of maps $f_k : \mathcal{O} \to \mathcal{O}$, defined by

$$f_k \equiv \phi_{k\tau}^{(k-1)\tau}.$$

It is instructive to note that when X does not depend on the time, the flow map ϕ has the property: $\phi_t^s = \phi_{t-s}^0$, and so $f_k = \phi_\tau^0 \equiv \phi_\tau$, for every k. Thus, the discrete dynamical system reduces to the autonomous one discussed in the previous exercise.

In this exercise you are to study the following nonautonomous, discrete dynamical systems.

(a) A sequence of logistic maps $f_k : \mathbb{R} \to \mathbb{R}$, given by

$$f_k(x) = r_k\, x(1 - x),$$

where $\{r_k\}_{k=1}^{\infty}$ is a sequence of positive numbers.

(b) A sequence of cubic maps $f_k : \mathbb{R} \to \mathbb{R}$, given by

$$f_k(x) = r_k\, x(1-x)(2-x),$$

where $\{r_k\}_{k=1}^{\infty}$ is a sequence of positive numbers.

(c) A sequence of Hénon maps $f_k : \mathbb{R}^2 \to \mathbb{R}^2$, given by

$$f(x,y) = (\, r_k - q_k\, y - x^2,\ x\,),$$

where $\{r_k\}_{k=1}^{\infty}$ and $\{q_k\}_{k=1}^{\infty}$ are sequences of numbers.

For each of the maps assigned to you for study, read the material on the Maple worksheets referenced by CDChapter 3 on the electronic component and use the special-purpose software to complete the exercises on the worksheet.

3.6 Summary

The most important concept discussed in this chapter is that of the flow (or flow map) ϕ generated by a vector field X. This geometrical concept can be thought of as the expression for the general solution of the system of differential equations associated with X—it contains the maximal integral curves for all initial value problems.

For an autonomous vector field $X : \mathcal{O} \subseteq \mathbb{R}^n \to \mathbb{R}^n$, the *flow* is a map: $\phi : \mathcal{D} \to \mathbb{R}^n$, defined on the open set $\mathcal{D} = \{\, (t,x) \in \mathbb{R} \times \mathbb{R}^n \,|\, x \in \mathcal{O}\}$ in $\mathbb{R} \times \mathbb{R}^n$, and has the properties

$$
\begin{aligned}
\frac{\partial \phi}{\partial t}(t,x) &= X(\phi(t,x)), &&\text{for all } t \in I_x \\
\phi(0,x) &= x, &&\text{for all } x \in \mathcal{O} \\
\phi(s, \phi(t,x)) &= \phi(s+t, x), &&\text{for all } t \in I_x \text{ and } s \in I_{\phi(t,x)}.
\end{aligned}
$$

A further concept introduced in the chapter is the notion of using iterates of a map to prove existence of a desired result. In our case, this map was the Picard map $T : \mathcal{C} \to \mathcal{C}$ and its iterates converge to the local solution of the initial value problem. This concept has been abstracted to many other situations and has become extremely useful in mathematics. Depending on your course of study, you might encounter this idea in other places in this book, for example, in the proof of the stability of periodic solutions (Chapter 7) and proof of the Hartman-Grobman Linearization Theorem (Appendix B).

Chapter 4

Linear Systems

In this chapter we study linear systems of differential equations and see how the general theory specializes to give us a much more complete description of the flow. As would be expected, the material here relies on many topics from linear algebra and so a good background in this subject will be helpful. (Appendix C has some review material on linear algebra and matrix analysis.)

Linear systems have the form

$$x' = A(t)x + b(t),$$

where $A(t) = \{a_{ij}(t)\}_{i,j=1}^{n}$ is an $n \times n$ matrix and $b(t)$ is a vector in $\mathbb{R}^n$. Thus, the vector field for the system is $X(t,x) = A(t)x + b(t)$ and, for a fixed t, is an affine transformation $X(t,\cdot) : \mathbb{R}^n \to \mathbb{R}^n$. Written out in terms of components, the linear system is:

$$
\begin{aligned}
x'_1 &= a_{11}(t)x_1 + a_{12}(t)x_2 + \cdots + a_{1n}(t)x_n + b_1(t) \\
x'_2 &= a_{21}(t)x_1 + a_{22}(t)x_2 + \cdots + a_{2n}(t)x_n + b_2(t) \\
&\;\;\vdots \\
x'_n &= a_{n1}(t)x_1 + a_{n2}(t)x_2 + \cdots + a_{nn}(t)x_n + b_n(t)
\end{aligned}
$$

Both A and b, in general, depend on the time t and we assume the interval of times is $I = (r_1, r_2)$. If we let $\mathcal{M}_n$ denote the collection of $n \times n$ matrices with real entries, then $A : I \to \mathcal{M}_n$ is a given matrix-valued function and $b : I \to \mathbb{R}^n$ is a given vector-valued function or curve in $\mathbb{R}^n$. If $b(t) = 0$, for every t, the system is called *homogeneous*, and if A is constant (independent of t), the system is said to have *constant coefficients*.

D. Betounes, *Differential Equations: Theory and Applications*, DOI 10.1007/978-1-4419-1163-6_4, 119
© Springer Science + Business Media, LLC 2010

Example 4.1 The following system does not represent any particular physical situation, but serves to illustrate various aspects of the theory throughout the next several sections.

$$\begin{aligned} x_1' &= -x_1 + 2tx_2 + \sin(t^2) \\ x_2' &= -2tx_1 + x_2 + \cos(t^2) \end{aligned}$$

In vector form the system is $x' = A(t)x + b(t)$, where

$$A(t) = \begin{bmatrix} -1 & 2t \\ -2t & -1 \end{bmatrix} \qquad \text{and} \qquad b(t) = \begin{bmatrix} \sin(t^2) \\ \cos(t^2) \end{bmatrix}.$$

The time interval I on which A and b are defined is $I = (-\infty, \infty)$ and the system is nonhomogeneous with nonconstant coefficients.

Linear systems arise naturally in a number of physical situations, as the following example shows. In addition to the motivations for their study in connection with physical phenomena, we shall also see, in the next chapter, that linear systems are useful in the study of nonlinear systems.

Example 4.2 (Coupled Masses) Suppose we have two bodies (metal balls) lying on a frictionless table as shown in Figure 4.1. The bodies are attached together by a spring and each is also attached to a rigid support by a spring. For simplicity, assume the masses of the bodies are the same: $m_1 = m_2 = 1$ and that the stiffnesses of the springs, indicated by their spring constants k_1, k_2, k_3, are all the same as well: $k_1 = k_2 = k_3 = 1$.

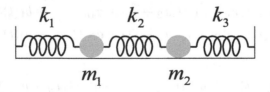

Figure 4.1: *Two bodies with springs attaching them together and to rigid supports.*

The figure shows the bodies and springs lying in a straight line, with the springs at their natural lengths, neither extended nor compressed, and the bodies at their equilibrium positions. To get the bodies to oscillate in this line. we displace each of them forward or backward of their equilibrium positions and impart some initial velocities to them. Their subsequent motion is modeled by a 2nd-order, linear system of DEs which is derived as follows.

By Newton's 2nd Law, the motion of each body is such that its mass times its acceleration is equal to the sum of all the forces on the body. The downward force of gravity is canceled by the upward supporting force of the table, so we can forget about those forces. There remains only the forces of the two springs acting on the body. By Hooke's Law, a spring exerts a force whose magnitude is the spring constant times the amount of displacement and whose direction is opposite the extension or compression. So suppose $x_1(t), x_2(t)$ denote the displacements of the bodies from equilibrium at time t. See Figure 4.2.

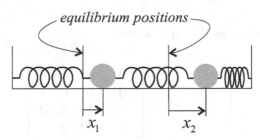

Figure 4.2: *Displacements of the bodies from their equilibrium positions.*

Referring to Figure 4.2, we see that body 1 has spring forces $-k_1 x_1 = -x_1$ and $k_2(x_2 - x_1) = x_2 - x_1$ acting on it. Note that in the situation depicted, the 2nd spring is extended and its force on body 1 is to the right (the positive x-direction). Its force on body 2, however, is toward the left, and thus, the two spring forces acting on body 2 are $-k_2(x_2 - x_1) = -(x_2 - x_1)$ and $-k_3 x_2 = -x_2$. Putting all of this together gives (mass times acceleration equals sum of forces) for each body:

$$
\begin{aligned}
x_1'' &= -x_1 + (x_2 - x_1) \\
x_2'' &= -(x_2 - x_1) - x_2
\end{aligned}
$$

or

$$
\begin{aligned}
x_1'' &= -2x_1 + x_2 \\
x_2'' &= x_1 - 2x_2
\end{aligned}
$$

In terms of vectors and matrices, we can write this system of DEs as

$$
x'' = Kx,
$$

where

$$K = \begin{bmatrix} -2 & 1 \\ 1 & -2 \end{bmatrix}.$$

To reduce this system to 1st-order, we introduce velocities: $v_1 = x_1'$, $v_2 = x_2'$ and get the system:

$$\begin{aligned}
x_1' &= v_1 \\
x_2' &= v_2 \\
v_1' &= -2x_1 + x_2 \\
v_2' &= x_1 - 2x_2
\end{aligned}$$

In matrix form this system is $z' = Az$, where

$$z = \begin{bmatrix} x_1 \\ x_2 \\ v_1 \\ v_2 \end{bmatrix}, \qquad A = \begin{bmatrix} 0 & 0 & 1 & 0 \\ 0 & 0 & 0 & 1 \\ -2 & 1 & 0 & 0 \\ 1 & -2 & 0 & 0 \end{bmatrix}.$$

Later in this chapter we will see how to solve this linear system of DEs. We will also study the general system of coupled masses.

Example 4.3 (Linear, nth-order DEs) The general form of an nth-order, nonhomogeneous, linear differential equation is typically given as

$$x^{(n)} + a_{n-1}(t)x^{(n-1)} + \cdots + a_2(t)x'' + a_1(t)x' + a_0(t)x = r(t), \qquad (4.1)$$

with the coefficient functions a_j, $j = 0, \ldots, N - 1$, defined and continuous on an interval I. In your undergraduate DE course you studied the theory for constructing the general solution of such an equation and learned how to solve, in closed form, particular cases of this type of equation. It is important for you to note that the discussion and theory presented here, which is for 1st-order systems of DEs, contains all of that undergraduate material. This is so because we can rewrite equation (4.1) as the 1st-order system

$$z' = A(t)z + b(t),$$

where

$$A = \begin{bmatrix} 0 & 1 & & & & \\ 0 & 0 & 1 & & & \\ & & \ddots & \ddots & & \\ 0 & 0 & 0 & \cdots & 0 & 1 \\ -a_0 & -a_1 & a_2 & \cdots & -a_{n-2} & -a_{n-1} \end{bmatrix}, \tag{4.2}$$

and $b = (0, 0, \ldots, r)$ (exercise).

As a particularly simple example of this, consider the 2nd-order linear homogeneous DE

$$x'' - \frac{2}{t-1} x' + \frac{2}{(t-1)^2} x = 0,$$

with t in the interval $I = (-\infty, 1)$. Letting $v = x'$, reduces this 2nd-order DE to the following 1st-order system

$$\begin{aligned} x' &= v \\ v' &= \frac{-2}{(t-1)^2} x + \frac{2}{t-1} v \end{aligned}$$

To write this system in matrix form, let $z = (x, v)$ and

$$A(t) = \begin{bmatrix} 0 & 1 \\ \frac{-2}{(t-1)^2} & \frac{2}{t-1} \end{bmatrix}.$$

Then the system of 1st-order DEs is $z' = A(t)z$.

Before embarking on a detailed study of linear systems of equations, we point out that linear systems have many special properties which are not generally present in nonlinear systems. In particular we mention the following

Properties of Linear Systems:

(1) The solutions (integral curves) exist for all times $t \in I$, i.e., each maximum interval of existence coincides with I (see Theorem 4.1 below).

(2) Any linear combination of solutions of the *homogeneous* system

$$x' = A(t)x$$

is also a solution. Specifically, if $\alpha_1, \ldots, \alpha_k : I \to \mathbb{R}^n$ are integral curves and $c_1, \ldots, c_k$ are constants, then the curve

$$\beta(t) = c_1\alpha_1(t) + \cdots + c_k\alpha_k(t),$$

for $t \in I$, is also an integral curve. This is called the *superposition principle*.

(3) There is an "explicit" formula for the general solution of the nonhomogeneous equation $x' = A(t)x + b(t)$. It is

$$\alpha(t) = G(t)c + G(t) \int_0^t G(s)^{-1}b(s)ds,$$

where G is the *fundamental matrix* for A and $c \in \mathbb{R}^n$ is an arbitrary constant vector. The fundamental matrix is discussed below and the above formula is explicit only to the extent that we are able to compute the fundamental matrix and its inverse and the definite integral involved.

4.1 Existence and Uniqueness for Linear Systems

Note: Since the domain of the vector field $X(t, x) = A(t)x + b(t)$ is $B = I \times \mathbb{R}^n$, we can always assume, without loss of generality that $0 \in I$ (exercise). This will be convenient and will simplify the notation.

From here on we also make the assumption that $A : I \to \mathcal{M}_n$ and $b : I \to \mathbb{R}^n$ are continuous functions. Continuity of a matrix-valued function $A = \{a_{ij}\}$ just means that each of its entries $a_{ij} : I \to \mathbb{R}$ is a continuous function.

We let

$$C := \{\, \beta : I \to \mathbb{R}^n \,|\, \beta \text{ is continuous} \,\}$$

denote the set of all continuous curves on I. For a given point $c \in \mathbb{R}^n$, the *Picard iteration map* $T : C \to C$ is defined by

$$T(\beta)(t) = c + \int_0^t [A(s)\beta(s) + b(s)]ds,$$

for $t \in I$. It is easy to see that $T(\beta)$ is continuous on I (indeed is differentiable everywhere, except possibly at the endpoints of I). Thus, T is a mapping from C into C. In proving the general Existence and Uniqueness Theorem (Chapter 3), we found that, beginning with any curve β, the sequence of iterates

$$\beta, \ T(\beta), \ T^2(\beta), \ T^3(\beta), \ \ldots \ ,$$

converges to the integral curve α that satisfies $\alpha(0) = c$. The transformation T we used there arose after judiciously picking certain constants and intervals in order to make T a contraction. Here, for linear systems, this is not needed and, further, the maximum interval existence for integral curves turns out to be the same as the given interval I on which A and b are continuous:

Theorem 4.1 (Existence and Uniqueness for Linear Systems) *If $A : I \to M_n$ and $b : I \to \mathbb{R}^n$ are continuous on I, then for each $c \in \mathbb{R}^n$, there is a unique solution $\alpha : I \to \mathbb{R}^n$ of the IVP:*

$$\begin{aligned} x' &= A(t)x + b(t) \\ x(0) &= c. \end{aligned}$$

Note: *What's new here is that the solution α is defined on* all *of the given interval I where A and b are continuous. The general existence and uniqueness theorem only guarantees us the existence of a solution defined on* some *interval about 0.*

Proof: This theorem is not a corollary of Theorem 3.1, since it yields a stronger result than we can get by applying that theorem to this special case. Thus, the proof proceeds a little differently and exploits the linearity of the system.

As in the previous proof, we will use the ℓ_1 norm on $\mathbb{R}^n$, i.e., for $x \in \mathbb{R}^n$

$$\|x\| = \sum_{i=1}^{n} |x_i|.$$

Since each of the entries a_{ij} of the matrix A is a continuous function on $I = [r_1, r_2]$, there is a constant K such that

$$|a_{ij}(t)| \leq K,$$

for all $i, j = 1, \ldots, n$ and $t \in I$. This constant K plays the same role here as in the general existence and uniqueness theorem. Namely, we would like to show that

$$\|X(x,t) - X(y,t)\| \leq nK\|x - y\|, \tag{4.3}$$

for all $x, y \in \mathbb{R}^n$ and $t \in I$ (Lipschitz condition). This is easy to do since the vector field X has the form: $X(x,t) = A(t)x + b(t)$ and so $X(x,t) - X(y,t) = A(t)(x-y)$. Thus, inequality (4.3) results from the following:

$$
\begin{aligned}
\|A(t)(x-y)\| &= \sum_{i=1}^{n} \left| \sum_{j=1}^{n} a_{ij}(t)(x_j - y_j) \right| \\
&\leq \sum_{i=1}^{n} \sum_{j=1}^{n} |a_{ij}(t)| |x_j - y_j| \\
&\leq \sum_{i=1}^{n} \sum_{j=1}^{n} K |x_j - y_j| \\
&= nK \|x - y\|
\end{aligned}
$$

for all $x, y \in \mathbb{R}^n$ and $t \in I$. This is the inequality we need.

As before, the set C is a Banach space when endowed with the sup norm:

$$
\|\beta\| = \sup\{ \|\beta(t)\| \,|\, t \in I \}.
$$

We proceed to show that for any $\beta \in C$, the sequence $\{T^k(\beta)\}_{k=0}^{\infty}$ converges to the solution we seek.

For this note that because of the special (affine) nature of the Picard iteration map T, we have

$$
\begin{aligned}
T^p(\beta)(t) - T^{p+1}(\beta)(t) &= T\Big(T^{p-1}(\beta)\Big)(t) - T\Big(T^p(\beta)\Big)(t) \\
&= \int_0^t A(s)\Big[T^{p-1}(\beta)(s) - T^p(\beta)(s) \Big] ds,
\end{aligned}
$$

for all $p = 1, 2, \ldots$. We use this successively in deriving the following estimates. For convenience we let

$$
M = \|\beta - T(\beta)\|.
$$

Then for any $t \in I$ with $t \geq 0$, we have

$$
\begin{aligned}
\|T(\beta)(t) - T^2(\beta)(t)\| &= \left\| \int_0^t A(s)\Big[\beta(s) - T(\beta)(s) \Big] ds \right\| \\
&\leq \int_0^t \left\| A(s)\Big[\beta(s) - T(\beta)(s) \Big] \right\| ds \\
&\leq \int_0^t nK \|\beta(s) - T(\beta)(s)\| ds \\
&\leq \int_0^t nKM \, ds = nKM \, t.
\end{aligned}
$$

Next, use this in the following (similar) derivation:

$$
\begin{aligned}
\|T^2(\beta)(t) - T^3(\beta)(t)\| &= \left\| \int_0^t A(s)\Big[T(\beta)(s) - T^2(\beta)(s)\Big] ds \right\| \\
&\leq \int_0^t nK\|T(\beta)(s) - T^2(\beta)(s)\| ds \\
&\leq \int_0^t nKnKM\, s\, ds = n^2 K^2 M\, t^2/2
\end{aligned}
$$

Continuing in this fashion, it's not hard to see that inductively we get

$$
\|T^k(\beta)(t) - T^{k+1}(\beta)(t)\| \leq n^k K^k M \frac{|t|^k}{k!},
$$

for any k and all $t \in I$ with $t \geq 0$. Similar reasoning gives the same inequality for $t < 0$. Letting $q = r_2 - r_1$ and taking the sup over $t \in I$ gives

$$
\|T^k(\beta) - T^{k+1}(\beta)\| \leq n^k K^k M \frac{q^k}{k!},
$$

for $k = 1, 2, 3, \ldots$. Using this and the triangle inequality for the sup norm, we easily get the following:

$$
\begin{aligned}
&\|T^k(\beta) - T^m(\beta)\| \\
={}& \|T^k(\beta) - T^{k+1}(\beta) + T^{k+1}(\beta) - T^{k+2}(\beta) + \cdots + T^{m-1}(\beta) - T^m(\beta)\| \\
\leq{}& \|T^k(\beta) - T^{k+1}(\beta)\| + \|T^{k+1}(\beta) - T^{k+2}(\beta)\| + \cdots + \|T^{m-1}(\beta) - T^m(\beta)\| \\
\leq{}& M \sum_{j=k}^{m-1} \frac{n^j K^j q^j}{j!} \\
={}& M(S_{m-1} - S_{k-1}),
\end{aligned}
$$

for all $k < m$. Here we have introduced S_N to stand for

$$
S_N \equiv \sum_{j=0}^{N} \frac{(nKq)^j}{j!}.
$$

Now since $\lim_{N \to \infty} S_N = e^{nKq}$, the sequence $\{S_N\}_{N=0}^\infty$ is Cauchy and hence by the above inequality, the sequence $\{T^k(\beta)\}_{k=0}^\infty$ is Cauchy (exercise). But then, since the latter sequence is a Cauchy sequence in the Banach space C of curves, it is a convergent sequence. Thus, there exists an $\alpha \in C$, such that

$$
\lim_{k \to \infty} T^k(\beta) = \alpha.
$$

That α is a fixed point: $T(\alpha) = \alpha$, and therefore is a solution of the IVP, follows as in the proof of Theorem 3.1. That theorem also shows that α is unique. $\square$

Exercises 4.1

1. Show that the nth-order linear system (4.1) can be rewritten as a 1st-order system $z' = A(t)z + b(t)$, with A as in (4.2).

2. Show that there is no loss of generality in assuming that 0 is in the interval I, which is the domain for the coefficient matrix: $A : I \rightarrow \mathcal{M}_n$ and forcing vector $b : I \rightarrow \mathbb{R}^n$. *Hint*: Let $\tilde{I} = I - t_*$, where $t_* \in I$ and define a new coefficient matrix $\tilde{A}$ and forcing vector $\tilde{b}$ on $\tilde{I}$. Make sure you formulate and prove a precise result. See also Exercise 2, Section 3.4.

3. (**Homogeneous, Constant Coefficient Systems**) For systems $x' = Ax$, which are homogeneous with constant matrix A, the theory simplifies considerably.

   (a) Show that the Picard iteration map: $T : C \rightarrow C$ is

   $$T(\beta)(t) = c + A \int_0^t \beta(s)ds.$$

   (b) For the following matrices A and initial points c, compute, by hand, the iterates: $\beta_1 = T(c), \beta_2 = T(\beta_1), \ldots, \beta_8 = T(\beta_7)$. Use a computer to plot these approximations $\beta_k, k = 1, \ldots, 8$, and the solution α to $x' = Ax$, on some appropriate time interval.

   (i) $A = \begin{bmatrix} 0 & 1 \\ -1 & 0 \end{bmatrix}$, $c = (1,0)$. Based on your calculations, determine a formula for β_k and find the limit: $\lim_{k \to \infty} \beta_k(t)$.

   (ii) $A = \begin{bmatrix} -1 & 1 \\ -1 & -1 \end{bmatrix}$, $c = (1,1)$.

   (c) Use Part (a) to show that the 1st and 2nd iterates of the constant map c are: $T(c)(t) = c + tAc$, $T^2(c)(t) = c + tAc + \frac{t^2}{2}A^2c$. Then show that, in general, the kth iterate is

   $$T^k(c)(t) = \left[I + tA + \frac{t^2}{2}A^2 + \cdots + \frac{t^k}{k!}A^k \right] c.$$

4.2   The Fundamental Matrix and the Flow

The explicit formula for the integral curves of the general linear system $x' = A(t)x + b(t)$ involves a matrix known as the fundamental matrix G for the system (more precisely, for the coefficient matrix A). The definition of G follows from the existence and uniqueness theorem above.

Definition 4.1 (The Fundamental Matrix) For each $j = 1, \ldots, n$, let $\gamma_j : I \to \mathbb{R}^n$ be the solution of the IVP

$$
\begin{aligned}
x' &= A(t)x \\
x(0) &= \varepsilon_j.
\end{aligned}
$$

Here $\varepsilon_1 = (1, 0, 0, \ldots, 0)$, $\varepsilon_2 = (0, 1, 0, \ldots, 0)$, $\ldots$, $\varepsilon_n = (0, 0, 0, \ldots, 1)$ are the standard unit vectors in $\mathbb{R}^n$. The *fundamental matrix* for A is the $n \times n$ matrix G whose columns are $\gamma_1, \gamma_2, \ldots, \gamma_n$. Symbolically we write this as

$$
G = [\gamma_1, \gamma_2, \ldots, \gamma_n].
$$

More precisely, if $\gamma_j = (\gamma_{1j}, \ldots, \gamma_{nj})$, then $G : I \to \mathcal{M}_n$ is the matrix-valued function defined by

$$
G(t) = \{\gamma_{ij}(t)\} = \begin{bmatrix} \gamma_{11}(t) & \cdots & \gamma_{1n}(t) \\ \vdots & & \vdots \\ \gamma_{n1}(t) & \cdots & \gamma_{nn}(t) \end{bmatrix}. \tag{4.4}
$$

Example 4.4 Consider the system $z' = A(t)z$ with

$$
A(t) = \begin{bmatrix} 0 & 1 \\ \frac{-2}{(t-1)^2} & \frac{2}{t-1} \end{bmatrix}.
$$

How do we find the fundamental matrix G for this matrix A? Generally this is difficult, but this system arose in Example 4.3 by reducing the DE

$$
x'' - \frac{2}{t-1} x' + \frac{2}{(t-1)^2} x = 0,
$$

to 1st-order: $z' = A(t)z$, with $z = (x, v)$ and $v = x'$. Furthermore, this 2nd-order DE is a Cauchy-Euler (equidimensional) DE and so we can determine its general solution. It is (see your undergraduate DE book):

$$
x = a(t-1) + b(t-1)^2,
$$

where a, b are arbitrary constants. Then $v = x'$ is

$$v = a + 2b(t - 1).$$

Consequently, the general solution of $z' = A(t)z$ is

$$z = (x, v) = \left(a(t - 1) + b(t - 1)^2, \ a + 2b(t - 1) \right).$$

We need to find two solutions z_1, z_2 which satisfy $z_1(0) = (1, 0), z_2(0) = (0, 1)$. This is where the arbitrary constants a, b in the general solution come in—we choose them to get z_1, z_2. Now in general

$$z(0) = (-a + b, \ a - 2b).$$

So to get z_1 we must choose a, b to satisfy

$$\begin{aligned} -a + b &= 1 \\ a - 2b &= 0 \end{aligned}$$

This gives $a = -2, b = -1$ and thus

$$z_1 = \left(-2(t - 1) - (t - 1)^2, \ -2 - 2(t - 1) \right) = \left(1 - t^2, -2t \right).$$

To get z_2 we must choose a, b so that

$$\begin{aligned} -a + b &= 0 \\ a - 2b &= 1 \end{aligned}$$

This gives $a = -1, b = -1$ and thus,

$$z_2 = \left(-(t - 1) - (t - 1)^2, \ -1 - 2(t - 1) \right) = \left(t - t^2, 1 - 2t \right).$$

The fundamental matrix is then formed by using these two vectors z_1, z_1 as columns:

$$G(t) = \begin{bmatrix} 1 - t^2 & t - t^2 \\ -2t & 1 - 2t \end{bmatrix}.$$

Note that the 2nd row of G comes from differentiating the first row, but this occurs only because of the special nature of this system (it arises by a reduction to first order). The other thing to note is that, as mentioned earlier, the fundamental matrix is used in formulating the general solution of linear systems of 1st-order DEs. But in this example the general solution can be found by other means (i.e., by using the general solution of $x'' - \frac{2}{t-1} x' + \frac{2}{(t-1)^2} x = 0$ as we did above).

The fundamental matrix G can be computed explicitly in many cases (for instance, the last example), but for the most part it is only given theoretically from the existence and uniqueness theorem for linear systems. Its primary use, in either case, is to delineate the special nature of the solutions to linear systems. The case $n = 1$, i.e., when the system consists of a single scalar DE

$$x' = p(t)x + q(t),$$

should be familiar to you from your undergraduate DE course. Recall that the solution of this, which satisfies the initial condition: $x(0) = c$ is

$$x = \mu(t)\left[c + \int_0^t \mu(s)^{-1}q(s)ds\right], \tag{4.5}$$

where

$$\mu(t) = e^{\int_0^t p(s)ds}$$

is an "integrating factor" for the equation. Since μ satisfies $\mu' = p(t)\mu$ and $\mu(0) = 1$, it follows that μ is indeed the fundamental matrix for the case $n = 1$. You should see now why the fundamental matrix is not always explicitly computable. In the 1-dimensional case here, the integral $\int_0^t p(s)ds$ is only given theoretically and while many examples and exercises are designed so that the integrals are expressible in terms of standard elementary functions, this need not always be the case.

Our goal in the next several theorems is to show that the form of the above solution (4.5) for $n = 1$ is precisely the same for $n > 1$. Thus, the fundamental matrix is analogous to the notion of an integrating factor.

Theorem 4.2 *If G is the fundamental matrix for A, then G satisfies:*

$$G'(t) = A(t)G(t) \qquad \forall t \in I \tag{4.6}$$
$$G(0) = I, \tag{4.7}$$

where I denotes the $n \times n$ identity matrix.

Proof: The proof is an easy application of the product rule for matrix and vector-valued functions (see Appendix C) and the above existence and uniqueness theorem. Thus, differentiating G columnwise and using the fact that the columns, by definition, satisfy $\gamma_j' = A(t)\gamma_j$, we get

$$\begin{aligned}
G'(t) &= [\gamma_1'(t), \ldots, \gamma_n'(t)] \\
&= [A(t)\gamma_1(t), \ldots, A(t)\gamma_n(t)] \\
&= A(t)[\gamma_1(t), \ldots, \gamma_n(t)] \\
&= A(t)G(t),
\end{aligned}$$

for all $t \in I$. Also that $G(0) = I$ is clear since the columns of the identity matrix I are $\varepsilon_j = \gamma_j(0)$, $j = 1, \ldots, n$. $\square$

Remark: An alternative way of defining the fundamental matrix G for A is to just declare that it is the unique solution to the matrix differential equation: $G' = A(t)G$, which satisfies $G(0) = I$. It's an easy exercise to extend the existence and uniqueness theorem from above to the case of matrix DEs and then to use elementary properties of matrix operations to show that this way of defining G is equivalent to the first way (exercise).

Example 4.5 Consider the system $x' = A(t)x$, where

$$A(t) = \begin{bmatrix} -1 & 2t \\ -2t & -1 \end{bmatrix}.$$

The fundamental matrix for A is

$$G(t) = e^{-t} \begin{bmatrix} \cos(t^2) & \sin(t^2) \\ -\sin(t^2) & \cos(t^2) \end{bmatrix}.$$

To check this, first note that

$$G(0) = e^0 \begin{bmatrix} \cos(0) & \sin(0) \\ -\sin(0) & \cos(0) \end{bmatrix} = \begin{bmatrix} 1 & 0 \\ 0 & 1 \end{bmatrix} = I.$$

And more importantly we calculate

$$
\begin{aligned}
G'(t) &= -e^{-t} \begin{bmatrix} \cos(t^2) & \sin(t^2) \\ -\sin(t^2) & \cos(t^2) \end{bmatrix} + e^{-t} \begin{bmatrix} -2t\sin(t^2) & 2t\cos(t^2) \\ -2t\cos(t^2) & -2t\sin(t^2) \end{bmatrix} \\
&= e^{-t} \begin{bmatrix} -\cos(t^2) - 2t\sin(t^2) & -\sin(t^2) + 2t\cos(t^2) \\ \sin(t^2) - 2t\cos(t^2) & -\cos(t^2) - 2t\sin(t^2) \end{bmatrix}
\end{aligned}
$$

We compare this with

$$
\begin{aligned}
A(t)G(t) &= \begin{bmatrix} -1 & 2t \\ -2t & -1 \end{bmatrix} e^{-t} \begin{bmatrix} \cos(t^2) & \sin(t^2) \\ -\sin(t^2) & \cos(t^2) \end{bmatrix} \\
&= e^{-t} \begin{bmatrix} -\cos(t^2) - 2t\sin(t^2) & -\sin(t^2) + 2t\cos(t^2) \\ \sin(t^2) - 2t\cos(t^2) & -\cos(t^2) - 2t\sin(t^2) \end{bmatrix}
\end{aligned}
$$

This shows that $G'(t) = A(t)G(t)$. Thus, G is the fundamental matrix for A.

While this example serves to illustrate the previous theorem, you may wonder how we got the fundamental matrix for A in the first place. Here's one way to do this. (This technique is discussed in the exercises for 2×2 and 3×3 coefficient matrices A in general.)

Start with the system written in the form:

$$x_1' = -x_1 + 2tx_2 \tag{4.8}$$
$$x_2' = -2tx_1 - x_2 \tag{4.9}$$

Rearrange this as

$$x_1' + x_1 = 2tx_2$$
$$x_2' + x_2 = -2tx_1$$

Now multiply both sides by e^t to get

$$e^t x_1' + e^t x_1 = 2te^t x_2$$
$$e^t x_2' + e^t x_2 = -2te^t x_1$$

This is the same as

$$\left(e^t x_1\right)' = 2te^t x_2$$
$$\left(e^t x_2\right)' = -2te^t x_1$$

Then if we let $u_1 = e^t x_1$ and $u_2 = e^t x_2$, the above system is

$$u_1' = 2tu_2 \tag{4.10}$$
$$u_2' = -2tu_1 \tag{4.11}$$

Now look for a solution of this system of the form

$$u_1(t) = P(t^2) \tag{4.12}$$
$$u_2(t) = P'(t^2), \tag{4.13}$$

where P is a twice-differentiable function. We determine P by substituting u_1, u_2 into (4.10)-(4.11). The first of these equations is satisfied for any choice of P and the second equation reduces to

$$P''(t^2) = -P(t^2),$$

for all t. Thus, any solution of $P''(s) = -P(s)$ will give a solution of (4.10)-(4.11) via definitions (4.12)-(4.13). One solution of $P'' = -P$ is $P(s) = \cos(s)$ and this gives a solution

$$(u_1(t), u_2(t)) = \Big(\cos(t^2), -\sin(t^2)\Big)$$

of (4.10)-(4.11) whose value at $t = 0$ is $(1,0)$. Another solution of $P'' = -P$ is $P(s) = \sin(s)$ and this gives a solution

$$(u_1(t), u_2(t)) = \Big(\sin(t^2), \cos(t^2)\Big)$$

of (4.10)-(4.11) whose value at $t = 0$ is $(0,1)$. Multiplying each of these by e^{-t} gives two solutions (x_1, x_2) of the original system (4.8)-(4.9). Finally, we can use these to get its fundamental matrix:

$$G(t) = e^{-t} \begin{bmatrix} \cos(t^2) & \sin(t^2) \\ -\sin(t^2) & \cos(t^2) \end{bmatrix}.$$

An important fact about the fundamental matrix is that it is invertible, i.e., $G(t)^{-1}$ exists for all $t \in I$. This is equivalent to saying that $\det(G(t)) \neq 0$, for all $t \in I$. This non obvious fact can be demonstrated using the following formula, which relates the determinant of G and the trace of A:

Proposition 4.1 (Liouville's Formula) *If G is the fundamental matrix for A, then*

$$\det(G(t)) = e^{\int_0^t \text{tr}(A(s))\,ds},$$

for all $t \in I$.

Proof: If we view the matrix $G = [R_1, \dots, R_n]$ in terms of its rows R_i, then the matrix differential equation that G satisfies, namely: $G' = AG$, can also be expressed as

$$R_i' = \sum_{j=1}^{n} a_{ij} R_j,$$

for $i = 1, \dots, n$. Now since the determinant of a matrix is a multilinear function of the rows of the matrix, it's not hard to show that the derivative of $\det(G)$ satisfies a type of product rule (exercise). This is exhibited in the first line of the following calculation:

$$
\begin{aligned}
\frac{d}{dt}\det(G) &= \sum_{i=1}^{n}\det([R_1,\ldots,R_i',\ldots,R_n]) \\
&= \sum_{i=1}^{n}\det([R_1,\ldots,\sum_{j=1}^{n}a_{ij}R_j,\ldots,R_n]) \\
&= \sum_{i=1}^{n}\sum_{j=1}^{n}a_{ij}\det([R_1,\ldots,R_j,\ldots,R_n]) \\
&= \sum_{i=1}^{n}\sum_{j=1}^{n}a_{ij}\delta_{ij}\det(G). \\
&= \sum_{i=1}^{n}a_{ii}\det(G) \\
&= \operatorname{tr}(A)\det(G).
\end{aligned}
$$

Note: In the first three equations, the quantities: R_i', $\sum_{j=1}^{n}a_{ij}R_j$, and R_j, occur in the ith row of the indicated matrix. In the above, δ_{ij} is the Kronecker delta function, and we have used the fact that

$$
\det([R_1,\ldots,R_j,\ldots,R_n]) = 0,
$$

if $j \neq i$ (since two rows of the determinant will be the same). Also, it is clear that $\det(G(0)) = \det(I) = 1$. Thus, altogether, we have shown that the real-valued function $x = \det(G(t))$ satisfies the IVP

$$
\begin{aligned}
x' &= \operatorname{tr}(A(t))x \\
x(0) &= 1
\end{aligned}
$$

But by the discussion prior to Theorem 4.2, we know that $x = e^{\int_0^t \operatorname{tr}(A(s))ds}$ is the unique solution of this IVP. Hence: $\det(G(t)) = e^{\int_0^t \operatorname{tr}(A(s))ds}$. $\square$

Remark: You should note that in the case when $n = 1, \det(G(t)) = G(t)$ and $\operatorname{tr}(A(t)) = A(t)$. Thus, the relationship $G(t) = e^{\int_0^t A(s)ds}$ in Liouville's formula is precisely what you would expect, based on the discussion prior to the theorem.

Example 4.6 An interesting homogeneous system which has periodic solutions is the system $x' = A(t)x$ with coefficient matrix

$$A(t) = \begin{bmatrix} -\sin t & \cos t \\ \cos t & -\sin t \end{bmatrix}.$$

You can use the technique described in the last example (or the general results in Section 4.2, Exercise 2 below) to determine that the fundamental matrix for A is

$$G(t) = e^{\cos t - 1} \begin{bmatrix} \cosh(\sin t) & \sinh(\sin t) \\ \sinh(\sin t) & \cosh(\sin t) \end{bmatrix}.$$

We check that Liouville's formula holds in this example. First, using the hyperbolic identity $\cosh^2 u - \sinh^2 u = 1$, we get

$$\det(G(t)) = \left(e^{\cos t - 1}\right)^2 \left[\cosh^2(\sin t) - \sinh^2(\sin t)\right] = e^{2\cos t - 2}.$$

Also

$$\mathrm{tr}(A(t)) = -2\sin t.$$

Thus,

$$e^{\int_0^t \mathrm{tr}(A(s))ds} = e^{\int_0^t (-2\sin(s))ds} = e^{2\cos t - 2} = \det(G(t)).$$

It is an easy exercise to verify that matrices for the systems in Examples 4.4 and 4.5 also satisfy Liouville's formula.

We next consider the nonhomogeneous system

$$x' = A(t)x + b(t),$$

and show how to express its general solution (i.e., the flow) in terms of the fundamental matrix G for A. To motivate the formula, we provide the following heuristic derivation of it. This should look similar to what you did as an undergraduate in solving 1st-order, linear DE's, since, indeed, that is what the following reduces to when $n = 1$.

First we take the DE and rewrite it as

$$x' - A(t)x = b(t).$$

Then multiply both sides by $G(t)^{-1}$, to get

$$G(t)^{-1}x' - G(t)^{-1}A(t)x = G(t)^{-1}b(t).$$

You can consider $G(t)^{-1}$ as an "integrating factor" since the left side of the last DE is now the exact derivative of $G(t)x$. To see this, just do the calculation:

$$\begin{aligned}
\frac{d}{dt}\left[G^{-1}x\right] &= G^{-1}x' + \left[\frac{d}{dt}G^{-1}\right]x \\
&= G^{-1}x' - G^{-1}G'G^{-1}x \\
&= G^{-1}x' - G^{-1}AGG^{-1}x \\
&= G^{-1}x' - G^{-1}Ax,
\end{aligned}$$

where we have used the product rule and the rule for differentiating the matrix inverse (exercise). Thus, the DE becomes:

$$\frac{d}{dt}\left[G(t)^{-1}x\right] = G(t)^{-1}b(t).$$

Integrating both sides of this from 0 to t and using the Fundamental Theorem of Calculus on the left side yields

$$G(t)^{-1}x - x(0) = \int_0^t G(s)^{-1}b(s)ds.$$

Let the generic initial condition be $x(0) = c$ and solve the last equation for x to get the solution of the initial-value problem expressed by

$$x = G(t)c + G(t)\int_0^t G(s)^{-1}b(s)ds.$$

With this as the motivation, we now proceed to show that this is in fact the solution of the general linear differential equation.

Theorem 4.3 (Solution of the Nonhomogeneous DE) *If G is the fundamental matrix for A, then the general solution of the nonhomogeneous, linear system $x' = A(t)x + b(t)$ is*

$$\alpha(t) = G(t)\left[c + \int_0^t G(s)^{-1}b(s)ds\right], \tag{4.14}$$

for $t \in I$ and c an arbitrary constant (vector).

Proof: We first show that for any $c \in \mathbb{R}^n$, the α given by equation (4.14) is a solution of the nonhomogeneous equation. This involves a straight-forward calculation of the derivative of α using the product rule, the Fundamental Theorem of Calculus, and the fact that $G' = AG$. The calculation is as follows:

$$
\begin{aligned}
\alpha'(t) &= G'(t)\left[c + \int_0^t G(s)^{-1}b(s)ds\right] + G(t)G(t)^{-1}b(t) \\
&= A(t)G(t)\left[c + \int_0^t G(s)^{-1}b(s)ds\right] + b(t) \\
&= A(t)\alpha(t) + b(t)
\end{aligned}
$$

Next we must show that formula (4.14) includes all solutions of the non-homogeneous equation. More precisely: if $\beta : J \to \mathbb{R}^n$ is a solution of the nonhomogeneous equation, then there is a constant $c \in \mathbb{R}^n$ such that $\beta = \alpha$ on J. To see this, choose any $t_0 \in J$ and let

$$
c = G(t_0)^{-1}\beta(t_0) - \int_0^{t_0} G(s)^{-1}b(s)ds. \tag{4.15}
$$

With this choice for c in (4.14), we get that $\alpha(t_0) = \beta(t_0)$. Hence by the uniqueness part of the Existence and Uniqueness Theorem, it follows that $\alpha = \beta$ on J. $\square$

Rewriting formula (4.14) slightly gives

$$
\alpha(t) = G(t)c + G(t)\int_0^t G(s)^{-1}b(s)ds = \alpha_c(t) + \alpha_p(t),
$$

and this expresses the general solution as the sum of two terms. The first term

$$
\alpha_c(t) = G(t)c,
$$

is the general solution of the homogeneous equation, while the second term

$$
\alpha_p(t) = G(t)\int_0^t G(s)^{-1}b(s)ds, \tag{4.16}
$$

is a particular solution of the nonhomogeneous equation (exercise). Alternatively, let γ_j, $j = 1, \ldots, n$ denote the columns of G and $c = (c, \ldots, c_n)$. Then the expression for general solution is

$$
\alpha(t) = \sum_{j=1}^n c_j\gamma_j(t) + \alpha_p(t).
$$

These are the familiar forms, which you studied as an undergraduate, of expressing the general solution of the nonhomogeneous equation as the sum of the general solution of the homogeneous equation and any one particular solution of the nonhomogeneous equation. The discussion here not only generalizes the special case you studied as an undergraduate, but also gives an explicit formula (4.16) for the construction of a particular solution. The exercises will show how this formula contains the variation-of-parameters method for constructing particular solutions of the nonhomogeneous equation.

The choice of the constant c in equation (4.15) in the proof above was necessary in order to specialize the general solution (4.6) so that it passes through the point $q \equiv \beta(t_0)$ at time t_0. Substituting this value of c into formula (4.14) gives

$$
\begin{aligned}
\alpha(t) &= G(t)\left[G(t_0)^{-1}q - \int_0^{t_0} G(s)^{-1}b(s)ds + \int_0^t G(s)^{-1}b(s)ds\right] \\
&= G(t)\left[G(t_0)^{-1}q + \int_{t_0}^t G(s)^{-1}b(s)\,ds\right].
\end{aligned}
$$

From this we get the formula for the flow of the general linear system of DEs:

Corollary 4.1 (The Flow for Linear Systems) *For the general linear system*

$$x' = A(t)x + b(t),$$

the flow $\phi : I \times I \times \mathbb{R}^n \to \mathbb{R}^n$ *is given by*

$$\phi_t^u(x) = G(t)\left[G(u)^{-1}x + \int_u^t G(s)^{-1}b(s)ds\right], \qquad (4.17)$$

for all $u, t \in I$, *and* $x \in \mathbb{R}^n$. *In particular, the flow for the homogeneous system* $x' = A(t)x$ *is simply*

$$\phi_t^u(x) = G(t)G(u)^{-1}x. \qquad (4.18)$$

For a fixed $u, t \in I$, it is easy to see that the map $\phi_t^u : \mathbb{R}^n \to \mathbb{R}^n$ is a linear map in the homogeneous case (when $b = 0$) and is an affine map in the general nonhomogeneous case ($b \neq 0$). It is also easy to prove directly from formula (4.17) that the semigroup property for the flow holds (exercise).

Example 4.7 We consider again the homogeneous system $x' = A(t)x$ in Example 4.5 with

$$A(t) = \begin{bmatrix} -1 & 2t \\ -2t & -1 \end{bmatrix}.$$

We found that the fundamental matrix for A is

$$G(t) = e^{-t} \begin{bmatrix} \cos(t^2) & \sin(t^2) \\ -\sin(t^2) & \cos(t^2) \end{bmatrix}.$$

To find the flow map for this system, first recall that $\det(G(u)) = e^{-2u}$, and then use the formula for computing inverses of 2×2 matrices to get

$$G(u)^{-1} = e^u \begin{bmatrix} \cos(u^2) & -\sin(u^2) \\ \sin(u^2) & \cos(u^2) \end{bmatrix}.$$

Thus, the flow map is

$$\begin{aligned}
\phi_t^u(c) &= G(t)G(u)^{-1}c \\
&= e^{-t} \begin{bmatrix} \cos(t^2) & \sin(t^2) \\ -\sin(t^2) & \cos(t^2) \end{bmatrix} e^u \begin{bmatrix} \cos(u^2) & -\sin(u^2) \\ \sin(u^2) & \cos(u^2) \end{bmatrix} \begin{bmatrix} c_1 \\ c_2 \end{bmatrix} \\
&= e^{u-t} \begin{bmatrix} \cos(t^2 - u^2) & \sin(t^2 - u^2) \\ -\sin(t^2 - u^2) & \cos(t^2 - u^2) \end{bmatrix} \begin{bmatrix} c_1 \\ c_2 \end{bmatrix}
\end{aligned}$$

Recall that with u fixed, the curve $t \mapsto \phi_t^u(c)$ gives the integral curve which starts at c at time u. For example, the integral curves that start at $c = (1, 0)$ at times $u = 0, 1, 5$ are shown in Figure 4.3. The figure indicates that, regardless of the starting time, the integral curves spiral in toward the origin. This is easily verified from the formula for the flow map given above. Namely, since $\lim_{t \to \infty} e^{u-t} = 0$, we see that $\lim_{t \to \infty} G(t)G(u)^{-1}c = 0$. Note also that the flow lines for integral curves $\phi_t^0(c), \phi_t^1(c), \phi_t^5(c)$ shown in Figure 4.3 cross each other at various points. This is something that cannot happen for autonomous systems—*their flow lines cannot intersect*. But for nonautonomous systems, the intersection of flow lines is not prohibited. If you look at the situation dynamically, with the nonautonomous flow lines here representing the paths of three particles, you will see that these particles are never at the same place at the same time. This intersecting of flow lines is also exhibited in Figure 4.4 showing the paths traced from nine different initial points, all starting at time $u = 0$. One further analysis that we can do

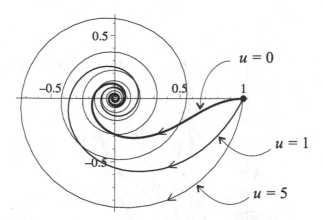

Figure 4.3: *Integral curves of $x_1' = -x_1 + 2tx_2$, $x_2' = -2tx_1 - x_2$ which start at $c = (1,0)$ at times $u = 0, 1, 5$.*

for this particular system is a study of how its direction field changes over time. The vector field involved is

$$X(t, x_1, x_2) = (-x_1 + 2tx_2, -2tx_1 - x_2).$$

The length of this vector is

$$|X(t, x_1, x_2)| = (1 + 4t^2)^{1/2}(x_1^2 + x_2^2)^{1/2} = (1 + 4t^2)^{1/2}|x|.$$

Thus, the length of X tends to infinity as $t \mapsto \infty$, and the normalized vector-field

$$\frac{X(x_1, x_2)}{|X(t, x_1, x_2)|} = \left(\frac{-x_1 + 2tx_2}{(1 + 4t^2)^{1/2}|x|}, \frac{-2tx_1 - x_2}{(1 + 4t^2)^{1/2}|x|} \right),$$

tends to an autonomous vector-field $Y(x_1, x_2)$:

$$\lim_{t \to \infty} \frac{X(t, x_1, x_2)}{|X(t, x_1, x_2)|} = \left(\frac{x_2}{|x|}, \frac{-x_1}{|x|} \right) \equiv Y(x_1, x_2).$$

The approach of $X/|X|$ to Y is fairly rapid, the two being almost indistinguishable for $t \geq 2$. Figure 4.5 shows three of the frames from such a movie, e.g., direction-field plots for $X(0, x)$, $X(0.2, x)$ and $X(1, x)$.

Example 4.8 (A Nonhomogeneous System) We conclude this section with the solution and discussion of the nonhomogeneous system in this Chapter's first example. Namely,

$$
\begin{aligned}
x_1' &= -x_1 + 2tx_2 + \sin(t^2) \\
x_2' &= -2tx_1 + x_2 + \cos(t^2)
\end{aligned}
$$

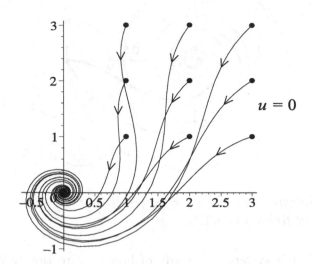

Figure 4.4: *Integral curves of* $x'_1 = -x_1 + 2tx_2$, $x'_2 = -2tx_1 - x_2$ *which start at* $c = (i, j)$, $i, j = 1, 2, 3$, *at time* $u = 0$.

In vector form this system is $x' = A(t)x + b(t)$, where

$$A(t) = \begin{bmatrix} -1 & 2t \\ -2t & -1 \end{bmatrix} \qquad \text{and} \qquad b(t) = \begin{bmatrix} \sin(t^2) \\ \cos(t^2) \end{bmatrix}.$$

The corresponding homogeneous system $x' = A(t)x$ was studied in the previous example, and its general solution is

$$\alpha_c(t) = G(t)c,$$

where

$$G(t) = e^{-t} \begin{bmatrix} \cos(t^2) & \sin(t^2) \\ -\sin(t^2) & \cos(t^2) \end{bmatrix}.$$

To get the general solution of the nonhomogeneous system, we have to add to this a particular solution α_p of the nonhomogeneous system. The theory shows us that we can get one such particular solution by computing

$$\alpha_p(t) = G(t) \int_0^t G(u)^{-1} b(u) du.$$

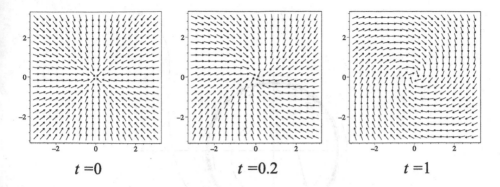

Figure 4.5: *Plots of the direction fields for the vector field* $X(t, x_1, x_2) = (-x_1 + 2tx_2, -2tx_1 - x_2)$ *at times* $t = 0, 0.2, 1$.

For this, we first calculate

$$G(u)^{-1}b(u) = e^u \begin{bmatrix} \cos(u^2) & -\sin(u^2) \\ \sin(u^2) & \cos(u^2) \end{bmatrix} \begin{bmatrix} \sin(u^2) \\ \cos(u^2) \end{bmatrix} = \begin{bmatrix} 0 \\ e^u \end{bmatrix}.$$

Then

$$\begin{aligned} \alpha_p(t) & = G(t) \int_0^t G(u)^{-1}b(u)du = e^{-t} \begin{bmatrix} \cos(t^2) & \sin(t^2) \\ -\sin(t^2) & \cos(t^2) \end{bmatrix} \begin{bmatrix} 0 \\ e^t - 1 \end{bmatrix} \\ & = (1 - e^{-t}) \begin{bmatrix} \sin(t^2) \\ \cos(t^2) \end{bmatrix}. \end{aligned}$$

It is clear from the above formula that the particular solution α_p asymptotically approaches the curve $\gamma(t) = (\sin(t^2), \sin(t^2))$, which is a curve that traces out a circle with radius 1, centered at the origin. The general solution of the nonhomogeneous DE is

$$\alpha(t) = \alpha_c(t) + \alpha_p(t) = G(t)c + \alpha_p(t),$$

where c is an arbitrary constant vector. As we have seen $\lim_{t \to \infty} G(t)c = 0$ for any c. Consequently, every integral curve of this nonhomogeneous system approaches the circle with radius 1, centered at the origin. See Figure 4.6

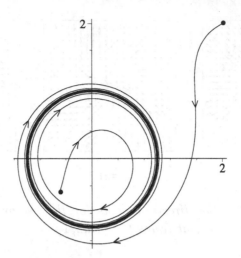

Figure 4.6: *Integral curves of the system:* $x_1' = -x_1 + 2tx_2 + \sin(t^2)$, $x_2' = -2tx_1 + x_2 + \cos(t^2)$ *which start at* $(2, 2)$ *and* $(-0.5, -05)$ *at time* $t = 0$.

Exercises 4.2

1. For each of the following 2nd-order DEs do the following:

   (i) Reduce the system to a 1st-order linear system: $z' = A(t)z$.

   (ii) Find the fundamental matrix G for A.

   (iii) Verify that Liouville's formula holds.

   (a) $x'' - \dfrac{1}{2(t + 1)} x' + \dfrac{1}{2(t + 1)^2} x = 0$, for $t \in [-1, \infty)$.

   Hint: This DE has two linearly independent solutions of the form $x = (t + 1)^k$ for appropriate choices of k.

   (b) $x'' - 3x' + 2x = 0$, for $t \in \mathbb{R}$.

   Hint: This DE has two linearly independent solutions of the form $x = e^{kt}$ for appropriate choices of k.

   (c) $x'' + (\tan t) x' = 0$, for $t \in (-\pi/2, \pi/2)$.

   Hint: In the 1st-order system, solve the second DE for v.

   (d) $x'' + p(t) x' = 0$, where p is continuous on an interval about 0.

   Hint: In the 1st-order system, solve the second DE for v.

2. In dimension $n = 2$ we can be more explicit about the fundamental matrix $G(t)$ for the coefficient matrix $A(t)$. If we can compute the integrals involved and find solutions of $P'' = cP$ in closed-form, then we can write an explicit closed-form expression for the fundamental matrix. This exercise shows you how this is done.

For convenience we write the general homogeneous system as

$$x_1' = h_1(t)x_1 + f_1(t)x_2 \qquad (4.19)$$
$$x_2' = f_2(t)x_1 + h_2(t)x_2 \qquad (4.20)$$

where h_1, h_2, f_1, f_2 are continuous on an interval about 0. We also assume that $f_1(0) \neq 0$ (which is needed in the construction below).

(a) Let $H_i(t) = \int_0^t h_i(s)ds$, and

$$u_i(t) = e^{-H_i(t)}x_i(t),$$

for $i = 1, 2$. Show that x_1, x_2 are solutions of (4.19)-(4.20) if and only if u_1, u_2 are solutions of

$$u_1' = \alpha_1(t)u_2 \qquad (4.21)$$
$$u_2' = \alpha_2(t)u_1 \qquad (4.22)$$

where

$$\alpha_i(t) = f_i(t)e^{(-1)^i[H_1(t)-H_2(t)]}, \qquad (i = 1, 2).$$

(b) Let $\beta(t) = \int_0^t \alpha_1(s)ds$. Then $\beta'(0) = \alpha_1(0) = f_1(0) \neq 0$, and so by the Inverse Function Theorem β is an invertible on a neighborhood of 0. We assume it is invertible on the whole interval I and let

$$c(s) = \frac{\alpha_2(\beta^{-1}(s))}{\alpha_1(\beta^{-1}(s))}.$$

Show that if P is a solution of $P'' = cP$, then

$$u_1(t) = P(\beta(t)) \qquad (4.23)$$
$$u_2(t) = P'(\beta(t)) \qquad (4.24)$$

is a solution of the system (4.21)-(4.22). The DE: $P'' = cP$ is called the *fundamental equation* for A (in the 2×2 case).

Next show that if P_1 and P_2 are solutions of $P'' = cP$ which satisfy $P_1(0) = 1, P_1'(0) = 0$ and $P_2(0) = 0, P_2'(0) = 1$, then

$$\begin{bmatrix} P_1(\beta(t)) & P_2(\beta(t)) \\ P_1'(\beta(t)) & P_2'(\beta(t)) \end{bmatrix} \qquad (4.25)$$

is the fundamental matrix for

$$\begin{bmatrix} 0 & \alpha_1(t) \\ \alpha_2(t) & 0 \end{bmatrix} \qquad (4.26)$$

(c) Finally, with the notation from above, show that

$$G(t) = \begin{bmatrix} e^{H_1(t)} P_1(\beta(t)) & e^{H_1(t)} P_2(\beta(t)) \\ e^{H_2(t)} P_1'(\beta(t)) & e^{H_2(t)} P_2'(\beta(t)) \end{bmatrix} \quad (4.27)$$

is the fundamental matrix for

$$A(t) = \begin{bmatrix} h_1(t) & f_1(t) \\ f_2(t) & h_2(t) \end{bmatrix} \quad (4.28)$$

3. Use the results of Exercise 2 above to find the fundamental matrix G for each of the following coefficient matrices A.

(a) $A(t) = \begin{bmatrix} 1 & 1 \\ 0 & -1 \end{bmatrix}$.

(b) $A(t) = \begin{bmatrix} 1 & 1 \\ 3 & -1 \end{bmatrix}$. Hint: The fundamental equation $P'' = cP$ is a Cauchy-Euler equation with $c(s) = 3(1 - 2s)^{-2}$. It has solutions of the form: $P(s) = (1 - 2s)^k$. Find two values of k that work and then use the general solution of the DE to find P_1, P_2.

(c) $A(t) = \begin{bmatrix} 0 & 2 \\ -2 & 0 \end{bmatrix}$.

(d) $A_\pm(t) = \begin{bmatrix} h(t) & f(t) \\ \pm f(t) & h(t) \end{bmatrix}$. Each choice of sign $(\pm)$ gives a different G.

4. Use the results of Exercise 2 above to find the fundamental matrix G for each of the following coefficient matrices A. Also, compute the flow matrix: $G(t)G(u)^{-1}$ and study the integral curves of the system as in Example 4.7.

(a) $A(t) = \begin{bmatrix} -1/4 & -\cos t \\ \cos t & -1/4 \end{bmatrix}$. Show that all the integral curves tend to 0 as $t \to \infty$.

(b) $A(t) = \begin{bmatrix} \sin t & 1 \\ -1 & \sin t \end{bmatrix}$. Show that all the integral curves are periodic of period 2π. Indeed, with $u \in [0, 2\pi]$ a given initial time, show that the curve $t \mapsto \phi_t^u(c)$ traces out an oval with minimum radius $r_{min} = e^{\cos u - 1}$ and maximum radius $r_{max} = e^{\cos u + 1}$. Show that the symmetry axis for the oval makes an angle u to the line from 0 to c.

(c) $A(t) = \begin{bmatrix} -1 & 1 \\ t & -1 \end{bmatrix}$. Hint: $P''(s) = sP(s)$ is the fundamental equation and will have to be solved by series methods: $P(s) = \sum_{n=0}^{\infty} a_n s^n$. The Airy functions Ai, Bi are two independent solutions, but not the ones needed for G.

5. A method for finding the fundamental matrix for any coefficient matrix $A(t)$ can, in principle, be developed along the lines of that indicated in Exercise 2 above (which is for 2×2 matrices $A(t)$). This exercise studies the situation for 3×3 matrices:

$$x_1' = a_{11}(t)x_1 + a_{12}(t)x_2 + a_{13}(t)x_3 \tag{4.29}$$
$$x_2' = a_{21}(t)x_1 + a_{22}(t)x_2 + a_{23}(t)x_3 \tag{4.30}$$
$$x_3' = a_{31}(t)x_1 + a_{32}(t)x_2 + a_{33}(t)x_3 \tag{4.31}$$

(a) Let $H_i(t) = \int_0^t a_{ii}(s)ds$, and

$$u_i(t) = e^{-H_i(t)}x_i(t),$$

for $i = 1, 2, 3$. Show that x_1, x_2, x_3 are solutions of (4.29)-(4.31) if and only if u_1, u_2, u_3 are solutions of

$$u_1' = \alpha_{12}(t)u_2 + \alpha_{13}(t)u_3 \tag{4.32}$$
$$u_2' = \alpha_{21}(t)u_1 + \alpha_{23}(t)u_3 \tag{4.33}$$
$$u_3' = \alpha_{31}(t)u_1 + \alpha_{32}(t)u_2 \tag{4.34}$$

where

$$\alpha_{ij}(t) = a_{ij}(t)e^{H_j(t)-H_i(t)}, \quad (i, j = 1, 2, 3, \ i \neq j).$$

(b) Part (a) reduces everything to the problem of finding the fundamental matrix for the system (4.32)-(4.34). For this, we can assume that the coefficients α_{ij} in one of the equations are not all zero. By reindexing if necessary, we can assume this is the case for the first equation: one of α_{12}, α_{13} is not zero. Reindexing again if necessary, we can assume that $\alpha_{12} \neq 0$ and that $\alpha_{12}(0) \neq 0$. Then $\beta(t) \equiv \int_0^t \alpha_{12}(s)ds$ is invertible on a neighborhood of 0. Below we use the notation

$$\bar{f}(s) = f(\beta^{-1}(s)).$$

Now look for solutions of the system with $u_1(t) = P(\beta(t)), u_1(t) = Q(\beta(t))$, where P, Q are two functions yet to be determined. The equations to determine P, Q are called the *fundamental equations* for A. Substituting these assignments of u_1, u_2 into the first two equations of the system gives

$$\alpha_{12}P'(\beta) = \alpha_{12}Q(\beta) + \alpha_{13}u_3 \tag{4.35}$$
$$\alpha_{12}Q'(\beta) = \alpha_{21}P(\beta) + \alpha_{23}u_3 \tag{4.36}$$

Show that eliminating u_3 and then letting $t = \beta^{-1}(s)$ gives

$$\bar{\alpha}_{12}\bar{\alpha}_{23}(P' - Q) = \bar{\alpha}_{13}(\bar{\alpha}_{12}Q' - \bar{\alpha}_{21}P) \tag{4.37}$$

This is fundamental equation 1. To get fundamental equation 2, divide into three cases:

(c) (CASE 1: $\alpha_{13} = 0, \alpha_{23} = 0$) In this case show that $Q = P'$. Further, show that

$$u_1 = P(\beta) \tag{4.38}$$
$$u_2 = P'(\beta) \tag{4.39}$$
$$u_3 = \int [\alpha_{31} P(\beta) + \alpha_{32} P'(\beta)] dt \tag{4.40}$$

is a solution of system (4.32)-(4.34) provided P satisfies

$$P'' = \frac{\bar{\alpha}_{21}}{\bar{\alpha}_{12}} P. \tag{4.41}$$

This is fundamental equation 2 when $\alpha_{13} = 0, \alpha_{23} = 0$.

(d) (CASE 2: $\alpha_{13} = 0, \alpha_{23} \neq 0$) In this case show that $Q = P'$. Further, show that

$$u_1 = P(\beta) \tag{4.42}$$
$$u_2 = P'(\beta) \tag{4.43}$$
$$u_3 = \frac{\alpha_{12}}{\alpha_{23}} P''(\beta) - \frac{\alpha_{21}}{\alpha_{23}} P(\beta) \tag{4.44}$$

is a solution of system (4.32)-(4.34) provided P satisfies

$$\frac{\bar{\alpha}_{12}^2}{\bar{\alpha}_{23}} P''' + \left(\frac{\alpha_{12}}{\alpha_{23}}\right)' P'' - \left[\bar{\alpha}_{32} + \frac{\bar{\alpha}_{12}\bar{\alpha}_{21}}{\bar{\alpha}_{23}}\right] P' - \left[\bar{\alpha}_{31} + \left(\frac{\alpha_{21}}{\alpha_{23}}\right)'\right] P = 0. \tag{4.45}$$

This is fundamental equation 2 when $\alpha_{13} = 0, \alpha_{23} \neq 0$.

(e) (CASE 3: $\alpha_{13} \neq 0$) In this case show that

$$u_1 = P(\beta) \tag{4.46}$$
$$u_2 = Q(\beta) \tag{4.47}$$
$$u_3 = \frac{\alpha_{12}}{\alpha_{13}} (P'(\beta) - Q(\beta)) \tag{4.48}$$

is a solution of system (4.32)-(4.34) provided P, Q satisfy

$$\frac{\bar{\alpha}_{12}^2}{\bar{\alpha}_{13}} P'' + \left[\left(\frac{\bar{\alpha}_{12}}{\alpha_{13}}\right)' - \frac{\bar{\alpha}_{12}^2 \bar{\alpha}_{23}}{\bar{\alpha}_{13}^2}\right] P' - \left[\bar{\alpha}_{31} + \frac{\bar{\alpha}_{12}\bar{\alpha}_{21}}{\bar{\alpha}_{13}}\right] P$$
$$= \left[\bar{\alpha}_{32} + \left(\frac{\alpha_{12}}{\alpha_{13}}\right)' - \frac{\bar{\alpha}_{12}^2 \bar{\alpha}_{23}}{\bar{\alpha}_{13}^2}\right] Q \tag{4.49}$$

This is fundamental equation 2 when $\alpha_{13} \neq 0$.

(f) Discuss the solutions of the fundamental equations (1)-(2) in each of the cases and show the general solutions P, Q involve three arbitrary constants a, b, c. These then give then general solution of (4.32)-(4.34) as $u_1 = P(\beta)$, $u_2 = Q(\beta)$, and $u_3 = R(\beta)$, where

$$R = \begin{cases} \frac{\bar{\alpha}_{12}}{\bar{\alpha}_{13}}(P' - Q) & \text{(Case 3)} \\ \frac{\bar{\alpha}_{12}}{\bar{\alpha}_{23}}P'' - \frac{\bar{\alpha}_{21}}{\bar{\alpha}_{23}}P & \text{(Case 2)} \\ \int [\bar{\alpha}_{31}(s)P(s) + \bar{\alpha}_{32}(s)Q(s)]\, ds & \text{(Case 1)} \end{cases} \qquad (4.50)$$

Show that by appropriate choices of a, b, c, there are solutions $P_i, Q_i, i = 1, 2, 3$, so that the fundamental matrix for

$$\widetilde{A} = \begin{bmatrix} 0 & \alpha_{12} & \alpha_{13} \\ \alpha_{21} & 0 & \alpha_{23} \\ \alpha_{31} & \alpha_{32} & 0 \end{bmatrix} \qquad (4.51)$$

is

$$\widetilde{G} = \begin{bmatrix} P_1(\beta) & P_2(\beta) & P_3(\beta) \\ Q_1(\beta) & Q_2(\beta) & Q_3(\beta) \\ R_1(\beta) & R_2(\beta) & R_3(\beta) \end{bmatrix} \qquad (4.52)$$

Finally, use this to determine the fundamental matrix for A.

6. Use Exercise 5 above to (a) verify that the fundamental equations for A are as indicated, (b) find the general solution P, Q (involving three arbitrary constants) of the fundamental equations, and (c) find the fundamental matrix G for A.

(a) $A(t) = \begin{bmatrix} 0 & 1 & 1 \\ 1 & 0 & 1 \\ 1 & 1 & 0 \end{bmatrix}$, $\quad \begin{aligned} P' - Q &= Q' - P \\ P'' - P' - 2P &= 0 \end{aligned}$

(b) $A(t) = \begin{bmatrix} 0 & 1 & -1 \\ 0 & 0 & -1 \\ 0 & 1 & 0 \end{bmatrix}$, $\quad \begin{aligned} P' - Q &= Q' \\ P'' - P' &= -2Q \end{aligned}$

(c) $A(t) = \begin{bmatrix} 0 & 1 & -1 \\ t & 0 & 1 \\ t & 1 & 0 \end{bmatrix}$, $\quad \begin{aligned} Q' - Q &= tP - P' \\ P'' + P' &= 0 \end{aligned}$

7. Use Exercise 5 above and series $P(t) = \sum_{n=0}^{\infty} a_n t^n$ to find the fundamental matrix G for A.

$A(t) = \begin{bmatrix} 0 & 1 & 0 \\ t & 0 & 1 \\ 1 & 0 & 0 \end{bmatrix}$, $\quad \begin{aligned} P' - Q &= 0 \\ P''' - tP' - 2P &= 0 \end{aligned}$

8. Use the results of Exercise 5 above to find the fundamental matrix G for the following matrices A. DO NOT try to find P and Q explicitly. Rather just express G in terms of P and Q and their derivatives. Then check that G has the correct form by computing AG and comparing with G'.

(a) $A(t) = \begin{bmatrix} 0 & 1 & 1 \\ 1 & 0 & t \\ -1 & t & 0 \end{bmatrix}$, $\begin{aligned} Q' + tQ &= tP' + P \\ P'' - tP' &= 0 \end{aligned}$

(b) $A(t) = \begin{bmatrix} 0 & 1 & 1 \\ 0 & 0 & 1 \\ t & 2 & 0 \end{bmatrix}$, $\begin{aligned} P' &= Q' + Q \\ P'' - P' - tP &= Q \end{aligned}$

9. Suppose $\tilde{A}$ is an $n \times n$ (constant) matrix and $f : I \to \mathbb{R}$ is a continuous function on an interval I. Define $A : I \to M_n$ by $A(t) = f(t)\tilde{A}$. Let $\beta(t) = \int_0^t f(s)ds$. Show that if $\tilde{G}$ is the fundamental matrix for $\tilde{A}$ then the fundamental matrix G for A is given by $G(t) = \tilde{G}(\beta(t))$.

10. This exercise further studies the system $x' = A(t)x$ from Example 4.6 with coefficient matrix
$$A(t) = \begin{bmatrix} -\sin t & \cos t \\ \cos t & -\sin t \end{bmatrix}.$$

If you wish you can use the results in Exercise 2 above to show that the fundamental matrix for A is
$$G(t) = e^{\cos t - 1} \begin{bmatrix} \cosh(\sin t) & \sinh(\sin t) \\ \sinh(\sin t) & \cosh(\sin t) \end{bmatrix}.$$

Do the following:

(a) Use Theorem 4.2 to verify that G is indeed the fundamental matrix for A. That is, check that $G' = AG$ and $G(0) = I$.

(b) Verify that Liouville's formula holds for this A and G.

(c) Show that the flow map is given by
$$\phi_t^u(c) = e^{\cos t - \cos u} \begin{bmatrix} \cosh(\sin t - \sin u) & \sinh(\sin t - \sin u) \\ \sinh(\sin t - \sin u) & \cosh(\sin t - \sin u) \end{bmatrix} c$$

(d) Show that each integral curve (other than the fixed point at the origin) is periodic of period 2π.

(e) The vector-field for this system,
$$X(t, x) = ((-\sin t)x_1 + (\cos t)x_2, (\cos t)x_1 - (\sin t)x_2)$$

is periodic of period 2π. So, to study its variation over time it suffices to consider times $t \in [0, 2\pi]$. Do an animation of the direction field for

X using 33 frames for the times $t = j\pi/16$, $j = 0, 1, \ldots, 32$. What happens along the lines $x_1 = x_2$ and $x_1 = -x_2$? What does this imply for the integral curves that start at points on these lines? (Prove your assertions.)

(e) Let $R_+, R_- : \mathbb{R}^2 \to \mathbb{R}^2$ be the maps defined by $R_\pm(x_1, x_2) = \pm(x_2, x_1)$. Geometrically, R_+ (respectively R_-) is the map that reflects points in $\mathbb{R}^2$ about the line $x_1 = x_2$ (respectively the line $x_1 = -x_2$). (Draw some pictures to convince yourself of this.) Show that

$$\phi_t^u(R_\pm(c)) = R_\pm(\phi_t^u(c)),$$

for every $c \in \mathbb{R}^2$. Interpret what this says about the integral curves of the system. Specifically, the lines $x_1 = x_2$ and $x_1 = -x_2$ divide the plane into fours sectors. If c is a point in one of these sectors, how does the integral curve $t \mapsto \phi_t^u(c)$ relate to integral curves in the other three sectors?

11. If $g : I \to \mathbb{R}$ is a differentiable function on an interval I, then an important special case of the quotient rule from calculus is

$$(g^{-1})' = -g^{-2}g' = -g^{-1}g'g^{-1}.$$

Show that the generalization of this to matrix-valued functions holds. That is, show that if $G : I \to \mathcal{M}_n$ is differentiable, with $\det(G(t)) \neq 0$ for all $t \in I$, then $G^{-1} : I \to \mathcal{M}_n$, defined by $G^{-1}(t) \equiv G(t)^{-1}$ is differentiable and

$$(G^{-1})' = -G^{-1}G'G^{-1}.$$

Note that G^{-1} and G' will not commute, in general, and so the derivative cannot be written as $-G^{-2}G'$. What do you think the formula for the derivative $(G^{-2})'$ should be? What about $(G^{-p})'$, for a positive integer p? Prove your conjectures.

12. Show that $\alpha_c(t) \equiv G(t)c$, for $t \in I$ and $c \in \mathbb{R}^n$, an arbitrary constant, is the *general* solution of the homogeneous equation $x' = A(t)x$. That is, any solution of the homogeneous equation coincides with γ_c for some choice of c. Show that $\alpha_p(t) \equiv G(t) \int_0^t G(s)^{-1}b(s)ds$, for $t \in I$, is a particular solution of the nonhomogeneous equation $x' = A(t)x + b(t)$.

13. Show, directly from formula (4.17), that the flow for $x' = A(t)x + b(t)$, satisfies the semigroup property.

14. Consider the general 2nd-order, nonhomogeneous, linear DE

$$u'' + p(t)u' + q(t)u = r(t), \tag{4.53}$$

for $u : I \to \mathbb{R}$. This problem is designed to show you how your undergraduate study of this equation is related to the general theory in the text. In particular, you should first review how the above DE is solved by *variation of parameters* in undergraduate DE books.

(a) Let $x = u, y = u'$ and write equation (4.53) as a system of 1st-order DEs:

$$\begin{bmatrix} x' \\ y' \end{bmatrix} = A(t) \begin{bmatrix} x \\ y \end{bmatrix} + b(t). \tag{4.54}$$

(b) Let u_1, u_2, be the two solutions of

$$u'' + p(t)u' + q(t)u = 0,$$

which satisfy the initial conditions $u_1(0) = 1, u_1'(0) = 0$ and $u_2(0) = 0, u_2'(0) = 1$, respectively. Express the fundamental matrix G in terms of these and show that

$$\det(G(t)) = W(u_1(t), u_2(t)),$$

where the right-hand side is the Wronskian of u_1, u_2.

(c) Let $f : I \to \mathbb{R}^2$ be the particular solution of equation (4.54) given by

$$f(t) = G(t) \int_0^t G(s)^{-1} b(s) ds. \tag{4.55}$$

Compute the components f_1, f_2 of f explicitly. Discuss how f_1 gives the formula for the particular solution of equation (4.53) found by the method of variation of parameters in undergraduate DE books.

(d) Show how the general solution

$$G(t)c + f(t) = G(t)c + G(t) \int_0^t G(s)^{-1} b(s) ds \tag{4.56}$$

of equation (4.54) gives the general solution of equation (4.53).

15. Discuss the general analog of Exercise 14. That is consider the general, nth-order, linear DE

$$x^{(n)} + a_{n-1}(t)x^{(n-1)} + \cdots + a_2(t)x'' + a_1(t)x' + a_0(t)x = r(t),$$

and write this in 1st-order linear form $z' = A(t)z + b(t)$, as in the first section. Show how the fundamental matrix and the general solution (4.14) involve the Wronskian and the method of variation of parameters.

16. **(Variation of Parameters)** Consider the general version of the variation of parameters technique discussed in the last two exercises. Namely, look at the problem of finding a particular solution of the nonhomogeneous system $x' = A(t)x + b(t)$. (Here $A(t)$ is general, not of the special forms in the last

two exercises.) For this we need to know the fundamental matrix G for A (i.e., we need to have n linearly independent solutions of the homogeneous DE $x' = A(t)x$). Then for any constant vector c, the curve $\alpha(t) = G(t)c$ is a solution of the homogeneous DE. However, if we let c be a function of t (i.e. a variable vector), then we can hope to find a particular solution of the nonhomogeneous DE of the form $\alpha_p(t) = G(t)c(t)$. Show that this is indeed the case and that c is given by the integral: $c(t) = \int G(t)^{-1}b(t)dt$. Compare the result here with that in Formula (4.16).

17. Prove directly from formula (4.17) that the semigroup property: $\phi_u^t \circ \phi_t^s = \phi_u^s$, for the flow holds.

18. (See worksheet: `fmatrix.mws`) Find the fundamental matrix G for the system

$$x_1' = -7x_1 - 10x_2 + e^t \tag{4.57}$$
$$x_2' = 4x_1 + 5x_2 - e^{-t} \tag{4.58}$$

and use it to compute, explicitly, the general solution of the system (You may use a computer algebra system if you wish). Determine solutions of the homogeneous and nonhomogeneous problems that satisfy the initial conditions $x_1(0) = 1, x_2(0) = 1$. Determine the solution of the nonhomogeneous problem that satisfies $x_1(0) = 0, x_2(0) = 0$. Plot all three solutions using a suitable range of t values.

19. **(Floquet Theory)** Many important linear systems have coefficient matrices that are periodic, and this exercise deals with some basic results for such systems (cf. [Wa 98, pp. 195-198], [Cr 94, pp. 93-105], for more discussion).

Suppose $A : \mathbb{R} \to \mathcal{M}_n$ is a continuous, matrix-valued function and that there is a least positive number p, called the *period* of A, such that $A(t+p) = A(t)$, for all $t \in \mathbb{R}$. Let $G : \mathbb{R} \to \mathcal{M}$ be the fundamental matrix for A.

(a) Show that G has the property

$$G(t + p) = G(t)G(p),$$

for all $t \in \mathbb{R}$. *Hint*: Define $H(t) \equiv G(t + p)G(p)^{-1}$ and then show that H is also a fundamental matrix for A. You may use the fact that fundamental matrices (as we have defined them) are unique.

(b) The eigenvalues of the matrix $G(p)$ are called the *characteristic multipliers* of A. Show that μ is a characteristic multiplier if and only if the homogeneous system $x' = A(t)x$ has a solution α such that $\alpha(t+p) = \mu\alpha(t)$ for all t. (Hence, the system has a periodic solution (of period p) if and only if $\mu = 1$ is a characteristic multiplier.)

4.3 Homogeneous, Constant Coefficient Systems

We now specialize the discussion of the general linear system to the case where $b(t) \equiv 0$ (a homogeneous system) and A does not depend on t (a constant coefficient system), i.e.,

$$x' = Ax.$$

This is the simplest of *all* systems of DEs, linear and nonlinear, and is a type of system for which complete and definitive theory exists. Such systems also play a role in the study of autonomous, *nonlinear* DEs, as we will see in the next chapter. As you might expect, the properties of A as a linear transformation $A : \mathbb{R}^n \rightarrow \mathbb{R}^n$, which you studied in linear algebra (eigenvalues, eigenvectors, direct sum decomposition of $\mathbb{R}^n$, etc.) will play a role in the study of the corresponding system of DEs.

The fundamental matrix G has a particularly nice form when A does not depend on t. To motivate what this form is, we consider the basic equation

$$G'(t) = AG(t)$$

which G satisfies and differentiate both sides of this successively to get:

$$
\begin{aligned}
G''(t) &= AG'(t) = AAG(t) = A^2 G(t) \\
G'''(t) &= A^2 G'(t) = A^3 G(t)
\end{aligned}
$$

$$\vdots$$

$$G^{(k)}(t) = A^{k-1} G'(t) = A^k G(t).$$

Thus, in particular, we get

$$G^{(k)}(0) = A^k$$

for the value of the kth derivative of G at $t = 0$. Hence, heuristically, the power series expansion of G, centered at $t = 0$ is

$$G(t) = \sum_{k=0}^{\infty} G^{(k)}(0) \frac{t^k}{k!} = \sum_{k=0}^{\infty} A^k \frac{t^k}{k!}$$

and consequently, we "conclude" that

$$G(t) = e^{At}. \tag{4.59}$$

This motivates why the fundamental matrix is given by the matrix exponential, as the following theorem asserts.

Theorem 4.4 (Constant Coefficient, Homogeneous Systems) *The fundamental matrix for the system*

$$x' = Ax$$

(with A a constant matrix) is

$$G(t) = e^{At},$$

and thus the general solution of the system is

$$\alpha(t) = e^{At}c,$$

for $t \in \mathbb{R}$ and $c \in \mathbb{R}^n$ an arbitrary constant. Furthermore the flow ϕ : $\mathbb{R} \times \mathbb{R}^n \to \mathbb{R}^n$, for the system is simply

$$\phi(t, x) = e^{At}x.$$

Proof: This follows from the discussion in Appendix C, which you should read when time permits. All the basic theory for series of real numbers extends to series of $n \times n$ matrices, and this enables one to extend most real-valued functions of a real variable to corresponding matrix-valued functions of a matrix variable. In particular, it is shown in Appendix C that for any matrix B in $\mathcal{M}_n$, the series of matrices

$$\sum_{k=0}^{\infty} \frac{B^k}{k!}$$

converges. This gives a function from $\mathcal{M}_n$ to $\mathcal{M}_n$, called the matrix exponential and denoted by

$$e^B \equiv \sum_{k=0}^{\infty} \frac{B^k}{k!}.$$

From this, we can define a function $G : \mathbb{R} \to \mathcal{M}_n$ by $G(t) \equiv e^{At}$. Since G is given by a power series, we can compute $G'(t)$ by differentiating the power series term by term and we arrive at the result: $G'(t) = AG(t)$ (see Appendix C for details). Also it is easily seen that $G(0) = I$. Hence G is the fundamental matrix for A and the other assertions in the theorem follow from this. $\square$

Note that since the system $x' = Ax$ is autonomous, only the time zero initial times are needed to describe the flow completely:

$$\phi_t \equiv \phi_t^0.$$

Also the flow gives a 1-parameter *group:* $\{\phi_t\}_{t\in\mathbb{R}}$, of linear maps $\phi_t : \mathbb{R}^n \to \mathbb{R}^n$. The fact that this set is an (Abelian) group follows from a property of the matrix exponential:

$$\phi_s \circ \phi_t = e^{As}e^{At} = e^{As+At} = e^{A(s+t)} = \phi_{s+t}$$

(exercise).

Also notice how the 1-dimensional system $x' = ax$, with well-known solution $x = ce^{at}$ is nicely generalized to the n-dimensional system $x' = Ax$ by the introduction of the matrix exponential.

Corollary 4.2 (Constant Coefficient, Linear Systems) *The flow map for the system*

$$x' = Ax + b(t)$$

(with A a constant matrix) is

$$\phi_t^u(x) = e^{A(t-u)}x + \int_u^t e^{A(t-s)}b(s)ds, \qquad (4.60)$$

for all $t, u \in I$, and $x \in \mathbb{R}^n$. This gives a two-parameter semigroup $\{\phi_t^u\}_{t,u\in\mathbb{R}}$ of affine maps $\phi_t^u : \mathbb{R}^n \to \mathbb{R}^n$.

For the homogeneous system $x' = Ax$, the explicit form of the flow, $\phi_t(x) = \phi(t,x) = e^{At}x$, gives us a wealth of information about the integral curves and qualitative properties of the system. In the ensuing sections we will explore these aspects and also develop techniques that will enable us to explicitly compute e^{At} for a certain matrices A. If A is a 2×2 matrix, one can always *try* to compute e^{At} directly from its power series definition.

Example 4.9 Suppose A is the upper triangular matrix

$$A = \begin{bmatrix} 1 & 1 \\ 0 & 3 \end{bmatrix}.$$

The direct approach to computing e^{At} amounts to computing $A^2, A^3, A^4, \ldots$, out to a sufficiently high power so that we can inductively determine the general formula for A^k. Then we can ascertain what matrix the series $\sum_{k=0}^{\infty} A^k t^k / k!$ converges to. This will be the matrix e^{At}. For the matrix in this example, we easily find

$$A^2 = \begin{bmatrix} 1 & 1+3 \\ 0 & 3^2 \end{bmatrix},$$

$$A^3 = \begin{bmatrix} 1 & 1+3+3^2 \\ 0 & 3^3 \end{bmatrix},$$

$$A^4 = \begin{bmatrix} 1 & 1+3+3^2+3^3 \\ 0 & 3^4 \end{bmatrix},$$

and inductively we get

$$A^k = \begin{bmatrix} 1 & 1+3+\cdots+3^{k-1} \\ 0 & 3^k \end{bmatrix} = \begin{bmatrix} 1 & \frac{1}{2}(3^k-1) \\ 0 & 3^k \end{bmatrix}.$$

Using this in the series we find

$$
\begin{aligned}
G = e^{At} &= \sum_{k=0}^{\infty} \frac{A^k t^k}{k!} \\
&= \sum_{k=0}^{\infty} \begin{bmatrix} 1 & \frac{1}{2}(3^k-1) \\ 0 & 3^k \end{bmatrix} \frac{t^k}{k!} \\
&= \begin{bmatrix} \sum_{k=0}^{\infty} \frac{t^k}{k!} & \frac{1}{2}\sum_{k=0}^{\infty}(3^k-1)\frac{t^k}{k!} \\ 0 & \sum_{k=0}^{\infty} 3^k \frac{t^k}{k!} \end{bmatrix} \\
&= \begin{bmatrix} e^t & \frac{1}{2}(e^{3t}-e^t) \\ 0 & e^{3t} \end{bmatrix}
\end{aligned}
$$

Note that the series summation sign in the second line above can be applied to each of the four entries to give the matrix whose entries are the series shown in the third line (see Appendix C).

We can check our work by computing

$$AG = \begin{bmatrix} 1 & 1 \\ 0 & 3 \end{bmatrix} \begin{bmatrix} e^t & \frac{1}{2}(e^{3t}-e^t) \\ 0 & e^{3t} \end{bmatrix} = \begin{bmatrix} e^t & \frac{3}{2}e^{3t}-\frac{1}{2}e^t) \\ 0 & 3e^{3t} \end{bmatrix},$$

and seeing that this is equal to

$$G' = \frac{d}{dt} \begin{bmatrix} e^t & \frac{1}{2}(e^{3t}-e^t) \\ 0 & e^{3t} \end{bmatrix} = \begin{bmatrix} e^t & \frac{3}{2}e^{3t}-\frac{1}{2}e^t) \\ 0 & 3e^{3t} \end{bmatrix}.$$

With $G = e^{At}$ determined, the general solution of the system

$$
\begin{aligned}
x_1' &= x_1 + x_2 \\
x_2' &= 3x_2
\end{aligned}
$$

can be expressed by

$$
\begin{aligned}
\alpha(t) &= e^{At}c \\
&= \begin{bmatrix} e^t & \frac{1}{2}(e^{3t} - e^t) \\ 0 & e^{3t} \end{bmatrix} \begin{bmatrix} c_1 \\ c_2 \end{bmatrix} \\
&= \left((c_1 - \tfrac{1}{2}c_2)e^t + \tfrac{1}{2}c_2 e^{3t}, \; c_2 e^{3t} \right).
\end{aligned}
$$

Of course, we can arrive at the same solution in a more elementary way by solving the second equation, $x_2' = 3x_2$, of the system for x_2, substituting this in the first equation of the system and solving the resulting DE for x_1. This technique, however, relies on the special form of the system (A is upper triangular) and so is only effective in that case.

Writing the general solution in terms of the matrix exponential e^{At} is the more general technique, but the computation of e^{At} can be difficult (even with the additional techniques developed below).

Exercises 4.3

1. As in Example 4.9, compute the fundamental matrix $G(t) = e^{At}$ for the system $x' = Ax$ and use this to write out explicitly the general solution $\alpha(t) = e^{At}c$ for the system. Check your answer for G by (i) computing AG and comparing to G' and (ii) evaluating $G(0)$.

(a) $A = \begin{bmatrix} -1 & 0 \\ 0 & -2 \end{bmatrix}$.

(b) $A = \begin{bmatrix} -1 & 1 \\ 0 & -2 \end{bmatrix}$.

(c) $A = \begin{bmatrix} 1 & 1 \\ 0 & -1 \end{bmatrix}$.

(d) $A = \begin{bmatrix} -1 & 1 & 0 \\ 0 & -2 & 0 \\ 0 & 0 & 5 \end{bmatrix}$.

(e) $A = \begin{bmatrix} 0 & 1 & 0 \\ 0 & 0 & 1 \\ 0 & 0 & 0 \end{bmatrix}$.

(f) $A = \begin{bmatrix} 0 & 1 & 0 & 0 \\ 0 & 0 & 1 & 0 \\ 0 & 0 & 0 & 1 \\ 0 & 0 & 0 & 0 \end{bmatrix}$.

(g) $A = \begin{bmatrix} 1 & 1 \\ 1 & 1 \end{bmatrix}$.

(h) $A = \begin{bmatrix} 0 & 0 & 1 \\ 0 & 1 & 0 \\ 1 & 0 & 0 \end{bmatrix}$.

(i) $A = \begin{bmatrix} 0 & 0 & 1 \\ 1 & 0 & 0 \\ 0 & 1 & 0 \end{bmatrix}$. In this case show that the fundamental matrix is

$$G = \begin{bmatrix} f'' & f & f' \\ f' & f'' & f \\ f & f' & f'' \end{bmatrix},$$

where f is the solution of $f''' = f$, subject to the initial conditions: $f(0) = 0$, $f'(0) = 0$, $f''(0) = 1$. You will have to use series since f is not a well-known function.

2. For the general 2×2, upper triangular matrix

$$A = \begin{bmatrix} a & b \\ 0 & c \end{bmatrix},$$

compute e^{At} and use this to write out explicitly the general solution of the system

$$\begin{aligned} x_1' &= ax_1 + bx_2 \\ x_2' &= cx_2. \end{aligned}$$

Can you use the system of DEs directly to determine what e^{At} is, rather than computing it from the power series?

3. The Maple worksheet `matexpo.mws` on the electronic component has a procedure for computing the approximations

$$G(t, N) \equiv \sum_{k=0}^{N} A^k \frac{t^k}{k!},$$

to the fundamental matrix $G(t) = e^{At}$. Study this material and work the exercises listed there.

4. **(Nilpotent Matrices)** An $n \times n$ matrix N is called *nilpotent* if $N^p = 0$, for some positive integer p. The least positive integer p for which this happens is called its *index of nilpotencey*. Compute the form of e^{Nt}, where N is nilpotent of index p. The matrices in Exercises 1(e) and 1(f) above are standard examples of nilpotent matrices in dimensions 3 and 4, and these are denoted by N_3 and N_4, respectively. Based on this, what would be the corresponding form for the $n \times n$, standard, nilpotent matrix N_n? Compute $e^{N_n t}$.

5. There are many properties of the matrix exponential that are useful for computing e^{At} and for other theoretical topics. Here are two properties that we will need later in the text (all matrices are assumed to be $n \times n$):

(1) If P is invertible, then $e^{P^{-1}AP} = P^{-1}e^A P$.

(2) If A and B commute, i.e., $AB = BA$, then $e^{A+B} = e^A e^B$.

Prove Property (1). This is easy to do directly from the series definition. Property (2) is not so easy. There is a proof of this in the Matrix Analysis section in Appendix C. By far, Property (2) is the most fundamental. It is the basic law of exponents, but beware that it is only guaranteed to work for commuting matrices. To see this do the following. Let

$$A = \begin{bmatrix} 1 & 0 \\ 0 & 3 \end{bmatrix}, \qquad B = \begin{bmatrix} 0 & 1 \\ 0 & 0 \end{bmatrix}.$$

Show that $AB \neq BA$. Compute e^{A+B} (see Example 5.3), $e^A e^B$, and verify that $e^{A+B} \neq e^A e^B$.

6. Floquet's theory for homogeneous systems $x' = A(t)x$ with periodic coefficient matrix A was introduced in Exercise 19, Section 4.2. We continue here with that discussion, showing how it involves the matrix exponential. We assume $A : \mathbb{R} \to \mathcal{M}_n$ is a continuous, matrix-valued function and that there is a least positive number p, called the *period* of A, such that $A(t+p) = A(t)$, for all $t \in \mathbb{R}$. Let $G : \mathbb{R} \to \mathcal{M}$ be the fundamental matrix for A.

(a) Exercise 19 of 4.2, Part (a), shows that the fundamental matrix G is not necessarily periodic when A is (unless $G(p) = I$). However, Floquet's Theorem says that G can be written as

$$G(t) = R(t)e^{Mt},$$

where R has period p, is invertible, and $R(0) = I$. Here M is a constant matrix. Prove Floquet's Theorem. You may use the fact that there is a matrix M such that $e^{Mp} = G(p)$. (Heuristically, $M = \frac{1}{p}\ln(G(p))$.) Hint: Define $R(t) \equiv G(t)e^{-Mt}$.

(b) To illustrate Floquet's Theorem do the following. For each of the periodic matrices A: (a) find the fundamental matrix G (for this, see Exercises 2 and 4(a) in Section 4.2), (b) verify that $G(t) = R(t)e^{Mt}$ for the given R and M, (c) find the characteristic exponents of the system and determine if the system has periodic solutions (using Floquet theory as well as the particular form of G).

(i) $A_{\pm}(t) = \begin{bmatrix} \sin t & 1 \\ \pm 1 & \sin t \end{bmatrix}$. In Floquet's theorem:

$$R(t) = e^{1-\cos t}I, \qquad M_{\pm} = \begin{bmatrix} 0 & 1 \\ \pm 1 & 0 \end{bmatrix},$$

where I is the 2×2 identity matrix. NOTE: Two famous formulas for the matrix exponentials are

$$e^{tM_+} = \begin{bmatrix} \cosh t & \sinh t \\ \sinh t & \cosh t \end{bmatrix}, \qquad e^{tM_-} = \begin{bmatrix} \cos t & \sin t \\ -\sin t & \cos t \end{bmatrix}.$$

Prove that these formulas are correct.

(ii) $A(t) = \begin{bmatrix} -1 & -\cos t \\ \cos t & -1 \end{bmatrix}$. In Floquet's theorem:

$$R(t) = \begin{bmatrix} \cos(\sin t) & \sin(\sin t) \\ -\sin(\sin t) & \cos(\sin t) \end{bmatrix}, \qquad M = -I,$$

where I is the 2×2 identity matrix.

4.4 The Geometry of the Integral Curves

Note: The remainder of this chapter is devoted to the study of the homogeneous, constant coefficient, linear system: $x' = Ax$.

The theory so far has shown us that the integral curve of $x' = Ax$ that starts at the point $c \in \mathbb{R}^n$ at time $t = 0$ is

$$\phi_t(c) = e^{At}c.$$

While such curves can be quite complicated (especially for $n > 3$), there are some simple, special cases that are easy to describe and visualize. The general case can often be built from these special ones by the superposition principle.

The simplest type of integral curve for any system, linear or nonlinear, is a fixed point (or more precisely, the constant curve corresponding to the fixed point). For the linear systems being considered here, the set of fixed points is well-known.

Fixed Points: The collection of fixed points (equilibrium points) for the system $x' = Ax$, is precisely the kernel, or null space, of the linear transformation A:

$$\ker(A) = \{\, x \in \mathbb{R}^n \mid Ax = 0 \,\}.$$

Thus, $0 = (0, \ldots, 0)$ is always a fixed point, and is the only fixed point, called a *simple fixed point*, precisely when $\det A \neq 0$. On the other hand when $\det A = 0$, there are infinitely many fixed points and any nonzero fixed point is also known as an eigenvector of A corresponding to eigenvalue zero. We will use the alternative notation

$$E_0 = \ker(A),$$

when we want to emphasize that the kernel of A is also an eigenspace of A, corresponding to eigenvalue 0.

Thus, E_0 is a subspace of $\mathbb{R}^n$, with dimension k, say, and geometrically E_0 is a single point 0 if $k = 0$, or is a line through the origin if $k = 1$, or is a plane through the origin if $k = 2$, etc. An integral curve that starts at a point $v \in E_0$ remains there for all time: $\phi_t(v) = v$ for all t.

Other Eigenvalues: As we shall see the other eigenvalues of A play an important role in the theory, and the corresponding eigenvectors v will provide us special types of integral curves $\phi_t(v)$, that are only slightly more complicated than those for fixed points.

By definition, an eigenvalue of A is a number λ (real or complex) for which there is a nonzero vector v, such that

$$Av = \lambda v.$$

Note that v is a complex vector: $v \in \mathbb{C}^n$, when λ is complex. You should recall that the eigenvalues are the roots of the *characteristic equation*:

$$\det(A - \lambda I) = 0,$$

and that for a given root λ of this equation, the eigenvectors corresponding to λ are the solutions of the system

$$(A - \lambda I)v = 0.$$

In general there are infinitely eigenvectors corresponding to a given eigenvalue; indeed, they comprise the subspace: $\ker(A - \lambda I)$.

Definition 4.2 For a *real* eigenvalue λ, the subspace:

$$
\begin{aligned}
E_\lambda &= \{\, v \in \mathbb{R}^n \mid Av = \lambda v \,\} \\
&= \ker(A - \lambda I)
\end{aligned}
$$

is called the *eigenspace* of A corresponding to λ.

The above definition applies to complex eigenvalues λ as well, except that then the eigenspace is a subspace of $\mathbb{C}^n$, i.e., consists of complex vectors. Since we are just interested in solutions of $x' = Ax$ that are real (i.e., $x_1, \ldots, x_n$ are real-valued functions), we will have to modify the above definition in the complex case. It will simplify the discussion if we just concentrate on real eigenvalues first, and deal with the complex case later.

4.4.1 Real Eigenvalues

Suppose λ is a real eigenvalue of A and $v \in E_\lambda$ is an eigenvector. Then from $Av = \lambda v$, we get $A^2 v = A(\lambda v) = \lambda^2 v$. Similarly we find that $A^3 v = \lambda^3 v$, and generally that $A^k v = \lambda^k v$, for any $k = 0, 1, 2, 3, \ldots$. From this it follows that

$$e^{At} v = e^{\lambda t} v,$$

for all $t \in \mathbb{R}^n$ (exercise). Consequently, the integral curve that starts at the point $v \in E_\lambda$ at time $t = 0$ is given by:

$$\phi_t(v) = e^{\lambda t} v. \tag{4.61}$$

Thus, this integral curve (when $v \neq 0$) is a half-line (or ray) in $\mathbb{R}^n$, and as t increases through positive times, $\phi_t(v)$ runs away from the origin if $\lambda > 0$, and runs toward the origin if $\lambda < 0$. In either case the entire ray through the origin, containing v, is traced out by this integral curve as t varies from $-\infty$ to ∞. This ray is contained in the eigenspace E_λ and geometrically E_λ is composed entirely of a bundle of such rays (straight-line integral curves) all of which are directed away from the origin if λ is positive or directed toward the origin if λ is negative. See Figure 4.7.

Thus, from the above discussion we see that the geometry of the integral curve $t \mapsto \phi_t(c)$ is quite simple if the initial point c is in one of the real eigenspaces: $c \in E_\lambda$. The integral curve (when $c \neq 0$) is either a point (when $\lambda = 0$) or a ray through the origin.

The next simplest case is when $c \in E_{\lambda_1} \oplus E_{\lambda_2}$ is the sum of two eigenvectors from different eigenspaces

$$c = k_1 v_1 + k_2 v_2,$$

where $v_i \in E_{\lambda_i}, i = 1, 2$ and $\lambda_1 \neq \lambda_2$ (Thus, v_1 and v_2 are linearly independent). The integral curve starting at c at time zero is

$$\begin{aligned} \phi_t(c) &= e^{At} c = k_1 e^{At} v_1 + k_2 e^{At} v_2 \\ &= k_1 e^{\lambda_1 t} v_1 + k_2 e^{\lambda_2 t} v_2. \end{aligned}$$

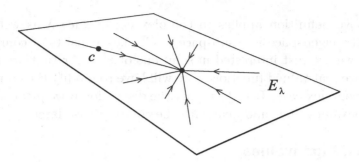

Figure 4.7: *The integral curve passing through a point $c \in E_\lambda$, is a half-ray, contained completely in E_λ. The ray is directed toward the origin if $\lambda < 0$ and away from the origin if $\lambda > 0$. E_λ is comprised entirely of such rays. The picture is for the case when E_λ is two-dimensional.*

The nature of this curve is easily analyzed as follows (for simplicity we assume that $k_1 \geq 0, k_2 \geq 0$). First note that it lies in the plane M determined by v_1 and v_2. Next, if $k_1 \neq 0, k_2 = 0$, or if $k_1 = 0, k_2 \neq 0$, the integral curve is a straight-line, integral curve along v_1 or v_2, directed toward/away from the origin, as expected from the above analysis. In the case when neither k_1 nor k_2 is zero, the curve $\phi_t(c)$ is a linear combination of the motions along the two straight-line integral curves just mentioned. Thus, if λ_1 and λ_2 are both negative, the motion of the curve is toward the origin (with $\lim_{t\to\infty} |\phi_t(c)| = 0$). If λ_1 and λ_2 are both positive, the motion of the curve is away from the origin (with $\lim_{t\to\infty} |\phi_t(c)| = \infty$). If λ_1 and λ_2 are of opposite signs, with, say λ_1 negative, the motion of the curve is away from the origin and is asymptotic to the straight-line integral curve along v_2 (i.e., $\lim_{t\to\infty} |\phi_t(c) - k_2 e^{\lambda_2 t} v_2| = 0$).

An easy way to visualize how the curve $\phi_t(c)$ looks is to consider the *canonical* form of it (this foreshadows a technique discussed later in detail). Thus, observe that (in the case at hand, but not in general):

$$\phi_t(c) = P\beta(t),$$

where P is the $n \times 2$ matrix: $P = [v_1, v_2]$, formed using the the vectors v_1, v_2 as columns of P, and β is the following curve in $\mathbb{R}^2$:

$$\beta(t) = (k_1 e^{\lambda_1 t}, k_2 e^{\lambda_2 t}).$$

Since the linear transformation $P : \mathbb{R}^2 \to M$ is 1-1 and maps the curve β onto the integral curve in question, this latter curve will be similar to β.

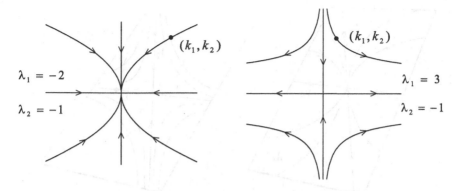

Figure 4.8: *Plots of the curves* $\beta(t) = (k_1 e^{\lambda_1 t}, k_2 e^{\lambda_2 t})$ *in the two cases:* **(left)** $\lambda_1 = -2, \lambda_2 = -1$ *giving the ratio* $r = 1/2$ *and* **(right)** $\lambda_1 = 3, \lambda_2 = -1$, *giving the ratio* $r = -1/3$.

However, the curve β lies on *part* of the graph of a power function $y = m|x|^r$ (which part depends on the sign of k_1). This is easily seen by taking the component expression for β,

$$x_1 = k_1 e^{\lambda_1 t}$$
$$x_2 = k_2 e^{\lambda_2 t},$$

and eliminating the parameter t (we assume $k_1 \neq 0$). We find that

$$x_2 = m|x_1|^r,$$

where $m = |k_2|/|k_1|^r$ and $r = \lambda_2/\lambda_1$. Thus, the nature of the integral curve $\phi_t(c)$ depends on the ratio r of the two eigenvalues. Also note that $\beta(0) = (k_1, k_2)$ and this helps locate the branch of $x_2 = m|x_1|^r$ that β lies on. So, for example, the integral curve $\phi_t(c)$ will be similar to one of the curves shown on the left in Figure 4.8 if $r = 1/2$ and will be similar to one of the curves shown on the right if $r = -1/3$. The direction of flow along the integral curve is determined by the signs of the two eigenvalues, as explained above.

This analysis of β, then gives us a qualitative understanding of what the integral curves $\phi_t(c) = P\beta(t)$ in the plane M, for several choices of $c \in M$, will look like. Note that β will change with c but P remains the same. In all cases P is 1-1 and so $\phi_t(c)$ will be similar to β. Figure 4.9 illustrates two possible phase portraits for the flow on the plane $M \subseteq \mathbb{R}^n$.

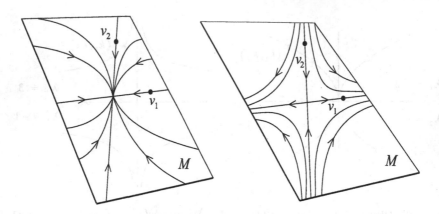

Figure 4.9: *Plots of several integral curves $\phi_t(c) = P\beta(t)$ in the plane $M = \text{span}\{v_1, v_2\}$. The integral curves that start at a point c on one of the axes determined by v_1 and v_2 are straight-line integral curves and run toward or away from the origin according to whether the eigenvalue is negative or positive. The other integral curves have their form determined by the ratio $r = \lambda_2/\lambda_1$. The figure shows the two cases:* (a) $r = 1/2$ *and* (b) $r = -1/3$.

We can extend the above discussion to the case where A has p linearly independent eigenvectors: $v_1, v_2, \ldots, v_p$,

$$Av_j = \lambda_j v_j \qquad \text{for } j = 1, \cdots, p.$$

Note: We are *not* assuming here that the λ_j's are distinct. We are assuming, however, that these eigenvalues, as well as the corresponding eigenvectors v_j, are *real*. Then suppose the initial point c is in the span of these eigenvectors, say

$$c = k_1 v_1 + \cdots + k_p v_p.$$

It is easy to see that the integral curve that starts at c at time zero is given by

$$\phi_t(c) = k_1 e^{\lambda_1 t} v_1 + \cdots + k_p e^{\lambda_p t} v_p. \tag{4.62}$$

The geometry of this integral curve can be analyzed as in the case above where $p = 2$ and is found to depend on the signs (positive/negative) of the eigenvalues and on the ratios: $\lambda_2/\lambda_1, \ldots, \lambda_p/\lambda_1$ (see the exercises). This then gives us an understanding of all the integral curves that start at a point $c \in \text{span}\{v_1, \ldots, v_p\}$. In the special case when the span of the v_i's is all of $\mathbb{R}^n$, i.e., $\text{span}\{v_1, \ldots, v_p\} = \mathbb{R}^n$, then necessarily $p = n$ and we obtain a description and understanding of every possible integral curve. This is recorded in the following theorem:

Theorem 4.5 (Real Eigenbasis Theorem) *Suppose A is an $n \times n$ matrix, and suppose that $\mathbb{R}^n$ has a basis: $\{v_1, \ldots, v_n\}$, consisting entirely of eigenvectors of A, say $Av_i = \lambda_i v_i$, with the λ_i's being the corresponding eigenvalues. Then for any $c \in \mathbb{R}^n$, the integral curve of the system $x' = Ax$, that passes through c at time $t = 0$, is given by:*

$$\phi_t(c) = k_1 e^{\lambda_1 t} v_1 + \cdots + k_n e^{\lambda_n t} v_n, \tag{4.63}$$

where the numbers: $k_1, \ldots, k_n$ are the components of c with respect to the basis $v_1, \cdots, v_n$. That is:

$$c = k_1 v_1 + \cdots + k_n v_n.$$

Note: In the theorem the eigenvalues $\lambda_1, \ldots, \lambda_n$, need not all be distinct. We also emphasize that the assumption of the theorem implies that all the eigenvalues of A are real, but the converse need not be true. That is: if all the eigenvalues of A are real, it does *not* follow that $\mathbb{R}^n$ has a basis of eigenvectors of A. We shall encounter many examples of this later. The next few examples illustrate the content of the eigenbasis theorem.

Example 4.10 Consider the system: $x' = Ax$, where A is the 2×2 matrix:

$$A = \begin{bmatrix} 1 & 2 \\ 2 & -2 \end{bmatrix}.$$

The equation to determine the eigenvalues of A (eigenvalue equation) is

$$
\begin{aligned}
\det(A - \lambda I) &= \det \begin{bmatrix} 1 - \lambda & 2 \\ 2 & -2 - \lambda \end{bmatrix} \\
&= (\lambda - 1)(\lambda + 2) - 4 = \lambda^2 + \lambda - 6 \\
&= (\lambda - 2)(\lambda + 3) = 0.
\end{aligned}
$$

Thus, the eigenvalues of A are $\lambda = 2, -3$. Corresponding eigenvectors can be found as follows (they are not unique):

(1) ($\lambda = 2$) The eigenvector equation $(A - 2I)v = 0$ is

$$\begin{bmatrix} -1 & 2 \\ 2 & -4 \end{bmatrix} \begin{bmatrix} x_1 \\ x_2 \end{bmatrix} = 0.$$

We can solve this system of equations by guesswork, i.e., clearly:

$$v_1 = \begin{bmatrix} 2 \\ 1 \end{bmatrix},$$

is one vector that works.

(2) ($\lambda = -3$) The eigenvector equation $(A + 3I)v = 0$ is

$$\begin{bmatrix} 4 & 2 \\ 2 & 1 \end{bmatrix} \begin{bmatrix} x_1 \\ x_2 \end{bmatrix} = 0.$$

Again by guesswork we get

$$v_2 = \begin{bmatrix} -1 \\ 2 \end{bmatrix},$$

is one possible solution.

You can readily verify that these two vectors v_1, v_2 are linearly independent, and therefore constitute a basis for $\mathbb{R}^2$. By the theorem, the general solution of the DE: $x' = Ax$, is given by the curve

$$\begin{aligned} \gamma(t) &= k_1 e^{2t} \begin{bmatrix} 2 \\ 1 \end{bmatrix} + k_2 e^{-3t} \begin{bmatrix} -1 \\ 2 \end{bmatrix} \\ &= \begin{bmatrix} 2k_1 e^{2t} - k_2 e^{-3t} \\ k_1 e^{2t} + 2k_2 e^{-3t} \end{bmatrix} \\ &= (2k_1 e^{2t} - k_2 e^{-3t}, k_1 e^{2t} + 2k_2 e^{-3t}). \end{aligned}$$

Note: We have written the column matrix in the second line as an ordered pair (or vector in $\mathbb{R}^2$) in the third line. We will use either notation interchangeably. In the above general solution, k_1, k_2 are arbitrary constants which can be chosen so that the above solution γ satisfies an initial condition: $\gamma(0) = c$. Of course, another way to express the general solution is:

$$\begin{aligned} x_1 &= 2k_1 e^{2t} - k_2 e^{-3t} \\ x_2 &= k_1 e^{2t} + 2k_2 e^{-3t}. \end{aligned}$$

We can sketch, by hand, the phase portrait of this system as follows (see Figure 4.10). Graph the basis eigenvectors v_1, v_2, with initial points at the origin. In the general solution:

$$\gamma(t) = k_1 e^{2t} v_1 + k_2 e^{-3t} v_2,$$

take $k_2 = 0$ to get a curve: $\gamma(t) = k_1 e^{2t} v_1$ which is on the line through the origin containing v_1. This line is the eigenspace E_2 and in fact γ is the half-ray in this space in the direction of v_1 (if $k_1 > 0$), or in the direction of $-v_1$ (if $k_2 < 0$). Note that $\gamma(t)$ runs off to infinity as $t \to \infty$. Similarly, taking $k_2 = 0$, we get a half-ray: $\gamma(t) = k_2 e^{-3t} v_2$, in the eigenspace E_{-3} for the corresponding solution curve. In this case however $\lim_{t \to \infty} \gamma(t) = 0$. These two special types of integral curves, along with several integral curves for which $k_1 \neq 0$ and $k_2 \neq 0$, are shown in Figure 4.10.

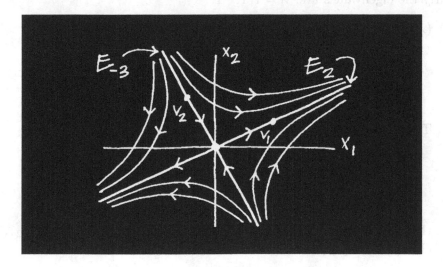

Figure 4.10: *Hand-drawn phase portrait for the linear system:* $x_1' = x_1 + 2x_2$, $x_2' = 2x_1 - 2x_2$.

The sketches of the integral curves that are not straight lines come from the general discussion above and Figure 4.8. In the example here, we can write the general solution as

$$\gamma(t) = P \begin{bmatrix} k_1 e^{2t} \\ k_2 e^{-3t} \end{bmatrix},$$

where P is the 2×2 matrix:

$$P = \begin{bmatrix} 2 & -1 \\ 1 & 2 \end{bmatrix}.$$

Thus, the graph of the curve $\beta(t) \equiv (k_1 e^{2t}, k_2 e^{-3t})$, which coincides with part of the graph of the function $x_2 = m|x_1|^{-3/2}$, is mapped by P onto the integral curve γ and so the graph of γ is similar to the graph of β.

Example 4.11 Consider the system $x' = Ax$ with

$$A = \begin{bmatrix} 1 & 0 & -1 \\ 0 & 1 & -1 \\ 0 & 0 & -1 \end{bmatrix}.$$

Clearly the eigenvalues are: $\lambda = 1, 1, -1$.

(1) ($\lambda = 1$) The eigenvector equation $(A - I)v = 0$ is

$$\begin{bmatrix} 0 & 0 & -1 \\ 0 & 0 & -1 \\ 0 & 0 & -2 \end{bmatrix} \begin{bmatrix} x_1 \\ x_2 \\ x_3 \end{bmatrix} = 0.$$

This system of equations has two obvious solutions:

$$v_1 = \begin{bmatrix} 1 \\ 0 \\ 0 \end{bmatrix}, \quad v_2 = \begin{bmatrix} 0 \\ 1 \\ 0 \end{bmatrix},$$

that are linearly independent. The eigenspace $E_1 = \mathrm{span}\{v_1, v_2\}$, is clearly the x-y plane.

(2) ($\lambda = -1$) The eigenvector equation $(A + I)v = 0$ is

$$\begin{bmatrix} 2 & 0 & -1 \\ 0 & 2 & -1 \\ 0 & 0 & 0 \end{bmatrix} \begin{bmatrix} x_1 \\ x_2 \\ x_3 \end{bmatrix} = 0.$$

Written out in non-matrix form, this system of equations is

$$\begin{aligned} 2x_1 - x_3 &= 0 \\ 2x_2 - x_3 &= 0. \end{aligned}$$

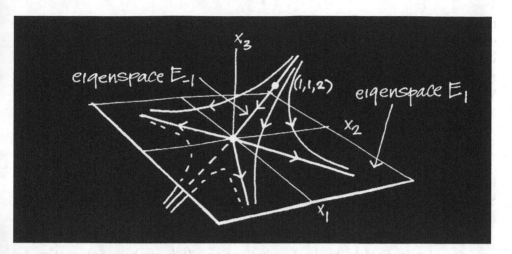

Figure 4.11: *Hand-drawn phase portrait for the linear system:* $x_1' = x_1 - x_3$, $x_2' = x_2 - x_3$, $x_3' = -x_3$.

If we take $x_3 = 2$, then $x_1 = 1, x_2 = 1$, and so

$$v_3 = \begin{bmatrix} 1 \\ 1 \\ 2 \end{bmatrix},$$

is an eigenvector corresponding to $\lambda = -1$. The eigenspace E_{-1}, being the subspace spanned by the vector v_3, is geometrically the straight line through the origin in the direction of v_3.

The general solution of the system $x' = Ax$ in this example is

$$\begin{aligned} \gamma(t) &= e^t(k_1 v_1 + k_2 v_2) + e^{-t} k_3 v_3 \\ &= \left(k_1 e^t + k_3 e^{-t}, \; k_2 e^t + k_3 e^{-t}, \; 2k_3 e^{-t} \right), \end{aligned} \tag{4.64}$$

where k_1, k_2, k_3 are arbitrary constants. The geometry of the integral curves for this system is fairly simple and is easy to plot by hand.

Consider a given initial point $c \in \mathbb{R}^3$, and let γ be the integral curve that starts at c at time zero: $\gamma(0) = c$. If c lies in the eigenspace $E_1 =$ the x-y plane, then we can write c as $c = k_1 v_1 + k_2 v_2$. Then $\gamma(t) = e^t(k_1 v_1 + k_2 v_2) = e^t c$ is just the ray from the origin through c, and $\gamma(t)$ tends to infinity as $t \to \infty$. This is illustrated in Figure 4.11. If the initial point c lies in the eigenspace $E_{-1} =$ the line through the origin in the direction $v_3 = (1, 1, 2)$, then c is a multiple of v_3, say $c = k_3 v_3$. So $\gamma(t) = k_3 e^{-t} v_3 = e^{-t} c$, is just a ray on this line, with $\lim_{t \to \infty} \gamma(t) = 0$. These two observations describe what happens for integral curves starting in one of the eigenspaces, and also enable us to discern what happens when c is not in one of the eigenspaces. When this is the case, choose k_1, k_2, k_3, such that $c = k_1 v_1 + k_2 v_2 + k_3 v_3$. Then the integral curve:

$$\gamma(t) = e^t (k_1 v_1 + k_2 v_2) + k_3 e^{-t} v_3,$$

is a curve that lies in the plane containing the vectors $k_1 v_1 + k_2 v_2$ and v_3. Since $\lim_{t \to \infty} k_3 e^{-t} v_3 = 0$, this curve becomes asymptotic to the x_1-x_2 plane as $t \to \infty$. This is shown in Figure 4.11.

Note that because of the particular form of the system in this example we can solve it directly by back substitution. It is perhaps instructive to look at this. Writing the system out in non-matrix form gives:

$$
\begin{align}
x_1' &= x_1 - x_3 & (4.65) \\
x_2' &= x_2 - x_3 & (4.66) \\
x_3' &= -x_3. & (4.67)
\end{align}
$$

The last equation readily gives: $x_3(t) = a_3 e^{-t}$, with a_3 an arbitrary constant. Substituting this in the first and second equations gives the DEs:

$$
\begin{align}
x_1' &= x_1 - a_3 e^{-t} \\
x_2' &= x_2 - a_3 e^{-t}.
\end{align}
$$

The solutions of these are: $x_2(t) = k_2 e^t + a_3 e^{-t}/2$ and $x_1(t) = k_1 e^t + a_3 e^{-t}/2$, which, if we relabel the constants by $k_3 = a_3/2$, is the same as before. Of course, the eigenvector method provides us with the extra geometric information needed to sketch integral curves. This would be difficult to do just from the above solutions alone.

The above example exhibits the decomposition $\mathbb{R}^3 = E_1 \oplus E_{-1}$ of $\mathbb{R}^3$ into a direct sum of the eigenspaces of A. Likewise in the very first example we had the decomposition $\mathbb{R}^2 = E_2 \oplus E_{-3}$. Something like this always occurs when $\mathbb{R}^n$ has a basis of eigenvectors of A. Furthermore, the geometric analysis of the integral curves will be similar to the above examples, except harder to visualize. Thus, suppose in the eigenbasis theorem above we relabel the λ's, just listing the *distinct* ones, say: $\mu_1, \ldots, \mu_p$, and relabel the eigenvectors, say: $v_1^1, \ldots, v_{r_1}^1$ is a basis for E_{μ_1}, etc., out to: $v_1^p, \ldots, v_{r_p}^p$ is a basis for E_{μ_p}. Then (4.63) has the form:

$$\phi_t(c) = e^{\mu_1 t}(k_1^1 v_1^1 + \cdots + k_{r_1}^1 v_{r_1}^1) + \cdots + e^{\mu_p t}(k_1^p v_1^p + \cdots + k_{r_p}^p v_{r_p}^p). \quad (4.68)$$

This is the general solution of $x' = Ax$ under the assumption that $\mathbb{R}^n$ has a basis of eigenvectors of A. We have the decomposition

$$\mathbb{R}^n = E_{\mu_1} \oplus \cdots \oplus E_{\mu_p},$$

of $\mathbb{R}^n$ into a direct sum of the eigenspaces of A. The geometric interpretation of the integral curve (4.68), is difficult even in small dimensions. However a special case of this is easy to visualize, and even though this has already been mentioned, we reiterate it here in this setting for emphasis.

Suppose we consider an integral curve γ that starts at an initial point in one of the eigenspaces, $c \in E_{\mu_i}$ say. Then we can write c as $c = b_i^i v_i^1 + \cdots + b_{r_i}^i v_{r_i}^1$ and so

$$\phi_t(c) = e^{\mu_i t}(k_i^i v_i^1 + \cdots + k_{r_i}^i v_{r_i}^1) = e^{\mu_i t} c.$$

This integral curve is a ray from the origin in the direction of c and it tends to infinity or to the origin as $t \to \infty$ depending on whether μ_i is greater than or less than zero. Geometrically, E_{μ_i} is either a line through 0, or a plane through 0, or a higher-dimensional subspace of $\mathbb{R}^n$, and is filled up with (comprised of) integral curves which are all rays directed away from or toward 0. This then is what the part of the phase portrait looks like on E_{μ_i}. **Note:** The special case when A has 0 as an eigenvalue is included in the above discussion. Then, as we have noted, the corresponding eigenspace $E_0 = \ker A$, consists entirely of fixed points; the integral curve starting at a point $c \in E_0$ stays there for all time.

Warning: The discussion to this point has been very limited and should not mislead you into thinking the general case will be similar. In general we can *not* express the integral curves of $x' = Ax$ solely in terms of the eigenvectors and eigenvalues of A, and we can *not* decompose $\mathbb{R}^n$ into a direct sum of eigenspaces. One reason for this is that A may have some eigenvalues that are complex (not real), or even when all the eigenvalues are real there may not be enough linearly independent eigenvectors to form a basis for $\mathbb{R}^n$. Thus, we need to look at these other possibilities.

4.4.2   Complex Eigenvalues

The geometry of integral curves that arise from complex eigenvalues is more interesting, in some respects, than for real eigenvalues. The discussion in the real case was based on the fact that the action of e^{At} on c is very simple if c is an eigenvector or a sum of eigenvectors. This same simplicity holds in the complex case except that the eigenvectors lie in $\mathbb{C}^n$ and not in $\mathbb{R}^n$. Transferring the integral curves from $\mathbb{C}^n$ back to $\mathbb{R}^n$ accounts for the only complication in the discussion here.

One important observation about complex eigenvalues is that they come in complex conjugate pairs, i.e., if $\lambda = a + bi$ is an eigenvalue of A, then so is its conjugate $\overline{\lambda} = a - bi$. This is so since the eigenvalue equation $\det(A - \lambda I) = 0$ is a polynomial equation with real coefficients. Another useful observation is that if $\gamma : \mathbb{R} \to \mathbb{C}^n$ is a solution of the corresponding system $z' = Az$ of complex DEs, then both its real part: $\alpha = \text{Re } \gamma$, and imaginary part: $\beta = \text{Im } \gamma$, are solutions of the original system $x' = Ax$. Thus, we could extend the discussion to the complex domain, obtain general results there, and then reduce these back to the real domain. We do not take this approach here (see [Arn 78a]), but rather proceed as follows.

Suppose $\lambda = a + bi$ (with $b \neq 0$) is a complex eigenvalue of A, and v is a corresponding eigenvector (obtained by solving $(A - \lambda I)v = 0$). Then necessarily v is a complex vector: $v \in \mathbb{C}^n$, and we can write it as $v = u + wi$, where $u, w \in \mathbb{R}^n$. You can verify that u and w are linearly independent (exercise). If we write out the equation $Av = \lambda v$ in terms of real and imaginary parts, we get

$$
\begin{aligned}
Au + iAw &= A(u + iw) = Av = \lambda v = (a + bi)(u + iw) \\
&= (au - bw) + i(bu + aw)
\end{aligned}
$$

Hence, if we equate real and imaginary parts of this, we obtain

$$
\begin{aligned}
Au &= au - bw \\
Aw &= bu + aw.
\end{aligned}
$$

This proves part of the following fundamental result:

Theorem 4.6 *Suppose A is a real, $n \times n$ matrix, $u, w \in \mathbb{R}^n$, and $a, b \in \mathbb{R}$ with $b \neq 0$. Let $v = u + iw$ and $\lambda = a + bi$. Then the following are equivalent:*

(1) $Av = \lambda v$.

(2) u, w satisfy

$$Au = au - bw \tag{4.69}$$
$$Aw = bu + aw. \tag{4.70}$$

(3) u, w satisfy

$$[(A - aI)^2 + b^2 I]u = 0 \tag{4.71}$$
$$w = -b^{-1}(A - aI)u. \tag{4.72}$$

If any one of these holds, then u and w are linearly independent.

Proof: This is left as an (easy but interesting) exercise.

Equations (4.71)-(4.72) suggest the following definition:

Definition 4.3 Suppose $\lambda = a + bi$, with $b \neq 0$, is a complex eigenvalue of A. The corresponding *pseudo-eigenspace* for the complex conjugate pair $a \pm bi$ is defined as

$$E_{a \pm bi} = \{\, u \in \mathbb{R}^n \mid [(A - aI)^2 + b^2 I]u = 0 \,\}.$$

It is important to note that the vectors in $E_{a \pm bi}$ are *not* eigenvectors of A, but rather comprise a subspace of $\mathbb{R}^n$ that corresponds to the pair $a \pm bi$ of eigenvalues of A. In essence the characteristic polynomial $p(x) = \det(A - xI)$ contains a quadratic factor, $q(x) = (x - a)^2 + b^2$, which is irreducible (over $\mathbb{R}$). As shown in Appendix C, the factorization of $p(x)$ completely into linear and irreducible quadratic factors is used to decompose $\mathbb{R}^n$ into a direct sum of (generalized) eigenspaces. The relation between $E_{a \pm bi}$ and the complex eigenspace for $\lambda = a + bi$ is described in the following theorem.

Theorem 4.7 *Suppose $\lambda = a + bi$, with $b \neq 0$, is a complex eigenvalue of A and let*

$$V_\lambda = \{\, v \in \mathbb{C}^n \mid Av = \lambda v \,\}$$

be the complex eigenspace of A corresponding to λ. Define a map $L : E_{a \pm bi} \rightarrow V_\lambda$ by

$$Lu = u - ib^{-1}(A - aI)u.$$

Then L is a real, linear isomorphism between the vector spaces $E_{a\pm bi}$ and V_λ considered as vector spaces over $\mathbb{R}$. Furthermore, the inverse of L is given by

$$L^{-1}(u + iw) = u.$$

Consequently, $E_{a\pm bi}$ is a even-dimensional subspace (over the reals). In particular, if V_λ has dimension r as a complex vector space and has basis

$$u_1 + iw_1, u_2 + iw_2, \ldots, u_r + iw_r,$$

then $E_{a\pm bi}$ has dimension $2r$ and basis

$$u_1, w_1, \ u_2, w_2, \ldots, u_r, w_r.$$

Proof: It is easy to see that L is linear and 1-1. By results in the last theorem one can show that L is onto (exercise). The formula for L^{-1} is clear. It is a general result from linear algebra that a complex vector space V of dimension r has real dimension $2r$ and more specifically, if $v_1, \ldots, v_r$ is a basis for V over the complex numbers, then $v_1, iv_1, \ldots, v_r, iv_r$ is a basis for V over the real field (exercise). Applying this to V_λ, using the fact that L is an isomorphism, and taking into account the formula for L^{-1}, gives the result (exercise). $\square$

Remark: The prescription for finding a basis for $E_{a\pm bi}$ suggested by the theorem is also the one that works best in practice. Namely, work in the complex domain. For each complex eigenvalue λ, solve the eigenvector equation $(A - \lambda I)v = 0$ for a set of linear independent (over $\mathbb{C}$) complex eigenvectors: $v_j = u_j + iw_j$, $j = 1, \ldots, r$. Then take real and imaginary parts of these vectors to get linearly independent real vectors: u_j, w_j, $j = 1, \ldots, r$.

With all this linear algebra out of the way, we can now more easily understand the nature of the integral curves $\phi_t(u)$ that start with initial point in

$$u \in E_{a\pm bi}.$$

We shall see that these integral curves are either ellipses (when $a = 0$) or spirals (when $a \neq 0$). Compare this with the situation where the initial point is in one of the real eigenspaces of A. Then the integral curves are either points (fixed points) or rays through the origin.

For a given $u \in E_{a\pm bi}$ let

$$w = -b^{-1}(A - aI)u.$$

Then we can use equations(4.69)-(4.70) above to compute $\phi_t(u) = e^{At}u$ (and also $\phi_t(w) = e^{At}w$ at the same time). This is rather lengthy and is relegated to the exercises (since it is nevertheless interesting). A much shorter way to get the same results is to work in the complex domain and use Euler's formula as follows.

We let $v = u + iw$ and $\lambda = a + bi$. Then $Av = \lambda v$ and consequently, in the complex domain, we get

$$e^{At}v = e^{\lambda t}v. \tag{4.73}$$

Now all we have to do is express each side of this equation in terms of real and imaginary parts. Using Euler's formula on the right-hand side gives: $e^{\lambda t}v = e^{at+bti}(u + wi) = e^{at}[\cos(bt) + i\sin(bt)](u + iw) = e^{at}[\cos(bt)u - \sin(bt)w] + ie^{at}[\sin(bt)u + \cos(bt)w]$. The left-hand side of equation(4.73) separates easily into real and imaginary parts since e^{At} is a real matrix. Thus, $e^{At}v = e^{At}u + ie^{At}w$ is the separation into real and imaginary parts. Consequently, by equating real and imaginary parts on both sides of equation (4.73) we arrive at

$$\phi_t(u) = e^{At}u = e^{at}[\cos(bt)u - \sin(bt)w]$$
$$\phi_t(w) = e^{At}w = e^{at}[\sin(bt)u + \cos(bt)w].$$

These are the formulas for the integral curves that start at u and w, respectively, at time zero. More generally, from these we can get the explicit formula for the integral curve that starts at any point in the plane:

$$M = \text{span}\{u, w\}$$

spanned by u and w. Thus, suppose that $c \in M$ and write c in terms of the basis: $c = ku + hw$. We easily get from the above pair of equations that

$$\phi_t(c) = e^{At}c = ke^{At}u + he^{At}w$$
$$= e^{at}[\cos(bt)(ku + hw) + \sin(bt)(hu - kw)]$$
$$= e^{at}[\cos(bt)c + \sin(bt)d] \tag{4.74}$$
$$= P\beta(t). \tag{4.75}$$

Here $d = hu - kw$, $P = [u, w]$ is the $n \times 2$ matrix with columns u, w, and β is the curve in $\mathbb{R}^2$ defined by

$$\beta(t) = e^{at} \begin{bmatrix} \cos(bt) & \sin(bt) \\ -\sin(bt) & \cos(bt) \end{bmatrix} \begin{bmatrix} k \\ h \end{bmatrix}.$$

You can easily see that if $a = 0$, then β is a circle of radius $(k^2 + h^2)^{1/2}$, centered at the origin and if $a \neq 0$, then β is a spiral that winds either toward the origin (if $a < 0$) or away from the origin (if $a > 0$). These two cases are shown in Figure 4.12.

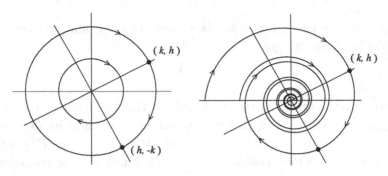

Figure 4.12: *Plots of the curve $\beta(t) = e^{at} R(bt)(k, h)$ for the two cases $a = 0$ and $a \neq 0$ and for various choices of the initial point (k, h). The graphs are for the case $b > 0$, so that the motion along the curves is clockwise. When $b < 0$ the motion is counterclockwise.*

In addition

$$
\begin{aligned}
\beta(0) &= (k, h) \\
\beta(\pi/2b) &= e^{a\pi/2}(h, -k) \\
\beta(\pi/b) &= e^{a\pi}(-k, -h) \\
\beta(3\pi/2b) &= e^{3a\pi/2}(-h, k),
\end{aligned}
$$

gives four points on the circle or spiral which have position vectors that make, in succession, angles of 90 degrees with each other. This should be expected since the curve β involves the *rotation matrix*:

$$
R(\theta) = \begin{bmatrix} \cos(\theta) & \sin(\theta) \\ -\sin(\theta) & \cos(\theta) \end{bmatrix}.
$$

By varying the initial point (k, h), we get either a series of concentric circles (when $a = 0$) or spirals (when $a \neq 0$). To interpret the corresponding integral curves in the plane M, consider P as a linear transformation $P : \mathbb{R}^2 \to \mathbb{R}^n$. Then P is 1-1 (since its columns u, w are linearly independent) and has M for its range. Thus, the integral curve $\alpha(t) \equiv \phi_t(c)$ is similar to the curve $\beta(t)$ in $\mathbb{R}^2$. It is thus an ellipse (exercise) when $a = 0$ and otherwise is a spiral.

To learn how to sketch a number of the integral curves α for various choices of $c = ku + hw$ in M, we look first at the case $a = 0$. Then all the integral curves are concentric ellipses, so we first plot the basic one: $k = 1, h = 0$, i.e., the initial point is u. This ellipse can be plotted easily by graphing the vectors $\pm u, \pm w$, and noting that $\alpha(0) = u, \alpha(\pi/2b) = -w, \alpha(\pi/b) = -u$, and $\alpha(3\pi/2b) = w$. Figure 4.13 shows this graph, along with one going through a generic initial point $c = ku + hw$. In the latter case the integral curve passes through the points $c, d, -c, -d$ at times $0, \pi/2b, \pi/b, 3\pi/2b$, respectively. *Note:* If $c = ku + hw$, then $d \equiv hu - kw$

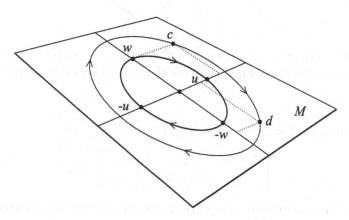

Figure 4.13: *Graphs of the curves $\phi_t(u) = \cos(bt)u - \sin(bt)w$ and $\phi_t(c) = \cos(bt)c + \sin(bt)d$ in the plane M spanned by u and w. Note, in the latter curve, if $c = ku + hw$, then $d = hu - kw$.*

This is the picture under the assumption that $a = 0$, i.e., the complex eigenvalue $\lambda = a + bi = bi$, is purely imaginary. If, however, $a \neq 0$, then the exponential factor e^{at} in the integral curve $\alpha(t) = e^{at}[(\cos bt)c + (\sin bt)d]$ causes the otherwise elliptical path of α to either spiral in toward the origin (when $a < 0$) or spiral out away from the origin (when $a > 0$). This is indicated in Figure 4.14.

Example 4.12 Consider the system $x' = Ax$, where A is the 2×2 matrix:

$$A = \begin{bmatrix} 1 & 5 \\ -1 & -1 \end{bmatrix}.$$

The eigenvalues of A are easily computed to be $\lambda = \pm 2i$. An eigenvector

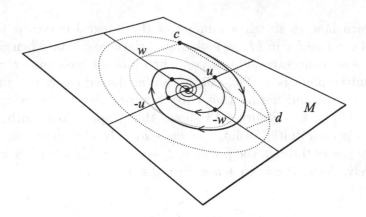

Figure 4.14: *Graphs of* $\phi_t(u) = e^{at}[\cos(bt)u - \sin(bt)w]$ *and* $\phi_t(c) = e^{at}[\cos(bt)c + \sin(bt)d]$, *for* $a < 0$. *Note that if* $c = ku + hw$, *then* $d \equiv hu - kw$.

corresponding to $\lambda = 2i$ can be found by solving

$$\begin{bmatrix} 1 - 2i & 5 \\ -1 & -1 - 2i \end{bmatrix} \begin{bmatrix} x_1 \\ x_2 \end{bmatrix} = 0.$$

You can solve this system of equations by guesswork, i.e., clearly

$$v = \begin{bmatrix} 5 \\ -1 + 2i \end{bmatrix},$$

is one vector that works (check this!). Taking the real and imaginary parts of v gives $u = (5, -1)$ and $w = (0, 2)$. Thus, we know from the above discussion that the general solution of this is $\alpha(t) = P\beta(t)$, where P is the matrix

$$P = \begin{bmatrix} 5 & 0 \\ -1 & 2 \end{bmatrix}$$

and β is the circle

$$\beta(t) = \begin{bmatrix} \cos(2t) & \sin(2t) \\ -\sin(2t) & \cos(2t) \end{bmatrix} \begin{bmatrix} k \\ h \end{bmatrix}.$$

To plot the approximate phase portrait by hand, we simply plot the vectors $\pm u, \pm w$, sketch in an ellipse passing through the endpoints of these vectors, and then draw a sequence of ellipses concentric to this one. The result is shown in Figure 4.15.

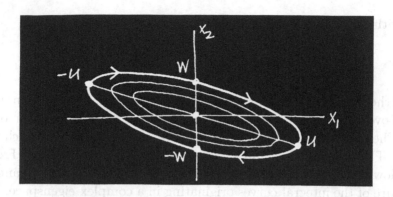

Figure 4.15: *Hand-drawn sketch of the phase portrait for the system* $x'_1 = x_1 + 5x_2$, $x'_2 = -x_1 - x_2$.

The above discussion for a single complex eigenvector $v = u + iw$ can easily be extended. Thus, suppose $v_j = u_j + iw_j$, $j = 1, \ldots, p$ are linearly independent complex eigenvectors of A corresponding to complex eigenvalues $\lambda_j = a_j + b_j i$ (not necessarily distinct). Then $u_1, w_1, \ldots, u_p, w_p$ are linearly independent vectors in $\mathbb{R}^n$ and for any initial point

$$c = \sum_{j=1}^{p}(k_j u_j + h_j w_j)$$

in the span of these vectors, we get

$$e^{At}c = \sum_{j=1}^{p} e^{a_j t}\left[\cos(b_j t)\Big(k_j u_j + h_j w_j\Big) + \sin(b_j t)\Big(h_j u_j - k_j w_j\Big)\right].$$

This yields the explicit form for the integral curve that starts at c at time zero. To interpret it geometrically, we introduce the following extension of the above notation. Define curves $\beta_j : \mathbb{R} \to \mathbb{R}^2$, and $\beta : \mathbb{R} \to \mathbb{R}^{2p}$ by

$$\beta_j(t) = e^{a_j t} R(b_j t)(k_j, h_j) \qquad j = 1, \ldots, p \qquad (4.76)$$

$$\beta(t) = \Big(\beta_1(t), \ldots, \beta_p(t)\Big). \qquad (4.77)$$

Also, for each $j = 1, \ldots, p$, let $P_j = [u_j, w_j]$ be the $n \times 2$ matrix with the indicated columns, and let $P = [u_1, w_1, \ldots, u_p, w_p]$ be the $n \times 2p$ matrix with the indicated columns. Then an alternative way of describing the integral curve with initial point $c \in M \equiv \text{span}\{u_1, w_1, \ldots, u_p, w_p\}$ is

$$\phi_t(c) = P\beta(t) = \sum_{j=1}^{p} P_j \beta_j(t).$$

Each of the curves $P_j \beta_j(t)$ is an ellipse/spiral in the plane $M_j = \text{span}\{u_j, w_j\}$ and the overall integral curve $\phi_t(c)$ is comprised of the superposition of all of these. The result is rather difficult to visualize except for $p = 1$ (which we did above). The case $p = 2$ requires four dimensions for viewing (see Example 4.13 below). Nevertheless, the result gives us a qualitative understanding of the nature of the integral curves originating in a complex eigenspace. In the special case when $n = 2p$, this includes every possible integral curve.

Theorem 4.8 (Complex Eigenbasis Theorem) *Suppose $n = 2p$ and A is an $n \times n$ matrix with real entries. If $v_j = u_j + iw_j$, $j = 1 \ldots , p$ are linearly independent (over $\mathbb{C}$) eigenvectors for A corresponding to the eigenvalues $\lambda_j = a_j + b_j i$, $j = 1, \ldots, p$, then $\{u_1, w_1, \ldots, u_p, w_p\}$ is a basis for $\mathbb{R}^n$ and for any $c \in \mathbb{R}^n$, the integral curve of the system $x' = Ax$, which passes through c at time $t = 0$, is given by*

$$\phi_t(c) = \sum_{j=1}^{p} e^{a_j t} \left[\cos(b_j t) \left(k_j u_j + h_j w_j \right) + \sin(b_j t) \left(h_j u_j - k_j w_j \right) \right] \quad (4.78)$$

where the real numbers k_j, h_j are the components of c with respect to the basis, i.e.,

$$c = \sum_{j=1}^{p} (k_j u_j + h_j w_j).$$

Furthermore, $\mathbb{R}^n$ decomposes into a direct sum of planes

$$\mathbb{R}^n = M_1 \oplus \cdots \oplus M_p,$$

with $M_j = \text{span}\{u_j, w_j\}$, $j = 1, \ldots, p$, and relative to this the integral curve is expressed as a sum

$$\phi_t(c) = \sum_{j=1}^{p} P_j \beta_j(t) \quad (4.79)$$

of curves $P_j \beta_j$ lying in M_j. Here

$$\beta_j(t) = e^{a_j t} R(b_j t)(k_j, h_j)$$

is a standard circle (when $a_j = 0$) or spiral (when $a_j \neq 0$) in $\mathbb{R}^2$ and $P_j :$ $\mathbb{R}^2 \to M_j$ is the linear transformation determined by the $n \times 2$ matrix $P_j = [u_j, w_j]$

A primary example of the Complex Eigenbasis Theorem is the system of masses coupled by springs as in Example 4.2.

Example 4.13 (Two Coupled Masses) For simplicity, assume that the masses are the same, say $m_1 = m_2 = 1$, and that the spring constants are all equal, say $k_1 = k_2 = k_3 = 1$. Then, from Example 4.2, the 2nd-order system for the displacements $x = (x_1, x_2)$ of the masses from equilibrium is $x'' = Kx$, where

$$K = \begin{bmatrix} -2 & 1 \\ 1 & -2 \end{bmatrix}.$$

The corresponding first-order system is $z' = Az$, where A is the 4×4 matrix

$$A = \begin{bmatrix} 0 & 0 & 1 & 0 \\ 0 & 0 & 0 & 1 \\ -2 & 1 & 0 & 0 \\ 1 & -2 & 0 & 0 \end{bmatrix}.$$

The computation of the eigenvalues of A via $\det(A - \lambda I) = 0$ is not hard if we use some properties of determinants and row reduce $A - \lambda I$ before taking the determinant. (What we do here works in general for the case of N coupled masses.) These properties are merely that the determinant is a antisymmetric, multilinear function of its rows: (1) If P is an $n \times n$ matrix and if each entry in one row of P is multiplied by a constant c to give the matrix $\tilde{P}$, then $\det(\tilde{P}) = c \det(P)$, and (2) the determinant of P does not change if one of its rows is added to another row.

With this understood, we multiply the 1st and 2nd rows of $A - \lambda I$ by λ. In the resulting matrix we add the 3rd row to the 1st row and the 4th row to the 2nd row. Now the resulting matrix is a block diagonal matrix and its determinant is easy to compute. Here is the implementation of this:

$$\begin{vmatrix} -\lambda & 0 & 1 & 0 \\ 0 & -\lambda & 0 & 1 \\ -2 & 1 & -\lambda & 0 \\ 1 & -2 & 0 & -\lambda \end{vmatrix} = \frac{1}{\lambda^2} \begin{vmatrix} -\lambda^2 & 0 & \lambda & 0 \\ 0 & -\lambda^2 & 0 & \lambda \\ -2 & 1 & -\lambda & 0 \\ 1 & -2 & 0 & -\lambda \end{vmatrix}$$

$$= \frac{1}{\lambda^2} \begin{vmatrix} -(\lambda^2 + 2) & 1 & 0 & 0 \\ 1 & -(\lambda^2 + 2) & 0 & 0 \\ -2 & 1 & -\lambda & 0 \\ 1 & -2 & 0 & -\lambda \end{vmatrix}$$

$$= \frac{1}{\lambda^2} \begin{vmatrix} -(\lambda^2+2) & 1 \\ 1 & -(\lambda^2+2) \end{vmatrix} \begin{vmatrix} -\lambda & 0 \\ 0 & -\lambda \end{vmatrix}$$

$$= (\lambda^2+2)^2 - 1$$

From this it's easy to see that the eigenvalues of A are $\lambda = \pm i, \pm\sqrt{3}\,i$. A straight-forward calculation gives complex eigenvectors

$$v_1 = (1, 1, i, i), \qquad v_2 = (1, -1, \sqrt{3}\,i, -\sqrt{3}\,i),$$

corresponding to i and $\sqrt{3}i$, respectively. These are linearly independent with respect to the complex number field $\mathbb{C}$ and so the Complex Eigenbasis Theorem applies. Taking real and imaginary parts of these vectors gives

$$u_1 = \begin{bmatrix} 1 \\ 1 \\ 0 \\ 0 \end{bmatrix}, \quad w_1 = \begin{bmatrix} 0 \\ 0 \\ 1 \\ 1 \end{bmatrix}, \quad u_2 = \begin{bmatrix} 1 \\ -1 \\ 0 \\ 0 \end{bmatrix}, \quad w_2 = \begin{bmatrix} 0 \\ 0 \\ \sqrt{3} \\ -\sqrt{3} \end{bmatrix},$$

and these vectors form a basis for $\mathbb{R}^4$. By the Complex Eigenbasis Theorem (with $a_j = 0$, $j = 1, 2$ and $b_1 = 1, b_2 = \sqrt{3}$) the general solution of $z' = Az$ is

$$\begin{aligned} \gamma(t) &= (\cos t)(k_1 u_1 + h_1 w_1) + (\sin t)(h_1 u_1 - k_1 w_1) \\ &\quad + (\cos\sqrt{3}t)(k_2 u_2 + h_2 w_2) + (\sin\sqrt{3}t)(h_2 u_2 - k_2 w_2) \end{aligned}$$

where the arbitrary constants k_1, h_1, k_2, h_2 are able to be chosen so that γ satisfies any $\gamma(0) = c$ for any given $c \in \mathbb{R}^4$. One merely solves the algebraic system

$$k_1 u_1 + h_1 w_1 + k_2 u_2 + h_2 w_2 = c.$$

In this example, the planes $M_1 = \text{span}\{u_1, w_1\}$ and $M_2 = \text{span}\{u_2, w_2\}$, are perpendicular to each other in $\mathbb{R}^4$ and the integral curves that start at a point $c \in M_1$ or $c \in M_2$ are circles or ellipses, respectively. For a point $c \in \mathbb{R}^4 = M_1 \oplus M_2$, not in one of these planes, the integral curve through c is more complicated.

Note that $\gamma(t)$ gives the *state* of the system of two masses at time t and this consists of the two positions and the two velocities. The first two components of $\gamma(t)$ give the positions and so the general solution of $x'' = Ax$ is

$$\alpha(t) = \left(k_1 \cos t + h_1 \sin t\right)\begin{bmatrix} 1 \\ 1 \end{bmatrix} + \left(k_2 \cos\sqrt{3}\,t + h_2 \sin\sqrt{3}\,t\right)\begin{bmatrix} 1 \\ -1 \end{bmatrix}.$$

The two vectors

$$v_1 = \begin{bmatrix} 1 \\ 1 \end{bmatrix}, \qquad v_2 = \begin{bmatrix} 1 \\ -1 \end{bmatrix},$$

are called the *normal mode vectors* of the coupled mass system and the numbers $1, \sqrt{3}$ are called the *normal frequencies* of vibration. The reason for this terminology is as follows. Suppose the masses are initially displaced from equilibrium and released with no initial velocities. Then the motion of the system is described by the curve

$$\alpha(t) = k_1 \cos(t)\, v_1 + k_2 \cos(\sqrt{3}t)\, v_2.$$

The case $k_1 = 1, k_2 = 0$, corresponds to initial displacements $x_1 = 1, x_2 = 1$ and the curve:

$$\eta_1(t) \equiv \cos(t)\, v_1,$$

is one of the *normal modes* of the system. In this mode of vibration the masses move in unison (i.e., in the same way) about their equilibrium positions with frequency 1. This is illustrated in Figure 4.16.

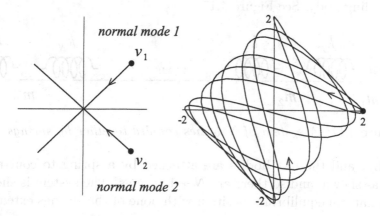

Figure 4.16: *The integral curves for the two normal modes are straight line segments. In this example, a superposition of the two normal modes gives a space-filling integral curve.*

The case $k_1 = 0, k_2 = 1$, corresponds to initial displacements $x_1 = 1, x_2 = -1$, and the curve:

$$\eta_2(t) \equiv \cos(\sqrt{3}t)\, v_2,$$

is the second *normal mode* of the system. In this mode of vibration both masses vibrate, oppositely, about equilibrium with frequency $\sqrt{3}$.

As shown in Figure 4.16, the motion of the masses is quite simple for either of the normal modes of vibration. However, the general motion of the masses, being a superposition

$$\alpha(t) = k_1\eta_1(t) + k_2\eta_2(t),$$

of the two normal modes of vibration, can be more complex and interesting. A typical plot of the curve α is shown in Figure 4.16 for $k_1 = 1 = k_2$, so that $\alpha(0) = (2, 0)$. There is additional study of this example and its generalization on the Maple worksheet oscillate.mws on the electronic component.

Example 4.14 (N-coupled Masses) The techniques used in the last example for 2-coupled masses generalizes. Here we derive the general equations of motion $x'' = Kx$ and the corresponding 1st-order system $z' = Az$. The rest of the general study is left for the exercises.

Consider a system of N bodies with masses $m_1, \ldots, m_N$ which lie on a frictionless table, are aligned in a straight line, and are connected by springs—in succession each body is connected by a spring to the preceding and succeeding body. See Figure 4.17.

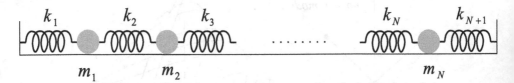

Figure 4.17: *A system of N bodies coupled together by springs.*

The first and the last bodies are attached by a spring to constraining brackets, as shown, and so there are $N + 1$ springs. The system is shown at rest in its natural equilibrium position with none of the springs extended or compressed.

Consider a motion of the system whereby each body is displaced from equilibrium and given an initial velocity, each either forward or backward in the line of the system. Let $x_j(t)$, $j = 1, \ldots, N$, be the position of the jth body, relative to its equilibrium position, at time t. Newton's 2nd Law gives the differential equation that x_j must satisfy: $m_j x_j'' = F_j$. All we have to do is figure out what F_j is. This is entirely similar to what we did for the case of two bodies. Figure 4.18 shows one possible position of the jth body at time t. Also shown are possible positions of the two bodies on each side of the jth body. For the situation depicted, you can see that each spring

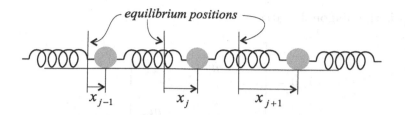

Figure 4.18: *A hypothetical position of the jth body at time t.*

attached to the jth body is stretched from its natural length—the jth spring is stretched by amount $x_j - x_{j-1}$ and the $(j + 1)$st spring is stretched by amount $x_{j+1} - x_j$. Each spring will attempt to contract to its natural length, thereby exerting forces to the left and right, respectively, on the jth body. Thus, from Hooke's Law, we get the equation of motion for the jth body:

$$m_j x_j'' = -k_j(x_j - x_{j-1}) + k_{j+1}(x_{j+1} - x_j). \tag{4.80}$$

While the argument leading to this equation was based on the positions shown in Figure 4.18, one can show that the same equation results regardless of the positions of the $(j - 1)$st, jth, and $(j + 1)$st bodies (exercise). The argument (and Figure 4.18) also assumes that the jth body is not the first or the last. However one can show that when $j = 1$ or $j = N$, equation (4.80) still results, *provided* we take $x_0 \equiv 0$ or $x_{N+1} \equiv 0$ (exercise). In summary, after rearranging equation (4.80) slightly, we get the equations of motion for the entire system of bodies:

Equations of Motion:

$$m_j x_j'' = k_j x_j - (k_j + k_{j+1})x_{j-1} + k_{j+1}x_{j+1}, \tag{4.81}$$

for $j = 1, \ldots, N$. This is a linear system of 2nd-order differential equations and can be written in matrix form as

$$x'' = Kx,$$

where $x = (x_1, \ldots, x_N)$ is the vector whose components are the displacements of the individual bodies from equilibrium, and K is an $N \times N$ matrix. It will be convenient to express K as a product

$$K = \mu^{-1}B,$$

where μ is the diagonal matrix

$$
\mu = \begin{bmatrix} m_1 & & & \\ & m_2 & & \\ & & \ddots & \\ & & & m_n \end{bmatrix}, \tag{4.82}
$$

called the *mass matrix*, and B is the tridiagonal matrix:

$$
\begin{bmatrix}
-(k_1 + k_2) & k_2 & 0 & & & \\
k_2 & -(k_2 + k_3) & k_3 & 0 & & \\
0 & k_3 & -(k_3 + k_4) & k_4 & 0 & \\
& \ddots & & \ddots & & \ddots & \\
& & & k_{n-1} & -(k_{n-1} + k_n) & k_n \\
& & & 0 & k_n & -(k_n + k_{n+1})
\end{bmatrix}
$$
$$\tag{4.83}$$

The matrix B is called a tridiagonal matrix because the only nonzero entries are the entries on the diagonal, superdiadiagonal, and subdiagonal. Such matrices arise in many other important applications and have been well-studied (cf. [Par 80], [Fi 86]). We shall see later that B has many interesting properties for the case when all the spring constants are the same, say $k_j = 1$, for $j = 1, \ldots, N$.

As we did above with two bodies, we use the standard procedure to reduce the 2nd-order system $x'' = Kx$ to a 1st-order system by introducing the velocities $v \equiv x'$. This gives

$$
\begin{bmatrix} x' \\ v' \end{bmatrix} = \begin{bmatrix} 0 & I \\ K & 0 \end{bmatrix} \begin{bmatrix} x \\ v \end{bmatrix}, \tag{4.84}
$$

where I and 0 denote the $N \times N$ identity matrix and zero matrix, respectively. In summary: the 1st-order linear system for the motion of the n-boies is $z' = Az$, where

$$
A = \begin{bmatrix} 0 & I \\ K & 0 \end{bmatrix}.
$$

The special form of the matrix A, which is typical of systems that arise in physical problems involving 2nd-order derivatives, will be studied in the exercises.

It's important to note that the Complex Eigenbasis Theorem, when slightly modified (in interpretation) is a theorem that contains both the real and complex eigenbasis theorems.

Corollary 4.3 (Real and Complex Eigenbasis Theorem) *Suppose A is an $n \times n$ real matrix with real eigenvalues $\lambda_j, j = 1, \ldots, q$ and complex eigenvalues $a_j \pm b_j i, j = 1, \ldots, p$. Suppose further that*

$$n = q + 2p.$$

Assume that there are corresponding real eigenvectors $v_j, j = 1, \ldots, q$ and complex eigenvectors $u_j + i w_j, j = 1, \ldots, p$, that are linearly independent over $\mathbb{R}$ and $\mathbb{C}$, respectively. Then

$$\{v_1, \ldots, v_q, u_1, w_1, \ldots, u_p, w_p\},$$

is a basis for $\mathbb{R}^n$ and for any $c \in \mathbb{R}^n$, the integral curve of the system $x' = Ax$ that passes through c at time $t = 0$ is given by

$$\phi_t(c) = \sum_{j=1}^{q} m_j e^{\lambda_j t} v_j \qquad (4.85)$$

$$+ \sum_{j=1}^{p} e^{a_j t} \left[\cos(b_j t) \left(k_j u_j + h_j w_j \right) + \sin(b_j t) \left(h_j u_j - k_j w_j \right) \right],$$

where the real numbers m_j, k_j, h_j are the components of c with respect to the basis, i.e.,

$$c = \sum_{j=1}^{q} m_j v_j + \sum_{j=1}^{p} (k_j u_j + h_j w_j).$$

When a system has real and complex eigenvalues, the geometry of its integral curves can be interesting and complicated. Combining the straight-line motion in a real eigenspace with the elliptical or spiral motion in a complex pseudo-eigenspace gives motions which are helical or spiraling-helical, as the next example shows.

Example 4.15 Consider the system $x' = Ax$ where

$$A = \begin{bmatrix} -1 & 5 & -2 \\ -5 & -3 & -2 \\ 3 & 3 & -2 \end{bmatrix}.$$

To find the eigenvalues λ we compute the determinant

$$|A - \lambda I| = \begin{vmatrix} -(\lambda + 1) & 5 & -2 \\ -5 & -(\lambda + 3) & -2 \\ 3 & 3 & -(\lambda + 2) \end{vmatrix}$$

$$= \left\{ \begin{array}{l} -(\lambda + 1) \begin{vmatrix} -(\lambda + 3) & -2 \\ 3 & -(\lambda + 2) \end{vmatrix} \\[18pt] -5 \begin{vmatrix} -5 & -2 \\ 3 & -(\lambda + 2) \end{vmatrix} - 2 \begin{vmatrix} -5 & -(\lambda + 3) \\ 3 & 3 \end{vmatrix} \end{array} \right.$$

$$= -(\lambda^3 + 6\lambda^2 + 48\lambda + 80)$$

We can guess at one root of $\lambda^3 + 6\lambda^2 + 48\lambda + 80 = 0$ (small integers) and get $\lambda = -2$. Then dividing $\lambda^3 + 6\lambda^2 + 48\lambda + 80$ by $\lambda + 2$ gives the factorization

$$\lambda^3 + 6\lambda^2 + 48\lambda + 80 = (\lambda + 2)(\lambda^2 + 4\lambda + 40).$$

Using the quadratic formula on $\lambda^2 + 4\lambda + 40 = 0$ gives the other two roots $\lambda = -2 \pm 6i$. The eigenvectors are found as follows.

For eigenvalue $\lambda = -2$, the eigenvector equation is

$$(A + 2I)v = \begin{bmatrix} 1 & 5 & -2 \\ -5 & -1 & -2 \\ 3 & 3 & 0 \end{bmatrix} \begin{bmatrix} x \\ y \\ z \end{bmatrix} = 0.$$

This gives three equations, the third of which is $3x_1 + 3x_2 = 0$. So we take $x_1 = 1, x_2 = -1$. Using this in one of the other equations gives $x_3 = -2$. Thus, an eigenvector is

$$v = \begin{bmatrix} 1 \\ -1 \\ -2 \end{bmatrix},$$

which you can easily check. A more systematic way of solving the above system would be to use Gaussian elimination.

For eigenvalue $\lambda = -2 + 6i$, the eigenvector equation is

$$[A + (2 - 6i)I]v = \begin{bmatrix} 1 - 6i & 5 & -2 \\ -5 & -1 - 6i & -2 \\ 3 & 3 & -6i \end{bmatrix} \begin{bmatrix} x \\ y \\ z \end{bmatrix} = 0$$

To find an eigenvector v, we use Gaussian elimination on the above matrix. For this, we first divide the last row by 3 to get:

$$\begin{bmatrix} 1-6i & 5 & -2 \\ -5 & -1-6i & -2 \\ 1 & 1 & -2i \end{bmatrix}.$$

Now, using elementary row operations, we create zeros in the first column where the entries -5 and 1 occur. We multiply the 2nd row by $1-6i$ and add to it 5 times the first row. Then we multiply the 3rd row by $1-6i$ and add to it -1 times the first row. This gives the reduced matrix

$$\begin{bmatrix} 1-6i & 5 & -2 \\ 0 & -12 & -12+12i \\ 0 & -4-6i & -10-2i \end{bmatrix}.$$

You will have to do some scratch work and complex arithmetic to verify this. Now we divide the 2nd row by 12 and the 3rd row by 2, giving

$$\begin{bmatrix} 1-6i & 5 & -2 \\ 0 & -1 & -1+i \\ 0 & -2-3i & -5-i \end{bmatrix}.$$

Finally, we multiply the 2nd row by $2+3i$ and add it to -1 times the 3rd row to get

$$\begin{bmatrix} 1-6i & 5 & -2 \\ 0 & -1 & -1+i \\ 0 & 0 & 0 \end{bmatrix}.$$

This is sufficiently reduced to get an eigenvector. The 2nd row of the above matrix corresponds to the equation $-x_2+(-1+i)x_3 = 0$. So we take $x_3 = 1$ and $x_2 = -1+i$. Using this in the equation corresponding to the 1st row gives $(1-6i)x_1+5(-1+i)-2 = 0$. Solving this gives $x_1 = (7-5i)/(1-6i) = 1+i$. In summary, we get an eigenvector

$$v = \begin{bmatrix} 1+i \\ -1+i \\ 1 \end{bmatrix} = \begin{bmatrix} 1 \\ -1 \\ 1 \end{bmatrix} + \begin{bmatrix} 1 \\ 1 \\ 0 \end{bmatrix} i,$$

with real and imaginary parts:

$$u = \begin{bmatrix} 1 \\ -1 \\ 1 \end{bmatrix}, \quad w = \begin{bmatrix} 1 \\ 1 \\ 0 \end{bmatrix}.$$

From this, and the work above, we get that the general solution of the system is

$$\alpha(t) \ = \ ae^{-2t}v + e^{-2t}\cos 6t\,(ku + hw) + e^{-2t}\sin 6t\,(hu - kw)$$

$$= \ e^{-2t}\begin{bmatrix} a \\ -a \\ -2a \end{bmatrix} + e^{-2t}\cos 6t \begin{bmatrix} k+h \\ -k+h \\ k \end{bmatrix} + e^{-2t}\sin 6t \begin{bmatrix} h-k \\ -h-k \\ h \end{bmatrix}$$

$$= \ \begin{bmatrix} ae^{-2t} + (k+h)e^{-2t}\cos 6t + (h-k)e^{-2t}\sin 6t) \\ -ae^{-2t} + (-k+h)e^{-2t}\cos 6t - (h+k)e^{-2t}\sin 6t) \\ -2ae^{-2t} + ke^{-2t}\cos 6t + he^{-2t}\sin 6t) \end{bmatrix}$$

where a, k, h are arbitrary constants. This is a complicated expression, but is easiest to analyze as written in the first line.

The vector v spans the eigenspace $E_{-2} = \text{span}\{v\}$, which is 1-dimensional (a line through the origin) and integral curves that start in E_{-2} run toward the origin on this line. The vectors u, w span the 2-dimensional pseudo-eigenspace $E_{-2\pm 6i} = \text{span}\{u, w\}$, which is plane through the origin, and integral curves that start in $E_{-2\pm 6i}$ spiral toward the origin, with the rotation determined by turning u onto $-w$.

For a general initial point $c \in E_{-2} \oplus E_{-2\pm 6i} = \mathbb{R}^3$, the integral curve $e^{At}c$ will spiral about the line E_{-2} as it approaches the plane $E_{-2\pm 6i}$ (and the origin). This is shown in the sketch in Figure 4.19. In this particular example the line $E_{-2} = \text{span}\{v\}$ is perpendicular to the plane $E_{-2\pm 6i} = \text{span}\{u, w\}$ since v is perpendicular to both u and w (as is easily checked). This makes the phase portrait easier to view. The worksheet linearDEplot.mws shows how to study the system in this example using a computer.

In the above corollary (the Real and Complex EigenBasis Theorem) the eigenvalues are *not* assumed to be distinct. If we relabel so that $\mu_1, \ldots, \mu_r$ is a list of the distinct real eigenvalues and $p_1 \pm q_1 i, \ldots, p_s \pm q_s i$ is a list of the distinct complex eigenvalues, then one can show, *with the assumptions in the corollary*, that $\mathbb{R}^n$ decomposes into a direct sum of eigenspaces and pseudo-eigenspaces:

$$\mathbb{R}^n = E_{\mu_1} \oplus \cdots \oplus E_{\mu_r} \oplus E_{p_1 \pm q_1 i} \oplus \cdots \oplus E_{p_s \pm q_s i}.$$

Caution: The setting for the corollary is still very limited and, while it does cover many important cases, you should not be misled into thinking

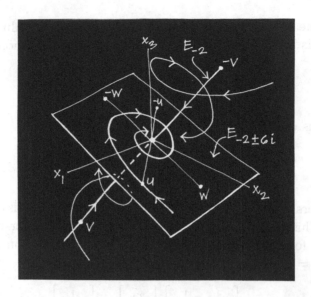

Figure 4.19: *A by-hand sketch of the phase portrait for a 3d-system* $x' = Ax$, *where A has eigenvalues* $\lambda = -2, -2 + \pm 6i$

the general case will be similar. In general we can *not* express the integral curves of $x' = Ax$ as in Formula (4.85), and we can *not* decompose $\mathbb{R}^n$ into a direct sum of eigenspaces and pseudo-eigenspaces. What we can do is decompose $\mathbb{R}^n$ into

$$\mathbb{R}^n = GE_{\mu_1} \oplus \cdots \oplus GE_{\mu_r} \oplus GE_{p_1 \pm q_1 i} \oplus \cdots \oplus GE_{p_s \pm q_s i},$$

a direct sum of *generalized* eigenspaces and pseudo-eigenspaces, each of which generally is a larger subspace containing the corresponding eigenspace or pseudo-eigenspace. This decomposition is discussed in Appendix C.

Example 4.16 A simple example of a generalized eigenspace occurs in the system $x' = Ax$, where

$$A = \begin{bmatrix} -3 & 1 \\ -1 & -5 \end{bmatrix}.$$

The eigenvalue equation is

$$\det(A - \lambda I) = \begin{vmatrix} -(\lambda + 3) & 1 \\ -1 & -(\lambda + 5) \end{vmatrix} = (\lambda + 4)^2 = 0.$$

So $\lambda = -4$ is the only eigenvalue. This, by itself, does not give any information about the eigenspace E_{-4}, which could be two-dimensional if there

ere two linearly independent eigenvectors for this eigenvalue. To determine whether this is the case, we look at the eigenvector equation

$$(A + 4I)v = \begin{bmatrix} 1 & 1 \\ -1 & -1 \end{bmatrix} \begin{bmatrix} x_1 \\ x_2 \end{bmatrix} = 0.$$

One obvious eigenvector is

$$v_1 = \begin{bmatrix} 1 \\ -1 \end{bmatrix},$$

and all others are multiples of this. Thus, $E_{-4} = \text{span}\{v_1\}$ is one-dimensional (a line). This gives part of the phase portrait. To understand the rest of it, we look for a *generalized* eigenvector v_2, i.e., a solution of the equation $(A + 4I)v_2 = v_1$. Written out explicitly, this equation is

$$\begin{bmatrix} 1 & 1 \\ -1 & -1 \end{bmatrix} \begin{bmatrix} x_1 \\ x_2 \end{bmatrix} = \begin{bmatrix} 1 \\ -1 \end{bmatrix}.$$

It is easy to see that

$$v_2 = \begin{bmatrix} 0 \\ 1 \end{bmatrix},$$

is a solution of this. Note that v_2 is not an eigenvector for $\lambda = -4$, since it doesn't satisfy $(A + 4I)v = 0$. However, from $(A + 4I)v_2 = v_1$, we do get that v_2 satisfies

$$(A + 4I)^2 v_2 = (A + 4I)(A + 4I)v_2 = (A + 4I)v_1 = 0.$$

This is why v_2 is called a generalized eigenvector. The generalized eigenspace is

$$GE_{-4} = \{v \in \mathbb{R}^2 \mid (A + 4I)^p v = 0, \text{ for some } p\}.$$

In this example, since $v_1, v_2 \in GE_{-4}$ and are linearly independent, we have that $GE_{-4} = \mathbb{R}^2$.

To understand the geometry of the other integral curves in the phase space, we compute $e^{At} v_2$. For this, we start with

$$Av_2 = v_1 + \lambda v_2,$$

(where $\lambda = -4$) and apply A to both sides:

$$A^2 v_2 = Av_1 + \lambda A v_2 = \lambda v_1 + \lambda(v_1 + \lambda v_2) = 2\lambda v_1 + \lambda^2 v_2.$$

Applying A to both sides of this gives

$$A^3 v_2 = 2\lambda A v_1 + \lambda^2 A v_2 = 2\lambda^2 v_1 + \lambda^2 (v_1 + \lambda v_2) = 3\lambda^2 v_1 + \lambda^3 v_2.$$

By induction we get

$$A^k v_2 = k\lambda^{k-1} v_1 + \lambda^k v_2,$$

for $k = 1, 2, \ldots$ Consequently

$$
\begin{aligned}
e^{At} v_2 &= \sum_{k=0}^{\infty} \frac{t^k}{k!} A^k v_2 \\
&= v_2 + \sum_{k=1}^{\infty} \frac{t^k}{k!} (k\lambda^{k-1} v_1 + \lambda^k v_2) \\
&= t e^{\lambda t} v_1 + e^{\lambda t} v_2 \\
&= e^{\lambda t} (t v_1 + v_2)
\end{aligned}
$$

The geometry of the integral curve $\alpha(t) = e^\lambda (t v_1 + v_2)$ is analyzed as follows. First note that the curve $\beta(t) = t v_1 + v_2$ is the straight line through the point v_2 in the direction of the vector v_1. This is shown in Figure 4.20.

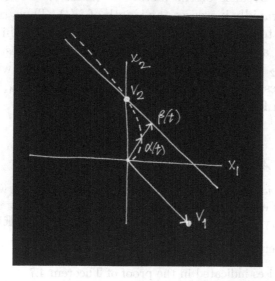

Figure 4.20: *The curves* $\alpha(t) = e^{\lambda t}(t v_1 + v_2)$ *and* $\beta(t) = t v_1 + v_2$

Now view $\alpha(t)$ and $\beta(t)$ as position vectors, with their tips directed toward points on the respective curves as they are traced out. Both vectors $\alpha(t), \beta(t)$ point to v_2 at time $t = 0$, but for positive times the length of $\alpha(t)$

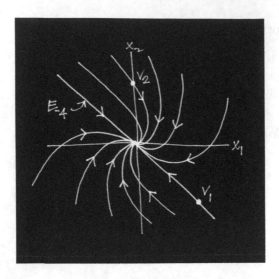

Figure 4.21: *The phase portrait of the system* $x_1' = -3x_1 + x_2$, $x_2' = -x_1 - 5x_2$.

decreases rapidly because of the factor $e^{\lambda t}$ (since $\lambda = -4$). For negative times this factor rapidly increases the length of $\alpha(t)$. Thus, $\alpha(t)$ traces out a curve like the dotted curve shown in Figure 4.20 above. This is the integral curve that starts at v_2 at time $t = 0$, and integral curves that start at a multiple kv_2 of v_2 will be similar. Sketching these, along with the straight line integral curves that arise from starting at point in E_{-4}, gives the phase portrait shown in Figure 4.21.

Exercises 4.4

1. Show that if $Av = \lambda v$, then $e^{At}v = e^{\lambda t}v$.

2. Suppose $v = u + iw$ is an eigenvector for the (real) matrix A which corresponds to a complex eigenvalue $\lambda = a + bi$, with $b \neq 0$. Show that u and w (the real and imaginary parts of v) are linearly independent (over $\mathbb{R}$).

3. Prove Theorem 4.6.

4. Do the exercises indicated in the proof of Theorem 4.7.

5. For each of the following linear systems, $x' = Ax$, find the eigenvalues and eigenvectors of A, and use these to write down the general solution of the system. Also sketch (by hand) the phase portrait, showing the eigenspaces E_λ, $E_{a \pm bi}$, and the directions along the integral curves. In each case draw enough integral curves to display how the system is behaving.

(a) $\begin{bmatrix} 1 & 0 \\ 0 & -1 \end{bmatrix}$, (b) $\begin{bmatrix} -1 & 2 \\ 2 & -1 \end{bmatrix}$, (c) $\begin{bmatrix} -1 & -2 \\ 2 & -1 \end{bmatrix}$

(d) $\begin{bmatrix} 2 & 0 \\ 0 & 1 \end{bmatrix}$, (e) $\begin{bmatrix} -1 & 0 \\ 0 & -2 \end{bmatrix}$, (f) $\begin{bmatrix} -1 & 0 \\ 0 & -1 \end{bmatrix}$

(g) $\begin{bmatrix} 0 & -2 \\ 2 & 0 \end{bmatrix}$, (h) $\begin{bmatrix} -3 & -2 \\ 17 & 3 \end{bmatrix}$, (i) $\begin{bmatrix} 1 & 3 \\ 2 & 6 \end{bmatrix}$.

(j) $\begin{bmatrix} -9 & 3 \\ 1 & -11 \end{bmatrix}$, (k) $\begin{bmatrix} 2 & 16 \\ 5 & 4 \end{bmatrix}$, (l) $\begin{bmatrix} 2 & 1 & 0 \\ -1 & 2 & 0 \\ 0 & 0 & 3 \end{bmatrix}$

(m) $\begin{bmatrix} -3 & -2 & 0 \\ 17 & 3 & 0 \\ 0 & 0 & -1 \end{bmatrix}$, (n) $\begin{bmatrix} -1 & -10 & -8 \\ 2 & 2 & 7 \\ 4 & -5 & -4 \end{bmatrix}$, (o) $\begin{bmatrix} 1 & -3 & -3 \\ -5 & -1 & 5 \\ 2 & 2 & -4 \end{bmatrix}$

Hints: For (n): One eigenvalue is $\lambda = -3$. Use a view $[-45, 60]$. Are the eigenspaces perpendicular? For (o): The eigenvalues are $\lambda = -2, -6, 4$. Use a view $[-135, 45]$. Do a sketch above the x_1-x_2 plane only. Sketch the integral curves in the planes $E_{-2} \oplus E_{-6}$, $E_{-2} \oplus E_4$, and $E_{-6} \oplus E_4$.

6. For each of the systems $x' = Ax$ in Exercise 5 above, use a computer to study and draw the phase portrait of the system. Make sure you draw all the integral curves that are straight lines and sufficiently many other integral curves to adequately delineate the main features of the phase portrait. On the printouts of your studies, annotate and mark all the features (eigenspaces, directions of flow on the integral curves, etc.)

7. Suppose K is any $n \times n$ matrix and let A be the matrix (given in block form):

$$A = \begin{bmatrix} 0 & I \\ K & 0 \end{bmatrix}, \tag{4.86}$$

where 0 and I denote the $n \times n$ zero matrix and identity matrix, respectively. Use the technique from Example 4.13 to show that

$$\det(A - \lambda I) = (-1)^n \det(K - \lambda^2 I).$$

Hence conclude that the eigenvalues of A are $\lambda_j = \pm\sqrt{\mu_j}$, $j = 1, \ldots, n$, where μ_j, $j = 1, \ldots, n$ are the eigenvalues of K. Also show that if u is an eigenvector of K corresponding to eigenvalue μ, then $(u, \sqrt{\mu}\, u)$ is an eigenvector of A corresponding to $\lambda = \sqrt{\mu}$. Show how these results give the eigenvalues and eigenvectors in Example 4.13.

8. **(Two Coupled Masses)** Use the work done in Example 4.13 to further study the dynamics of the system of two equal masses coupled by three identical springs. In particular, suppose α is the curve:

$$\alpha(t) = \left(k_1 \cos t + h_1 \sin t \right) v_1 + \left(k_2 \cos \sqrt{3}\, t + h_2 \sin \sqrt{3}\, t \right) v_2,$$

that gives the two positions x_1, x_2 of each mass at time t.

   (a) Use a computer to plot the curve α for each of the following sets of initial conditions:

   (i) $x_1(0) = .5$, $x_2(0) = 1$, $x_1'(0) = 0$, $x_2'(0) = 0$.
   (ii) $x_1(0) = -.5$, $x_2(0) = 1$, $x_1'(0) = 0$, $x_2'(0) = 0$.
   (iii) $x_1(0) = .5$, $x_2(0) = 1$, $x_1'(0) = 0.2$, $x_2'(0) = 0$.
   (iv) $x_1(0) = 0$, $x_2(0) = 0$, $x_1'(0) = 0.5$, $x_2'(0) = 1$.

   In each case use the following time intervals: $[0, 10]$, $[0, 50]$, $[0, 100]$.

   (b) For each of the ones assigned in part (a) plot, in the same figure, x_1 as a function of t and x_2 as a function of t.

   (c) Annotate and label your drawings and use these to analyze and describe the motion of the system of masses. Also use the code on the worksheet oscillate.mws to animate both the motion of α and the actual motion of the two masses. This should help in writing your description.

9. **(Three Coupled Masses)** This exercise is to study the system of three masses: $m_j = 1$, $j = 1, 2, 3$, coupled by four springs with equal spring constants $k_j = 1$, $j = 1, 2, 3$. Using Example 4.13 as a guide, do the following part in your study:

   (a) Write done matrices K and A for the 2nd- and 1st-order systems $x'' = Kx$ and $z' = Az$, that govern the motion. Determine all the eigenvalues and corresponding eigenvectors of K and then use the result in Exercise 9 above to find the eigenvectors/eigenvalues of A. Use this to write the explicit formula for the general solution of both $x'' = Kx$ and $z' = Az$.

   (b) Determine the three normal modes η_1, η_2, η_3 of vibration for the system of masses. Plot each of these space curves and describe what each of the corresponding motions of the three masses is like. Look at the projections on the coordinate planes.

   (c) Study the motion of the three masses that results from the initial displacements $x_1(0) = 1$, $x_2(0) = 0.5$, $x_3(0) = -1$, and initial velocities all zero $x_j'(0) = 0$, $j = 1, 2, 3$.

11. **(N Coupled Masses)** In the case of N equal masses, say $m_j = 1$, $j = 1, \ldots, N$, and identical springs, say $k_j = 1$, $j = 1, \ldots, N + 1$, it is possible to

explicitly determine the formula for the general solution of the equations of motion. This exercise shows you how. The crux of the matter is finding the eigenvalues and eigenvectors of the tridiagonal matrix

$$
K = \begin{bmatrix}
-2 & 1 & 0 & & & & \\
1 & -2 & 1 & 0 & & & \\
0 & 1 & -2 & 1 & 0 & & \\
& & \ddots & \ddots & \ddots & & \\
& & & 1 & -2 & 1 \\
& & & 0 & 1 & -2
\end{bmatrix}. \tag{4.87}
$$

A direct approach to this will be difficult, so take the following indirect approach. First write K as

$$
K = -2I + M,
$$

where I is the $N \times N$ identity matrix and $M \equiv K + 2I$. Thus, M has the same form as K except that it has zeros on its diagonal. To find eigenvalues and eigenvectors for M is easier. For this show that the vector equation

$$
Mv = \mu v,
$$

is equivalent to the system of equations:

$$
\begin{align}
v_2 &= \mu v_1 \tag{4.88} \\
v_{p-1} + v_{p+1} &= \mu v_p \quad (p = 2, \ldots, N-1) \tag{4.89} \\
v_{N-1} &= \mu v_N, \tag{4.90}
\end{align}
$$

where $v = (v_1, \ldots, v_N)$. To determine numbers $\mu, v_1, \ldots, v_N$ that satisfy this, we make the key observation that equation (4.89) has the same form as the trig identity

$$
\sin(\psi + \theta) + \sin(\psi - \theta) = 2 \cos \theta \sin \psi.
$$

Use this to show that for any θ, the $N + 1$ numbers

$$
\begin{align}
\mu &= 2 \cos \theta \tag{4.91} \\
v_p &= \sin p\theta, \quad (p = 1, \ldots, N), \tag{4.92}
\end{align}
$$

satisfy equations (4.88)-(4.89). Then show that in order for equation (4.90) to be satisfied, it is necessary and sufficient that θ be a number of the form

$$
\theta_j = \frac{j\pi}{N+1}, \quad (j = 1, \ldots, N). \tag{4.93}
$$

Use all of this to complete the proof of the following theorem.

Theorem 4.9 *The eigenvalues* $\mu_1, \ldots, \mu_N$ *of the* $N \times N$, *tridiagonal matrix*

$$M = \begin{bmatrix} 0 & 1 & 0 & & & \\ 1 & 0 & 1 & 0 & & \\ 0 & 1 & 0 & 1 & 0 & \\ & \ddots & \ddots & \ddots & & \\ & & 1 & 0 & 1 \\ & & & 0 & 1 & 0 \end{bmatrix}$$

are

$$\mu_j = 2\cos\theta_j, \qquad (j = 1, \ldots, N), \tag{4.94}$$

where

$$\theta_j = \frac{j\pi}{N+1}, \qquad (j = 1, \ldots, N). \tag{4.95}$$

Furthermore, corresponding eigenvectors $v^{(1)}, \ldots, v^{(N)}$ *are given by*

$$v^{(j)} = (\sin\theta_j, \sin 2\theta_j, \ldots, \sin N\theta_j), \tag{4.96}$$

for $j = 1, \ldots, N$.

Next since $A = -2I + M$, it is easy to see that

$$Mv = \mu v \quad \Longrightarrow \quad Av = (-2 + \mu)v.$$

Using this and the result in Exercise 9 above, prove the following theorem.

Theorem 4.10 *Let* K *be the* $N \times N$, *tridiagonal matrix in equation* (4.87). *Then the general solution of the system* $x'' = Kx$ *is*

$$\alpha(t) = \sum_{j=1}^{N} (k_j \cos b_j t + \ell_j \sin b_j t) v^{(j)}, \tag{4.97}$$

where

$$b_j = \sqrt{2} \left(1 - \cos\frac{j\pi}{N+1} \right)^{1/2}, \tag{4.98}$$

for $j = 1, \ldots, N$, *and*

$$v^{(j)} = \left(\sin\frac{j\pi}{N+1}, \sin\frac{2j\pi}{N+1}, \ldots, \sin\frac{Nj\pi}{N+1} \right), \tag{4.99}$$

for $j = 1, \ldots, N$.

12. Use the techniques from Exercise 11 to find the eigenvalues and corresponding eigenvectors of the $N \times N$, tridiagonal matrix

$$K = \begin{bmatrix} a & b & 0 & & & & \\ c & a & b & 0 & & & \\ 0 & c & a & b & 0 & & \\ & & \ddots & \ddots & \ddots & & \\ & & & c & a & b \\ & & & 0 & c & a \end{bmatrix} \qquad (4.100)$$

where a, b, c are real numbers and $bc > 0$. *Hint:* Write $K = aI + M$ and write out the equations like equations (4.88)-(4.90) for $Mv = \mu v$. Then look for solutions of these equations of the form

$$\mu = 2g \cos \theta \qquad (4.101)$$
$$v_p = h^p \sin p\theta, \qquad (p = 1, \ldots, N), \qquad (4.102)$$

where θ, g and h are numbers to be determined.

13. **(Generalized Eigenvectors)** Suppose A is a 3×3 matrix with real eigenvalue λ of multiplicity three and corresponding eigenvector v_1. Assume that the eigenspace is one-dimensional, $E_\lambda = \text{span}\{v_1\}$, and suppose that v_2, v_3 are vectors such that

$$(A - \lambda I)v_2 = v_1$$
$$(A - \lambda I)v_3 = v_2$$

As in Example 4.16, one can show that $e^{At}v_2 = e^{\lambda t}(tv_1 + v_2)$. Using this and $(A - \lambda I)v_1 = 0$, show that

$$e^{At}v_3 = e^{\lambda t}\left(\frac{1}{2}t^2 v_1 + tv_2 + v_3\right).$$

Assuming v_1, v_2, v_3 are linearly independent, find a formula for the flow $\phi_t(c) = e^{At}c$, where $c = k_1 v_1 + k_2 v_2 + k_3 v_3$ is any point in $\mathbb{R}^3 = \text{span}\{v_1, v_2, v_3\} = GE_\lambda$.

4.5 Canonical Systems

One could argue that the chapter is closed on our study of the linear system $x' = Ax$, since, after all, we have explicitly exhibited its general solution: $\phi(t, c) = e^{At}c$, constructed from the matrix exponential. However, this matrix exponential is only given theoretically, and in most cases it is impossible to compute the entries of the matrix e^{At} effectively. Even if one is satisfied that, *in theory*, e^{At} is known, and, *in practice*, is often computable, nevertheless there is a good deal more to understand and clarify about how these solutions of $x' = Ax$ behave.

Additionally, as we have seen in the theorems from the previous section, the qualitative nature of the phase portrait, consisting of straight-line, elliptical, and spiral integral curves (and combinations thereof) was derived only for the case where there enough linearly independent, complex eigenvectors to get a (real) basis for $\mathbb{R}^n$.

One important further object of study in this regard deals with canonical systems $y' = Jy$, and the theory for transforming a given system: $x' = Ax$, into one of the canonical systems. The motive here should be quite clear: canonical systems $y' = Jy$ are particularly simple in form, and therefore we can:

(1) easily compute e^{Jt}, and

(2) easily understand the qualitative nature of the canonical phase portrait.

Then, as we shall see, the phase portrait of the given system, $x' = Ax$, will be similar to the portrait of its corresponding canonical system, only just a linear distortion of it. In essence, the idea is this: rather than studying all possible systems $x' = Ax$, we need only study a few of the simplest type: the canonical systems $y' = Jy$.

Before discussing what canonical systems are, we look at the transformation technique (not the general technique, but rather what we need here for linear systems).

Definition 4.4 The linear system: $x' = Ax$ is called *linearly equivalent* to the system $y' = By$, if the matrix A is *similar* to the matrix B, i.e., if there exists an invertible linear transformation $P : \mathbb{R}^n \to \mathbb{R}^n$, such that

$$P^{-1}AP = B. \tag{4.103}$$

If $X(x) = Ax$ and $Y(y) = By$ are the two vector fields associated with the linear systems, then one can view (4.103) as a transformation law between the respective vector fields. Later when we study nonlinear systems, we will find that (4.103) is really a special case of a more general way of transforming one vector field into another.

The importance of this notion of linear equivalence is that linearly equivalent systems have similar phase portraits. This is made precise by the following theorem which gives an alternative, more geometric, condition for linear equivalence.

Theorem 4.11 *The two systems $x' = Ax$ and $y' = By$ are linearly equivalent if and only if there is an invertible linear transformation P that maps the integral curves of the one system into integral curves of the other system. More precisely: for every integral curve $\beta : I \to \mathbb{R}^n$ of $y' = By$, the curve α defined by*

$$\alpha(t) = P\beta(t)$$

is an integral curve of $x' = Ax$.

Proof: Suppose first that the systems are linearly equivalent, so that $P^{-1}AP= B$ for some invertible P. Then for any integral curve $\beta : I \to \mathbb{R}^n$ of $y' = By$, the curve defined by $\alpha(t) = P\beta(t)$, for $t \in I$, is an integral curve of $x' = Ax$. This follows from the elementary computation:

$$\begin{aligned} \alpha'(t) &= P\beta'(t) = PB\beta(t) \\ &= PBP^{-1}\alpha(t) = A\alpha(t), \end{aligned}$$

for every $t \in I$.

On the other hand, suppose P is an invertible linear transformation that maps each integral curve β of $y' = By$ into an integral curve of $x' = Ax$ according to the prescription: $\alpha(t) = P\beta(t)$. From this we wish to show that $P^{-1}AP = B$. To get at this, note that for any $b \in \mathbb{R}^n$, the curve $\alpha(t) \equiv Pe^{Bt}b$ must be an integral curve of $x' = Ax$, and in fact an integral curve that satisfies $\alpha(0) = Pb$. But this is also true of the curve $\phi(t) \equiv e^{At}Pb$. Thus, by the Existence and Uniqueness Theorem: $\alpha(t) = \phi(t)$ for all t, i.e.,

$$Pe^{Bt}b = e^{At}Pb,$$

for all t. Since this holds for all vectors b, it follows that

$$Pe^{Bt} = e^{At}P,$$

for all t. Differentiating both sides of this identity gives

$$PBe^{Bt} = Ae^{At}P,$$

and taking $t = 0$ in this yields $PB = AP$, or equivalently $B = P^{-1}AP$, as desired. ☐

The theorem gives us a reason to look for a matrix B in the similarity class:

$$[A] = \{\, P^{-1}AP \mid P \text{ is invertible}\,\},$$

which has a particularly simple form. The study of the phase portrait of $x' = Ax$ is reduced to the simpler case $y' = By$, where $B = P^{-1}AP$. The correspondence between the respective phase portraits is given by

$$x = Py.$$

Also note that since A and B are related by $A = PBP^{-1}$, results from Appendix C give

$$
\begin{aligned}
e^{At}c &= e^{PBP^{-1}t}c \\
&= Pe^{Bt}P^{-1}c,
\end{aligned}
$$

for every $t \in \mathbb{R}$ and $c \in \mathbb{R}^n$. This gives us a way to compute the general solution: $\phi_t(c) = e^{At}c$, of $x' = Ax$ provided we know how to compute e^{Bt}.

Out of all the matrices similar to a given matrix A we will choose the one known as the Jordan canonical form J for A. This is our choice of *simplest* form in the similarity classes, even though one could use some other choice as well. In this case the computation of e^{Jt} is quite simple and the corresponding canonical system $y' = Jy$ is has the simplest possible phase portrait.

The description of the Jordan canonical form for a matrix A is somewhat complicated. As a motivation, we first discuss two special classes of matrices where the description is easy.

4.5.1 Diagonalizable Matrices

When A is a diagonalizable matrix, its Jordan canonical form is, as we shall see, an appropriate diagonal matrix $J = D$, with $P^{-1}AP = D$. This certainly seems desirable, since a diagonal matrix is most simple and the corresponding dynamical system $y' = Dy$ is as simple as one can ask for. Let's look at this in more detail.

Suppose A is an $n \times n$ matrix satisfying the hypotheses of the Real Eigenbasis Theorem, i.e., there is a basis: $v_1, \ldots, v_n$, of $\mathbb{R}^n$ consisting of eigenvectors of A: say, $Av_j = \lambda_j v_j$. This of course implies that all the eigenvalues of A, namely $\lambda_1, \ldots, \lambda_n$, are all real. Note, we are *not* assuming all the eigenvalues are distinct. From such a basis of eigenvectors $v_1, \ldots, v_n$, we can form a matrix P whose columns are the respective vectors v_j. We indicate this by:

$$P = [v_1, \ldots, v_n].$$

Note that P is an invertible matrix since $v_1, \ldots, v_n$ is a basis (exercise). Now using properties of matrix multiplication, one sees that

$$
\begin{aligned}
AP &= A[v_1, \ldots, v_n] = [Av_1, \ldots, Av_n] \\
&= [\lambda_1 v_1, \ldots, \lambda_n v_n] = [v_1, \ldots, v_n]D \\
&= PD,
\end{aligned}
\tag{4.104}
$$

where D is the diagonal matrix

$$
D = \mathrm{diag}(\lambda_1, \ldots, \lambda_n) =
\begin{bmatrix}
\lambda_1 & 0 & \cdots & 0 \\
0 & \lambda_2 & \cdots & 0 \\
\vdots & & & \vdots \\
0 & 0 & \cdots & \lambda_n
\end{bmatrix}
\tag{4.105}
$$

This shows that A is diagonalizable: $P^{-1}AP = D$, and gives an explicit construction for finding a matrix P that diagonalizes A. Conversely if A is diagonalizable, i.e., if there exists an invertible matrix P such that $P^{-1}AP = D$, where D is a diagonal matrix, then necessarily the columns of P constitute a basis of eigenvectors of A, and the diagonal entries of D are precisely the eigenvalues of A (convince yourself of this!).

The above discussion motivates the notion of the Jordan form in the very special case when A is diagonalizable. The Jordan form is just a diagonal matrix made up of the eigenvalues of A. Note, by not specifying how to order the eigenvalues along the diagonal of D, we are being rather nebulous about what D is, and in fact we get a different $J = D$ for each ordering. Otherwise said, in the diagonalizable case, a Jordan form (rather than *the* Jordan form) for A is just any one of these diagonal matrices.

Example 4.17 The matrix

$$
A =
\begin{bmatrix}
11 & -10 \\
10 & -14
\end{bmatrix}
$$

is diagonalizable, since its characteristic polynomial $p_A(\lambda) = \lambda^2 + 3\lambda - 54$
has two distinct real roots: $-9, 6$. Two corresponding eigenvectors are $v_1 = (1, 2), v_2 = (2, 1)$. It's easy to verify that if we let

$$P = \begin{bmatrix} 1 & 2 \\ 2 & 1 \end{bmatrix},$$

then

$$P^{-1}AP = \begin{bmatrix} -9 & 0 \\ 0 & 6 \end{bmatrix} \equiv J.$$

The canonical system $y' = Jy$ is as simple as possible. Its phase portrait is
shown on the right in Figure 4.22. The phase portrait of the original system:
$x' = Ax$ is shown on the left in Figure 4.22 and is readily seen to be similar
to the canonical system.

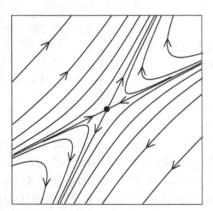 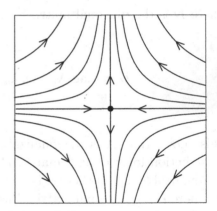

Figure 4.22: **(Left):** *Phase portrait for a linear system.* **(Right):** *Phase
portrait of the corresponding canonical system.*

The linear transformation $x = Py$ transforms the integral curves of the
canonical system into integral curves of the original system. The matrix
exponential for the canonical system is easy to compute since

$$J^k = \begin{bmatrix} -9 & 0 \\ 0 & 6 \end{bmatrix}^k = \begin{bmatrix} (-9)^k & 0 \\ 0 & 6^k \end{bmatrix}$$

for every k. From this we get

$$e^{Jt} = \sum_{k=0}^{\infty} \frac{t^k}{k!} J^k$$

$$= \sum_{k=0}^{\infty} \frac{t^k}{k!} \begin{bmatrix} (-9)^k & 0 \\ 0 & 6^k \end{bmatrix}$$

$$= \begin{bmatrix} \sum_{k=0}^{\infty} \frac{(-9t)^k}{k!} & 0 \\ 0 & \sum_{k=0}^{\infty} \frac{(6t)^k}{k!} \end{bmatrix}$$

$$= \begin{bmatrix} e^{-9t} & 0 \\ 0 & e^{6t} \end{bmatrix}.$$

Now we can compute e^{At}, using: $A = PJP^{-1}$ and an identity from the Appendix:

$$
\begin{aligned}
e^{At} &= e^{PJP^{-1}t} = Pe^{Jt}P^{-1} \\
&= \begin{bmatrix} 1 & 2 \\ 2 & 1 \end{bmatrix} \begin{bmatrix} e^{-9t} & 0 \\ 0 & e^{6t} \end{bmatrix} \begin{bmatrix} -\frac{1}{3} & \frac{2}{3} \\ \frac{2}{3} & -\frac{1}{3} \end{bmatrix} \\
&= \begin{bmatrix} -\frac{1}{3}e^{-9t} + \frac{4}{3}e^{6t} & \frac{2}{3}e^{-9t} - \frac{2}{3}e^{6t} \\ -\frac{2}{3}e^{-9t} + \frac{2}{3}e^{6t} & \frac{4}{3}e^{-9t} - \frac{1}{3}e^{6t} \end{bmatrix}.
\end{aligned}
$$

This computation is readily seen to generalize to the computation of e^{At} for any diagonalizable matrix A. The general solution of $x' = Ax$ can now be written as

$$
\begin{aligned}
\phi_t(c) &= e^{At}c \\
&= \frac{1}{3}\begin{bmatrix} (-c_1 + 2c_2)e^{-9t} + (4c_1 - 2c_2)e^{6t} \\ (-2c_1 + 4c_2)e^{-9t} + (2c_1 - c_2)e^{6t} \end{bmatrix}.
\end{aligned}
$$

Note: While it is instructive to write out the general solution in this form, it is sometimes better to write the general solution as we did before in terms of the eigenvectors. To see how this is actually contained in the above computation, we relabel the arbitrary constants as $b = (b_1, b_2) \equiv P^{-1}c$. Then

$$
\begin{aligned}
\gamma(t) &= e^{At}c = Pe^{Jt}P^{-1}c = Pe^{Jt}b \\
&= P\begin{bmatrix} b_1 e^{-9t} \\ b_2 e^{6t} \end{bmatrix} \\
&= b_1 e^{-9t}v_1 + b_2 e^{6t}v_2
\end{aligned}
$$

4.5.2 Complex Diagonalizable Matrices

As the complex analog of what we did in the last section, suppose the hypotheses of the Complex Eigenbasis Theorem hold. That is, suppose $n = 2p$ and that A is an $n \times n$ real matrix with complex (nonreal) eigenvalues $\lambda_j = a_j + b_j i$, $b_j \neq 0$, for $j = 1, \ldots, p$ and corresponding complex eigenvectors $v_j = u_j + i w_j$, $j = 1, \ldots, p$, which are linearly independent over $\mathbb{C}$. Then while it is possible to diagonalize A over the complex numbers (exercise), it is not possible to find a real matrix P, so that $P^{-1} A P$ is a diagonal matrix. However, what we can do is the following.

According to Theorem 5.6, the equations: $A v_j = \lambda_j v_j$, $j = 1, \ldots, p$, are equivalent to

$$
\begin{aligned}
A u_j &= a_j u_j - b_j w_j \\
A w_j &= b_j u_j + a_j w_j,
\end{aligned}
$$

for $j = 1, \ldots, p$. We can write this in matrix form as

$$
A [u_j, w_j] = [u_j, w_j] \begin{bmatrix} a_j & b_j \\ -b_j & a_j \end{bmatrix} \qquad (j = 1, \ldots, p),
$$

where $[u_j, w_j]$ is the $n \times 2$ matrix with columns u_j, w_j. Thus, if we let $P = [u_1, w_1, \ldots, u_p, w_p]$ be the $n \times n$ matrix with the indicated vectors as columns, then the matrix equations above can be combined into a single matrix equation:

$$
\begin{aligned}
A P &= A[u_1, w_1, \ldots u_p, w_p] \\
&= \begin{bmatrix} | & | & & | & | \\ u_1 & w_1 & \cdots & u_p & w_p \\ | & | & & | & | \end{bmatrix} \begin{bmatrix} a_1 & b_1 & & & \\ -b_1 & a_1 & & & \\ & & \ddots & & \\ & & & a_p & b_p \\ & & & -b_p & a_p \end{bmatrix}.
\end{aligned}
$$

For later use we let $C(a_j, b_j)$ is the 2×2 matrix:

$$
C(a_j, b_j) \equiv \begin{bmatrix} a_j & b_j \\ -b_j & a_j \end{bmatrix}.
$$

Consequently, P is an invertible matrix such that

$$
P^{-1} A P = J,
$$

where J is the $n \times n$, block diagonal, matrix

$$J = \begin{bmatrix} C(a_1, b_1) & 0 & \cdots & 0 \\ 0 & C(a_2, b_2) & \cdots & 0 \\ \vdots & & & \\ 0 & 0 & \cdots & C(a_p, b_p) \end{bmatrix},$$

with the 2×2 matrices $C(a_j, b_j)$ on the diagonal. While the matrix J is block diagonal, it is not diagonal since the blocks are not 1×1 matrices. But this is the best we can do with the assumptions on A, and so we take J to be the Jordan canonical form for A.

Example 4.18 As we saw in Example 4.13 for two coupled masses, the 4×4 matrix

$$K = \begin{bmatrix} 0 & 0 & 1 & 0 \\ 0 & 0 & 0 & 1 \\ -2 & 1 & 0 & 0 \\ 1 & -2 & 0 & 0 \end{bmatrix}$$

has complex eigenvalues $\lambda_1 = i, \lambda_2 = \sqrt{3}\,i$, with corresponding complex eigenvectors $v_1 = (1, 1, i, i), v_2 = (1, -1, \sqrt{3}\,i, -\sqrt{3}\,i)$, which are linearly independent over $\mathbb{C}$. According to the above discussion, the matrix

$$P \equiv \begin{bmatrix} 1 & 0 & 1 & 0 \\ 1 & 0 & -1 & 0 \\ 0 & 1 & 0 & \sqrt{3} \\ 0 & 1 & 0 & -\sqrt{3} \end{bmatrix}$$

can be used to bring A to Jordan form: $P^{-1}AP = J$, where

$$J = \begin{bmatrix} C(0, 1) & 0 \\ 0 & C(0, \sqrt{3}) \end{bmatrix} = \begin{bmatrix} 0 & 1 & 0 & 0 \\ -1 & 0 & 0 & 0 \\ 0 & 0 & 0 & \sqrt{3} \\ 0 & 0 & -\sqrt{3} & 0 \end{bmatrix}.$$

It is left as an exercise to compute P^{-1} and verify this result directly.

4.5.3 The Nondiagonalizable Case: Jordan Forms

We have just seen that when A is diagonalizable, a convenient choice for a corresponding canonical system is $y' = Jy$, where J the diagonal matrix obtained by diagonalizing A. In addition, when A is complex diagonalizable, a sensible choice of canonical system involves a J which is block diagonal with the special 2×2 matrices $C(a, b)$ on the diagonal. To generalize these results to other cases requires the theory of Jordan forms.

This theory says that (1) any given $n \times n$ matrix A is similar to a matrix J with a very special form, a Jordan form, and (2) when A is diagonalizable, J is a diagonal matrix obtained by diagonalizing A. In this subsection we describe what an $n \times n$ Jordan form is. The association of a given matrix with its corresponding Jordan form is more difficult to describe. This is done in Appendix C.

Definition 4.5 (Jordan Forms) An $n \times n$ *Jordan form* (or *Jordan canonical form*) J is a special type of $n \times n$ matrix which is built out of Jordan blocks (which are matrices of smaller size). We first describe what a Jordan block looks like. There are two types: real and complex. A *real Jordan block* (of size k) is a $k \times k$ matrix of the form:

$$J_k(\lambda) = \begin{bmatrix} \lambda & 1 & 0 & \cdots & 0 & 0 \\ 0 & \lambda & 1 & \cdots & 0 & 0 \\ \vdots & & \ddots & \ddots & & \vdots \\ 0 & 0 & 0 & \cdots & \lambda & 1 \\ 0 & 0 & 0 & \cdots & 0 & \lambda \end{bmatrix}. \tag{4.106}$$

The matrix indicated by (4.106) has λ's down the diagonal and 1's on the supradiagonal. It's not hard to see that $J_k(\lambda)$ has λ as its only eigenvalue (repeated k times), and that the corresponding eigenspace E_λ is one-dimensional, being spanned by the vector $e_1 = (1, 0, \ldots, 0)$. Furthermore, we can decompose $J_k(\lambda)$ into a sum of *commuting* matrices:

$$J_k(\lambda) = \lambda I_k + N_k, \tag{4.107}$$

where I_k is the $k \times k$ identity matrix, and N_k is a nilpotent matrix, i.e., some power of it is the zero matrix, in this case the kth power: $N_k^k = 0$. You should verify these assertions.

On the other hand, a *complex Jordan block* is a matrix of the form

$$C_{2m}(a, b) = \begin{bmatrix} C(a, b) & I_2 & 0 & 0 & \cdots & 0 \\ 0 & C(a, b) & I_2 & 0 & \cdots & 0 \\ \vdots & & \vdots & & & \vdots \\ 0 & 0 & 0 & C(a, b) & I_2 \\ 0 & 0 & 0 & & 0 & C(a, b) \end{bmatrix} \tag{4.108}$$

The matrix indicated by (4.108) is a $2m \times 2m$ matrix, given in block form in terms of the 2×2 matrices:

$$C(a, b) = \begin{bmatrix} a & b \\ -b & a \end{bmatrix}, \tag{4.109}$$

and I_2 is the 2×2 identity matrix. In (4.108), the $C(a, b)$'s are arranged along the diagonal, and the I_2 on the supradiagonal. Note that in the above notation $C(a, b) = C_2(a, b)$. You should verify the following properties of the complex Jordan block $C_{2m}(a, b)$. It has only $a \pm bi$ as eigenvalues (each repeated m times). One can decompose it into a sum of commuting matrices:

$$C_{2m}(a, b) = D_{2m}(a, b) + M_{2m}, \tag{4.110}$$

where M_{2m} is nilpotent: $M_{2m}^m = 0$.

With these preliminaries out of the way, we now define what is meant by a *Jordan canonical form*, sometimes called a *Jordan form*, or a *Jordan matrix*. It is a matrix which is either (1) a single real Jordan block $J_k(\lambda)$, or (2) a single complex Jordan block $C_{2m}(a, b)$, or (3) a matrix built up from real and complex blocks (of varying size) arranged along the diagonal, viz.:

$$J = \begin{bmatrix} J_{k_1}(\lambda_1) & & & & & \\ & \ddots & & & & \\ & & J_{k_r}(\lambda_r) & & & \\ & & & C_{2m_1}(a_1, b_1) & & \\ & & & & \ddots & \\ & & & & & C_{2m_s}(a_s, b_s) \end{bmatrix}. \tag{4.111}$$

It is understood in this description of J that there might be no real Jordan blocks, or there might be no complex Jordan blocks, but if both types are present, the real ones are listed first along the diagonal. We also agree to arrange the real blocks in order of decreasing size: $k_1 \geq \cdots \geq k_r$, and likewise for the complex blocks: $m_1 \geq \cdots \geq m_s$. Further, the λ's, a's, and b's in (4.111) are *not* assumed to be distinct. Note that if the general Jordan matrix J indicated in (4.111) is $n \times n$, then

$$n = k_1 + \cdots + k_r + 2m_1 + \cdots + 2m_s.$$

Also it is clear that the eigenvalues of J are $\lambda_1, \ldots, \lambda_r$, and $a_1 \pm b_1 i, \ldots, a_s \pm b_s i$.

That concludes the definition of what a Jordan canonical form is. Note that one possible Jordan form is a diagonal matrix. This is where there are only real blocks in (4.111) and all these blocks are of size 1: $J_1(\lambda_1), \ldots, J_1(\lambda_n)$.

The major theorem concerning Jordan forms is the following:

Theorem 4.12 (Jordan Canonical Form) *If A is any $n \times n$ matrix, then there exists an invertible matrix P, such that:*

$$P^{-1}AP = J,$$

where J is a Jordan canonical form.

Example 4.19 (2 × 2 Jordan Forms) What are all the possible 2×2 Jordan forms J? To enumerate these, consider the type (real/complex) and number of Jordan *blocks* that J could be composed of. Thus, if J has any complex blocks, there can be only one and this block must be all of J. So $J = C(a, b)$. The other possible forms of J then must consist entirely of real blocks, and J can have either (a) one such block: $J = J_2(\lambda)$, or (b) two real blocks: $J = [J_1(\lambda), J_1(\mu)]$. Thus, we find only *three* possible 2×2 Jordan forms:

$$(1) \begin{bmatrix} \lambda & 1 \\ 0 & \lambda \end{bmatrix}, \quad (2) \begin{bmatrix} \lambda & 0 \\ 0 & \mu \end{bmatrix}, \quad (3) \begin{bmatrix} a & b \\ -b & a \end{bmatrix}.$$

Note that λ, μ, a, b can be any real numbers (zero included) and that $\lambda = \mu$ in form (2) is also a possibility. Of course $b \neq 0$. Thus, while we are only listing three distinct Jordan forms in the enumeration, any one of the three can have a drastically different phase portrait depending on the signs of the constants λ, μ, a, b and whether some of them are zero or not. If we limit ourselves to the cases where the origin is a simple fixed point, i.e., $\det(J) \neq 0$, so that 0 is the only fixed point, then necessarily $\lambda \neq 0$ and $\mu \neq 0$. Figure 4.23 shows the phase portrait for Case (1), which is called an *improper node*. This name is based on the fact the matrix of type (1) has only one eigenvalue λ and the corresponding eigenspace is 1-dimensional.

The Jordan form (2), which is diagonal, has three different types of phase portraits associated with it and these depend on the nature of the eigenvalues λ, μ. See Figure 4.24. These possibilities are (2a) $\lambda = \mu$ equal, or (2b) $\lambda \neq \mu$ and have the same sign, or (2c) $\lambda \neq \mu$ and have opposite signs.

The pictures for forms (1), (2a), and (2b) are for the case where λ, μ are negative, and as you can see the flow is directed toward the fixed point at the origin. For this reason the origin is called a *stable* fixed point (we discuss

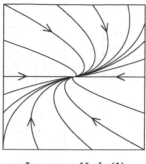

Improper Node (1)

Figure 4.23: *Canonical phase portrait for an improper node, with $\lambda < 0$.*

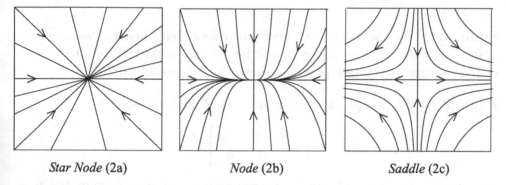

Star Node (2a)   *Node* (2b)   *Saddle* (2c)

Figure 4.24: *Canonical phase portraits for a star node ($\mu = \lambda < 0$), node ($\mu < 0, \lambda < 0$), and saddle point ($\mu < 0, \lambda > 0$).*

stability of systems later in more detail). The names *improper node, star node*, and *node* are used to refer to the three different behaviors exhibited in (1), (2a), and (2b). The fixed point that occurs in case (2c) (real eigenvalues of opposite sign), is called a *saddle* (or *hyperbolic*) point. It is an *unstable* fixed point since any integral curve starting near it will move away from it off to infinity.

The remaining two canonical phase portraits, shown in Figure 4.25, are for the complex eigenvalue cases (3a)-(3b). The fixed point in case (3a) is called a *focus* and is stable in the picture shown, which is drawn assuming $a < 0$. In the other case (3b), the fixed point is referred to as a *center* and is stable regardless of whether b is negative (case shown in the picture) or positive.

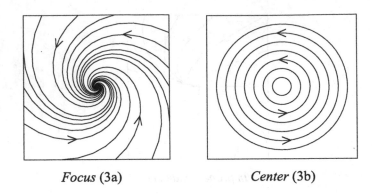

Focus (3a)   *Center* (3b)

Figure 4.25: *The canonical phase portraits corresponding to complex eigenvalues are a focus (when $a \neq 0$) and center (when $a = 0$).*

Summary: By the theory of linear equivalence any linear system in the plane: $x' = Ax$, with $\det(A) \neq 0$, will have a phase portrait similar to one of the six canonical phase portraits shown in Figures 4.23-4.25.

Example 4.20 In a previous example we saw that the matrix

$$A = \begin{bmatrix} -3 & -2 \\ 17 & 3 \end{bmatrix},$$

had eigenvalues $\lambda = \pm 5i$ and a corresponding complex eigenvector $v = (-2, 3 + 5i)$. Taking the real part $u = (-2, 3)$, imaginary part $w = (0, 5)$, and forming a matrix P with these vectors as columns gives

$$P = \begin{bmatrix} -2 & 3 \\ 0 & 5 \end{bmatrix}.$$

For this choice of P, it's easy to compute that

$$P^{-1}AP = \begin{bmatrix} 0 & 5 \\ -5 & 0 \end{bmatrix} = C(0, 5),$$

which shows that $J = C(0, 5)$ is the Jordan form for A. Thus, we know that the phase portrait for $x' = Ax$ will be similar to the center point phase portrait in Figure 4.25.

Example 4.21 (3 × 3 Jordan Forms) To enumerate the possible 3×3 Jordan forms J, first consider the case where J consists entirely of real

Jordan blocks. These blocks can have sizes ranging from 3 down to 1, and thus the possibilities are: $J = J_3(\lambda), J = [J_2(\lambda), J_1(\mu)]$, and $J = [J_1(\lambda), J_1(\mu), J_1(\nu)]$. Next, if J has any complex blocks it can have only one and it must be a $C(a, b)$ (since complex blocks come only in even sizes). Thus, $J = [J_1(\lambda), C(a, b)]$ is the only such possibility. Our reasoning leads us to conclude that J must have one of the following four forms:

$$(1) \begin{bmatrix} \lambda & 1 & 0 \\ 0 & \lambda & 1 \\ 0 & 0 & \lambda \end{bmatrix}, \qquad (2) \begin{bmatrix} \lambda & 1 & 0 \\ 0 & \lambda & 0 \\ 0 & 0 & \mu \end{bmatrix},$$

$$(3) \begin{bmatrix} \lambda & 0 & \\ 0 & \mu & 0 \\ 0 & 0 & \nu \end{bmatrix}, \qquad (4) \begin{bmatrix} \lambda & 0 & 0 \\ 0 & a & -b \\ 0 & b & a \end{bmatrix}.$$

In this enumeration the constants λ, μ, ν, a, b can have any real values with the exception that $b \neq 0$. As in the last example, the four Jordan forms listed here can have distinctly different phase portraits depending on the values of these constants. We will not discuss here the various possibilities.

Final Remark: From a mathematical point of view, this section has shown that understanding the types of behavior of a given linear systems $x' = Ax$, is reduced to understanding what can occur for any of the canonical systems $y' = Jy$. Knowing which of these canonical behaviors is exhibited by the given system is another matter.

The determination of the Jordan canonical form J for a given matrix A can be difficult, if not impossible, to do in practice. Since the eigenvalues of A are needed in the Jordan form, the first step is to determine them. Then $\mathbb{R}^n$ can be decomposed as a direct sum of generalized eigenspaces and pseudo-eigenspaces (cf. Appendix C), and then an algorithm can be developed for choosing bases of cyclic vectors for these subspaces. Maple has a built-in procedure, based on this approach, for not only determining J, but also a matrix P that brings A to Jordan form: $P^{-1}AP = J$. See the worksheet `jorforms.mws` on the electronic component. As with all computer algorithms, there can be problems in using this procedure due to numerical error, memory overload, and run times.

Exercises 4.5

1. Show that the notion of similarity is an equivalence relation on the set $\mathcal{M}_n$ of all $n \times n$ matrices. Thus, linear equivalence is an equivalence relation on the set of all linear, homogeneous, constant coefficient systems.

2. Suppose $v_1, \ldots, v_n$ is a basis for $\mathbb{R}^n$ consisting of eigenvectors of A and let $P = [v_1, \ldots, v_n]$ be the matrix formed with these vectors as columns. Show that P is invertible. We used this result in showing that A is diagonalizable when $\mathbb{R}^n$ has a basis of eigenvectors for A. Show that, conversely, if A is diagonalizable, then $\mathbb{R}^n$ has a basis of eigenvectors of A.

3. **(Complex Diagonalizability)** Suppose $n = 2p$ and that A is an $n \times n$ real matrix with complex (nonreal) eigenvalues $\lambda_j = a_j + b_j i$, $b_j \neq 0$, for $j = 1, \ldots, p$ and corresponding complex eigenvectors $v_j = u_j + i w_j$, $j = 1, \ldots, p$, which are linearly independent over $\mathbb{C}$. Show that there is an $n \times n$, complex matrix P, which is invertible and

$$P^{-1}AP = \operatorname{diag}(\lambda_1, \overline{\lambda}_1, \ldots, \lambda_p, \overline{\lambda}_p),$$

where the matrix on the right is the diagonal matrix with the λ_j's and their complex conjugates on the diagonal.

4. For each of the possible 2×2 and 3×3 Jordan forms J:

   (a) Compute $e^{Jt} = \sum_{k=0}^{\infty} \frac{t^k}{k!} J^k$ explicitly. *Hint*: In some cases it will be convenient to split J into the sum of two matrices: $J = M + N$, where M and N commute.

   (b) Write out the general solution: $y = e^{tJ}c$, of the canonical system, $y' = Jy$, explicitly in the forms:

$$
\begin{aligned}
y_1 &= \ldots \\
y_2 &= \ldots
\end{aligned}
\quad \text{or} \quad
\begin{aligned}
y_1 &= \ldots \\
y_2 &= \ldots \\
y_3 &= \ldots
\end{aligned}
$$

5. Write down all the possible 4×4 Jordan canonical forms J.

6. Calculate e^{tJ}, where J is the matrix:

$$
J =
\begin{bmatrix}
-1 & 2 & 1 & 0 & & & & \\
-2 & -1 & 0 & 1 & & & & \\
0 & 0 & -1 & 2 & & & & \\
0 & 0 & -2 & -1 & & & & \\
& & & & -3 & 1 & 0 & 0 \\
& & & & 0 & -3 & 1 & 0 \\
& & & & 0 & 0 & -3 & 1 \\
& & & & 0 & 0 & 0 & -3
\end{bmatrix}.
$$

7. In a previous exercise you drew the phase portrait for two of the 3×3 canonical systems. Examples of the remaining two types of 3×3 systems are:

   (a)

   $$A = \begin{bmatrix} -1 & 1 & 0 \\ 0 & -1 & 0 \\ 0 & 0 & -1/2 \end{bmatrix}$$

   (b)

   $$A = \begin{bmatrix} -2 & 1 & 0 \\ 0 & -2 & 1 \\ 0 & 0 & -2 \end{bmatrix}$$

   Use a computer to draw the phase portraits for these two examples. Use a window size: $-10 \le x, y, z \le 10$. Mark the directions on the integral curves and label the eigenspaces (if any). You will have to use your judgment on how best to display the main features of the phase portrait.

8. For each of the following systems: $x' = Ax$:

   (i) Determine the Jordan form J for A.

   (ii) Find a matrix P such that $P^{-1}AP = J$. To check your answer, Compute P^{-1} and then $P^{-1}AP$.

   (iii) Explicitly compute $G(t) = e^{At} = Pe^{Jt}P^{-1}$. Check your answer by (i) computing $AG(t)$ and comparing with $G'(t)$ and (ii) evaluating $G(0)$.

   (iv) Draw, by hand and by computer, the phase portraits for the system $x' = Ax$ and $y' = Jy$.

   (a) $A = \begin{bmatrix} 7 & -6 \\ 4 & -7 \end{bmatrix}$ (b) $A = \begin{bmatrix} 1 & -1 \\ 2 & -1 \end{bmatrix}$

   (c)

   $$A = \begin{bmatrix} 4 & -5 & -5 \\ 4 & -5 & -2 \\ 1 & 1 & -2 \end{bmatrix}$$

   For this problem: find an equation for the plane in $\mathbb{R}^3$, with the property that all integral curves that start at a point in this plane, remain in this plane for all time. Can you rotate so that your view is looking perpendicularly down on this plane ?

   (d)

   $$A = \begin{bmatrix} 5 & -4 & -4 \\ 4 & -5 & -2 \\ 0 & 0 & -3 \end{bmatrix}$$

9. Show that equation (4.107) holds and that the matrix N_k there is nilpotent: $N_k^k = 0$. Do a similar thing for equation (4.110).

10. Explain why the eigenvalues of the matrix (4.111) are $\lambda_1, \ldots, \lambda_r$, and $a_1 \pm b_1 i, \ldots, a_s \pm b_s i$.

11. Study the material on the Maple worksheet jorforms.mws and work the exercises listed there.

12. Read the material in CDChapter 5 on the electronic component that pertains to linear, discrete dynamical systems and work the exercises there.

4.6 Summary

This chapter discussed the theory for linear systems of differential equations, which in general have the form

$$x' = A(t)x + b(t),$$

where $A : I \to \mathcal{M}_n$ and $b : I \to \mathbb{R}^n$ are given continuous functions on an interval I, which we can assume contains 0. The key ingredient for understanding the integral curves and geometric properties of such systems is the fundamental matrix G. This is the differentiable, matrix-valued function: $G : I \to \mathcal{M}_n$, which satisfies

$$G'(t) = A(t)G(t) \quad \text{(for all } t \in I),$$
$$G(0) = I.$$

In terms of the fundamental matrix, the general solution of the linear system is given by the formula

$$\alpha(t) = G(t)c + G(t) \int_0^t G(s)^{-1}b(s)\, ds \quad \text{(for } t \in I),$$

where $c = (c_1, \ldots, c_n) \in \mathbb{R}^n$ is an arbitrary vector. The first term gives the general solution $\beta(t) \equiv G(t)c$ of the corresponding homogeneous equation $x' = A(t)x$, while the second term gives an integral curve γ which is a particular solution of the nonhomogenous equation.

The chapter predominantly concentrated on the study of homogeneous, constant coefficient, systems $x' = Ax$. For such systems the fundamental matrix is given by

$$G(t) = e^{At} = \sum_{k=0}^{\infty} A^k \frac{t^k}{k!},$$

and this is can be effectively computed either exactly, by using the Jordan form for A (see Chapter 6), or approximately, by using the terms in its Taylor series expansion (see the Maple worksheet `matexpo.mws` on the electronic component).

The flow map for the system $x' = Ax$ is given simply by

$$\phi_t = e^{At},$$

and this formula indicates why properties of the matrix A are reflected in corresponding properties for the integral curves $\phi_t(c) = e^{At}c$ of the system. The formula $\phi_t(c) = e^{At}c$ for the general solution can be calculated by several methods.

(1) *The eigenvalue/eigenvector method*: This requires decomposing c:

$$c \in \mathbb{R}^n = GE_{\mu_1} \oplus \cdots \oplus GE_{\mu_r} \oplus GE_{p_1 \pm q_1 i} \oplus \cdots \oplus GE_{p_s \pm q_s i},$$

in terms of *generalized* eigenvectors and then determining the action of e^{At} on the respective generalized eigenvectors. The chapter only pursued this method for the case when A is diagonalizable over $\mathbb{C}$. Even with this limited setting, the derivation of the formula for $e^{At}c$ in terms of eigenvectors and eigenvalues helps one understand the basic geometry of the integral curves and lays the foundation for the stability results in Chapter 6.

(2) *The transformation method*: This method computes the desired formula

$$\phi_t(c) = e^{At}c = Pe^{Jt}P^{-1}c,$$

from the Jordan canonical form J for A and the matrix P that brings A to Jordan form: $P^{-1}AP = J$. This method actually is just the eigenvalue/eigenvector method in disguise: P is made up of all the (generalized) eigenvectors and J contains the eigenvalues and basic structure of A (Cf. Exercise 4 in Section 2 of Chapter 6).

Chapter 5

Linearization & Transformation

In this chapter we present two basic techniques that are useful for analyzing nonlinear systems. One technique consists of *linearizing* about the fixed points to obtain local, qualitative pictures of the phase portrait via the corresponding linear systems. Thus, our previous work on linear systems has direct bearing on nonlinear systems. The validity of this technique is contained in the Linearization Theorem, which we present later on, after first applying it in numerous examples and exercises.

The other technique studied here, which has broader significance, involves the idea of *transforming* one system of DEs into another, perhaps simpler, system. You have already studied this for linear systems where the transformation was via a linear transformation and the resulting simpler system was the linear system determined by the Jordan form. The general technique uses nonlinear transformations and is motivated by the example of transforming to polar coordinates which you studied in Chapter 2. The transformation theory also motivates the notion of topological equivalence of systems of DEs, which is the basis for the Linearization Theorem.

5.1 Linearization

The linearization technique is quite easy to describe. Thus, suppose

$$x' = X(x)$$

is a given system of DEs, determined by a vector field $X : \mathcal{O} \to \mathbb{R}^n$ on an open set $\mathcal{O} \subseteq \mathbb{R}^n$. By algebraic or numerical means we look for the fixed points of the system: points $c \in \mathcal{O}$ where the vector field X vanishes: $X(c) = 0$. These give us constant solutions: $\alpha(t) \equiv c, \forall t \in \mathbb{R}$. Such solutions

D. Betounes, *Differential Equations: Theory and Applications*, DOI 10.1007/978-1-4419-1163-6_5,
© Springer Science + Business Media, LLC 2010

are not *per se* very interesting, but the theory (with some exceptional cases) is that near each fixed point $c \in \mathcal{O}$, the phase portrait of the nonlinear system resembles the phase portrait of the corresponding linear system:

$$y' = Ay,$$

where A is the Jacobian matrix, or derivative, of X at c:

$$A \equiv X'(c).$$

The precise nature of this similarity between the phase portraits will be established shortly.

   We emphasize that this is only a *local* similarity of phase portraits (namely, similarity in a small neighborhood of the fixed point). Thus, if we find five fixed points for a system, we get five corresponding linear systems, and a rough idea of what the nonlinear system looks like near each fixed point. When we back off and look at the *global* picture, the theory does not tell us how to connect up the integral curves in the five local pictures to get the global integral curves.

   We also emphasize that some types of fixed points c form exceptional cases to the linearization technique. That is, the phase portrait of the nonlinear system near c *need not* be similar to that of $y' = Ay$, near zero. These cases occur when either $\det(A) = 0$ or when A has a purely imaginary eigenvalue. The following is some special terminology that is in common use for the linearization technique:

Definition 5.1 (Hyperbolic and Simple Fixed Points) A fixed point c of a vector field $X : \mathcal{O} \to \mathbb{R}^n$ is called a *hyperbolic* fixed point if each eigenvalue $\lambda = a + bi$, of the matrix

$$A \equiv X'(c),$$

has nonzero real part: $a \neq 0$. If $\det A \neq 0$, then c is called a *simple* fixed point. Note that a hyperbolic fixed point is simple and that a nonhyperbolic fixed point is either nonsimple or A has a purely imaginary eigenvalue. A hyperbolic fixed point is further classified by the real parts of the eigenvalues of A. If every real part is negative, it is called a *sink*; (2) if every real part positive it is called a *source*; (3) if at least two of the real parts have opposite signs, it is called a *saddle*.

   We work through a number of examples illustrating the linearization technique, before going into the theory in more detail.

Example 5.1 (Predator-Prey) The general form for the predator-prey system is

$$x' = (a - by)x$$
$$y' = (cx - d)y,$$

where a, b, c, d are parameters (nonnegative constants). With x representing the number of prey (say rabbits) and y the number of predators (say foxes), the interpretation of the form of the predator-prey system comes from the observation that if there are only a small number of foxes ($y \approx 0$), then the first equation in the system is approximately: $x' = ax$, so the rabbit population increases exponentially. On the other hand for a relatively large number of foxes $(a - by) < 0$, the first equation of the system is approximately exponential decay of the rabbit population. Similarly, in the second equation of the system, if there are only a small number of rabbits ($x \approx 0$), then the DE is approximated by: $y' = -dy$, i.e., the fox population exponentially decays to 0 (there is very little for them to eat).

For clarity in applying the linearization technique, we specialize to the case where the parameters $a = 10, b = 1, c = 1, d = 50$ (see the exercises for the general case):

$$x' = (10 - y)x \qquad (5.1)$$
$$y' = (x - 50)y, \qquad (5.2)$$

The vector field here is given by

$$X(x, y) = \left((10 - y)x, (x - 50)y \right),$$

and the fixed points are easily seen to be:

$$(0, 0), \quad (50, 10).$$

Next, we compute the derivative of X and get

$$X'(x, y) = \begin{bmatrix} 10 - y & -x \\ y & x - 50 \end{bmatrix}.$$

Evaluating this at the two fixed points yields the matrices:

$$A = X'(0, 0) = \begin{bmatrix} 10 & 0 \\ 0 & -50 \end{bmatrix}, \quad B = X'(10, 50) = \begin{bmatrix} 0 & -50 \\ 10 & 0 \end{bmatrix}.$$

The point $(0,0)$ is a hyperbolic fixed point and the theory is that near the point $(0,0)$, the nonlinear phase portrait looks like that of the linear system $v' = Av$ at $(0,0)$, and thus should resemble a saddle point. This is why such fixed points are called hyperbolic: the linearized system has integral curves that are hyperbolas (and their asymptotes).

The other fixed point $(50,10)$ is nonhyperbolic and has a corresponding linear system $v' = Bv$, which has a center at the origin. As we shall see the Linearization Theorem is not strong enough to guarantee us that the nonlinear system looks like a center near $(50,10)$. However other techniques tell us this is the case. Thus, to sketch the phase portrait by hand we draw a small saddle portrait around $(0,0)$ and small center portrait around $(50,10)$ and take a stab at joining the integral curves to form a global picture. This is shown in Figure 5.1.

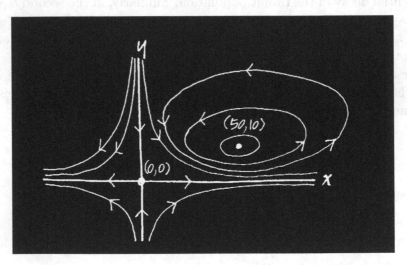

Figure 5.1: *Rough sketch of a predator-prey phase portrait modeled by:* $x' = (10 - y)x$, $y' = (x - 50)y$, *based on the calculations that* $(0,0)$ *is a saddle point and* $(50,10)$ *is a center. The straight line integral curves (separatrices) are also shown.*

We have also shown integral curves in the sketch that are straight lines, the separatrices for the saddle point, and which just happen to coincide with coordinate axes. Otherwise said, the lines with equations $x = 0$ and $y = 0$ are *invariant lines* for the system, since they are invariant under the flow map ϕ for the system. For an initial point c on one of these lines, the integral curve $t \mapsto \phi_t(c)$ remains on this line for all time, $t \in I_c$. The

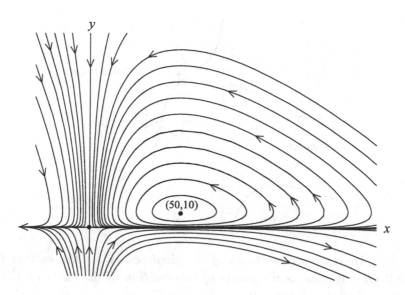

Figure 5.2: *Computer-drawn sketch of the same predator-prey phase portrait for $x' = (10 - y)x$, $y' = (x - 50)y$.*

invariant lines for this system are easily determined by inspection of the two DEs comprising the system (as was done in Chapter 2). For $x = 0$ the first equation is automatically satisfied and the second equation reduces to $y' = -50y$. Thus, the flow on the y-axis is toward the origin. On the other hand, for $y = 0$ the second equation is automatically satisfied and the first equation reduces to $x' = 10x$. Thus, the flow on the x-axis is away from the origin.

A much more accurate sketch of the phase portrait is the computer drawn picture shown in Figure 5.2. **Note:** Since the predator-prey model represents population sizes, we normally are interested only in the part of the phase portrait in the 1st quadrant (where $x \geq 0, y \geq 0$). However from just a mathematical point of view the complete phase portrait for this system covers the whole plane.

We emphasize that the linearization technique is a valuable tool especially if the system is very complex or in a higher dimensional phase space.

Example 5.2 Here's an abstract dynamical system, not pertaining to anything in particular:

$$
\begin{aligned}
x' &= xy^3 + x^3 + y - 4 \\
y' &= x^3y - y^2 + 1.
\end{aligned}
$$

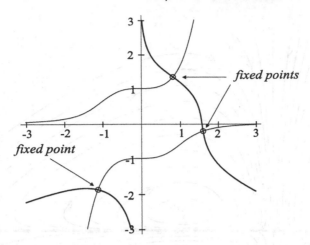

Figure 5.3: *Approximate locations of the fixed points for the system in Example 5.2 are obtained as the points of intersection of the curves $xy^3 + x^3 + y - 4 = 0$ and $x^3y - y^2 + 1 = 0$.*

The fixed points for $X(x,y) = (xy^3 + x^3 + y - 4, x^3y - y^2 + 1)$ must be found numerically. The plots of the curves $xy^3 + x^3 + y - 4 = 0$ and $x^3y - y^2 + 1 = 0$ are shown in Figure 5.3 and illustrate the approximate locations of the three fixed points. Using a computer, we find the fixed points, approximated to ten decimal places, are:

$$
\begin{aligned}
c_1 &= (0.844104506, 1.344954022)\\
c_2 &= (1.618800027, -0.2239139508)\\
c_3 &= (-1.101303898, -1.870387369).
\end{aligned}
$$

It is also convenient to use a computer to compute the values $A_j = X'(c_j)$, $j = 1, 2, 3$ of the Jacobian matrix at the fixed points and then to find the eigenvalues of these matrices. Thus, for example, we find that

$$
A_1 = X'(c_1) = \begin{bmatrix} 4.570424558 & 5.580703340 \\ 2.874886902 & -2.088473862 \end{bmatrix},
$$

and that the eigenvalues of this matrix are

$$
\lambda = 6.449537816, -3.967587120.
$$

Hence the linear system has a *saddle point* at the origin, and so the phase portrait of the nonlinear system will look qualitatively like a saddle point

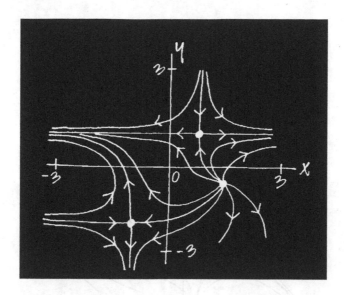

Figure 5.4: *Hand-drawn sketch of the phase portrait for the system in Example 5.2, based on the linearization about the three fixed points.*

in a small neighborhood of c_1. With a similar analysis we find that the eigenvalues for the matrices A_2 and A_3 are

$$\lambda = 6.825189775, 5.715039601$$

and

$$\lambda = -9.132572544, 8.632951735,$$

respectively. Thus, in the nonlinear system the phase portrait is similar to an *unstable node* in a neighborhood of c_2 and is similar to a *saddle point* in a neighborhood of c_3. Figure 5.4 shows a qualitative sketch of the nonlinear phase portrait. Note that we have drawn the three local portraits to resemble their linear counterparts. However, unlike the last example, the separatrices for the two saddle points cannot be expected to be straight lines in general. These will be distorted in the similarity between the linear and nonlinear systems. However, the stability/instability and type of fixed points will be the same in both systems. Note also that knowing the qualitative look of the phase portrait near each fixed point is often not enough to get a *global* qualitative sketch of the entire phase portrait. In this example it's not too hard to connect the three local pictures to get a global picture as indicated in Figure 5.4. A more exact representation of the global phase portrait is shown in the computer-generated picture in Figure 5.5.

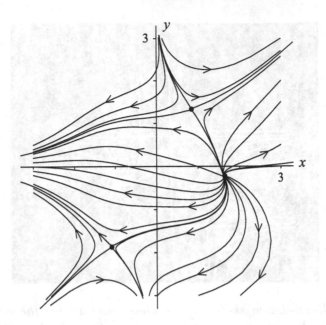

Figure 5.5: *Phase portrait of $x' = xy^3 + x^3 + y - 4$ and $y' = x^3y - y^2 + 1$.*

Comment: For systems in the plane, the nature of the fixed points (saddle, node, or center) and the stability can usually be discerned from the direction field plots. Thus, the linearization analysis is not so useful in dimension two, except in theoretical work (see the Exercises). However in higher dimensions, linearization is an essential tool for determining the nature of the fixed points in any particular example (and for the theory too). This is true even in dimension three where the direction field plot is visualizable, but mostly difficult to read and analyze for information on the fixed points. The next example exhibits this.

Example 5.3 This example is of an abstract dynamical system in $\mathbb{R}^3$.

$$\begin{align} x' &= x(1 - y^2) & (5.3) \\ y' &= (x + 3)z & (5.4) \\ z' &= y(z - 2). & (5.5) \end{align}$$

The fixed points are the solutions of the system:

$$\begin{align} x(1 - y^2) &= 0 \\ (x + 3)z &= 0 \\ y(z - 2) &= 0 \end{align}$$

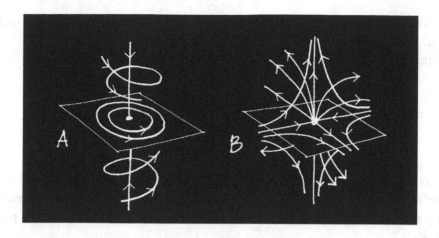

Figure 5.6: *A sketch of the phase portraits for the linear systems: $v' = Av$ and $v' = Bv$, that arise from linearizing the system (5.3)-(5.5) at the fixed points $(-3, \pm 1, 2)$.*

and are easily determined to be the three points

$$(0, 0, 0), \quad (-3, \pm 1, 2).$$

The derivative of the vector field for this system is

$$X'(x, y, z) = \begin{bmatrix} 1 - y^2 & -2xy & 0 \\ z & 0 & x + 3 \\ 0 & z - 2 & y \end{bmatrix}.$$

Evaluating this at the three fixed points gives the matrices for the corresponding linear systems, which decompose into block submatrices. Thus,

$$A = X'(-3, -1, 2) = \begin{bmatrix} 0 & -6 & 0 \\ 2 & 0 & 0 \\ 0 & 0 & -1 \end{bmatrix},$$

has a 2×2 sub block corresponding to a center at the origin (since its eigenvalues are $\pm 2\sqrt{3}\, i$). The 1×1 sub block gives an eigenspace E_{-1} which is the z-axis. The phase portrait for the linear system $v' = Av$, with $v = (x, y, z)$, is familiar to us from our previous work. A sketch of the linear system's phase portrait is shown in Figure 5.6. Since A has pure imaginary eigenvalues, the Linearization Theorem does not guarantee that the linear and nonlinear systems will look similar at this fixed point. However, other investigations will reveal that this is the case.

Similarly, at the fixed point $(-3, 1, 2)$, we find that

$$B = X'(-3, 1, 2) = \begin{bmatrix} 0 & 6 & 0 \\ 2 & 0 & 0 \\ 0 & 0 & 1 \end{bmatrix}.$$

Since the 2×2 upper sub block of B has eigenvalues $\pm 2\sqrt{3}$, the linear system, $v' = Bv$, has a saddle point at the origin and the integral curves in the x-y plane are hyperbolic-like. The system also has straight line integral curves along the z-axis. Figure 5.6 shows a sketch of this linear system's phase portrait.

Finally, examining the remaining fixed point we find:

$$C = X'(0, 0, 0) = \begin{bmatrix} 1 & 0 & 0 \\ 0 & 0 & 3 \\ 0 & -2 & 0 \end{bmatrix}.$$

This is entirely similar to the first case: the upper 1×1 sub block of C gives the eigenspace E_1 which is the x-axis, with integral curve lying on this axis directed away from the origin. The lower 2×2 sub block of C has purely imaginary eigenvalues $\pm\sqrt{6}i$, and gives that the integral curves starting in the y-z plane are ellipses in this plane. Sketching the linear system's portrait gives the picture shown in Figure 5.7.

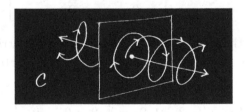

Figure 5.7: *A sketch of the phase portrait for the linear system: $v' = Cv$, that arises from linearizing the system (5.3)-(5.5) at the fixed point $(0, 0, 0)$.*

To get a first, rough sketch of the nonlinear system's phase portrait we combine the three pictures shown in Figures 5.6-5.7 together into one picture with the origins there located now at the respective three fixed points. In doing this it is helpful to know that the nonlinear system has some invariant lines (straight-line integral curves) in its phase portrait. In fact, there are four such lines and they can be determined by examining the form of the system

$$x' = x(1 - y^2)$$
$$y' = (x + 3)z$$
$$z' = y(z - 2).$$

If $y = \pm 1$ and $x = -3$, then the 1st and 2nd equations are automatically satisfied and the 3rd equation reduces to $z' = \pm(z-2)$. Thus, the two vertical lines through the fixed points $(-3, \pm 1, 2)$ are invariant under the flow and for $y = +1$, the flow is away from the fixed point, while for $y = -1$ the flow is toward the fixed point. A third invariant line comes from $y = 0$ and $z = 0$. Then the 2nd and 3rd equations are satisfied and the first equation becomes $x' = x$. This invariant line goes through the fixed point $(0, 0, 0)$ and the flow on it is away from the fixed point. A fourth invariant line, one that does not pass through any of the fixed points, is the line $x = 0, z = 2$. On this line the 1st and 3rd equations are satisfied and the 2nd equation is $y' = 6$. Thus, the flow on this line is uniformly in one direction.

Putting together all this information from the linearization and the analysis of invariant lines gives the sketch shown in Figure 5.8.

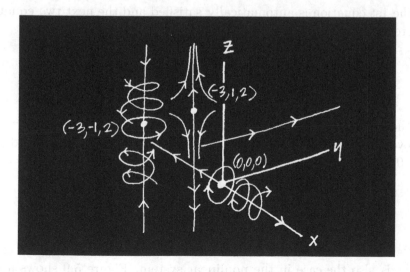

Figure 5.8: *Rough sketch of the phase portrait for the system:* $x' = x(1 - y^2), y' = (x + 3)z, z' = y(z - 2)$, *near each of its three fixed points.*

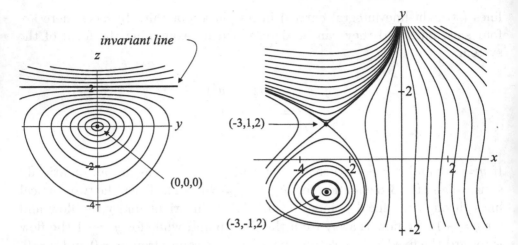

Figure 5.9: *The flow of the system* (5.3)-(5.5) *in the invariant plane* $x = 0$ (*on the left*) *and in the invariant plane* $z = 2$ (*on the right*).

For this example there is also some additional information about the phase portrait that comes from looking for *invariant planes*. These are planes in $\mathbb{R}^3$ that are invariant under the flow: if c is in the plane, then $\phi_t(c)$ is in the plane for all $t \in I_c$. There are three such planes here and they can easily be found by inspection of the system of DEs (5.3)-(5.5). Thus, for $x = 0$ the 1st equation is automatically satisfied and the next two equations reduce to the 2×2 system:

$$\begin{aligned} y' &= 3z \\ z' &= y(z - 2). \end{aligned}$$

Thus, the plane $x = 0$ is invariant and the flow in this plane is governed by the above 2×2 system. This system has an invariant line $z = 2$, fixed point at $y = 0, z = 0$, and corresponding linear system at this point, $w' = Mw$, where

$$M = \begin{bmatrix} 0 & 3 \\ -2 & 0 \end{bmatrix}.$$

Since the eigenvalues of M are $\pm\sqrt{6}\, i$, the origin is a center and we expect that this is also the case in the nonlinear system. Figure 5.9 shows a computer plot of the phase portrait for this system. The Figure also shows the phase portrait for the system in the other invariant plane $z = 2$.

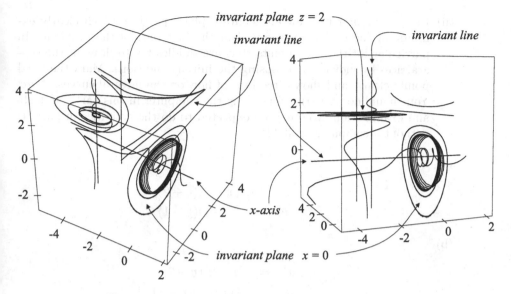

Figure 5.10: *Two views of the phase portrait.*

In this plane the original system (5.3)-(5.5) reduces to

$$x' = x(1 - y^2)$$
$$y' = 2(x + 3),$$

which has fixed points at $x = -3, y = 1$ and $x = -3, y = -1$, with corresponding linear systems $w' = Pw$ and $w' = Qw$, where

$$P = \begin{bmatrix} 0 & 6 \\ 2 & 0 \end{bmatrix}, \qquad Q = \begin{bmatrix} 0 & -6 \\ 2 & 0 \end{bmatrix}.$$

Thus, the two fixed points are a saddle and a center, respectively.

All of this analysis makes the study of the 3-D phase portrait somewhat easier. Figure 5.10 shows the result.

Exercises 5.1

1. For each of the following nonlinear systems in the plane:

   (i) Find all the fixed points and classify them by the Linearization Theorem (saddle, node, possible center, etc.). Find all invariant lines, if any. Make a sketch, by hand, of the phase portrait based on this information. To help discern what the global picture is like, determine the direction of the flow at a few selected points. On the finished sketch mark the direction of flow on all integral curves.

(ii) Use a computer to draw an accurate phase portrait which clearly exhibits all the features you found in the first part of the problem. In
particular for the fixed points that are saddles try to draw in the separatrices by judiciously choosing the initial conditions. Mark the fixed
points clearly and show directions of flow on the integral curves. *Note*:
You should do part (i) before using the computer in this part. On tests
and examinations you may be expected to do the activities in part (i)
without a computer available.

(a)

$$
\begin{aligned}
x' &= (x-2)(y+2) \\
y' &= (x-1)(y-1)
\end{aligned}
$$

(b)

$$
\begin{aligned}
x' &= (x-2)(y+2) \\
y' &= -(x-1)(y-1)
\end{aligned}
$$

(c)

$$
\begin{aligned}
u' &= -u - 2v\tfrac{1}{2}(u^2 - v^2) - 3 \\
v' &= -u + 2v + 5
\end{aligned}
$$

(d)

$$
\begin{aligned}
x' &= x(1 - y^2) \\
y' &= (x^2 - 1)y
\end{aligned}
$$

(e)

$$
\begin{aligned}
x' &= (x+y)/2 + x^2 \\
y' &= (3y - x)/2
\end{aligned}
$$

(f)

$$
\begin{aligned}
x' &= (x-1)(y-2) \\
y' &= (y-x)(y+x)
\end{aligned}
$$

(g) In this problem make two computer drawings: one with window size
about $-3 \le x, y \le 3$, and the other with a larger window size.

$$
\begin{aligned}
x' &= y^2 - 3x + 2 \\
y' &= x^2 - y^2
\end{aligned}
$$

2. For the following systems use a computer to find the fixed points and the eigenvalues of the Jacobian matrix at each fixed. Classify the fixed points and make a rough sketch (by hand) of the phase portrait.

   (a)

   $$x' = x^3 - xy^2 + y^3 - 1$$
   $$y' = x^3 - x - y^2 + 1$$

   (b)

   $$x' = x^2y + x^3 + xy^3 + y + 1$$
   $$y' = x^3y - x - x^2y^2 + 1$$

3. Consider the system

   $$x' = k(x - a)(y - b) \qquad (5.6)$$
   $$y' = m(x - c)(y - d), \qquad (5.7)$$

   where k, m are nonzero constants and a, b, c, d are any constants with $a \neq c$, $b \neq d$. Show that this system has two fixed points, two invariant lines, and only two types of phase portraits. For the latter note that each of the two fixed points *could* be a node, saddle, center, or focus, and so there technically could be eight types of phase portrait possible when the constants m, k, a, b, c, d are chosen appropriately. Show that in fact there are only two types and determine what they are. Sketch each type and label the drawing with the constants a, b, c, d.

   Show that the system (5.6)-(5.7) can be transformed into a system of the form:

   $$u' = ku(v - r) \qquad (5.8)$$
   $$v' = mv(u - s), \qquad (5.9)$$

   by the change of variables (translation) $u = x - a$, $v = y - d$. Describe this geometrically and explain the relation between the phase portraits for the two systems.

4. Consider a system of the form

   $$x' = A(x)B(y) \qquad (5.10)$$
   $$y' = C(x)D(y),, \qquad (5.11)$$

   where A, B, C, D are polynomial functions, with *real* roots $\{a_i\}_{i=1}^{\alpha}$, $\{b_j\}_{j=1}^{\beta}$, $\{c_k\}_{k=1}^{\gamma}$, and $\{d_m\}_{m=1}^{\delta}$ respectively.

(a) Determine the fixed points, the invariant lines, and numbers of each in terms of the roots of A, B, C, D.

(b) Show that a fixed point is nonsimple if and only if either A, C have a common root or B, D have a common root, or one of A, B, C, D has a repeated root.

(c) Determine which types of fixed points are possible (node, saddle, center, or focus) and which combinations of these can occur in the phase portrait, e.g., is it possible to choose A, B, C, D so that the phase portrait has two saddles and a node in it?

5. **(Predator-Prey Model):** For the general predator-prey model:

$$
\begin{aligned}
x' &= (a - by)x \\
y' &= (cx - d)y,
\end{aligned}
$$

find all the fixed points and classify them according to the linearization theorem. In the model one usually assumes the parameters a, b, c, d are all positive, so you can assume this here.

6. For each of the following nonlinear systems in $\mathbb{R}^3$:

(i) Find all the fixed points and classify them by the Linearization Theorem. Find all invariant lines and invariant planes, if any. Make a sketch, by hand, of the phase portrait based on this information.

(ii) Use a computer to draw an accurate phase portrait that clearly exhibits all the features you found in the first part of the problem. If there are invariant planes, create separate computer plots of the flow in these planes.

(a)

$$
\begin{aligned}
x' &= (1 - x^2)(z - 1) \\
y' &= x(y - 2) \\
z' &= (y + 1)z.
\end{aligned}
$$

(b)

$$
\begin{aligned}
x' &= -y(z - 1) \\
y' &= x(z + 1) \\
z' &= z(y + 2).
\end{aligned}
$$

(c)

$$
\begin{aligned}
x' &= y(z - 1) \\
y' &= x(z + 1) \\
z' &= (x - 1)z(y + 2).
\end{aligned}
$$

(d)

$$\begin{aligned}
x' &= x(y-4)(z-2) \\
y' &= (x+2)y(z-3) \\
z' &= (x+1)(y-3)z.
\end{aligned}$$

7. The famous Lorenz system of DE's is:

$$\begin{aligned}
x' &= -sx + sy \\
y' &= rx - y - xz \\
z' &= -bz + xy
\end{aligned}$$

where s, r, b are positive parameters. This system was introduced by E. N. Lorenz in the paper: "Deterministic Nonperiodic Flow," *J. of the Atmospheric Sciences*, **20** (1963) 130-141. This exercise studies some aspects of this system.

(a) Show that if $r \leq 1$, then there is only one fixed point, while if $r > 1$ there are three fixed points. In each case linearize about the fixed point and classify the fixed point.

(b) For the case $r = 28, s = 10$, and $b = 8/3$, sketch by hand the local phase portrait near each fixed point and then use a computer to draw the global phase portrait (or at least several integral curves which start near each fixed point).

(c) Get a copy of Lorenz's paper and (i) explain how he derives the Lorenz system from a system of PDEs in Section 5, (ii) explain the details of his analysis in Section 6, and (iii) use a computer to reproduce Figures 1 and 2 on page 137 of his paper.

5.2 Transforming Systems of DEs

The key idea behind the Linearization Theorem is the possibility of relating the integral curves of two different systems: $x' = X(x)$ and $y' = Y(y)$, by some sort of transformation. When this transformation has suitable properties, we can also directly transform one system into the other, which means transforming the vector field X into the vector field Y (a much-used technique in differential geometry). The polar coordinate example discussed in Chapter 2 and the transformation to canonical form for linear systems in the last chapter are good motivations for the general technique we discuss here. Before looking at the general technique, we return to the polar coordinate transformation as a means of motivation.

Transforming by the Polar Coordinate Map: We suppose

$$x' = X^1(x, y) \tag{5.12}$$
$$y' = X^2(x, y), \tag{5.13}$$

is any dynamical system in the plane (i.e., $\mathcal{O} \subseteq \mathbb{R}^2$). Transforming this system by the polar coordinate map amounts to, in simplistic terms, looking for solutions of the system that have the form

$$x = r \cos\theta \tag{5.14}$$
$$y = r \sin\theta, \tag{5.15}$$

where $r = r(t)$ and $\theta = \theta(t)$ are two unknown functions of t. To interpret things geometrically we introduce the following definition.

Definition 5.2 The *polar coordinate map* is the map $g : \mathbb{R}^2 \to \mathbb{R}^2$, defined by:
$$g(r, \theta) = (r \cos\theta, r \sin\theta),$$
for $(r, \theta) \in \mathbb{R}^2$.

Using this, we then assume that $(x(t), y(t)) = g(r(t), \theta(t))$ is a solution of the x-y system (5.12)-(5.13) and we derive a r-θ system of DEs that $(r(t), \theta(t))$ must satisfy.

As we have seen before, differentiating both sides of (5.14)-(5.15) gives:

$$x' = r' \cos\theta - r\theta' \sin\theta \tag{5.16}$$
$$y' = r' \sin\theta + r\theta' \cos\theta. \tag{5.17}$$

Using these relations between the respective functions and their respective derivatives, we can rewrite the original system entirely in terms of r and θ to get

$$r' \cos\theta - r\theta' \sin\theta = X^1(g(r, \theta)) \tag{5.18}$$
$$r' \sin\theta + r\theta' \cos\theta = X^2(g(r, \theta)). \tag{5.19}$$

Solving these for r' and θ', as we did in Chapter 2, yields the desired form of the polar version of the system:

Polar Coordinate Version:

$$r' = \cos\theta \cdot X^1(g(r, \theta)) + \sin\theta \cdot X^2(g(r, \theta)) \tag{5.20}$$
$$\theta' = \frac{-\sin\theta}{r} \cdot X^1(g(r, \theta)) + \frac{\cos\theta}{r} \cdot X^2(g(r, \theta)), \tag{5.21}$$

With the aim of generalizing the above procedure, we recast it in terms in matrix form as follows. The solutions of the x-y system are related to the solutions of the r-θ system by

$$(x, y) = g(r, \theta).$$

By the chain rule, the respective derivatives are related by

$$\begin{bmatrix} x' \\ y' \end{bmatrix} = g'(r, \theta) \begin{bmatrix} r' \\ \theta' \end{bmatrix},$$

where the Jacobian matrix of the polar coordinate map is

$$g'(r, \theta) = \begin{bmatrix} \cos \theta & -r \sin \theta \\ \sin \theta & r \cos \theta \end{bmatrix}.$$

Thus, the matrix form of the transformed system (5.18)-(5.19) is

$$g'(r, \theta) \begin{bmatrix} r' \\ \theta' \end{bmatrix} = X(g(r, \theta)).$$

Inverting the Jacobian matrix of the polar coordinate map, allows us to rewrite this last equation as:

$$(r', \theta') = g'(r, \theta)^{-1} X(g(r, \theta)), \tag{5.22}$$

which is the matrix version of (5.20)-(5.21). Thus, we see that the vector field for the transformed system is

$$Y(r, \theta) \equiv g'(r, \theta)^{-1} X(g(r, \theta)).$$

This provides a motivation for the general method of transforming one vector field into another, which is presented below. The definition, for technical reasons, is phrased in terms of $f \equiv g^{-1}$ rather than in terms of g.

The meaning, then, of the polar system (5.22) is this: Any solution curve $r = r(t), \theta = \theta(t)$, of the system gives rise to (is mapped onto) a solution curve, $(x, y) = g(r, \theta)$, of the original system (5.12)-(5.13). Thus, if we can plot the phase portrait in the r-θ plane of the polar system (5.22), then g will transform this into the phase portrait of the original system in the x-y plane. See Figure 5.11.

As with the canonical form for a linear system, the polar system is generally simpler than the original Cartesian system. However, it is important to

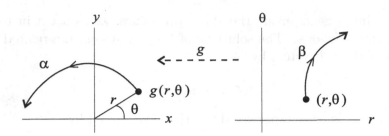

Figure 5.11: *The action of the polar coordinate map, transforming integral curves of the polar system into integral curves of the original Cartesian system.*

note that the transformation to the polar system uses a *nonlinear* transformation and so the polar system will bear little resemblance to the Cartesian system. Nevertheless, we can use the information in the polar system to discern the features in the phase portrait of the original system. This, as you learned in Chapter 2, relies upon interpreting θ as an angle and r as a distance. Thus, in Figure 5.11, the integral curve β shows that r and θ increase with time and so the corresponding integral curve α has a trajectory that flows away from the origin in a counterclockwise movement.

With this as motivation, we now look at the general definition of transforming vector fields by diffeomeorphisms.

Definition 5.3 (Transforming Vector Fields) We suppose that $\mathcal{O}$ and $\overline{\mathcal{O}}$ are open sets in $\mathbb{R}^n$.

(1) A map $f : \mathcal{O} \to \overline{\mathcal{O}}$ is called a *diffeomorphism* between $\mathcal{O}$ and $\overline{\mathcal{O}}$ if it is differentiable, 1-1, onto, and its inverse f^{-1} is also differentiable. In the sequel we will often denote the inverse of f by $g = f^{-1}$.

(2) If $X : \mathcal{O} \to \mathbb{R}^n$ is a vector field on $\mathcal{O}$, and $f : \mathcal{O} \to \overline{\mathcal{O}}$ is a diffeomorphism, then we get a *transformed vector field*: $f_*(X)$ on $\overline{\mathcal{O}}$, defined by

$$f_*(X)(y) = f'(f^{-1}(y))X(f^{-1}(y)), \qquad (5.23)$$

for each $y \in \overline{\mathcal{O}}$. In differential geometry, this vector field is called the *push-forward* of X by f. For $y \in \overline{\mathcal{O}}$, if we let $x \equiv f^{-1}(y)$, then definition (5.23) says the value of the vector field $f_*(X)$ at y is obtained by transforming the vector $X(x)$ by the Jacobian matrix $f'(x)$, i.e., $f_*(X)(y) = f'(x)X(x)$. Figure 5.12 shows an abstract picture of how

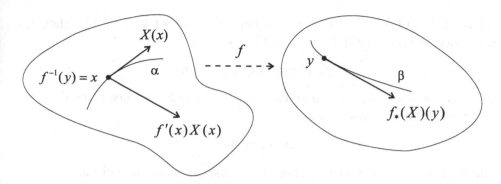

Figure 5.12: *Hypothetical picture of the transformation of the vector field X into the vector field $f_*(X)$ by a diffeomorphism f.*

this transformation of the vector field works. We also mention a word about the notation. As a transformation, or map, f transforms points $x \in \mathcal{O}$ into points $y = f(x) \in \overline{\mathcal{O}}$, while f_* denotes a different type of transformation (one induced by f), which transforms vector fields X on $\mathcal{O}$ into vector fields $Y = f_*(X)$ on $\overline{\mathcal{O}}$.

(3) Suppose $X : \mathcal{O} \to \mathbb{R}^n$ and $Y : \overline{\mathcal{O}} \to \mathbb{R}^n$ are two vector fields. We say that the two systems: $x' = X(x)$ and $y' = Y(y)$, are *differentiably equivalent* if there exists a diffeomorphism $f : \mathcal{O} \to \overline{\mathcal{O}}$, such that

$$Y = f_*(X).$$

We also say that the vector fields X and Y are differentiably equivalent when this happens.

We consider a few examples to illustrate how this transformation works and to verify it is the same as what we encountered previously.

Example 5.4 (Transforming by a Linear Transformation) Suppose Q is an invertible $n \times n$ matrix, and let $f : \mathbb{R}^n \to \mathbb{R}^n$ denote the linear transformation

$$f(x) = Qx,$$

(so that $f^{-1}(x) = Q^{-1}x$). The derivative of f at any $x \in \mathbb{R}^n$ is the same matrix, namely Q:

$$f'(x) = Q \qquad \forall x \in \mathbb{R}^n.$$

Thus, if $X : \mathcal{O} \to \mathbb{R}^n$ is any vector field, and we let $\overline{\mathcal{O}} = f(\mathcal{O})$, then the transformed vector field $f_*(X)$ on $\overline{\mathcal{O}}$ is given by

$$f_*(X)(y) = QX(Q^{-1}y),$$

for every $y \in \overline{\mathcal{O}}$. A particular case of this, which we encountered in the study of linear systems, is when

$$X(x) = Ax,$$

where A is an $n \times n$ matrix. Then the transformed vector field is

$$f_*(X)(y) = QAQ^{-1}y.$$

As we saw in the chapter on linear systems, the matrix Q can be chosen so that $QAQ^{-1} = J$ is the Jordan canonical form for A. In this case the transformed system is the canonical system, and according to the above definition, the two systems $x' = Ax$ and $y' = Jy$ are differentiably equivalent. Thus, the notion of differentiable equivalence is a generalization of the previous notion of linear equivalence.

Example 5.5 (Polar Coordinate Transformation) While the polar coordinate map $g(r, \theta) = (r \cos \theta, r \sin \theta)$, as a map from $\mathbb{R}^2$ onto $\mathbb{R}^2$, is infinitely differentiable, it is not a diffeomorphism (or even 1-1) unless we restrict its domain. There are many possible restrictions, but suppose we choose $\overline{\mathcal{O}} = \{(r, \theta) | 0 < r, -\pi/2 < \theta < \pi/2\} = (0, \infty) \times (-\pi/2, \pi/2)$. Then one can check that g is 1-1 on $\overline{\mathcal{O}}$ and has inverse:

$$f(x, y) = \left((x^2 + y^2)^{1/2}, \tan^{-1}(y/x) \right),$$

defined on $\mathcal{O} = g(\overline{\mathcal{O}}) = (0, \infty) \times (-\infty, \infty)$. Thus, f (and also g) is a diffeomorphism and

$$f'(x, y) = \begin{bmatrix} \dfrac{x}{(x^2+y^2)^{1/2}} & \dfrac{y}{(x^2+y^2)^{1/2}} \\ \dfrac{-y}{x^2+y^2} & \dfrac{x}{x^2+y^2} \end{bmatrix}. \tag{5.24}$$

Hence, the transformation formula for a vector field X on $\mathcal{O}$ is

$$f_*(X)(r, \theta) = f'(g(r, \theta))X(g(r, \theta))$$

$$= \begin{bmatrix} \cos \theta & \sin \theta \\ -\dfrac{\sin \theta}{r} & \dfrac{\cos \theta}{r} \end{bmatrix} \begin{bmatrix} X^1(r \cos \theta, r \sin \theta) \\ X^2(r \cos \theta, r \sin \theta) \end{bmatrix}. \tag{5.25}$$

The corresponding transformed system of DEs is

$$(r', \theta') = f_*(X)(r, \theta),$$

and you can see this is the same as we found in equation (5.22) previously.

There are a number of interesting and useful transformations in $\mathbb{R}^3$ that are analogous to the polar coordinate transformation. First and foremost is the spherical coordinate transformation which, as was the case for polar coordinates, will enable us to sketch by hand the integral curves for some systems in $\mathbb{R}^3$ by looking at the corresponding system in spherical coordinates. Visualization of phase portraits in $\mathbb{R}^3$ is usually difficult and so any additional aids like this are welcome.

5.2.1 The Spherical Coordinate Transformation

The transformation to spherical coordinates is traditionally given as

$$\begin{aligned}
x &= \rho \sin \phi \cos \theta \\
y &= \rho \sin \phi \sin \theta \\
z &= \rho \cos \phi.
\end{aligned}$$

As with the polar coordinate map, we formally get from this the *spherical coordinate map* $g : \mathbb{R}^3 \to \mathbb{R}^3$ defined by

$$g(\rho, \theta, \phi) = (\rho \sin \phi \cos \theta, \rho \sin \phi \sin \theta, \rho \cos \phi).$$

This map is differentiable with its derivative given by

$$g'(\rho, \theta, \phi) = \begin{bmatrix} \sin \phi \cos \theta & -\rho \sin \phi \sin \theta & \rho \cos \phi \cos \theta \\ \sin \phi \sin \theta & \rho \sin \phi \cos \theta & \rho \cos \phi \sin \theta \\ \cos \phi & 0 & -\rho \sin \phi \end{bmatrix}. \tag{5.26}$$

It is easy to calculate that $\det(g'(\rho, \theta, \phi)) = -\rho^2 \sin \phi$ (exercise) and so, by the Inverse Function Theorem, g has an inverse $f = g^{-1}$ on a neighborhood of any point (ρ, θ, ϕ) with $\rho \neq 0$ and ϕ not a multiple of π. It is also easy to find some explicit formulas for f (exercise). There are several choices depending on the choice of domain. Having a formula for f, we could use this to explicitly compute the transformed vector field $f_*(X)$ for any vector field on the domain of f.

However it is important to note that we can compute $f_*(X)$ directly from g without using f (and this works in general; see Exercise 2 below). Just observe that by the formula for the derivative of an inverse function

$$f'(g(\rho,\theta,\phi)) = (g^{-1})'(g(\rho,\theta,\phi)) = g'(\rho,\theta,\phi)^{-1},$$

and thus

$$\begin{aligned} f_*(X)(\rho,\theta,\phi) &= f'(g(\rho,\theta,\phi))X(g(\rho,\theta,\phi)) \\ &= g'(\rho,\theta,\phi)^{-1}X(g(\rho,\theta,\phi)). \end{aligned}$$

Thus, all we need is the inverse of the Jacobian matrix of g. Applying the formula for computing the inverse of matrix from its adjoint matrix ($A^{-1} = (A^\dagger)^T$), we easily find that

$$g'(\rho,\theta,\phi)^{-1} = \begin{bmatrix} \sin\phi\cos\theta & \sin\phi\sin\theta & \cos\phi \\ \dfrac{-\sin\theta}{\rho\sin\phi} & \dfrac{\cos\theta}{\rho\sin\phi} & 0 \\ \dfrac{\cos\phi\cos\theta}{\rho} & \dfrac{\cos\phi\sin\theta}{\rho} & \dfrac{-\sin\phi}{\rho} \end{bmatrix}. \tag{5.27}$$

Thus, with X given, we can define Y by

$$Y(\rho,\theta,\phi) = g'(\rho,\theta,\phi)^{-1}X(g(\rho,\theta,\phi)),$$

and get the corresponding system of DEs in spherical coordinates:

$$(\rho',\theta',\phi') = Y(\rho,\theta,\phi).$$

Alternatively, we could arrive at the same result by an argument like the one we used for polar coordinates in deriving equation (5.22) (exercise)

In general the corresponding spherical coordinate system will be quite complicated; however, there certain systems of differential equations that are transformed into particularly simple spherical coordinate systems. These are described in the following proposition.

Proposition 5.1 *Suppose $A, B, C : [0, \infty) \to \mathbb{R}$ are differentiable functions. Consider the system of DEs in $\mathbb{R}^3$ of the following special form:*

$$\begin{aligned} x' &= A(\rho)x - B(\rho)y + C(\rho)xz & (5.28) \\ y' &= A(\rho)y + B(\rho)x + C(\rho)yz & (5.29) \\ z' &= A(\rho)z - C(\rho)(x^2 + y^2), & (5.30) \end{aligned}$$

where, for convenience, we have let $\rho \equiv (x^2 + y^2 + z^2)^{1/2}$. *Then this system in Cartesian coordinates transforms into the following system in spherical coordinates:*

$$\rho' = A(\rho)\rho \tag{5.31}$$
$$\theta' = B(\rho) \tag{5.32}$$
$$\phi' = C(\rho)\rho \sin \phi. \tag{5.33}$$

Proof: Exercise.

It is easy to see that we could allow A, B, C to be more general expressions of (x, y, z) and obtain an extension of the result in the proposition.

Example 5.6 Consider the case where $A = 1 - \rho$, $B = 0, C = 1$. Then the Cartesian system in the proposition is

$$x' = (1 - \rho)x + xz \tag{5.34}$$
$$y' = (1 - \rho)y + yz \tag{5.35}$$
$$z' = (1 - \rho)z - (x^2 + y^2), \tag{5.36}$$

and the corresponding spherical coordinate system is

$$\rho' = (1 - \rho)\rho \tag{5.37}$$
$$\theta' = 0 \tag{5.38}$$
$$\phi' = \rho \sin \phi. \tag{5.39}$$

Since the spherical coordinate system is quite simple, we can use it to analyze and sketch, by hand, the original system (5.34)-(5.36). As with polar coordinate systems we take advantage of what the spherical coordinates mean geometrically in the Cartesian coordinate system.

Thus, suppose $\beta(t) = (\rho(t), \theta(t), \phi(t))$, for $t \in I$ is a solution of the spherical coordinate system (5.37)-(5.39). Then

$$\alpha(t) \equiv \Big(\rho(t) \sin \phi(t) \cos \theta(t), \ \rho(t) \sin \phi(t) \sin \theta(t), \ \rho(t) \cos \phi(t) \Big), \tag{5.40}$$

is a solution of the original system.

Now since $\theta' = 0$, it follows that $\theta(t) = \theta_0$ is constant, and we conclude that the integral curve α lies in the vertical plane through the origin that makes an angle θ_0 with the x-axis.

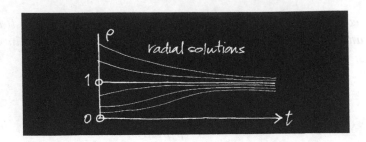

Figure 5.13: *Graphs of the solutions ρ of $\rho' = (1 - \rho)\rho$.*

The radial equation $\rho' = (1 - \rho)\rho$ indicates how the distance $\rho(t)$ of $\alpha(t)$ from the origin is changing. This distance is increasing when $0 < \rho < 1$ and is decreasing when $\rho > 1$. To be more specific, assume (without loss of generality) that $\rho(0) = \rho_0 \geq 0$. Using the techniques from Chapter 4 (or by just solving the radial equation directly) it is easy to see that the function ρ has graph like one of those shown in Figure 5.13.

Case 1: $(\rho_0 = 0, 1)$ The two fixed points $\rho_0 = 0, 1$ of the radial equation give special solutions α of the original system. Corresponding to $\rho_0 = 0$ is a fixed point at the origin $(0, 0, 0)$. However, for $\rho_0 = 1$, we get that α remains on the sphere of radius 1, centered at the origin. Since α is also on the plane $\theta = \theta_0$, it thus remains on the circle of radius 1, centered at the origin, in this plane. If we take into account the polar angle equation $\phi' = \sin \phi$, then we can conclude that $\alpha(t) \rightarrow (0, 0, -1)$ as $t \rightarrow \infty$, when $\rho_0 = 1$, and ϕ_0 is not a multiple of π. See Figure 5.14.

The solutions ϕ in this figure (which are similar in form to those in Figure 5.13) tend to π asymptotically as $t \rightarrow \infty$. The exceptions are those with $\phi(0) = 0, \pi$, which correspond to fixed points, in the original system (5.34)-(5.36), at the north and south poles $(0, 0, \pm 1)$ of the unit sphere.

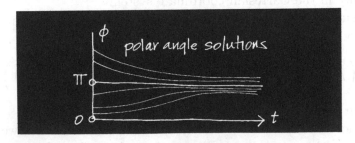

Figure 5.14: *Graphs of the solutions ϕ of $\phi' = \sin \phi$.*

Case 2: $(\rho_0 \neq 0, 1)$ In this case the function ρ is either decreasing (if $\rho_0 > 1$) or increasing (if $0 < \rho_0 < 1$) and tends asymptotically to 1 as $t \to \infty$. The polar angle equation $\phi' = \rho \sin \phi$, will have solutions that are qualitatively similar to those in Figure 5.14, and consequently

$$\lim_{t \to \infty} \alpha(t) = (0, 0, -1)$$

for all initial conditions except those with ϕ_0 a multiple of π. When $\phi_0 = 0, \pi$, etc., α lies on the z-axis and approaches one of the three fixed points $(0, 0, 0), (0, 0, \pm 1)$ of the original system. Figure 5.15 shows a sketch of the integral curves in the plane $\theta_0 = \theta_0$ and a sketch of the phase portrait in 3-D.

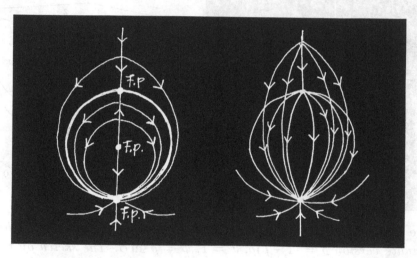

Figure 5.15: *Sketches of the integral curves of the system (5.34)-(5.36). The sketch on the left shows the integral curves in a typical plane $\theta = \theta_0$, which is representative. The sketch on the right gives a spatial view of the phase portrait.*

Example 5.7 An interesting extension of the last example, but one that is not too complex, arises from the choice of $A = 1 - \rho, B = 1, C = 1$. Then the spherical coordinate version of this system is

$$
\begin{aligned}
\rho' &= (1 - \rho)\rho \\
\theta' &= 1 \\
\phi' &= \rho \sin \phi.
\end{aligned}
$$

The only difference is that now the azimuthal angle equation $\theta' = 1$ has solution $\theta(t) = t + \theta_0$. So each integral curve α (cf. equation (5.40)) that starts at a point not on the z-axis, will wind continually around the z-axis as θ increases with t. At the same time α will approach the south pole of the unit sphere: $\rho \to 1$, $\phi \to \pi$ (as before). Thus, α has a spiral-like appearance and if $\rho_0 = 1$, this spiral will lie completely on the unit sphere. If α starts at a point on the z-axis, then it will remain on this axis and approach one of the three fixed points $(0, 0, 0), (0, 0, \pm 1)$ of the system (just as before). Figure 5.16 shows a hand-drawn sketch of the phase portrait together with one drawn by Maple.

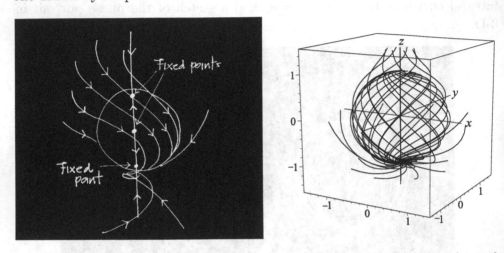

Figure 5.16: *Sketches of the integral curves of the system with spherical coordinate version* $\rho' = (1 - \rho)\rho$, $\theta' = 1$, $\phi' = \rho\sin\phi$. *The sketch on the left was done by hand while the one on the right was done by a computer.*

5.2.2   Some Results on Differentiable Equivalence

An important interpretation of differentiable equivalence is contained in the next theorem. It essentially gives an alternative definition of this concept, namely two vector fields X and Y are differentiably equivalent if and only if there is a diffeomorphism that maps the integral curves of X onto integral curves of Y.

Theorem 5.1 (Differentiable Equivalence) *Suppose* $X : \mathcal{O} \to \mathbb{R}^n$ *and* $Y : \overline{\mathcal{O}} \to \mathbb{R}^n$ *are vector fields on open subsets* $\mathcal{O}$ *and* $\overline{\mathcal{O}}$ *of* $\mathbb{R}^n$, *and suppose* $f : \mathcal{O} \to \overline{\mathcal{O}}$ *is a diffeomorphism. Then the following are equivalent:*

(a) $f_*(X) = Y$.

(b) *For each integral curve* $\alpha : I \to \mathbb{R}^n$ *of* X, *the curve*

$$\beta(t) = f(\alpha(t)),$$

$t \in I$, *is an integral curve of* Y.

Proof: [(a) $\Rightarrow$ (b)] Assume that (a) holds, and suppose that $\alpha : I \to \mathbb{R}^n$ is an integral curve of X. To see that $\beta(t) \equiv f(\alpha(t))$ is an integral curve of Y, we just use the chain rule, the definition of what $f_*(X) = Y$ means, and the assumption that

$$\alpha'(t) = X(\alpha(t)),$$

for all $t \in I$. Thus,

$$
\begin{aligned}
\beta'(t) &= \frac{d}{dt}(f \circ \alpha)(t) = f'(\alpha(t))\alpha'(t) \\
&= f'(\alpha(t))X(\alpha(t)) = f'(f^{-1}(\beta(t)))X(f^{-1}(\beta(t))) \\
&= f_*(X)(\beta(t)) = Y(\beta(t))
\end{aligned}
$$

[(b) $\Rightarrow$ (a)] Assuming (b) holds, we need to show that if $y \in \overline{\mathcal{O}}$ then

$$Y(y) = f_*(X)(y) = f'(f^{-1}(y))X(f^{-1}(y)).$$

For this, let $x = f^{-1}(y)$, and choose an integral curve $\alpha : I \to \mathcal{O}$ of X with $0 \in I$ and such that $\alpha(0) = x$. The Existence and Uniqueness Theorem guarantees that such a curve exists. Then assumption (b) says that $\beta(t) \equiv f(\alpha(t))$, is an integral curve of Y. Note that $\beta(0) = f(\alpha(0)) = f(x) = y$. But then by the chain rule and the fact that α and β are integral curves, we find

$$
\begin{aligned}
Y(y) &= Y(\beta(0)) = \beta'(0) = \frac{d}{dt}f(\alpha(t))\Big|_{t=0} \\
&= f'(\alpha(0))\alpha'(0) = f'(x)X(\alpha(0)) \\
&= f'(x)X(x) = f'(f^{-1}(y))X(f^{-1}(y)).
\end{aligned}
$$

This completes the proof. $\square$

The original definition of differentiable equivalence is most convenient to use in any given example, where the form of the system $x' = X(x)$ being studied often suggests a transformation that will give a simpler system: $y' = f_*(X)(y)$. The alternative definition of differentiable equivalence,

Property (b) of the theorem, is more useful in geometrically interpreting the relationship between the two systems. It just says the phase portraits of the original system and the transformed system will look qualitatively the same, one being a diffeomorphic distortion of the other (this distortion often is quite dramatic). Also note that since fixed points are integral curves, Property (b) says that f establishes a 1-1 correspondence between the fixed points of the original system and of the transformed system.

The other result we include here is rather straightforward to prove as well, and is important and inherent in our terminology:

Theorem 5.2 *The notion of differentiable equivalence is an equivalence relation on the set of all vector fields* $X : \mathcal{O} \rightarrow \mathbb{R}^n$ *on open subsets of* $\mathbb{R}^n$.

The proof is left as an exercise. The theorem allows us to identify all differentiably equivalent dynamical systems:

$$x' = X(x),\ y' = Y(y),\ z' = Z(z),\ \ldots,$$

with one another, and think of them all as essentially the same.

Exercises 5.2

1. Prove that differentiable equivalence is an equivalence relation on the set of all vector fields with domains in $\mathbb{R}^n$. More particularly, prove that

    (a) $f_*(X) = Y \implies (f^{-1})_*(Y) = X$.

    (b) $f_*(X) = Y,\ g_*(Y) = Z \implies (g \circ f)_*(X) = Z$.

    These follow easily from the formula for the derivative of an inverse function and the chain rule (see Appendix A).

2. As a generalization of the technique for transforming one vector field into another, consider the following. Suppose $X : \mathcal{O} \rightarrow \mathbb{R}^n$ and $Y : \overline{\mathcal{O}} \rightarrow \mathbb{R}^n$ are two vector fields, and $g : \overline{\mathcal{O}} \rightarrow \mathcal{O}$ is a differentiable map (we do *not* require g to be a diffeomorphism). We say that g *transforms* Y *into* X if

    $$g'(y)Y(y) = X(g(y)),\ \forall y \in \overline{\mathcal{O}}.$$

    Based on this concept, do the following:

    (a) Prove that the following statements are equivalent:

        (1) g transforms Y into X,

        (2) g transforms each integral curve of Y into an integral curve of X. That is: if $\beta : I \rightarrow \overline{\mathcal{O}}$ is an integral curve of Y, then $\alpha(t) \equiv g(\beta(t))$, for $t \in I$ is an integral curve of X.

(b) Suppose $\det(g'(y)) \neq 0$, for every $y \in \overline{\mathcal{O}}$. Prove that for each vector field X on $\mathcal{O}$, there is a unique vector field Y on $\overline{\mathcal{O}}$ that g transforms into X. This vector field Y is called *the g version* of X and is defined by

$$Y(y) = g'(y)^{-1}X(g(y)), \tag{5.41}$$

for $y \in \overline{\mathcal{O}}$. In addition, show that if g restricted to some open set $\overline{U} \subseteq \overline{\mathcal{O}}$ is a diffeomorphism $g : \overline{U} \to U$ and if $f \equiv g^{-1}$ denotes the inverse map, then

$$f_*(X) = Y.$$

As we have seen in the polar and spherical coordinate examples, it is often easier to compute the transformed vector field Y using equation (5.41) than it is to compute a local inverse $f = g^{-1}$ and then transform with that.

(c) A primary example that motivates the need for the generalization discussed in this exercise is the polar coordinate map:

$$g(r, \theta) = (r \cos \theta, r \sin \theta).$$

As a map, $g : \mathbb{R}^2 \to \mathbb{R}^2$ is *not* a diffeomorphism, since it is not even 1-1. By restricting the domain of g we can make it a diffeomorphism, but this is often inconvenient. The best alternative is to use the results from (a) and (b) above. Specifically let

$$\overline{\mathcal{O}} = \{ (r, \theta) \,|\, r > 0 \}.$$

Show that $g : \overline{\mathcal{O}} \to \mathbb{R}^2$ satisfies the condition in part (b), and explain how this reduces the study of the phase portrait for a vector field X to that of its g version (i.e., polar coordinate version) Y.

3. Suppose $A, B : [0, \infty) \to \mathbb{R}$ are differentiable functions, and consider the planar system:

$$
\begin{aligned}
x' &= A(r)x - B(r)y & (5.42) \\
y' &= B(r)x + A(r)y, & (5.43)
\end{aligned}
$$

where for convenience we have let $r = (x^2 + y^2)^{1/2}$. Show that this system transforms into the following system in polar coordinates

$$
\begin{aligned}
r' &= A(r)r & (5.44) \\
\theta' &= B(r) & (5.45)
\end{aligned}
$$

4. For each of the following systems $(x', y') = X(x, y)$, in the plane:

(i) Compute the polar coordinate version of the system:

$$r' = Y^1(r, \theta)$$
$$\theta' = Y^2(r, \theta).$$

Here $Y = f_*(X)$ is the vector field obtained by transforming to polar coordinates.

(ii) Use the polar coordinate version of the system (its particular form and its relation to the original system) to sketch *by hand* the phase portrait of $(x', y') = X(x, y)$.

(iii) Solve the polar coordinate system for r and θ as functions of t. (*Hint*: In (d) you might want to use the identity: $(1 - r)^2 - 1 = r(2 - r)$.)

(a) $X(x, y) = (x - y, x + y)$.

(b) $X(x, y) = (r^{-1}x - y, x + r^{-1}y)$.

(c) $X(x, y) = ((1 - r)x - y, x + (1 - r)y)$.

(d) $X(x, y) = ((1 - r)(2 - r)x - y, x + (1 - r)(2 - r)y)$.

In the above: $r = (x^2 + y^2)^{1/2}$. *Hint*: Each of the vector fields in (a)-(d) has the form $X(x, y) = (Ax - y, Ay + x)$, so you may use Exercise 3 if you wish.

5. Prove that $\det(g'(\rho, \theta, \phi)) = -\rho^2 \sin \phi$, where g is the spherical coordinate map. Find an explicit formula for a local inverse $f = g^{-1}$ of the spherical coordinate map and specify the domain of f.

6. Prove Proposition 5.1.

7. **(Spherical Coordinates)** For each of the following systems: (i) transform the system to spherical coordinates and use this to analyze the phase portrait of the given system, (ii) find all fixed points, and (iii) sketch the phase portrait showing all the pertinent features. (You should be able to do this easily by hand, but you may augment the study by using the computer if you absolutely must.)

(a) In the following system $\rho = (x^2 + y^2 + z^2)^{1/2}$.

$$x' = (\rho - 1)x - y + xz$$
$$y' = (\rho - 1)y + x + yz$$
$$z' = (\rho - 1)z - (x^2 + y^2),$$

(b) In the following system $\rho = (x^2 + y^2 + z^2)^{1/2}$.

$$x' = (1 - \rho)x - y$$
$$y' = (1 - \rho)y + x$$
$$z' = (1 - \rho)z,$$

(c) Here do both cases: $B = 0$ and $B = 1$.

$$
\begin{aligned}
x' &= (1 - \rho)(2 - \rho)x - By + xz \\
y' &= (1 - \rho)(2 - \rho)y + Bx + yz \\
z' &= (1 - \rho)(2 - \rho)z - (x^2 + y^2),
\end{aligned}
$$

8. **(Cylindrical Coordinates)** The cylindrical coordinate map $g : \mathbb{R}^3 \to \mathbb{R}^3$ is

$$
g(r, \theta, z) = (\, r \cos \theta, \; r \sin \theta, \; z \,).
$$

Suppose $A, B, C, D : [0, \infty) \to \mathbb{R}$ are differentiable functions. Consider the system in Cartesian coordinates:

$$
\begin{aligned}
x' &= A(r)x - B(r)y + C(r)xz \\
y' &= B(r)x + A(r)y + C(r)yz \\
z' &= D(r)z,
\end{aligned}
$$

where $r = (x^2 + y^2)^{1/2}$. Show that g transforms this system into the cylindrical coordinate system:

$$
\begin{aligned}
r' &= A(r)r + C(r)rz \\
\theta' &= B(r) \\
z' &= D(r)z,
\end{aligned}
$$

Use this to study the following system:

$$
\begin{aligned}
x' &= (1 - r^2)x - by \\
y' &= bx + (1 - r^2)y \\
z' &= (1 - r^2)z,
\end{aligned}
$$

In particular, do the following: (a) Show that the unit cylinder U $x^2 + y^2 = 1$ is invariant under the flow and, in fact, consists entirely of circular flows $\gamma_c(t) = (\cos bt, \sin bt, c)$, where c is any constant. (b) Show that, in the cylindrical coordinate system, $r(t), z(t)$ lie on the line $z = (z_0/r_0)r$ in the r-z plane and that $\lim_{t \to \infty} r(t) = z_0/r_0$. (c) Show that each integral curve α of the Cartesian system with initial point (x_0, y_0, z_0) having $x_0^2 + y_0^2 \ne 1, 0$, will spiral toward the circle γ_{z_0/r_0} on the unit cylinder U while remaining on the cone $z^2 = (z_0^2/r_0^2)(x^2 + y^2)$. (d) Discuss what happens on the z-axis. (e) Sketch, by hand, the phase portrait. Use this to construct a good computer plot of the phase portrait.

9. **(Hyperbolic Coordinates)** As you know, polar coordinates are closely connected with the geometry of the circle: for a fixed $r \ge 0$, the polar coordinate map $g(r, \theta) = (r \cos \theta, r \sin \theta)$ gives a parametrization of the circle

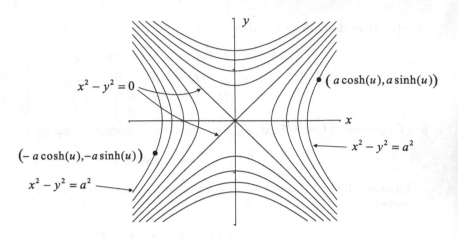

$$x^2 - y^2 = 0$$

$$(a\cosh(u), a\sinh(u))$$

$$x^2 - y^2 = a^2$$

$$(-a\cosh(u), -a\sinh(u))$$

$$x^2 - y^2 = a^2$$

Figure 5.17: *Hyperbolic coordinates on the plane.*

$x^2 + y^2 = r^2$, and every point in the plane lies on one of these circles. In an analogous, fashion we can introduce *hyperbolic coordinates* on the plane. Note that every point in the plane lies on one and only one of the hyperbolas with equation $x^2 - y^2 = \pm a^2$ (for $a = 0$, this is a pair of straight lines). See Figure 5.17.

A *hyperbolic coordinate map* is (naturally enough) given by

$$g(a, u) = (a\cosh u, a\sinh u),$$

and for a fixed a this parametrizes a branch of the hyperbola $x^2 - y^2 = a^2$. *Note:* On the other hand the map $h(a, u) = (a\sinh u, a\cosh u)$ parametrizes the hyperbola $x^2 - y^2 = -a^2$. If we restrict attention to the points in the region $\{(x, y \mid x \geq 0, |y| \leq x\}$ in Figure 5.17, then the *hyperbolic coordinates* of such a point (x, y) are the numbers a, u for which

$$x = a\cosh u$$
$$y = a\sinh u.$$

(a) Invert the above relationships, i.e., solve for a, u and thus get a locally defined inverse map f to the map g. This map can be used to transform a system $(x', y') = X(x, y)$ into one in hyperbolic coordinates.

(b) Compute the derivative of f. For this, it is helpful to know that:

$$\tanh^{-1}\left(\frac{y}{x}\right) = \tfrac{1}{2}\ln|x + y| - \tfrac{1}{2}\ln|x - y|.$$

Use this to calculate the matrix $f'(g(a, u))$. As in equation (5.23), compute, the hyperbolic transformation of a vector field X in the x-y plane into the vector field $f_*(X)$ in the a-u plane.

(c) Apply the results of Exercise 2 to g. Compute the derivative of g and determine the largest domain $\overline{\mathcal{O}}$ possible such that $\det(g'(a,u)) \neq 0$, for $(a,u) \in \overline{\mathcal{O}}$. Thus for each vector field X on $\mathcal{O} \subseteq \mathbb{R}^2$ there is a unique vector field Y on $\overline{\mathcal{O}}$ that g transforms into X. The vector field is the *hyperbolic coordinate version* of X.

(d) A simple illustration of the hyperbolic transformation technique is the basic saddle point system:

$$x' = y$$
$$y' = x.$$

The corresponding vector field is $X(x,y) = (y,x)$. Show that the hyperbolic coordinate version of this system is

$$a' = 0$$
$$u' = 1.$$

The integral curves for this latter system are vertical lines as shown in Figure 5.18.

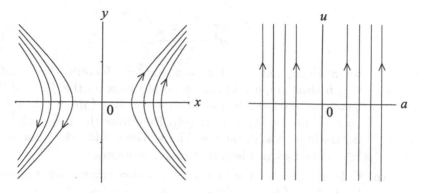

Figure 5.18: *The integral curves for $x' = y, y' = x$, and its version: $a' = 0, u' = 1$, in hyperbolic coordinates.*

Of course such a high-powered technique is not needed for so simple an example as this. However, this example serves to illustrate the content of the Flow Box Theorem in the next section—in a neighborhood of a nonfixed point there is a diffeomorphism f that transforms the integral curves into straight lines. In this example $f = g^{-1}$ is a local inverse of the hyperbolic coordinate map. The hyperbolic flow lines between the asymptotes in the saddle point system correspond to straight lines in the hyperbolic coordinate version of that system.

10. **(Hyperboloidal Coordinates)** A natural analog of spherical coordinates
    is what arises when you replace each trig function by the corresponding hyperbolic function:

$$x = a \sinh v \cosh u \qquad (5.46)$$

$$y = a \sinh v \sinh u \qquad (5.47)$$

$$z = a \cosh v. \qquad (5.48)$$

Here $a, u, v \in \mathbb{R}$ are called the *hyperboloidal coordinates* of the point (x, y, z) with Cartesian coordinates x, y, z. The corresponding *hyperboloidal map* $g : \mathbb{R}^3 \to \mathbb{R}^3$ is

$$g(a, u, v) = (\, a \sinh v \cosh u, \; a \sinh v \sinh u, \; a \cosh v \,). \qquad (5.49)$$

Analyze the nature of these coordinates and the corresponding transformation of systems of DEs from Cartesian coordinates to hyperboloidal coordinates by doing the following.

(a) For $(a, u, v) \in \mathbb{R}^3$, with $a \geq 0$, show that $(x, y, z) \equiv g(a, u, v)$ lies on the three surfaces with Cartesian equations

$$z^2 - x^2 + y^2 = a^2$$

$$y = mx$$

$$y^2 + \frac{k^2}{1 + k^2} z^2 = x^2,$$

where $m = \tanh u$ and $k = a^2 \sinh^2 v$. Identify these surfaces and sketch their graphs. On these graphs indicate the parts of these surfaces corresponding to $\{g(a, u, v) | u, v \in \mathbb{R}\}$, $\{g(a, u, v) | a, v \in \mathbb{R}\}$, and $\{g(a, u, v) | a, u \in \mathbb{R}\}$, respectively. Determine the image $g(\mathbb{R}^3)$ of $\mathbb{R}^3$ under the hyperboloidal map g. This will determine which points $(x, y, z) \in \mathbb{R}^3$ can be assigned hyperboloidal coordinates.

(b) Calculate $g'(a, u, v)$, $g'(a, u, v)^{-1}$, and determine at what points (a, u, v) the inverse of the Jacobian matrix fails to exist.

(c) State and prove an analog of Proposition 5.1 for hyperboloidal coordinates.

11. When you studied homogeneous DEs and Bernoulli DEs in an introductory differential equations course, the techniques you used to solve these involved transforming the homogeneous DE into a separable DE and transforming the Bernoulli DE into a linear DE. This is actually the same as the transformation technique discussed in this chapter and this exercise is to illustrate this.

(a) **(Homogeneous DEs)** Recall that a homogeneous DE has the form

$$x' = h\left(\frac{x}{t}\right),$$

where h is a given function. To apply the transformation theory in this chapter, we must first write this as an autonomous system

$$t' = 1$$
$$x' = h\left(\frac{x}{t}\right).$$

The corresponding vector field is $X(t, x) = (1, h(x/t))$, defined on $B = \{(t, x)|t \neq 0, x/t \in U\}$, where U is the domain of h. Let

$$f(t, x) \equiv \left(t, \frac{x}{t}\right),$$

which is defined for all (t, x) with $t \neq 0$. Show that f is a diffeomorphism and compute a formula for its inverse $g = f^{-1}$. Then show that

$$f_*(X)(t, y) = \left(1, \frac{-y + h(y)}{t}\right),$$

and so the transformed DE is the separable DE

$$y' = \frac{-y + h(y)}{t}.$$

Compare this with the method for solving homogeneous DEs in your undergraduate textbook.

(b) **(Bernoulli DEs)** A Bernoulli DE has the form

$$x' = a(t)x + b(t)x^r,$$

where $a, b : I \to \mathbb{R}$ are continuous functions and $r \neq 1$ is any real number. (When $r = 1$, the equation is linear, so this case is excluded from the discussion.) Reducing this nonautonomous DE to an autonomous one gives a system with associated vector field $X : I \times \mathcal{O} \to \mathbb{R}^2$ given by $X(t, x) = (1, a(t)x + b(t)x^r)$. For simplicity we take $\mathcal{O} = (0, \infty)$, even though for certain exponents r, the vector field X can be defined on a larger domain (see Chapter 4). Define $f : I \times \mathcal{O} \to \mathbb{R}^2$ by

$$f(t, x) = (t, x^{1-r}).$$

Show that f is a diffeomorphism and compute a formula for its inverse $g = f^{-1}$. Then show that

$$f_*(X)(t, y) = \left(1, (1 - r)\left[a(t)y + b(t)\right]\right),$$

and so the corresponding transformed DE is the linear DE

$$y' = (1 - r)\left[a(t)y + b(t)\right].$$

Compare this with the method for solving Bernoulli DEs in your undergraduate textbook.

5.3   The Linearization and Flow Box Theorems

Recall that in our study of linear systems $y' = Ay$, the notion of linear equivalence (of linear systems) led to the search among all the linear systems equivalent to $y' = Ay$ for one that was particularly simple in form. This was the canonical system: $z' = Jz$, where J is the Jordan form for A. By analogy, we would now like to do something similar for nonlinear systems $x' = X(x)$. This is essentially what the Linearization Theorem does for us (which you would well expect from all your work on applying it to specific examples). There are, however, several new features (or difficulties) that arise in looking for canonical forms for nonlinear systems:

- The canonical form for $x' = X(x)$ is only given locally on a neighborhood of each fixed point and nonfixed point, and varies from point to point. For fixed points c (with some exceptions to be explained shortly), the canonical form is $y' = Ay$, where $A = X'(c)$. To get the equivalence we must restrict X to a neighborhood $U \subseteq \mathcal{O}$ of c. For nonfixed points the canonical system is also a linear system and the equivalence is local.

- The local equivalence between $x' = X(x)$ and $y' = Ay$ is *not always* differentiable equivalence, but rather *topological equivalence*. This latter notion is more general.

Definition 5.4 (Topological Equivalence)

(1) A map $f : \mathcal{O} \to \overline{\mathcal{O}}$ is called a *homeomorphism* if it is continuous, 1-1, onto, and its inverse f^{-1} is also continuous.

(2) The system $x' = X(x)$ is said to be *topologically equivalent* to the system $y' = Y(y)$ if there exists a homeomorphism $f : \mathcal{O} \to \overline{\mathcal{O}}$, such that: (i) for each integral curve $\alpha : I \to \mathcal{O}$ of X, the curve $\beta = f \circ \alpha : I \to \overline{\mathcal{O}}$ is an integral curve of Y, and (ii) for each integral curve $\beta : I \to \overline{\mathcal{O}}$ of Y, the curve $\alpha = f^{-1} \circ \beta : I \to \mathcal{O}$ is an integral curve of X.

We now come to the Linearization Theorem. We do not prove this here, but refer the interested reader to Appendix B. The proof, while quite long, has many interesting aspects to it and connections to other important topics in analysis. Thus, it is worthy of your study and is highly recommended, if you have time.

Theorem 5.3 (Linearization Theorem) *Suppose that* $X : \mathcal{O} \to \mathbb{R}^n$ *is a vector field and* $c \in \mathcal{O}$ *is a fixed point of* X. *Let* $A = X'(c)$, *and let* $Y(y) = Ay$ *be the corresponding vector field. Assume that* c *is hyperbolic fixed point (i.e.,* $\det A \neq 0$ *and* A *has no purely imaginary eigenvalues). Then there exist neighborhoods* U *of* c *and* $\overline{U}$ *of* 0, *such that the restrictions of* X *to* U *and* Y *to* $\overline{U}$ *are topologically equivalent.*

To elaborate the content of the theorem, we mention that under the stated conditions, we can find a homeomorphism $f : U \to \overline{U}$ between neighborhoods $U, \overline{U}$ of $c, 0$, that maps each integral curve $\alpha : I \to U$ of X into an integral curve $f \circ \alpha : I \to \overline{U}$ of Y. The theorem is existential, and in any particular example it might be impossible to construct the homeomorphism that works. However, knowing that one exists allows us to study the linear system $y' = Ay$ near its fixed point 0, with the confidence that each of its integral curves will look similar to one of $x' = X(x)$.

The Linearization Theorem is based on an idea that occurs frequently in mathematics. This idea is that the coefficients of the terms in the Taylor series expansion

$$F(x) = F(0) + F'(0)x + \tfrac{1}{2}F''(0)(x, x) + \cdots,$$

describe the geometric nature, or behavior, of a function $F : U \subseteq \mathbb{R}^n \to \mathbb{R}^n$, near $x = 0$ (cf. Appendix A). If we assume that the fixed point c for the vector field X is the origin $c = 0$ (there is no loss of generality in this since we can always transform X by a translation), then $X(0) = 0$, $X'(0) = A$, and the Taylor series expansion of X has the form

$$X(x) = Ax + R(x).$$

Heuristically $R(x)$ stands for all the remaining terms in the Taylor series, i.e., the remainder term. In reality, since X is usually only assumed to be C^1, one *defines* R by $R(x) \equiv X(x) - Ax$. This then is a motivation for the Linearization Theorem and the starting point for its proof (see Appendix B).

The two conditions on the fixed point c in the theorem: that it be simple and hyperbolic, are necessary in its proof and when these conditions do not hold then the theorem gives no information about the nature of the phase portrait near c. Thus if c is nonsimple ($\det(A) = 0$) or if A has a pure imaginary eigenvalue, then the phase portrait may or may not resemble the linear system. In some cases it does and in other cases it does not. Here are two examples to illustrate what can happen.

Example 5.8 (A Nonsimple Fixed Point) The nonlinear system

$$x' = x^2$$
$$y' = y,$$

has $(0,0)$ as its only fixed point. The derivative of $X(x,y) = (x^2, y)$ is

$$X'(x,y) = \begin{bmatrix} 2x & 0 \\ 0 & 1 \end{bmatrix},$$

and so:

$$A = X'(0,0) = \begin{bmatrix} 0 & 0 \\ 0 & 1 \end{bmatrix}.$$

Since $\det A = 0$, the fixed point is nonsimple. As you can see from the phase portraits in Figure 5.19 the global phase portraits for the nonlinear and linear systems are not similar at all.

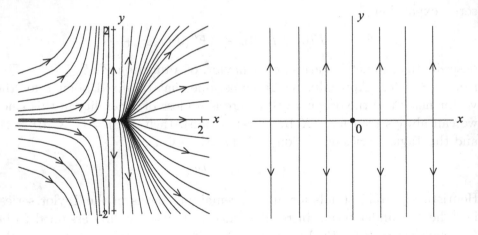

Figure 5.19: *Phase portraits for the system $x' = x^2, y' = y$, and its lineariza-tion $x' = 0, y' = y$.*

Locally on a small enough neighborhood of the origin, the phase portraits are, however somewhat similar, even though quantitatively their behaviors are quite different: the x-axis in the linear system is a line of fixed points, but in the nonlinear system it consists of two straight-line integral curves separated by the fixed point at the origin.

Example 5.9 (A Nonhyperbolic Fixed Point) Consider the system

$$x' = -y - x^3 - xy^2$$
$$y' = x - y^3 - x^2y.$$

If we linearize about the fixed point $(0, 0)$, we find that

$$X'(x, y) = \begin{bmatrix} -3x^2 - y^2 & -1 - 2xy \\ 1 - 2xy & -3y^2 - x^2 \end{bmatrix}$$

and

$$A = X'(0, 0) = \begin{bmatrix} 0 & -1 \\ 1 & 0 \end{bmatrix}.$$

Thus, the fixed point $(0, 0)$ is nonhyperbolic. The corresponding linear system is a center with the origin being stable but not asymptotically stable. However the nonlinear system has a phase portrait (see Figure 5.20), where the origin appears to be asymptotically stable, i.e., nearby integral curves tend to the origin in the limit. Compare the results here with those in the predator-prey example, where the linearized version was a center that *did* correspond qualitatively to the phase portrait of the nonlinear system.

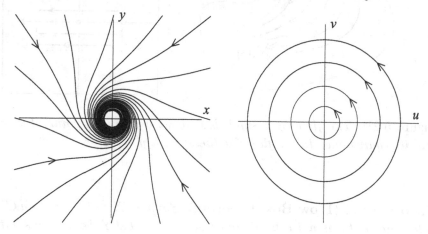

Figure 5.20: *Phase portraits for the system* $x' = -y - x^3 - xy^2, y' = x - y^3 - x^2y$, *and its linearization* $u' = -w, w' = u$.

Near a point c that is not a fixed point of X, the phase portrait has a standard (and rather uninteresting) look. This is described in the Flow Box heorem below which says that, in a small box (neighborhood) about c, the flow for X is differentiably equivalent to the straight-line flow for the system $y' = Y(y)$, with $Y : \mathbb{R}^n \to \mathbb{R}^n$ the *constant* vector field:

$$Y(y) = (1, 0, 0, ..., 0) = e_1,$$

for every $y \in \mathbb{R}^n$. The flow for ψ for Y is as simple as possible, namely:

$$\psi_t(c) = (c_1 + t, c_2, \ldots, c_n),$$

which is just a uniform flow in the direction: $e_1 = (1, 0, 0, ..., 0)$. According to the theorem the phase portrait for X on the box about c has a flow pattern that is the "same" (up to a distortion caused by a diffeomorphism). This is illustrated hypothetically in Figure 5.21, which illustrates the case for planar flow.

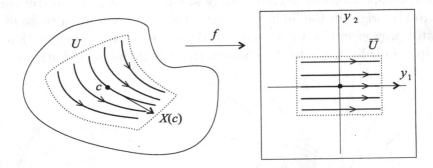

Figure 5.21: *The flow in a small box about a nonfixed point of* $x' = X(x)$ *and its equivalence to the flow for the system* $y' = e_1$.

Theorem 5.4 (Flow Box Theorem) *Suppose* $X : \mathcal{O} \to \mathbb{R}^n$ *is a* C^1 *vector field, and that* c *is not a fixed point:* $X(c) \neq 0$. *Let* Y *be the constant vector field:*

$$Y(y) = (1, 0, 0, \ldots, 0) = e_1,$$

for every $y \in \mathbb{R}^n$. *Then there exist neighborhoods* U *of* c *and* $\overline{U}$ *of* 0, *and a diffeomorphism* $f : U \to \overline{U}$, *such that*

$$f_*(X) = Y.$$

Thus the system $x' = X(x)$ is differentiably equivalent to the system:

$$
\begin{aligned}
y_1' &= 1 \\
y_2' &= 0 \\
&\ \ \vdots \\
y_n' &= 0,
\end{aligned}
$$

on the respective neighborhoods.

Proof: Technically, it will be easier to construct a diffeomorphism g such that $g_*(Y) = X$, locally. Then we can take $f \equiv g^{-1}$.

Let $v_1 = X(c)$ and $V = \{\, v \in \mathbb{R}^n \mid v \cdot v_1 = 0 \,\}$ be the subspace of $\mathbb{R}^n$ that is orthogonal to v_1. Choose any basis $\{v_2, \ldots, v_n\}$ for V and let $\phi : \mathcal{D} \to \mathbb{R}^n$ be the flow for X. Select any neighborhood $I \times B(c, r) \subseteq \mathcal{D}$ of $(0, c)$ and define

$$
W = \{\, (a_2, \ldots, a_n) \in \mathbb{R}^{n-1} \mid \sum_{i=2}^{n} |a_i| |v_i| < r \,\}.
$$

Then $I \times W \subseteq \mathbb{R}^n$ is a neighborhood of $0 \in \mathbb{R}^n$ and it is easy to see that if $(a_2, \ldots, a_n) \in W$, then $c + a_2 v_2 + \cdots + a_n v_n \in B(c, r)$. Thus, defining $g : I \times W \to \mathbb{R}^n$ by

$$
g(t, a_2, \ldots, a_n) = \phi_t(c + a_2 v_2 + \cdots + a_n v_n),
$$

makes sense and will give us the transformation we need. Note that

$$
g(0, 0, \ldots, 0) = \phi_0(c) = c.
$$

Geometrically, for fixed $a_2, \ldots, a_n$, the map $t \mapsto g(t, a_2, \ldots, a_n)$ is an integral curve passing through the point $c + a_2 v_2 + \cdots + a_n v_n$ on the hyperplane $V + c$. This is illustrated in Figure 5.22 for $n = 2$ and indicates why we would expect g to be a diffeomorphism locally on a neighborhood of $0 = (0, 0, \ldots, 0)$.

To show this we appeal to the Inverse Function Theorem (Appendix A), by showing that the Jacobian matrix $g'(0)$ is invertible. First we calculate the Jacobian at a general point $(t, a_2, \ldots, a_n) \in I \times W$. For the ease of notation we let $p = c + a_2 v_2 + \cdots + a_n v_n$, be the corresponding point in $B(c, r)$. Now use the chain rule and the property

$$
\frac{\partial \phi}{\partial t}(t, x) = X(\phi(t, x)),
$$

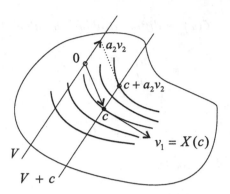

Figure 5.22: *The subspace V orthogonal to $v_1 = X(c)$ and its translate to a hyperplane $V + c$ through the point c.*

of the flow to easily calculate that

$$g'(t, a_2, \ldots, a_n) = \begin{bmatrix} X^1(\phi(t,p)) & \frac{\partial\phi^1}{\partial x_i}(t,p)v_2^i & \cdots & \frac{\partial\phi^1}{\partial x_i}(t,p)v_n^i \\ \vdots & \vdots & & \vdots \\ X^n(\phi(t,p)) & \frac{\partial\phi^n}{\partial x_i}(t,p)v_2^i & \cdots & \frac{\partial\phi^n}{\partial x_i}(t,p)v_n^i \end{bmatrix}. \quad (5.50)$$

Here, we are using implied summation on repeated indices in order to shorten the entries in the last $n-1$ columns. That is, $\frac{\partial\phi^k}{\partial x_i}(t,p)v_m^i \equiv \sum_{i=1}^n \frac{\partial\phi^k}{\partial x_i}(t,p)v_m^i$. Now note that since $\phi(0,x) = \phi_0(x) = x$, for all $x \in \mathcal{O}$, it follows that

$$\frac{\partial\phi^k}{\partial x_i}(0, x) = \lim_{h\to 0} \frac{\phi^k(0, x + he_i) - \phi^k(0, x)}{h} = \delta_{ik},$$

where δ_{ik} is the Kronecker delta function: $\delta_{ii} = 1$ and $\delta_{ik} = 0$ for $i \neq k$. Thus, taking $t = 0$ in equation (5.50) will reduce the expression for the last $n-1$ columns to just $v_2, \ldots, v_n$. Also $p = c$ for $a_2 = 0, \ldots, a_n = 0$ and so we get

$$g'(0, 0, \ldots, 0) = \begin{bmatrix} X^1(c) & v_2^1 & \cdots & v_n^1 \\ \vdots & \vdots & & \vdots \\ X^n(c) & v_2^n & \cdots & v_n^n \end{bmatrix}$$

$$= [v_1, v_2, \ldots, v_n].$$

This is the matrix with $v_1, v_2, \ldots, v_n$ as its columns. Since these vectors are linearly independent, it follows that $g'(0)$ is invertible. Thus, by the Inverse

Function Theorem, there is a neighborhood $\overline{U}$ of 0 and a neighborhood U of c, such that $g : \overline{U} \to U$ is a diffeomorphism.

Now all that is left is to show that $g_*(Y) = X$, where Y is the constant vector field: $Y(y) = e_1$ for all $y \in \overline{U}$. But this is easy. If $x \in U$, then

$$x = g(t, a_2, \ldots, a_n) = \phi_t(c + a_2 v_2 + \cdots + a_n v_n),$$

for some $t \in I$ and $a_2, \ldots, a_n \in W$. Then by definition of $g_*(Y)$ and equation (5.50), we get that

$$
\begin{aligned}
g_*(Y)(x) &= g'(g^{-1}(x))Y(g^{-1}(x)) \\
&= g'(t, a_2, \ldots, a_n)e_1 \\
&= \text{the first column of } g'(t, a_2, \ldots, a_n) \\
&= X\Big(\phi_t(c + a_2 v_2 + \cdots + a_n v_n)\Big) \\
&= X(x)
\end{aligned}
$$

The theorem is now complete if we take $f = g^{-1}$ and use one of the results from Exercise 1 in the preceding exercise set. $\square$

Exercises 5.3

1. Suppose $X : \mathcal{O} \to \mathbb{R}^n$ and $Y : \overline{\mathcal{O}} \to \mathbb{R}^n$ are vector fields on open sets $\mathcal{O}, \overline{\mathcal{O}}$ in $\mathbb{R}^n$ and that $f : \mathcal{O} \to \overline{\mathcal{O}}$ is a homeomorphism between $\mathcal{O}$ and $\overline{\mathcal{O}}$. Denote the flows for X and Y by ϕ^X and ϕ^Y. Show that the following statements are equivalent

   (a) The systems $x' = X(x)$ and $y' = Y(y)$ are topologically equivalent via the map f.

   (b) For every $y \in \overline{\mathcal{O}}$ and $t \in I_y$:

   $$\phi_t^Y = f \circ \phi_t^X \circ f^{-1}. \tag{5.51}$$

   Thus, in particular, whenever f is a diffeomorphism and $Y = f_*(X)$ is the push-forward of X, the flows for X and Y are related by formula (5.51).

2. The following systems dramatically illustrate that the Linearization Theorem does *not* apply at nonsimple fixed points. For each system, find all the fixed points, write down the linear system at the fixed points, and use a computer to draw the phase portrait for the nonlinear system.

   (a)

   $$
   \begin{aligned}
   x' &= x^2 + 2xy \\
   y' &= 2xy + y^2
   \end{aligned}
   $$

(b)

$$x' = x^2 - 2xy$$
$$y' = -2xy + y^2$$

(c)

$$x' = -5y^5$$
$$y' = x + y^2$$

3. The vector field:

$$X(x,y) = \left(\frac{-3y^2}{1+2y}, \frac{1}{1+2y}\right),$$

has no fixed points. As an illustration of the Flow Box Theorem, do the following:

(a) Let $f : \mathbb{R}^2 \to \mathbb{R}^2$ be the transformation defined by

$$f(x,y) = (x + y^3, y + y^2),$$

and consider the open sets:

$$\mathcal{O} = \{(x,y) \in \mathbb{R}^2 \,|\, y \geq -1/2\},$$

$$\overline{\mathcal{O}} = \{(u,v) \in \mathbb{R}^2 \,|\, v \geq -1/4\},$$

Show that $f : \mathcal{O} \to \overline{\mathcal{O}}$ is a diffeomorphism. Find an explicit formula for $g(u,v) \equiv f^{-1}(u,v)$. Show that $f_*(X) = e_2 = (0,1)$.

(b) Use a computer to draw the phase portrait for $(x',y') = X(x,y)$. Be careful with the integral curves starting near $y = -1/2$

4. Read the material in CDChapter 5 on the electronic component that pertains to the Linearization Theorem for discrete dynamical systems and work the exercises there.

Chapter 6

Stability Theory

In this chapter we study the topic of stability for dynamical systems. There are number of different concepts and definitions of stability and these apply to various types of integral curves: fixed points, periodic solutions, etc., for dynamical systems (cf. [Ha 82], [Rob 95], [RM 80], [AM 78], [Co 65], [Bel 53], [Mer 97]). This chapter provides an introduction to the subject, giving first a few results about stability of fixed points and then a brief discussion of stability of periodic solutions (also called cycles or closed integral curves).

The question of whether a given motion of a dynamical system is stable or not is a natural one, and we have already used the terminology—stable/unstable fixed point—throughout the text in numerous examples. The definitions of stability are given precisely below, but the basic idea in these definitions is whether the integral curves starting near a given fixed point (or more generally near a given integral curve) will stay near it (*stability*), and perhaps tend toward it asymptotically in time (*asymptotic stability*).

A classical example, studied in Chapter 1, is a ball rolling on a circular hoop as shown in Figure 6.1. The 2nd-order system governing the motion of ball, when reduced to a 1st-order system, gives

$$\begin{aligned} \theta' &= v \\ v' &= -k\sin(\theta), \end{aligned}$$

(this is assuming no frictional resistance). The geometry and physics of the setup shown in Figure 6.1 suggest there are two fixed points of the system. Placing the ball in the equilibrium position at the bottom of the hoop with no initial velocity corresponds to a fixed point (0,0) of the system which, from our experience with motion, should be a stable fixed point. That is, we

D. Betounes, *Differential Equations: Theory and Applications*, DOI 10.1007/978-1-4419-1163-6_6,

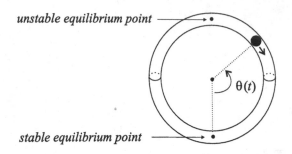

Figure 6.1: *Ball on a hoop.*

know that by placing the ball slightly to one side of equilibrium and releasing it with some (small) velocity produces a motion (integral curve) whereby the ball rolls back and forth about the equilibrium position.

If we add resistance due to frictional forces to the model, the corresponding system becomes

$$\theta' = v$$
$$v' = -k\sin(\theta) - bv,$$

and the equilibrium position becomes an *asymptotically stable* fixed point. This is so since for small displacements of the ball from equilibrium and small initial velocities, the ball rolls back and forth about the equilibrium position, but the amplitude of its displacement becomes less and less over time due to the friction. Eventually the motion ceases and the ball comes to rest at the equilibrium position (in theory, it takes infinitely long for this to occur). This type of motion is exhibited in the phase portrait shown in Figure 6.2, where the asymptotic stability of the stable equilibrium points is exhibited by the integral curves which spiral in toward the fixed points at $\theta = 0, \pm 2\pi, \ldots$. Compare this phase portrait with the one for motion with no frictional resistance shown in Figure 1.6.

The ball on the hoop example also has an *unstable* fixed point, namely the top of the hoop as shown in Figure 6.1. Placing the ball exactly on the equilibrium position shown results in the ball remaining there, but placing it slightly to one side or the other results in a motion away from the equilibrium point. The instability here is well known from our experience of the difficulty of balancing objects subject to the force of gravity.

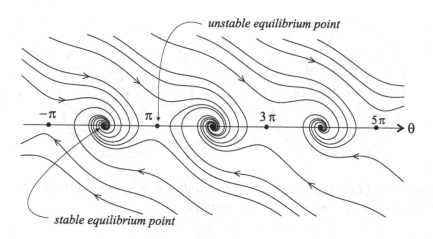

Figure 6.2: *Phase portrait for the ball on the hoop example with frictional resistance. Friction makes the stable equilibrium points into asymptotically stable ones.*

6.1 Stability of Fixed Points

For physical systems, like the ball in the hoop, and for systems in the plane, it often easy to discern the stability or asymptotic stability of fixed points from physical principles and pictures of the phase portraits. However, a formal, precise definition of these concepts is needed to clarify exactly what the concept is and to enable us to determine stability in more complex situations.

Recall: $B(c, \delta) = \{ x \in \mathbb{R}^n \mid |x - c| < \delta \}$ is the ball about c of radius δ.

Definition 6.1 (Stability of Fixed Points) Suppose $X : \mathcal{O} \to \mathbb{R}^n$ is a vector field on an open set $\mathcal{O}$ in $\mathbb{R}^n$, and let ϕ be the flow corresponding to the system $x' = X(x)$. Recall that for $x \in \mathcal{O}$, the maximum interval of existence for the integral curve passing through x at time $t = 0$ is denoted by I_x and $a_x < b_x$ denote the left- and right-hand endpoints of I_x, respectively.

A fixed point c is called *stable* if for each $\varepsilon > 0$ there exists a $\delta > 0$, such that for every $x \in B(c, \delta)$,

$$|\phi_t(x) - c| < \varepsilon \qquad \text{for all } t \in [0, b_x).$$

This definition tacitly assumes that δ is small enough so that $B(c, \delta) \subseteq \mathcal{O}$.

Figure 6.3 exhibits schematically what is involved in the definition. Generally if we are given an ε neighborhood of c as shown, then we will have to

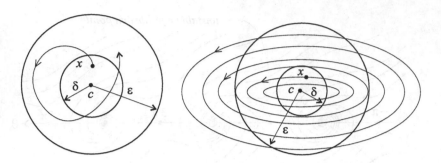

Figure 6.3: **(Left):** *Illustration of the stability definition.* **(Right):** *A choice of δ for a given ε in a particular example.*

choose a smaller δ neighborhood for the initial points x. The necessity of sometimes having to choose $\delta < \varepsilon$ is also illustrated in Figure 6.3.

A fixed point c is unstable if it is not stable, which is equivalent to saying that there exists an $\varepsilon > 0$, a sequence $\{x_k\}_{k=1}^{\infty}$ of points in $\mathcal{O}$, and a sequence $\{t_k\}_{k=1}^{\infty}$, of times, with $t_k \in [0, b_{x_k})$, such that $\lim_{k \to \infty} x_k = c$ and

$$|\phi_{t_k}(x_k) - c| \geq \varepsilon$$

(exercise). Thus, no matter how close x_k is to c, the flow $\phi_t(x_k)$ will eventually, at time t_k, carry it outside the epsilon neighborhood of c.

The definition of stability says, roughly, that we can make $\phi_t(x)$ remain near c for all $t \in [0, b_x)$, if x is near c. The next proposition says that this implies that $b_x = \infty$, and thus we get a stronger definition of stability.

Proposition 6.1 *Suppose c is a fixed point of X. Then c is stable if and only if for every $\varepsilon > 0$ there is a $\delta > 0$ such that for all $x \in B(c, \delta)$,*

(1) $b_x = \infty$, *i.e.,* $\phi_t(x)$ *is defined for all $t \geq 0$, and*

(2) $|\phi_t(x) - c| < \varepsilon$ *for all $t \geq 0$.*

Proof: Suppose c is stable and $\varepsilon > 0$ is given. Choose $\varepsilon_0 < \varepsilon$ such that $\overline{B}(c, \varepsilon_0) \subseteq \mathcal{O}$. Then choose $\delta > 0$ such that if $x \in B(c, \delta)$, then $|\phi_t(x) - c| < \varepsilon_0$, for all $t \in [0, b_x)$. But this says the $\phi_t(x)$ remains in the compact, convex set $\overline{B}(c, \varepsilon_0)$ for all $t \in [0, b_x)$. Hence by Theorem 3.8, we have $b_x = \infty$. Thus, (1) and (2) hold. Conversely it is clear that the criteria in (1) and (2) imply that c is stable. $\square$

Henceforth, we will use conditions (1) and (2) from the proposition as the criteria for stability. This makes it easier to state the condition for asymptotic stability.

Definition 6.2 (Asymptotic Stability) A fixed point c is called *asymptotically stable* if it is stable and if there exists a $\delta > 0$ such that

$$\lim_{t \to \infty} \phi_t(x) = c,$$

for every $x \in B(c, \delta)$. Note that because of the stability, there is a $\delta_0 > 0$ such that if $x \in B(c, \delta_0)$, then $\phi_t(x)$ is defined for all $t \geq 0$.

Applying the definition *directly* to determine stability in any particular example can be difficult. Thus, we need some theorems that will reduce the determination to checking other quantities—such as the eigenvalues of $X'(c)$ at a fixed point c or the Hessian of a certain function Λ (Liapunov function) on a neighborhood of c. We begin the description of these theorems in the next section with some results on stability of the origin 0 for linear systems.

Exercises 6.1

1. Show that a fixed point c is unstable if and only if there exists an $\varepsilon > 0$, a sequence $\{x_k\}_{k=1}^{\infty}$ of points in $\mathcal{O}$, and a sequence $\{t_k\}_{k=1}^{\infty}$, of times, with $t_k \in [0, b_{x_k})$, such that $\lim_{k \to \infty} x_k = c$ and

$$|\phi_{t_k}(x_k) - c| \geq \varepsilon,$$

for all k.

2. For each of the following systems, determine the stability of the fixed point $(0, 0)$ and prove that your determination is correct by transforming the system to polar coordinates and solving explicitly. In each case $r^2 = x^2 + y^2$.

   (a) $\quad x' = (1 - r^2)x - y$
   $\quad\quad\; y' = x + (1 - r^2)y$

   (b) $\quad x' = (r^2 - 1)x - y$
   $\quad\quad\; y' = x + (r^2 - 1)y$

6.2 Linear Stability of Fixed Points

For a linear, homogeneous, constant coefficient system $x' = Ax$, with $\det A \neq 0$, the origin 0 is the only fixed point and its stability is easy to determine in theory (but is often difficult to do in practice). This determination comes directly from the types of eigenvalues of the matrix A and a computation of the matrix exponential in the flow

$$\phi_t(c) = e^{tA}c,$$

for the system. The latter computation is facilitated by first computing e^{tJ}, where J is the Jordan canonical form for A. This computation is a generalization what you did in Exercise 4 from Section 4.5.

6.2.1 Computation of the Matrix Exponential for Jordan Forms

If J is a Jordan form, say

$$
J = \begin{bmatrix}
J_{k_1}(\lambda_1) & & & & & & \\
& \ddots & & & & & \\
& & J_{k_r}(\lambda_r) & & & & \\
& & & C_{2m_1}(a_1, b_1) & & & \\
& & & & \ddots & & \\
& & & & & C_{2m_s}(a_s, b_s)
\end{bmatrix}, \qquad (6.1)
$$

it is quite easy to compute e^{tJ} directly from its definition via the series

$$e^{tJ} = \sum_{k=0}^{\infty} \frac{t^k}{k!} J^k.$$

This so because J has block matrix form with square blocks, Jordan blocks, down its diagonal and zeros elsewhere. In addition, each Jordan block, whether real or complex, has a special form that makes its matrix exponential easy to compute. Most of the computations rely on the following basic result.

Proposition 6.2 (Block-diagonal Matrices) *Suppose A and B are $n \times n$ block-diagonal matrices with the same block structure, i.e.,*

$$
A = \begin{bmatrix}
A_1 & & & \\
& A_2 & & \\
& & \ddots & \\
& & & A_p
\end{bmatrix}, \qquad
B = \begin{bmatrix}
B_1 & & & \\
& B_2 & & \\
& & \ddots & \\
& & & B_p
\end{bmatrix}, \qquad (6.2)
$$

where A_j, B_j are $m_j \times m_j$ matrices. Then the product AB has the same block-diagonal structure and, indeed, is given by

$$AB = \begin{bmatrix} A_1 B_1 & & & \\ & A_2 B_2 & & \\ & & \ddots & \\ & & & A_p B_p \end{bmatrix}. \tag{6.3}$$

From this it follows that the matrix exponential of the block-diagonal matrix A has the same block structure and is given by

$$e^A = \begin{bmatrix} e^{A_1} & & & \\ & e^{A_2} & & \\ & & \ddots & \\ & & & e^{A_p} \end{bmatrix}. \tag{6.4}$$

Proof: First verify that the product AB has the stated form and then use this in the series definition of e^A when you compute $A^2, A^3, \ldots$. The details are left as an exercise. $\square$

Note that a special case of the proposition is when A is a diagonal matrix, i.e., each of the blocks A_j is a 1×1 matrix (a number). In particular, when $A = \lambda I$, is a multiple of the identity matrix I, we have

$$e^{\lambda I} = e^\lambda I.$$

The proposition applies directly to the Jordan form in equation (6.1) to give the following:

Exponential of a Jordan Form:

$$e^{tJ} = \begin{bmatrix} e^{tJ_{k_1}(\lambda_1)} & & & & & \\ & \ddots & & & & \\ & & e^{tJ_{k_r}(\lambda_r)} & & & \\ & & & e^{tC_{2m_1}(a_1,b_1)} & & \\ & & & & \ddots & \\ & & & & & e^{tC_{2m_s}(a_s,b_s)} \end{bmatrix}. \tag{6.5}$$

Thus, to complete the explicit computation of e^{tJ}, we need to compute the exact forms for the matrix exponentials of real and complex Jordan blocks,

$J_k(\lambda)$ and $C_{2m}(a, b)$, respectively. This involves decomposing $J_k(\lambda)$ and $C_{2m}(a, b)$ into a sum of commuting matrices, one of which is a nilpotent matrix. We look at the real case first.

Suppose $J_k(\lambda)$ is a real Jordan block

$$J_k(\lambda) = \begin{bmatrix} \lambda & 1 & 0 & \cdots & 0 & 0 \\ 0 & \lambda & 1 & \cdots & 0 & 0 \\ \vdots & & \ddots & \ddots & & \vdots \\ 0 & 0 & 0 & \cdots & \lambda & 1 \\ 0 & 0 & 0 & \cdots & 0 & \lambda \end{bmatrix}. \tag{6.6}$$

In the Chapter 4 exercises you were asked to show that $J_k(\lambda)$ decomposes into a sum of *commuting* matrices

$$J_k(\lambda) = \lambda I_k + N_k, \tag{6.7}$$

where I_k is the $k \times k$ identity matrix, and N_k is the $k \times k$ matrix with ones on the superdiagonal and zeros elsewhere, i.e.,

$$N_k = \begin{bmatrix} 0 & 1 & & & \\ & 0 & 1 & & \\ & & \ddots & & \\ & & & 0 & 1 \\ & & & & 0 \end{bmatrix}. \tag{6.8}$$

Further, N_k is a nilpotent matrix, i.e., some power of it is the zero matrix. In this case it is the kth power: $N_k^k = 0$. If you did not already verify these assertions, you should do this now. For example, when $k = 4$, the nilpotent matrix N_4 has the form

$$N_4 = \begin{bmatrix} 0 & 1 & 0 & 0 \\ 0 & 0 & 1 & 0 \\ 0 & 0 & 0 & 1 \\ 0 & 0 & 0 & 0 \end{bmatrix},$$

and it is easy to see that

$$N_4^2 = \begin{bmatrix} 0 & 0 & 1 & 0 \\ 0 & 0 & 0 & 1 \\ 0 & 0 & 0 & 0 \\ 0 & 0 & 0 & 0 \end{bmatrix}, \quad N_4^3 = \begin{bmatrix} 0 & 0 & 0 & 1 \\ 0 & 0 & 0 & 0 \\ 0 & 0 & 0 & 0 \\ 0 & 0 & 0 & 0 \end{bmatrix}, \quad N_4^4 = 0.$$

Because of the nilpotency of N_4, the series giving its matrix exponential terminates after the fourth term:

$$e^{tN_4} = I + tN_4 + \frac{t^2}{2!}N_4^2 + \frac{t^3}{3!}N_4^3 = \begin{bmatrix} 1 & t & \frac{t^2}{2!} & \frac{t^3}{3!} \\ 0 & 1 & t & \frac{t^2}{2!} \\ 0 & 0 & 1 & t \\ 0 & 0 & 0 & 1 \end{bmatrix}.$$

It is now easy to make an inductive guess about the case for a general k and get the following result.

Proposition 6.3 (Exponential of Real Jordan Blocks) *Suppose $J_k(\lambda)$ is a real Jordan block (see equation (6.6)). Then*

$$e^{tJ_k(\lambda)} = \begin{bmatrix} e^{\lambda t} & te^{\lambda t} & \frac{t^2}{2!}e^{\lambda t} & \cdots & \frac{t^{k-2}}{(k-2)!}e^{\lambda t} & \frac{t^{k-1}}{(k-1)!}e^{\lambda t} \\ 0 & e^{\lambda t} & te^{\lambda t} & \cdots & \frac{t^{k-3}}{(k-3)!}e^{\lambda t} & \frac{t^{k-2}}{(k-2)!}e^{\lambda t} \\ \vdots & & \ddots & \ddots & & \vdots \\ 0 & 0 & 0 & \cdots & e^{\lambda t} & te^{\lambda t} \\ 0 & 0 & 0 & \cdots & 0 & e^{\lambda t} \end{bmatrix}. \tag{6.9}$$

Proof: We use a result from Appendix C which says that if matrices A and B commute, i.e., $AB = BA$, then $e^{A+B} = e^A e^B$. Then in the case at hand we see that

$$\begin{aligned} e^{tJ_k(\lambda)} &= e^{\lambda tI_k + tN_k} = e^{\lambda tI_k}e^{tN_k} = e^{\lambda t}e^{tN_k} \\[2mm] &= e^{\lambda t}\begin{bmatrix} 1 & t & \frac{t^2}{2!} & \cdots & \frac{t^{k-2}}{(k-2)!} & \frac{t^{k-1}}{(k-1)!} \\ 0 & 1 & t & \cdots & \frac{t^{k-3}}{(k-3)!} & \frac{t^{k-2}}{(k-2)!} \\ \vdots & & \ddots & \ddots & & \vdots \\ 0 & 0 & 0 & \cdots & 1 & t \\ 0 & 0 & 0 & \cdots & 0 & 1 \end{bmatrix} \end{aligned}$$

Completing the above product gives the result (6.9) of the theorem. $\square$

The computation of the matrix exponential of a complex Jordan block is similar in many respects, but now bear in mind that the block form will naturally involve 2×2 matrices. Thus, suppose

$$C_{2m}(a,b) = \begin{bmatrix} C(a,b) & I_2 & 0 & 0 & \cdots & 0 & 0 \\ 0 & C(a,b) & I_2 & 0 & \cdots & 0 & 0 \\ \vdots & \vdots & \vdots & \vdots & \vdots & \vdots & \vdots \\ 0 & 0 & 0 & 0 & \cdots & C(a,b) & I_2 \\ 0 & 0 & 0 & 0 & \cdots & 0 & C(a,b) \end{bmatrix} \quad (6.10)$$

is a $2m \times 2m$, complex Jordan block with I_2 the 2×2 identity matrix and

$$C(a,b) = \begin{bmatrix} a & b \\ -b & a \end{bmatrix}.$$

We can decompose $C_{2m}(a,b)$ into a the sum of commuting matrices:

$$C_{2m}(a,b) = D_{2m}(a,b) + M_{2m},$$

where $D_{2m}(a,b)$ is the block-diagonal matrix with 2×2 blocks $C(a,b)$ down the diagonal and zeros elsewhere:

$$D_{2m}(a,b) = \begin{bmatrix} C(a,b) & 0 & 0 & 0 & \cdots & 0 & 0 \\ 0 & C(a,b) & 0 & 0 & \cdots & 0 & 0 \\ \vdots & \vdots & \vdots & \vdots & \vdots & \vdots \\ 0 & 0 & 0 & 0 & \cdots & C(a,b) & 0 \\ 0 & 0 & 0 & 0 & \cdots & 0 & C(a,b) \end{bmatrix} \quad (6.11)$$

and M_{2m} is the block matrix with 2×2 identity matrices on the superdiagonal and zeros elsewhere

$$M_{2m} = \begin{bmatrix} 0 & I_2 & 0 & 0 & \cdots & 0 & 0 \\ 0 & 0 & I_2 & 0 & \cdots & 0 & 0 \\ \vdots & \vdots & \vdots & \vdots & \vdots & \vdots & \vdots \\ 0 & 0 & 0 & 0 & \cdots & 0 & I_2 \\ 0 & 0 & 0 & 0 & \cdots & 0 & 0 \end{bmatrix}. \quad (6.12)$$

It is easy to compute (exercise) that the matrix exponential of $tC(a,b)$ is a multiple of the 2×2 rotation matrix $R(bt)$:

$$e^{tC(a,b)} = e^{at} \begin{bmatrix} \cos bt & \sin bt \\ -\sin bt & \cos bt \end{bmatrix} = e^{at} R(bt).$$

Consequently, by Proposition 6.2, we get that

$$
e^{tD_{2m}(a,b)} =
\begin{bmatrix}
e^{at}R(bt) & 0 & 0 & 0 & \cdots & 0 & 0 \\
0 & e^{at}R(bt) & 0 & 0 & \cdots & 0 & 0 \\
\vdots & \vdots & \vdots & \vdots & \vdots & \vdots & \vdots \\
0 & 0 & 0 & 0 & \cdots & e^{at}R(bt) & 0 \\
0 & 0 & 0 & 0 & \cdots & 0 & e^{at}R(bt)
\end{bmatrix}.
$$

The computation of the powers $M_{2m}^2, M_{2m}^3, \ldots, M_{2m}^m = 0$ is similar to the computation of the powers of N_k above. The *form* of these powers is exactly the same and we get the following similar form for the corresponding matrix exponential

$$
e^{tM_{2m}} =
\begin{bmatrix}
I_2 & tI_2 & \frac{t^2}{2!}I_2 & \cdots & \frac{t^{m-2}}{(m-2)!}I_2 & \frac{t^{m-1}}{(m-1)!}I_2 \\
0 & I_2 & tI_2 & \cdots & \frac{t^{m-3}}{(m-3)!}I_2 & \frac{t^{m-2}}{(m-2)!}I_2 \\
\vdots & \vdots & \ddots & \ddots & \vdots & \vdots \\
0 & 0 & 0 & \cdots & I_2 & tI_2 \\
0 & 0 & 0 & \cdots & 0 & I_2
\end{bmatrix}.
\tag{6.13}
$$

Putting these two results together and using $e^{tD_{2m}(a,b)+tM_{2m}} = e^{tD_{2m}(a,b)}e^{tM_{2m}}$ (by commutativity) we get the following result.

Proposition 6.4 (Exponential of Complex Jordan Blocks) *If $C_{2m}(a,b)$ is a $2m \times 2m$ complex Jordan Block (cf. equation (6.10)), then*

$$
e^{tC_{2m}(a,b)} =
\tag{6.14}
$$

$$
\begin{bmatrix}
e^{at}R(bt) & te^{at}R(bt) & \frac{t^2}{2!}e^{at}R(bt) & \cdots & \frac{t^{m-2}}{(m-2)!}e^{at}R(bt) & \frac{t^{m-1}}{(m-1)!}e^{at}R(bt) \\
0 & e^{at}R(bt) & te^{at}R(bt) & \cdots & \frac{t^{m-3}}{(m-3)!}e^{at}R(bt) & \frac{t^{m-2}}{(m-2)!}e^{at}R(bt) \\
\vdots & \vdots & \ddots & \ddots & \vdots & \vdots \\
0 & 0 & 0 & \cdots & e^{at}R(bt) & te^{at}R(bt) \\
0 & 0 & 0 & \cdots & 0 & e^{at}R(bt)
\end{bmatrix}
$$

Proof: This is clear from the discussion prior to the proposition. $\square$

From the explicit form of $e^{tJ_k(\lambda)}$ in equation (6.9), it is easy to see that if $\lambda < 0$, then each entry in this matrix tends to zero as t tends to infinity and thus $\lim_{t \to \infty} e^{tJ_k(\lambda)} = 0$. Here the limit is with respect to a certain matrix norm $\| \cdot \|$ on the collection $\mathcal{M}_n$ of all $n \times n$ matrices. For convenience we will use the following norm. If $B = \{b_{ij}\}$ is an $n \times n$ matrix, then its *norm* is

$$\|B\| = \max\{ |b_{ij}| \,|\, i, j \in \{1, \ldots, n\} \}. \tag{6.15}$$

Using this norm on matrices, the usual (Euclidean) norm $| \cdot |$ on vectors $x \in \mathbb{R}^n$, and Schwarz's inequality, it is easy to show that the following identity holds:

$$|Ax| \leq n\|A\|\,|x|$$

(exercise). See the Appendix C for more details on matrix norms and matrix analysis.

In a similar fashion, it's easy to see from the form of $e^{tC_{2m}(a,b)}$ in equation (6.14) that if $a < 0$, then $\lim_{t \to \infty} e^{tC_{2m}(a,b)} = 0$.

In summary, these two results say the exponentials of the Jordan blocks $e^{tJ_k(\lambda)}, e^{tC_{2m}(a,b)}$ tend to zero as t tends to infinity, provided $\lambda < 0$ and $a < 0$. This is the essence of the linear stability theorem.

Theorem 6.1 (Linear Stability) *Suppose A is an $n \times n$ matrix, all of whose eigenvalues have negative real parts. Then there are positive constants t_0, K, m such that*

$$\|e^{tA}\| \leq Ke^{-mt}, \tag{6.16}$$

for all $t \geq t_0$. Consequently, with $L = nK$, it follows that

$$|e^{tA}x| \leq Le^{-mt}|x|, \tag{6.17}$$

for all $x \in \mathbb{R}^n$ and all $t \geq 0$. Hence, the origin 0 is an asymptotically stable fixed point of the linear system $x' = Ax$.

Proof: Let J be the Jordan form for A and suppose J has the structure shown in equation (6.1). Then a list of the eigenvalues of A (with possible repeats) is

$$\lambda_1, \ldots, \lambda_r, a_1 \pm b_1, \ldots, a_s \pm b_s i,$$

and by assumption: $\lambda_1, \ldots, \lambda_r, a_1, \ldots, a_s < 0$. Since these numbers are negative, we can choose a small positive constant $m > 0$ so that also

$$m + \lambda_1, \ldots, m + \lambda_r, m + a_1, \ldots, m + a_s < 0.$$

Next we choose t_0 as follows. Using the computation of e^{tJ} in equations (6.5), (6.9), and (6.14), we see that the entries of the matrix $e^{mt}e^{tJ}$ are either 0 or have one of the following forms

$$\frac{t^p}{p!}e^{(m+\lambda_j)t}, \quad \pm\frac{t^p}{p!}e^{(m+a_j)t}\cos b_j t, \quad \pm\frac{t^p}{p!}e^{(m+a_j)t}\sin b_j t.$$

Since each of these types has limit zero as $t \to \infty$, we can choose a $t_0 > 0$ so that all the entries of $e^{mt}e^{tJ}$ have absolute value less than 1, for all $t \geq t_0$. Hence by definition of the matrix norm: $\|e^{mt}e^{tJ}\| \leq 1$, for all $t \geq t_0$. Otherwise said,

$$\|e^{tJ}\| \leq e^{-mt},$$

for all $t \geq t_0$. Now by the Jordan Canonical Form Theorem, there is an invertible matrix P such that $P^{-1}AP = J$. The constant K in the statement of the theorem can be taken to be $K \doteq n^2\|P\|\|P^{-1}\|$.

To get the required inequality (6.16), use the fact that $A = PJP^{-1}$, the identity $e^{PBP^{-1}} = Pe^B P^{-1}$ (see Chapter 4 and Appendix C), and the property $\|BC\| \leq n\|B\|\,\|C\|$ of the matrix norm (exercise). These together with the above inequality give

$$
\begin{aligned}
\|e^{tA}\| &= \|Pe^{tJ}P^{-1}\| \\
&\leq n\|P\|\,\|e^{tJ}P^{-1}\| \\
&\leq n^2\|P\|\,\|e^{tJ}\|\,\|P^{-1}\| \\
&\leq n^2\|P\|\,\|P^{-1}\|\,e^{-mt} \\
&= Ke^{-mt},
\end{aligned}
$$

for all $t \geq t_0$. This proves inequality (6.16).

We can use the above inequality to show that 0 is an asymptotically stable fixed point of $x' = Ax$. Thus, suppose $\varepsilon > 0$ is given, and let

$$\delta = \frac{\varepsilon}{n^2 K\|e^{-t_0 A}\|}.$$

If $|x - 0| = |x| < \delta$, then for all $t \geq 0$ we have

$$
\begin{aligned}
|e^{tA}x - 0| = |e^{tA}x| &\leq n\|e^{tA}\|\,|x| \\
&= n\|e^{-t_0 A}e^{(t+t_0)A}\|\,|x| \\
&\leq n^2\|e^{-t_0 A}\|\,\|e^{(t+t_0)A}\|\,|x| \\
&\leq n^2\|e^{-t_0 A}\|\,Ke^{-m(t+t_0)}\,\delta \\
&= \varepsilon e^{-m(t+t_0)} < \varepsilon.
\end{aligned}
$$

This proves stability of 0 as a fixed point. It is easy to see that $\lim_{t\to\infty} e^{At}x = 0$, and so the origin is also asymptotically stable. $\square$

Corollary 6.1 (Linear Instability) *Suppose A is an $n \times n$ matrix, all of whose eigenvalues have positive real parts. Then there are positive constants t_0, K, m such that*

$$Ke^{mt}|x| \le |e^{tA}x|, \tag{6.18}$$

for all $x \in \mathbb{R}^n$ and all $t \ge t_0$. Consequently, $\lim_{t\to\infty} |e^{tA}x| = \infty$, for all $x \ne 0$. Hence, the origin 0 is an unstable fixed point of the linear system $x' = Ax$.

Proof: Exercise.

The strong hypothesis in the above corollary is needed for the first two results there, but the instability follows from the weaker assumption that A have *at least* one eigenvalue with positive real part. For instance, suppose the eigenvalue is real: $\lambda > 0$, and that v is a corresponding eigenvector: $Av = \lambda v$. Then $\phi_t(v) = e^{At}v = e^{\lambda t}v$ and so $|\phi_t(v)| = e^{\lambda t}|v| \to \infty$, as $t \to \infty$. And more generally, any multiple of v flows off to infinity: $\lim_{t\to\infty} |\phi_t(bv)| = \infty$, for any b. We can use this to prove that the origin 0 is unstable by taking $\epsilon = 1$ and letting $x_k = v/k$, $t_k = \ln(2k/|v|)/\lambda$, for $k = 1, 2, 3, \ldots$ Then $\lim_{k\to\infty} x_k = 0$ and $|\phi_{t_k}(x_k)| = 2 > 1$, for all k. This proves part of the following proposition.

Proposition 6.5 (Linear Instability) *Suppose A has at least one eigenvalue with positive real part. Then the origin 0 is an unstable fixed point of the linear system $x' = Ax$.*

The standard example of this is in two-dimensions when the origin is a saddle point, say,

$$A = J = \begin{bmatrix} -1 & 0 \\ 0 & 3 \end{bmatrix}.$$

And in three-dimensions one could consider

$$A = J = \begin{bmatrix} -1 & 2 & 0 \\ -2 & -1 & 0 \\ 0 & 0 & 3 \end{bmatrix},$$

where the eigenvalues are $\lambda = -1 \pm 2i$, 3. Integral curves in the eigenspace E_3 (which is the z-axis) run off to infinity, while those in the pseudo-eigenspace $E_{-1 \pm 2i}$ (which is the x-z plane) spiral in toward the origin.

There are a few other results, based on the eigenvalues of A, that one can derive about the stability/instability of the linear system $x' = Ax$. However, a total understanding knowledge of the system's stability/instability requires more that knowing its eigenvalues. One must know its Jordan form J. From J we can construct the stable, unstable, and center subspaces for $x' = Ax$. This is described in the following exercise set.

Exercises 6.2

1. Prove Proposition 6.2.

2. Prove that if A is an $n \times n$ matrix, then

$$|Ax| \leq n\|A\| \, |x|, \tag{6.19}$$

for every $\in \mathbb{R}^n$. Here $|x| = (\sum_{i=1}^n x_i^2)^{1/2}$ is the Euclidean norm on vectors and $\|A\|$ is the maximum norm on matrices (as defined in the text). *Hint*: Use the *Schwarz inequality*:

$$\sum_{j=1}^n |v_j x_j| \leq (\sum_{j=1}^n v_j^2)^{1/2} (\sum_{i=j}^n x_j^2)^{1/2},$$

for all $v, x \in \mathbb{R}^n$. Prove that the same inequality (6.19) results if we use the ℓ_1 norm, $\|x\| = \sum_{i=1}^n |x_i|$, on vectors.

3. Prove Corollary 6.1. *Hint*: Theorem 6.1 applies to the matrix $-A$.

4. **(Stable, Unstable, and Center Subspaces)** The results on stability of the origin in Theorem 6.1 or on instability of the origin in Corollary 6.1 rely on the eigenvalues of A having real parts that are either all negative or all positive. In the general case, with only the restriction $\det(A) \neq 0$ on A, it is possible to extend these results as follows.

(a) By the Jordan Form Theorem, there is an invertible matrix $\tilde{P}$ such that $\tilde{P}^{-1}A\tilde{P} = \tilde{J}$ is a Jordan canonical form with $\tilde{J}$ having the structure shown in equation (6.1). Argue that by using permutation matrices, it is possible to find an invertible matrix P, such that $P^{-1}AP = J$, where J is the rearrangement of $\tilde{J}$ having all the Jordan blocks involving eigenvalues with negative, positive, and zero real parts come first, second and third in the ordering. That is, for a suitable P,

$$P^{-1}AP = J = \begin{bmatrix} J_1 & 0 & 0 \\ 0 & J_2 & 0 \\ 0 & 0 & J_3 \end{bmatrix},$$

where

$$
J_i = \begin{bmatrix}
J_{k_1^i}(\lambda_1^i) & & & & & \\
& \ddots & & & & \\
& & J_{k_{r_i}^i}(\lambda_{r_i}^i) & & & \\
& & & C_{2m_1^i}(a_1^i, b_1^i) & & \\
& & & & \ddots & \\
& & & & & C_{2m_{s_i}^i}(a_{s_i}^i, b_{s_i}^i)
\end{bmatrix},
$$

(6.20)

for $i = 1, 2$, and

$$
J_3 = \begin{bmatrix}
C_{2m_1^3}(0, b_1^3) & & \\
& \ddots & \\
& & C_{2m_{s_3}^3}(0, b_{s_3}^3)
\end{bmatrix}.
$$

(6.21)

Here $\lambda_1^1, \ldots, \lambda_{r_1}^1, a_1^1, \ldots, a_{s_1}^1$ are all negative, $\lambda_1^2, \ldots, \lambda_{r_2}^2, a_1^2, \ldots, a_{s_2}^2$ are all positive, and $b_1^3, \ldots, b_{s_3}^3$ are nonzero.

For studying stability of differential equations, this ordering of the Jordan blocks is more suitable than the standard ordering in (6.1) used in linear algebra.

(b) The matrices J_i are square matrices, say $\ell_i \times \ell_i$ matrices, with $\ell_1 + \ell_2 + \ell_3 = n$. Partition the matrix P into three submatrices

$$
P = [P_1, P_2, P_3],
$$

where P_i is an $n \times \ell_i$ matrix. Show that, as a linear map $P_i : \mathbb{R}^{\ell_i} \to \mathbb{R}^n$, each P_i is 1-1 (i.e., injective). Use this to show that $\mathbb{R}^n$ is direct sum of three subspaces

$$
\mathbb{R}^n = W_1 \oplus W_2 \oplus W_3,
$$

where

$$
W_i \equiv P_i(\mathbb{R}^{\ell_i}).
$$

The subspaces W_1, W_2 and W_3 are called the *stable*, *unstable*, and *center* subspaces of A, respectively.

(c) Show that

$$
AP_i = P_i J_i,
$$

for $i = 1, 2, 3$, and thus each subspace W_i is invariant under A, i.e., $A(W_i) \subseteq W_i$. Use this to show that

$$
e^{tA} P_i = P_i e^{tJ_i},
$$

for $i = 1, 2, 3$, and thus each subspace W_i is invariant under the flow for $x' = Ax$.

(d) **(Generalized Eigenbasis Theorem)** Prove that for any $c \in \mathbb{R}^n$, the integral curve of the system $x' = Ax$, which passes through c at time $t = 0$, is given by

$$\phi_t(c) = P_1 e^{tJ_1} u_1 + P_2 e^{tJ_2} u_2 + P_3 e^{tJ_3} u_3, \qquad (6.22)$$

where the vectors $u_i \in \mathbb{R}^{\ell_i}$ are the parameters of c in its decomposition

$$c = P_1 u_1 + P_2 u_2 + P_3 u_3,$$

into stable, unstable, and center components. This result can be considered as a generalization of the real/complex eigenbasis theorems in Chapter 4.

(e) Use Theorem 6.1 and its corollary (applied to the J_i's), to show that if $c \neq 0$ is in the stable subspace W_1 for A, then $\lim_{t \to \infty} \phi_t(c) = 0$, while if $c \neq 0$ is in the unstable subspace W_2, then $\lim_{t \to \infty} |\phi_t(c)| = \infty$. Show that for some choices of $c \neq 0$ in the center subspace W_3, the integral curve $\phi_t(c)$ remains bounded for all t, while for other choices $\phi_t(c)$ runs off to infinity as $t \to \infty$. *Note:* We are assuming here that $\ell_i \neq 0$, $i = 1, 2, 3$.

5. Study the material in CDChapter 6 on the electronic component that pertains to linear stability of discrete dynamical systems and work the exercises there.

6.3 Nonlinear Stability

For a simple fixed point c of a nonlinear system $x' = X(x)$, stability is easy to determine whenever the Linearization Theorem applies, i.e., whenever the matrix $A = X'(c)$ has no purely imaginary eigenvalues. Then the phase portrait near c is similar to the phase portrait of the linear system $y' = Ay$ near the origin 0, and thus the type of stability of c and 0 are the same. More generally, the following proposition shows that fixed points in topologically equivalent systems have the same type of stability.

Proposition 6.6 (Invariance of Stability) *Suppose $X : \mathcal{O} \to \mathbb{R}^n$ and $Y : \overline{\mathcal{O}} \to \mathbb{R}^n$ are vector fields on open sets $\mathcal{O}, \overline{\mathcal{O}}$ in $\mathbb{R}^n$. Suppose $f : \mathcal{O} \to \overline{\mathcal{O}}$ is a homeomorphism that makes the systems $x' = X(x)$ and $y' = Y(y)$ topologically equivalent. If $c \in \mathcal{O}$ is a fixed point, let $b = f(c)$. Then c is stable if and only if b is stable, and c is asymptotically stable if and only if b is asymptotically stable.*

Proof: Let ϕ^X and ϕ^Y denote the respective flows for X and Y. Then (by Exercise 1, Section 5.3) for every $y \in \overline{\mathcal{O}}$, we have that $I_y = I_{f^{-1}(y)}$ and

$$\phi_t^Y(y) = f \circ \phi_t^X \circ f^{-1}(y),$$

for all $t \in I_y$.

First suppose that c is stable. To show that $b = f(c)$ is stable, let $\varepsilon > 0$ be given. Then choose $\varepsilon_0 > 0$ such that

$$B(c, \varepsilon_0) \subseteq f^{-1}(B(b, \varepsilon)).$$

By stability of c, there is a $\delta_0 > 0$, such that

$$\phi_t^X (B(c, \delta_0)) \subseteq B(c, \varepsilon_0),$$

for all $t \geq 0$. Now since f is a diffeomorphism, $f(B(c, \delta_0))$ is an open set, so we can choose an $\delta > 0$ such that $B(b, \delta) \subseteq f(B(c, \delta_0))$. Thus, for all $t \geq 0$, we have

$$
\begin{aligned}
\phi_t^Y (B(b, \delta)) &= f\left(\phi_t^X \left(f^{-1}(B(b, \delta))\right)\right) \\
&\subseteq f\left(\phi_t^X (B(c, \delta_0))\right) \\
&\subseteq f(B(c, \varepsilon_0)) \\
&\subseteq B(b, \varepsilon).
\end{aligned}
$$

This establishes the stability of b. Conversely, assume that b is stable. Then showing that c is stable is exactly the same as above (just interchange c with b and f with f^{-1}).

Next assume that c is asymptotically stable. Then c is stable and there exists a $\delta_0 > 0$, such that $B(c, \delta_0) \subseteq \mathcal{O}$ and

$$\lim_{t \to \infty} \phi_t^X (x) = c,$$

for all $x \in B(c, \delta_0)$. Now choose $\delta > 0$ such that $B(b, \delta) \subseteq f(B(c, \delta_0))$. Then by continuity of f and the above relation between the flows for X and Y, we get that

$$\lim_{t \to \infty} \phi_t^Y (y) = \lim_{t \to \infty} f(\phi_t^X (f^{-1}(y))) = f(c) = b,$$

for all $y \in B(b, \delta)$. This establishes that b is asymptotically stable. The argument is entirely similar for showing that c is asymptotically stable when b is. $\square$

Corollary 6.2 (Nonlinear Stability) *Suppose c is a simple fixed point of X. If all the eigenvalues of $X'(c)$ have negative real parts, then c is asymptotically stable. If one of the eigenvalues of $X'(c)$ has positive real part, then c is unstable.*

Proof: Let $Y(y) = Ay$, where $A = X'(c)$. By the Linearization Theorem in Chapter 5, there are neighborhoods U of c and $\overline{U}$ of 0, such that $x' = X(x)$ and $y' = Y(y)$ are topologically equivalent when restricted to U and $\overline{U}$, respectively. By the last proposition, c and 0 have the same stability type. Thus, by the Linear Stability Theorem and Proposition 6.5, the results here follow. $\square$

Exercises 6.3

1. **(Stable, Center, and Unstable Manifolds)** For nonlinear systems there is a generalization of the stability result in Corollary 6.2, to allow for the case when the real parts of the eigenvalues of $X'(c)$ are partly negative, partly positive, and some are zero. This is analogous to the extension for linear systems $x' = Ax$, using stable, unstable, and center subspaces: W_1, W_2, W_3 (see Exercise 4 in the previous section). Now the respective spaces $M_1(c), M_2(c)$, and $M_3(c)$ are manifolds with W_1, W_2, W_3 as tangent spaces (when translated to c). The theory for this can be found in various sources in the literature (cf. [AM 78, p. 525], [Ha 82, p. 238], [Irw 80, p. 151], [Per 91, p. 104]). This exercise is to look at a *very* limited version of the theory.

   Suppose that $c \in \mathcal{O}$ is a hyperbolic fixed point of $X : \mathcal{O} \to \mathbb{R}^n$ and let $A = X'(c)$. Then by the Linearization Theorem from Chapter 5, there are neighborhoods U of c and $\overline{U}$ of 0 such that the system $x' = X(x)$ on U is topologically equivalent to the linear system $y' = Ay$ on $\overline{U}$. If $f : U \to \overline{U}$ is a homeomorphism that establishes this equivalence, let

   $$M_1(c) \equiv f^{-1}(W_1 \cap \overline{U}) \quad \text{and} \quad M_2(c) \equiv f^{-1}(W_2 \cap \overline{U}),$$

   where W_1, W_2 are the stable and unstable subspaces for A introduced in Exercise 4 in the previous section. The sets $M_1(c), M_2(c)$ are called the *stable manifold* and *unstable manifold* for X at c. Show that these manifolds are invariant under the flow: If $x \in M_i(c)$, then $\phi_t(x) \in M_i(c)$, for all $t \in I_x$. Further show that if $x \in M_1(c)$, then $|\phi_t(x)|$ is bounded for all $t \in [0, b_x)$, and thus $b_x = \infty$. Then show that $\lim_{t\to\infty} \phi_t(x) = c$, for all x in the stable manifold $M_1(c)$.

2. Read the material in CDChapter 6 on the electronic component on nonlinear stability of discrete dynamical systems and work the exercises there.

6.4 Liapunov Functions

Another valuable tool in analyzing the stability and asymptotic stability of a fixed point c, especially when the Linearization Theorem fails to apply, is the use of a certain real-valued function $\Lambda : U \to \mathbb{R}$, defined on a neighborhood U of c, and having certain easily stated properties. Such functions are called Liapunov functions, and while there is not an algorithm for discerning whether a Liapunov function exists for a given fixed point, when one does, we are guaranteed stability for the fixed point.

Definition 6.3 (Liapunov Functions) Suppose $X : \mathcal{O} \to \mathbb{R}^n$ is a vector field on $\mathcal{O} \subseteq \mathbb{R}^n$.

(1) If $F : \mathcal{O} \to \mathbb{R}$ is a scalar field on $\mathcal{O}$, i.e., a real-valued, C^1 function, its *covariant derivative* along X is the scalar field $\nabla_X F : \mathcal{O} \to \mathbb{R}$ defined by

$$\nabla_X F(x) \equiv \nabla F(x) \cdot X(x) = \sum_{j=1}^{n} X^j(x) \frac{\partial F}{\partial x_j}(x). \qquad (6.23)$$

(2) Suppose $c \in \mathcal{O}$ is a fixed point of X. A real-valued, C^1 function Λ defined on a open set $U \subseteq \mathcal{O}$ with $c \in U$ is called a *Liapunov function* for c if it satisfies the following two conditions:

  (a) $\Lambda(c) < \Lambda(x)$, for all $x \in U \setminus \{c\}$

  (b) $\nabla_X \Lambda(x) \leq 0$, for all $x \in U \setminus \{c\}$.

If in addition Λ satisfies the strict inequality $\nabla_X \Lambda(x) < 0$, for all $x \in U \setminus \{c\}$, then it is called a *strict Liapunov function* for the fixed point c.

The covariant derivative operator ∇_X is an important and fundamental operator in differential geometry and it has extensions to actions on vector fields Y and, more generally, on tensor fields T, giving vector and tensor fields $\nabla_X Y$ and $\nabla_X T$, respectively. For a scalar field F, the covariant derivative $\nabla_X F$ is geometrically interpreted in a number of ways. Let $\phi : \mathcal{D} \to \mathbb{R}^n$ be the flow generated by X. Then it is easy to see that

$$(\nabla_X F)(\phi_t(x)) = \frac{d}{dt}\Big(F(\phi_t(x)) \Big), \qquad (6.24)$$

for all $x \in \mathcal{O}$ and $t \in I_x$ (exercise). This says that the covariant derivative gives the rate of change of F along the flow ϕ. Taking $t = 0$ in the above equation gives that $\nabla_X F(x)$ is the directional derivative of F at x in the direction of $X(x)$. An auxiliary geometric interpretation involves the level sets, or hypersurfaces, of F:

$$S_F^k = \{ x \in \mathcal{O} \mid F(x) = k \},$$

$k \in \mathbb{R}$. One can show that the gradient $\nabla F(x)$ at each point x is perpendicular to the level hypersurface through x and points in the direction of greatest increase of F. In addition,

$$(\nabla_X F)(x) = \nabla F(x) \cdot X(x) = |\nabla F(x)|\,|X(x)|\,\cos(\theta(x)), \qquad (6.25)$$

for all $x \in \mathcal{O}$ and $t \in I_x$, where $\theta(x)$ is the angle between $\nabla F(x)$ and $X(x)$.

Thus, if Λ is a Liapunov function, then the condition $\nabla_X \Lambda(x) \leq 0$, says that the angle between $X(x)$ and $\nabla \Lambda(x)$ is either a right angle ($\theta = 90$) or is obtuse ($\theta > 90$). In the former case $X(x)$ is tangent to the hypersurface S_Λ^k (where $k \equiv \Lambda(x)$), and in the latter case $X(x)$ points toward the "inside" of this hypersurface. See Figure 6.4. The figure is for $n = 2$ and so each S_Λ^k is a level curve of Λ. The sequence of level curves shown indicates how Λ decreases in value toward the fixed point c. The plots of X at various points indicate why each integral curve, when crossing a level curve, must either remain tangent to the level curve or pass toward the inside of the curve (the side on which c is on).

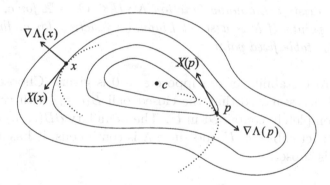

Figure 6.4: *If Λ is a Liapunov function for a fixed point c of X, then near c the integral curves of X are either tangent to the level hypersurfaces $S_\Lambda^k = \{\, x \in \mathcal{O} \mid \Lambda(x) = k \,\}$ or cross these hypersurfaces toward their interiors (direction of decreasing values of Λ).*

For emphasis, we state again, in a slightly different way, the ideas in the previous paragraphs. Condition (a) in the definition of a Liapunov function Λ says that Λ has a local minimum value at the fixed point c. Condition (b) of the definition says that for any $x \in U$, the values $\{\Lambda(\phi_t(x))\}_{t \geq 0}$ of Λ along the forward flow through x decrease with increasing t (or strictly decrease with increasing t, if Λ is a strict Liapunov function). This follows from the fact that the derivative of the function

$$g(t) \equiv \Lambda(\phi_t(x)),$$

for $t \in I_x$, is given by

$$g'(t) = \frac{d}{dt}\Big(\Lambda(\phi_t(x))\Big) = (\nabla_X \Lambda)(\phi_t(x)) \le 0,$$

and so g is a decreasing function (or strictly decreasing when Λ is a strict Liapunov function). This is the main idea used in the proof of the next theorem.

The geometrical interpretation of a Liapunov function for a fixed point c indicates why we would expect c to be a stable fixed point. The next theorem gives the details of the proof of this fact.

Theorem 6.2 (Liapunov Stability Theorem) *Suppose $X : \mathcal{O} \to \mathbb{R}^n$ is a vector field on an open set $\mathcal{O} \subseteq \mathbb{R}^n$ and $c \in \mathcal{O}$ is a fixed point of X. Assume there exists a Liapunov function $\Lambda : U \subseteq \mathcal{O} \to \mathbb{R}$ for c. Then c is a stable fixed point. If Λ is a strict Liapunov function for c, then c is an asymptotically stable fixed point.*

Proof: To prove stability of c, suppose $\varepsilon > 0$ is given. Choose a smaller epsilon, ε_0, if necessary, so that the closed ball $\overline{B}(c, \varepsilon_0)$ with center c and radius ε_0 is completely contained in U. The boundary $\partial B(c, \varepsilon_0)$ of this ball is then a compact subset of U and since Λ is continuous, it has a minimum value μ on this subset:

$$\mu \equiv \inf\{\, \Lambda(z) \,|\, z \in \partial B(c, \varepsilon_0) \,\}.$$

Since this value is attained at some point $z_0 \in \partial B(c, \varepsilon_0)$, we have, using condition (a) of the definition, that $\mu = \Lambda(z_0) > \Lambda(c)$.

Now choose $\delta > 0$ so that

$$B(c, \delta) \subseteq \{\, x \in B(c, \varepsilon_0) \,|\, \Lambda(x) < \mu \,\} = B(c, \varepsilon_0) \cap \Lambda^{-1}(-\infty, \mu).$$

See Figure 6.5. For an $x \in B(c, \delta)$, we have to show that (1) the maximal integral curve $t \mapsto \phi_t(x)$, $t \in I_x$, remains in $B(c, \varepsilon_0)$ for all $t \ge 0$, and (2) the maximal interval of existence $I_x = (a_x, b_x)$ has $b_x = +\infty$. For the rest of the proof, x will be fixed.

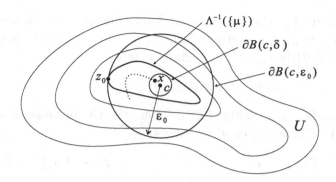

Figure 6.5: *Choice of the δ, for the given ε_0, necessary to prove stability of the fixed point c.*

Now $\Lambda(x) < \mu$, by definition of μ and choice of x. Since Λ is a Liapunov function we also have $\Lambda(\phi_t(x)) \leq \Lambda(x)$, for all $t \in [0, b_x)$. Thus,

$$\Lambda(\phi_t(x)) < \mu, \qquad\qquad \forall t \in [0, b_x). \qquad\qquad (6.26)$$

For convenience, define a real-valued function $f : I_x \to \mathbb{R}$, by

$$f(t) \equiv |\phi_t(x) - c|.$$

In terms of this notation, we would like to establish the following:

Claim: $f(t) < \varepsilon_0$ for all $t \in [0, b_x)$.

To prove the claim, assume to the contrary that it is not true, say $f(t_1) \geq \varepsilon_0$ for some $t_1 \in [0, b_x)$. See Figure 6.6. We get a contradiction as follows.

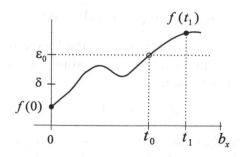

Figure 6.6: *Graph of the function f and application of the Intermediate Value Theorem.*

By continuity of f, the Intermediate Value Theorem guarantees us a $t_0 \in [0, t_1]$ such that $f(t_0) = \varepsilon_0$. Otherwise said, at time t_0, we have $|\phi_{t_0}(x) - c| = \varepsilon_0$, i.e., $\phi_{t_0}(x) \in \partial B(c, \varepsilon_0)$. So by definition of μ, we have

$$\mu \leq \Lambda(\phi_{t_0}(x)).$$

But this contradicts inequality (6.26). Thus, the Claim must be true.

Writing the Claim out explicitly gives

$$|\phi_t(x) - c| < \varepsilon_0 \leq \varepsilon, \qquad\qquad \forall t \in [0, b_x). \qquad (6.27)$$

This establishes stability once we show that $b_x = +\infty$. But this follows from Theorem 3.8 in Chapter 3, since the above says that $\phi_t(x)$ remains in the compact set $\overline{B}(x, \varepsilon_0) \subseteq \mathcal{O}$ for all $t \in [0, b_x)$.

Suppose that, in addition, Λ is a strict Liapunov function. We want to show that, in addition to the above, we also have $\lim_{t \to \infty} \phi_t(x) = c$. If this were not the case, then there would exist an $\varepsilon_1 < \varepsilon_0$ and a strictly increasing sequence $\{t_j\}_{j=1}^\infty$ of times such that

$$|\phi_{t_j}(x) - c| \geq \varepsilon_1, \qquad \text{for } j = 1, 2, 3, \ldots. \qquad (6.28)$$

Since $\{\phi_{t_j}(x)\}_{j=1}^\infty$ is a sequence in the compact set $\overline{B}(c, \varepsilon_0)$, some subsequence $\{\phi_{t_{j_k}}(x)\}_{k=1}^\infty$ converges to a point $z \in \overline{B}(c, \varepsilon_0)$. For convenience of notation let $s_k = t_{j_k}$. Then

$$\lim_{k \to \infty} \phi_{s_k}(x) = z, \qquad\qquad (6.29)$$

with $\{s_k\}_{k=1}^\infty$, a strictly increasing sequence of times. Then by (6.28), we have $|z - c| \geq \varepsilon_1$, and so $z \neq c$.

Now let $g(t) = \Lambda(\phi_t(z))$ for $t \in I_z = (a_z, \infty)$. Since Λ is a strict Liapunov function, it follows that

$$g'(t) = (\nabla_X \Lambda)(\phi_t(z)) < 0,$$

for every $t \in I_z$. Thus, g is a strictly decreasing function. In particular this says that z is not a fixed point (otherwise g would be a constant function). Also, if we pick some $s > 0$, then

$$\Lambda(\phi_s(z)) < \Lambda(z).$$

With s fixed, we note that the function $y \mapsto \Lambda(\phi_s(y))$ is continuous at z. and its value at z is less than $\Lambda(z)$. Thus, there exists an $0 < r < \varepsilon_0$ such that

$$\Lambda(\phi_s(y)) < \Lambda(z),$$

for all y with $|y - z| < r$. Now according to (6.29), we can choose a K such that $|\phi_{s_k}(x) - z| < r$ for all $k \geq K$. Consequently,

$$\Lambda(\phi_{s+s_k}(x)) = \Lambda(\phi_s(\phi_{s_k}(x)) < \Lambda(z), \tag{6.30}$$

for all $k \geq K$. Now since $\{s_k\}_{k=1}^{\infty}$ is a strictly increasing sequence, we can choose a subsequence $\{s_{k_j}\}_{j=1}^{\infty}$, that is also strictly increasing and satisfies $s_{k_j} > s + s_j$ for all j. Then because $f(t) = \Lambda(\phi_t(x))$ is a strictly decreasing function, we get from (6.30) that

$$\Lambda(\phi_{s_{k_j}}(x)) < \Lambda(z),$$

for every j. Taking $j \to \infty$ gives

$$\Lambda(z) = \lim_{j \to \infty} \Lambda(\phi_{s_{k_j}}(x)) < \Lambda(z),$$

which of course is a contradiction. This completes the proof. $\square$

Example 6.1 Consider the system in $\mathbb{R}^3$ with vector field

$$X(x, y, z) = \Big(y(z - 1), \ x(z + 1), -2xy \Big).$$

It is clear that $c = (0, 0, 0)$ is a fixed point of X. However, the Jacobian matrix of X is

$$X'(x, y, z) = \begin{bmatrix} 0 & z - 1 & y \\ z + 1 & 0 & x \\ -2y & -2x & 0 \end{bmatrix},$$

and so at the fixed point c

$$X'(0, 0, 0) = \begin{bmatrix} 0 & -1 & 0 \\ 1 & 0 & 0 \\ 0 & 0 & 0 \end{bmatrix}.$$

Thus, the Linearization Theorem does not apply (since c is not a simple fixed point). However, we claim that the function $\Lambda : \mathbb{R}^3 \to \mathbb{R}$, defined by

$$\Lambda(x, y, z) = \tfrac{1}{2}(x^2 + y^2 + z^2),$$

is a Liapunov function for c. Because of its simplicity, a function like this is often a standard choice, in all dimensions, for a Liapunov function. Namely, it is readily apparent that $\Lambda(x, y, z) \geq 0 = \Lambda(0, 0, 0)$ for all $(x, y, z) \in \mathbb{R}^3$, i.e., Λ has an absolute minimum at c. Also since $\nabla\Lambda(x, y, z) = (x, y, z)$, it is easy to calculate that

$$\nabla_X \Lambda = \nabla\Lambda \cdot X = xy(z - 1) + yx(z + 1) - 2xyz = 0,$$

for all $(x, y, z) \in \mathbb{R}^3$. Thus, Λ is a Liapunov function for c, and so by the theorem c is a stable fixed point.

Finding and verifying that a function Λ is a strict Liapunov function requires more work. In essence one must verify that Λ has a local minimum at c and that $\nabla_X \Lambda$ has a local maximum at c. The next example illustrates this.

Example 6.2 For vector fields on the plane, stability and asymptotic stability of fixed points are usually easy to ascertain from a direction field plot. However, this method is only an *experimental conjecture* and proof of the actual stability only comes from applying the theory. Thus, for example, consider the vector field

$$X(x, y) = \left(-y + xy - x^3 - \tfrac{1}{2}xy^2, \ -3y + xy + x^2 y - \tfrac{1}{2}xy^2 \right),$$

which has the origin $c = (0, 0)$ as a fixed point. Again the Linearization Theorem does not apply, since

$$X'(0, 0) = \begin{bmatrix} 0 & -1 \\ 0 & -3 \end{bmatrix}.$$

However examining the phase portrait in Figure 6.7 leads us to conjecture that the origin is asymptotically stable. Also shown in the figure are plots of several level curves for the function

$$\Lambda(x, y) = \tfrac{1}{2}(3x^2 - 2xy + y^2).$$

The graphical evidence suggests that Λ is a strict Liapunov function for $c = (0, 0)$, since the angle between $X(x, y)$ and $\nabla\Lambda(x, y)$ appears to be always larger than 90 degrees. However we need to verify this analytically. For this, first note that Λ can be written in the form

$$\Lambda(x, y) = \tfrac{1}{2}\left[2x^2 + (x - y)^2 \right],$$

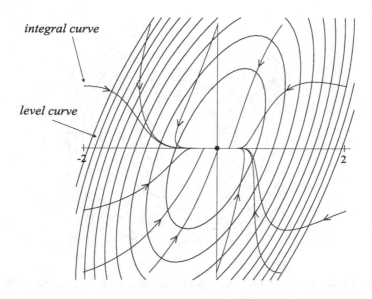

Figure 6.7: *Phase portrait for the system in Example 6.2 along with plots of level curves of a strict Liapunov function for $c = (0,0)$.*

from which it is obvious that Λ has an absolute minimum at $c = (0,0)$. Next, the gradient of Λ is

$$\nabla\Lambda = (\,3x - y, -x + y\,),$$

and so

$$\nabla_X \Lambda$$
$$= (3x - y)(-y + xy - x^3 - \tfrac{1}{2}xy^2) + (-x + y)(-3y + xy + x^2y - \tfrac{1}{2}xy^2)$$
$$= 2x^2y - 3x^4 - 2y^2$$
$$= -\left[(x^2 - y)^2 + 2x^4 + 2y^2\right].$$

In the last line, we manipulated the expression for $\nabla_X\Lambda$, so that it has a form that readily shows that $\nabla_X\Lambda(x,y) \leq 0$ for all $(x,y) \in \mathbb{R}^2$ and $\nabla_X\Lambda(x,y) = 0$ only for $(x,y) = (0,0) = c$. Hence Λ is a strict Liapunov function. Note that here we were able to use algebraic manipulation to discern that $\nabla_X\Lambda$ is strictly negative on $\mathbb{R}^2 \setminus \{c\}$. In general, you might have to use the Hessian (cf. Appendix A) to check that c is a local maximum of $\nabla_X\Lambda$.

Since Liapunov functions are not unique, you might wonder whether we could use a simpler function like

$$F(x,y) = \tfrac{1}{2}(x^2 + y^2)$$

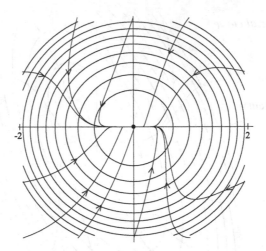

Figure 6.8: *Phase portrait for the system in Example* 6.2 *along with plots of level curves of the function* $F(x, y) = \frac{1}{2}(x^2 + y^2)$.

as a Liapunov function in this example. The graphical evidence in Figure 6.8 suggests that F might be a strict Liapunov function. Clearly F has an absolute minimum at $c = (0, 0)$ and a straightforward calculation gives

$$
\begin{aligned}
\nabla_X F &= x(-y + xy - x^3 - \tfrac{1}{2}xy^2) + y(-3y + xy + x^2 y - \tfrac{1}{2}xy^2) \\
&= -xy + x^2 y - x^4 - 3y^2 + xy^2
\end{aligned}
$$

Now $\nabla_X F(0, 0) = 0$, and so to see if F is a strict Liapunov function, we need to determine if $c = (0, 0)$ is a local maximum of the function $\nabla_X F$. Graphical evidence for this can always be obtained by graphing $\nabla_X F$ on a neighborhood of c. Figure 6.9 shows such a plot and seems to indicate that $\nabla_X F$ does have a local maximum at the origin. But, indeed, the graphical evidence leads us to a *false* conclusion. You can *prove* that the origin is *not* a local maximum for $\nabla_X F$ by calculating the Hessian $\mathcal{H}_{\nabla_X F}$ and using this to show that the origin is, in fact, a saddle point for the function $\nabla_X F$ (exercise). This does not show in Figure 6.9 because we did not use a small enough neighborhood of the origin. You can get a saddle-shaped graph by using a smaller neighborhood. Also note that $\nabla_X F(.05, .005) = -.0003174375$, while $\nabla_X F(.05, -.005) = .0001575625$. Each of these lends evidence to the assertion that $c = (0, 0)$ is not a local maximum for $\nabla_X F$, and thus F is not a Liapunov function for c.

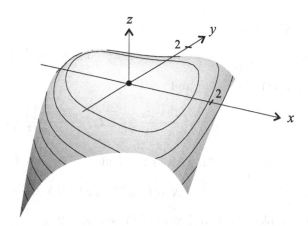

Figure 6.9: *Graph of* $\nabla_X F = -xy + x^2y - x^4 - 3y^2 + xy^2$ *on a neighborhood of* $(0,0)$.

Exercises 6.4

1. Suppose $\phi : \mathcal{D} \to \mathbb{R}^n$ be the flow generated by X. Show that

$$(\nabla_X F)(\phi_t(x)) = \frac{d}{dt}\Big(F(\phi_t(x)) \Big), \qquad (6.31)$$

for all $x \in \mathcal{O}$ and $t \in I_x$.

2. For each of the functions Λ and vector fields X assigned to you from the list below, do the following:

   (i) Prove that Λ is either a Liapunov function or a strict Liapunov function for the fixed point $c = (0,0)$. Also show that the Linearization Theorem does not apply.

   (ii) (For planar systems only) Plot, in the same figure, the phase portrait for the system and a number of level curves for the Liapunov function. Mark directions of flow on the integral curves. For several points c of intersection of an integral curve and a level curve, indicate approximately the directions of $X(c)$ and $\nabla\Lambda(c)$ and label the angle between these directions with its approximate degree measure (you may use a protractor for this).

   (a) $\Lambda = x^2 + 2y^2$ and
   $$X = (-x + 2xy^2, -x^2y).$$

   (b) $\Lambda = \frac{1}{2}(x^2 + x^2y^2 + y^4)$ and
   $$X = (-x^3 - 2xy^2, x^2y - y^3).$$

(c) $\Lambda = \frac{1}{2}(x^2 + y^2)$ and

$$X = (-y - x^3 - xy^2, \ x - y^3 - x^2 y).$$

(d) $\Lambda = x^2/(1 + x^2) + y^2$ and

$$X = \left(\frac{-2x}{(1 + x^2)^2} + 2y, \ \frac{-2(x + y)}{(1 + x^2)^2} \right).$$

(e) $\Lambda = \frac{1}{2m}(x^{2m} + y^{2m})$, with $m > 1$ and

$$X = (-y^{2m-1}, \ x^{2m-1}).$$

Do the plots for two choices of m, say $m = 2, 5$.

(f) $\Lambda = \frac{1}{2}(x^2 + 2y^2 + z^2)$ and

$$X = (-2y + yz, \ x - xz, \ xy).$$

(g) $\Lambda = \frac{1}{2}(x^2 + y^2)$ and

$$X = -y(z + 2), \ x(z + 2), \ x(y - 1)).$$

3. Show that $c = (0, 0)$ is a saddle point for the function

$$\nabla_X F = -xy + x^2 y - x^4 - 3y^2 + xy^2,$$

by calculating its Hessian (Appendix A). Also verify this experimentally by plotting its graph on a small enough neighborhood of c so that the graph has saddle shape. See Example 6.2 and Figure 6.9.

6.5   Stability of Periodic Solutions

Stability is a concept that applies not only to fixed points of $x' = X(x)$, but also to any integral curve in general. If $\gamma : I \to \mathbb{R}^n$ is an integral curve of X, we can always assume, without loss of generality, that $0 \in I$.

Recall that for $x \in \mathcal{O}$, the maximum interval of existence for the integral curve passing through x at time $t = 0$ is denoted by I_x and $a_x < b_x$ denote the left- and right-hand endpoints of I_x.

Definition 6.4 (Stability of Integral Curves) Suppose $X : \mathcal{O} \to \mathbb{R}^n$ is a vector field on an open set $\mathcal{O}$ in $\mathbb{R}^n$ and let ϕ be the flow corresponding to the system $x' = X(x)$. Suppose $\gamma : I \to \mathbb{R}^n$ is an integral curve of X and let $c = \gamma(0)$. Then γ is called *stable* if for each $\varepsilon > 0$ there exists a $\delta > 0$, such that $B(c, \delta) \subseteq \mathcal{O}$ and if $x \in B(c, \delta)$, then

(a) $[0, b_c) \subseteq [0, b_x)$ and

(b) $|\phi_t(x) - \gamma(t)| < \varepsilon$ for all $t \in [0, b_c)$.

If in addition to (a) and (b), one has

$$\lim_{t \to \infty} |\phi_t(x) - \gamma(t)| = 0,$$

then γ is called an *asymptotically stable* integral curve.

Figure 6.10 illustrates the definition of stability for an integral curve γ.

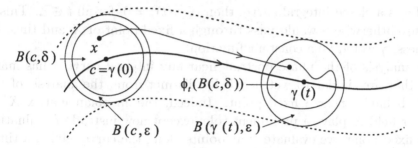

Figure 6.10: *Stability for an integral curve γ requires that flow lines can be made to remain in the moving ball $B(\gamma(t), \varepsilon)$, for any $\varepsilon > 0$.*

Note that the definition of stability for γ is, roughly speaking, more stringent than just requiring that the flow lines through points x near to $\gamma(0)$ stay near γ for times $t > 0$. Rather, stability requires that $\phi_t(x)$ remain in the ball $B(\gamma(t), \varepsilon)$. Furthermore, in the case when $b_c = \infty$, the definition can be rephrased in the more geometrical way: for each $\varepsilon > 0$, there is a $\delta > 0$, such that $b_x = \infty$ for all $x \in B(\gamma(0), \delta)$ and

$$\phi_t(B(\gamma(0), \delta)) \subseteq B(\gamma(t), \varepsilon),$$

for all $t > 0$. Also note that when γ is a constant function, i.e., a fixed point, the above definition reduces to the prior one for fixed points.

While one can attempt to analyze the stability of any integral curve, good results have, so far, only been obtained for fixed points (constant solutions) and periodic solutions of the system. Limit cycles, as discussed in Chapter 2, are examples of the latter type of integral curve. To proceed with the stability analysis, we need to make precise the concept of a "closed integral curve" (or cycle).

Definition 6.5 (Closed Integral Curves: Periodic Solutions)

(1) A function $f : \mathbb{R} \to S$ is called *periodic* if there exists a positive number $p > 0$ such that $f(t + p) = f(t)$, for all $t \in \mathbb{R}$. The number p is called a *period* of the function f.

(2) An integral curve γ of a vector field $X : \mathcal{O} \subseteq \mathbb{R}^n \to \mathbb{R}^n$, is called a *closed integral curve* (or *cycle*) if it is defined on all of $\mathbb{R}$, i.e., $\gamma : \mathbb{R} \to \mathcal{O}$, and is periodic. Otherwise said, γ is a *periodic solution* of the system of DEs: $x' = X(x)$. In the sequel when we refer to periodic solutions, we will assume they are non constant, i.e., not fixed points.

Note: If γ is a closed integral curve, then $X(\gamma(t)) \neq 0$, for all $t \in \mathbb{R}$. This is so because otherwise γ would pass through a fixed point of X and thus, by uniqueness, γ would be a constant function.

The analysis of the phase portrait near any integral curve γ has many aspects that are similar to, yet decidedly distinct from, the analysis of the system's behavior near a fixed point. Indeed, the Jacobian matrix X' of the vector field X plays a prominent role, except now instead of evaluating X' at a fixed point, we evaluate it at points along the curve, giving a time-dependent matrix, which is periodic when γ is. The motivation for using the Jacobian matrix here is the same as it was in the discussion of the Linearization Theorem for fixed points.

Heuristically, we consider a Taylor series expansion of X:

$$X(x + h) = X(x) + X'(x)h + \cdots,$$

and suppose that $\gamma : I \to \mathbb{R}^n$ is some integral curve. Assume a "nearby" integral curve α has the form $\alpha = \gamma + \xi$, where ξ is the *variation* and is thought of as being small in magnitude. See Figure 6.11. Using the fact that $\alpha'(t) = X(\alpha(t))$ and the above Taylor series expansion, we get (heuristically)

$$\gamma'(t) + \xi'(t) = X(\gamma(t) + \xi(t)) = X(\gamma(t)) + X'(\gamma(t))\xi(t) + \cdots.$$

Hence neglecting the higher-order terms (indicated by the ellipsis $\cdots$) and recalling that γ is an integral curve too, we get that the variation ξ satisfies

$$\xi'(t) = X'(\gamma(t))\xi(t).$$

This is a linear system of equations, called the *variational equations* by Poincaré, who also is responsible for first analyzing stability of periodic solutions in this fashion (cf. [Po 57]).

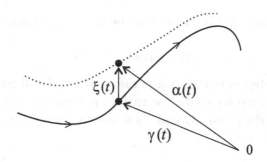

Figure 6.11: *An integral curve α near a given integral curve γ.*

Definition 6.6 (Characteristic Multipliers) Suppose $\gamma : I \to \mathbb{R}^n$ is an integral curve of X with $0 \in I$. The matrix-valued function $A : I \to \mathcal{M}_n$ defined by

$$A(t) = X'(\gamma(t)), \tag{6.32}$$

is called the *variational matrix* for γ. In the literature this matrix is also called the *monodromy matrix*. The corresponding linear system: $x' = A(t)x$, is called the system of *variational equations* for γ. The fundamental matrix $G : I \to \mathcal{M}_n$ for A is called the *characteristic matrix* for γ. If γ is a closed curve with period p, then the *characteristic multipliers* of γ are defined to be the eigenvalues of characteristic matrix at p, i.e., the eigenvalues of $G(p)$.

Example 6.3 (A 3-Dimensional System) Consider the system

$$
\begin{aligned}
x' &= -y + xz \\
y' &= x + yz \\
z' &= -z(x^2 + y^2)
\end{aligned}
$$

with corresponding vector-field $X : \mathbb{R}^3 \to \mathbb{R}^3$,

$$X(x, y, z) = (-y + xz, \ x + yz, -z(x^2 + y^2)).$$

Later we will analyze the integral curves of this system completely, but for now observe that restricting X to the x-y plane (the plane $z = 0$) we get $X(x, y, 0) = (-y, x, 0)$ and so X is tangent to the x-y plane and the integral curves that start in this plane satisfy

$$
\begin{aligned}
x' &= -y \\
y' &= x
\end{aligned}
$$

Thus, we get a plane full of circles for the integral curves. We choose one of these, say

$$\gamma(t) = (\cos t, \sin t, 0),$$

which is closed integral curve with period $p = 2\pi$. To illustrate the concepts introduced so far, we compute the variational matrix $A(t) = X'(\gamma(t))$, the fundamental matrix G for A, and then the characteristic multipliers for γ. An easy calculation gives

$$X'(x, y, z) = \begin{bmatrix} z & -1 & x \\ 1 & z & y \\ -2xz & -2yz & -(x^2 + y^2) \end{bmatrix},$$

and so

$$A(t) = X'(\cos t, \sin t, 0) = \begin{bmatrix} 0 & -1 & \cos t \\ 1 & 0 & \sin t \\ 0 & 0 & -1 \end{bmatrix}.$$

While it is generally difficult to explicitly compute the fundamental matrix G for a given time-dependent matrix A, we can do so in this case. The linear system corresponding to A is

$$\begin{aligned} x' &= -y + (\cos t)z \\ y' &= x + (\sin t)z \\ z' &= -z \end{aligned}$$

Solving the last equation gives $z = z_0 e^{-t}$, and substituting this in the first and second equations yields

$$\begin{aligned} x' &= -y + z_0 e^{-t}\cos t \\ y' &= x + z_0 e^{-t}\sin t \end{aligned}$$

The homogeneous part of this system has solution

$$\alpha_c(t) = R(t)v_0 = \begin{bmatrix} \cos t & -\sin t \\ \sin t & \cos t \end{bmatrix}\begin{bmatrix} x_0 \\ y_0 \end{bmatrix}.$$

And with $b(t) = (z_0 e^{-t}\cos t, z_0 e^{-t}\sin t)$, one can easily compute a particular solution of the nonhomogeneous system:

$$\alpha_p(t) = R(t)\int_0^t R(s)^{-1}b(s)\,ds = \begin{bmatrix} z_0(1 - e^{-t})\cos t \\ z_0(1 - e^{-t})\sin t \end{bmatrix}.$$

Using all of this gives the general solution of the linear system for A. Then the fundamental matrix is determined to be

$$G(t) = \begin{bmatrix} \cos t & -\sin t & (1 - e^{-t})\cos t \\ \sin t & \cos t & (1 - e^{-t})\sin t \\ 0 & 0 & e^{-t} \end{bmatrix}.$$

This is the characteristic matrix for the curve γ. Since γ has period 2π, we get

$$G(2\pi) = \begin{bmatrix} 1 & 0 & 1 - e^{-2\pi} \\ 0 & 1 & 0 \\ 0 & 0 & e^{-2\pi} \end{bmatrix}.$$

Hence the characteristic multipliers of γ, being the eigenvalues of the above matrix, are $1, 1, e^{-2\pi}$. We will see later that these multipliers are connected with the stability of γ.

Example 6.4 (A 2-Dimensional System) The following system has a form that simplifies considerably when written in polar coordinates. In Cartesian coordinates (with $r^2 = x^2 + y^2$) the system is

$$\begin{aligned} x' &= (r^2 - 1)x - by = -x - by + x^3 + xy^2 \\ y' &= bx + (r^2 - 1)y = bx - y + x^2y + y^3. \end{aligned}$$

Here $b > 0$ is a constant. It is easy to see that this system has a single cycle $(r = 1)$ which is the circle

$$\gamma(t) = (\cos bt, \sin bt, 0).$$

It has period $p = 2\pi/b$. The vector field for the system is

$$X(x, y) = (-x - by + x^3 + xy^2, \ bx - y + x^2y + y^3),$$

and has Jacobian matrix

$$X'(x, y) = \begin{bmatrix} -1 + 3x^2 + y^2 & -b + 2xy \\ b + 2xy & -1 + x^2 + 3y^2 \end{bmatrix}.$$

Then we get the variational matrix for γ is

$$A(t) = X'(\gamma(t)) = \begin{bmatrix} 2\cos^2 bt & -b + 2\cos bt \sin bt \\ b + 2\cos bt \sin bt & 2\sin^2 bt \end{bmatrix}.$$

It will be convenient to use some standard trig identities and write this as

$$A(t) = \begin{bmatrix} 1 + \cos 2bt & -b + \sin 2bt \\ b + \sin 2bt & 1 - \cos 2bt \end{bmatrix}.$$

Finding the fundamental matrix here involves a few tricks (see the Exercises). We get

$$G(t) = \begin{bmatrix} e^{2t} \cos bt & -\sin bt \\ e^{2t} \sin bt & \cos bt \end{bmatrix},$$

as is easily checked. This is the characteristic matrix for the curve γ. Next, since

$$G(2\pi/b) = \begin{bmatrix} e^{4\pi/b} & 0 \\ 0 & 1 \end{bmatrix},$$

the characteristic multipliers of γ are $e^{4\pi/b}, 1$. These values being ≥ 1 will indicate the instability of γ according to the theory to be introduced. In this example however, the instability is understood geometrically by using the polar coordinate version of the system. We will return to this in a moment.

For the theory, we will need the concept of the deformation matrix:

Definition 6.7 (Deformation Matrix) *If $\phi : \mathcal{D} \subseteq \mathbb{R} \times \mathbb{R}^n \to \mathbb{R}^n$ is the flow generated by a vector field X, then the* deformation matrix H *is the matrix-valued function $H : \mathcal{D} :\to \mathcal{M}_n$, defined by*

$$H(t, x) = \phi'(t, x) = \begin{bmatrix} \frac{\partial \phi^1}{\partial x_1}(t, x) & \cdots & \frac{\partial \phi^1}{\partial x_n}(t, x) \\ \vdots & & \vdots \\ \frac{\partial \phi^n}{\partial x_1}(t, x) & \cdots & \frac{\partial \phi^n}{\partial x_n}(t, x) \end{bmatrix}. \tag{6.33}$$

The deformation matrix H arises in continuum mechanics, where X represents the velocity vector field for a fluid, gas, or solid U that undergoes motion and deformation via the flow ϕ, with $\phi_t(U)$ representing the position of the continuum at time t. The connection between H and the characteristic matrix is given in the next proposition and this will be helpful in describing how G and the characteristic exponents transform under diffeomorphisms.

Proposition 6.7 *Suppose $\gamma : I \to \mathbb{R}^n$ is an integral curve of X, with $0 \in I$, and let $c = \gamma(0)$. Then the characteristic matrix G of γ coincides with the deformation matrix H at c. Specifically,*

$$G(t) = H(t, c),$$

for all $t \in I$. Furthermore,

$$G(t)X(c) = X(\gamma(t)), \tag{6.34}$$

for all $t \in I$. Hence if γ is a closed integral curve with period p, then

$$G(p)X(c) = X(c),$$

i.e., $G(p)$ has $\mu = 1$ as an eigenvalue and $X(c)$ as a corresponding eigenvector. Consequently, 1 is always a characteristic multiplier for any closed integral curve.

Proof: It is a straightforward calculation to show that the deformation matrix satisfies the matrix differential equation:

$$\frac{\partial H}{\partial t}(t, x) = X'(\phi(t, x))H(t, x), \tag{6.35}$$

for all $(t, x) \in \mathcal{D}$ and that $H(0, x) = I$, the identity matrix, for all $x \in \mathcal{O}$. (Exercise). Now since $\gamma(t) = \phi(t, c)$, for all $t \in I$, it is clear $\tilde{G}(t) \equiv H(t, c)$ is a fundamental matrix for $A(t) = X'(\gamma(t)) = X'(\phi(t, c))$. However, fundamental matrices (as we have defined them) are unique since they are solutions of matrix initial value problems. Hence, we have $\tilde{G} = G$, and the first assertion of the proposition follows.

Next, γ is an integral curve of X and so

$$\gamma'(t) = X(\gamma(t)),$$

for all $t \in I$. Differentiating both sides of this equation and using the chain rule and the above definitions gives

$$\begin{aligned} \gamma''(t) &= X'(\gamma(t))\gamma'(t) \\ &= A(t)\gamma'(t), \end{aligned}$$

for all $t \in I$. This says the curve γ' is a solution of the variational equations $x' = A(t)x$. Also $\gamma'(0) = X(\gamma(0)) = X(c)$ and so γ' satisfies the initial condition $x(0) = X(c)$. But since G is the fundamental matrix for A, we can express the solution γ' of this initial value problem in terms of it:

$$\gamma'(t) = G(t)X(c),$$

for all $t \in I$. However $\gamma'(t) = X(\gamma(t))$, for all t and so substituting this in the last equation gives $X(\gamma(t)) = G(t)X(c)$, for all $t \in I$. This proves the second assertion. It also easily yields the third assertion, since if γ is closed integral curve of period p, then taking $t = p$ in the last equation and using $\gamma(p) = \gamma(0) = c$, gives $X(c) = G(p)X(c)$. $\square$

The next proposition will be of use later when we will find it convenient to translate and rotate the coordinate system to more easily study the stability of γ. The proposition says, among other things, that the characteristic multipliers of γ are invariant under these types of transformations.

Notation: Let ϕ^X denote the flow generated by X.

Proposition 6.8 (Invariance) *Suppose* $f : \mathcal{O} \to \overline{\mathcal{O}}$ *is a diffeomorphism between two open sets* $\mathcal{O}$ *and* $\overline{\mathcal{O}}$ *in* $\mathbb{R}^n$. *For a vector field* $X : \mathcal{O} \to \mathbb{R}^n$ *on* $\mathcal{O}$, *let*

$$Y = f_*(X),$$

be the push-forward of X *to a vector field on* $\overline{\mathcal{O}}$. *Let* H^X *and* H^Y *denote the deformation matrices for the flows generated by* X *and* Y, *respectively. Then*

$$H^Y(t, y) = f'\left(\phi_t^X(f^{-1}(y))\right) H^X\left(t, f^{-1}(y)\right) f'\left(\phi_t^X(f^{-1}(y))\right)^{-1}, \quad (6.36)$$

for all (t, y) *in the domain of the flow for* Y. *Consequently, if* $\gamma : I \to \mathbb{R}^n$ *is an integral curve for* X *and*

$$\beta \equiv f \circ \gamma,$$

is the corresponding integral curve for Y, *then the characteristic matrices* G^X, G^Y *for* γ, β *are similar. Specifically*

$$G^Y(t) = f'(\gamma(t)) G^X(t) f'(\gamma(t))^{-1}, \quad (6.37)$$

for all $t \in I$. *In particular, if* γ *is a closed integral curve of period* p, *then* β *is a closed integral curve of period* p *and the characteristic multipliers for* γ *and* β *are the same.*

Proof: By the result of Exercise 1, Section 6.3, the flows for X and Y are conjugate

$$\phi_t^Y = f \circ \phi_t^X \circ f^{-1},$$

for all $t \in I_{f^{-1}(y)}$. Equation (6.36) for the relation between the deformation matrices follows directly by taking derivatives of both sides of this last equation and using the chain rule. Now $\gamma(0) = c$ and so $\beta(0) = f(c)$. Also $\gamma(t) = \phi_t^X(c)$, for all $t \in \mathbb{R}$. Taking $y = f(c)$ in equation (6.36) gives the equation (6.37) relating the characteristic matrices for γ and β. For $t = p$ this equation shows that $G^Y(p)$ and $G^X(p)$ are similar, and hence they have the same eigenvalues. $\square$

Proposition 6.7 gives us a start on the analysis of the stability of a closed integral curve γ. We know that one of its characteristic multipliers is $\mu_1 = 1$ and that $X(c)$ is a corresponding eigenvector. As we shall see below, if the remaining characteristic multipliers $\mu_2, \ldots, \mu_n$ have modulus less than one: $|\mu_j| < 1$, $j = 2, \ldots, n$, then the matrix $G(p)$, viewed as a linear map, will be a contraction when restricted to the subspace V_c perpendicular to $X(c)$. To motivate how this leads to the stability proof below, we return to the heuristic argument prior to Definition 6.6.

Integral curves α that are initially near the periodic solution γ can be written, to the first approximation, in the form $\alpha \approx \gamma + \xi$, where ξ, the variation, is a solution of the variational equations and initially $\xi(0) = v_0 \in V_c$. Here $V_c = \{ v \mid v \cdot X(c) = 0 \}$, is the subspace orthogonal to $X(c)$. Since G is the fundamental matrix for the variational equations $x' = A(t)x$, we have $\xi(t) = G(t)v_0$, and consequently

$$\alpha(t) \approx \gamma(t) + G(t)v_0,$$

for all t. For instance, in Example 6.3, $c = (1,0,0)$ and $X(c) = (0,1,0)$. Consequently, V_c is the x-z plane, and for $v_0 = (x_0, 0, z_0)$ in V_c, we have

$$\xi(t) = G(t)v_0 = \left([x_0 + (1 - e^{-t})z_0] \cos t, \; [x_0 + (1 - e^{-t})z_0] \sin t, \; z_0 e^{-t} \right).$$

Thus, the variation ξ asymptotically approaches

$$\beta(t) = \left([x_0 + z_0] \cos t, \; [x_0 + z_0] \sin t, \; 0 \right).$$

So for $|v_0|$, small we would expect α to remain near γ (but not approach it asymptotically).

On the other hand, In Example 6.4, $c = (1,0)$ and $X(c) = (0,1)$. So V_c is the x-axis, and for $v_0 = (x_0, 0)$ in V_c, we have

$$\xi(t) = G(t)v_0 = \left(x_0 e^{2t} \cos bt, \; x_0 e^{2t} \sin bt \right).$$

So the variation ξ spirals out to infinity and α does not remain near γ.

We can also use G to analyze the situation discretely, i.e., by looking at the times $t = p, 2p, 3p, \ldots$, where p is the period of γ. Since the variational matrix A is periodic: $A(t + p) = A(t)$, we can use the Floquet theory (see Exercise 19, Section 4.2), to conclude that the characteristic matrix G satisfies

$$G(t + p) = G(t)G(p),$$

for all $t \in \mathbb{R}$. By repeated application of this property we find that

$$G(kp) = G(p)^k,$$

for $k = 1, 2, 3, \ldots$ Using these discrete time steps in the above and noting that $\gamma(kp) = \gamma(0) = c$, for all k, we get

$$\alpha(kp) \approx c + G(p)^k v_0,$$

for $k = 1, 2, 3, \ldots$. For instance, in Example 6.3, a short computation shows that

$$G(2\pi)^k v_0 = \begin{bmatrix} 1 & 0 & 1 - e^{-2k\pi} \\ 0 & 1 & 0 \\ 0 & 0 & e^{-2k\pi} \end{bmatrix} \begin{bmatrix} x_0 \\ 0 \\ z_0 \end{bmatrix} = \begin{bmatrix} x_0 + z_0 - z_0 e^{-2k\pi} \\ 0 \\ z_0 e^{-2k\pi} \end{bmatrix}.$$

Consequently

$$\lim_{k \to \infty} \alpha(kp) \approx c + \lim_{k \to \infty} G(p)^k v_0 = \begin{bmatrix} 1 + x_0 + z_0 \\ 0 \\ 0 \end{bmatrix}.$$

On the other hand, in Example 6.4,

$$G(4\pi/b)^k v_0 = \begin{bmatrix} e^{4k\pi/b} & 0 \\ 0 & 1 \end{bmatrix} \begin{bmatrix} x_0 \\ 0 \end{bmatrix} = \begin{bmatrix} x_0 e^{4k\pi/b} \\ 0 \end{bmatrix},$$

and

$$\lim_{k \to \infty} \alpha(kp) \approx c + \lim_{k \to \infty} G(p)^k v_0 = \text{does not exist.}$$

since the latter quantity grows without bound.

However, if the restriction of $G(p)$ to V_c is a contraction (which it is not in either of the above examples), then the variation tends to zero $\lim_{k \to \infty} G(p)^k v_0 = 0$ and α tends (discretely) to c, i.e., $\lim_{k \to \infty} \alpha(kp) = c$ (see Appendix B,

Section 3). This rough argument indicates why we would expect γ to be asymptotically stable when $n-1$ of its characteristic multipliers have modulus less than 1.

To formulate a precise result, we need to first introduce another type of asymptotic stability. The reason for this is that periodic solutions generally cannot be asymptotically stable (exercise), but with suitable conditions they can be "orbitally asymptotically stable."

Definition 6.8 (Orbital Stability)

1. If $M \subseteq \mathbb{R}^n$ is a closed subset and $x \in \mathbb{R}^n$, let

$$d(x, M) \equiv \inf\{\,|x - m|\,|\,m \in M\,\},$$

denote the *distance* from x to the set M. A *neighborhood* of M is an open set that contains M. One can show that the set

$$B(M, \varepsilon) \equiv \{\,x \in \mathbb{R}^n\,|\,d(x, M) < \varepsilon\,\},$$

is a neighborhood of M. It consists of the points in $\mathbb{R}^n$ that have distance to M less than ε.

2. Suppose $\gamma : \mathbb{R} \to \mathbb{R}^n$ is periodic solution of $x' = X(x)$ and let $\Gamma \equiv \gamma(\mathbb{R})$ be the image of $\mathbb{R}$ under γ. Then γ is called *orbitally stable* if for every $\varepsilon > 0$, there is a $\delta > 0$, such that

$$d(\phi_t(x), \Gamma) < \varepsilon,$$

for all $t \geq 0$ and all x with $d(x, \Gamma) < \delta$. Equivalently, for each neighborhood Ω of Γ, there is a (smaller) neighborhood Ω_0 of Γ such that

$$\phi_t(x) \in \Omega,$$

for all $x \in \Omega_0$ and all $t \geq 0$.

If in addition

$$\lim_{t \to \infty} d(\phi_t(x), \Gamma) = 0,$$

then γ is called *orbitally asymptotically stable*.

In essence, orbital stability says that $\phi_t(x)$ will stay ε close to the set Γ, but not necessarily keep pace with the motion along γ, i.e., stay in the moving ball $B(\gamma(t), \varepsilon)$, as shown in Figure 6.10.

To prove the theorem below on orbital asymptotic stability of certain periodic solutions, we need some additional results, the first of which introduces the Poincaré map. The Poincaré map P is another tool to use in conjunction with the characteristic matrix for γ. We shall see that the Jacobian matrix $P'(c)$ coincides with a submatrix of $G(p)$, and so its eigenvalues coincide with $n-1$ of the characteristic values of γ.

Proposition 6.9 (The Poincaré Map) *Suppose $\gamma : \mathbb{R} \to \mathbb{R}^n$ is a closed integral curve of period p for X. Let $c = \gamma(0)$ and let*

$$M_c = \{\, x \in \mathbb{R}^n \mid (x - c) \cdot X(c) = 0 \,\}$$

be the hyperplane through c, perpendicular to $X(c)$. Then there is a neighborhood U of c and a C^1 map $\tau : U \to \mathbb{R}$, such that $\tau(c) = p$ and

$$\phi_{\tau(x)}(x) \in M_c,$$

for all $x \in U$. The map $P : U \cap M_c \to M_c$ defined by

$$P(x) = \phi_{\tau(x)}(x), \tag{6.38}$$

is called the Poincaré map *for γ at c.*

Proof: The proof is an easy consequence of the Implicit Function Theorem. To apply it, define a function $F : \mathcal{D} \to \mathbb{R}$, on the domain $\mathcal{D}$ of the flow map by

$$F(t, x) = (\phi_t(x) - c) \cdot X(c),$$

for $(t, x) \in \mathcal{D}$. Then F is C^1 and since $\gamma(t) = \phi_t(c)$ has period p, we get

$$F(p, c) = (\phi_p(c) - c) \cdot X(c) = (c - c) \cdot X(c) = 0.$$

Further, since the flow satisfies $d\phi_t(x)/dt = X(\phi_t(x))$, we have

$$\frac{\partial F}{\partial t}(t, x) = X(\phi_t(x)) \cdot X(c),$$

for all $(t, x) \in \mathcal{D}$. Hence in particular

$$\frac{\partial F}{\partial t}(p, c) = X(c) \cdot X(c) = |X(c)|^2 > 0.$$

Thus, the hypotheses of the Implicit Function Theorem apply and so we can solve the equation

$$F(t, x) = 0$$

explicitly for t as a function of x. More precisely, there is a C^1 function $\tau : U \to \mathbb{R}$, defined on a neighborhood U of c, such that $\tau(c) = p$ and

$$F(\tau(x), x) = 0,$$

for all $x \in U$. The latter equation is equivalent to saying $\phi_{\tau(x)}(x) \in M_c$, and this proves the result. $\square$

The Poincaré map $P : U \cap M_c \to M_c$, is also called the *first return map*, since for each $x \in U \cap M_c$, the integral curve that starts at x at time 0 returns to intersect the hyperplane M_c at the point $P(x)$ at time $\tau(x)$. See Figure 6.12.

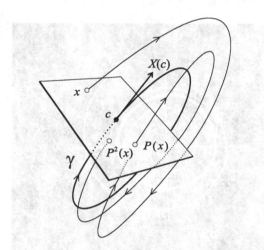

Figure 6.12: *The Poincaré map P for the periodic solution γ.*

The function τ is for this reason called the *time of return function*. The figure also indicates why the Poincaré map is expected to play a role in the stability analysis. The integral curve shown in the figure, which starts at $x \in U \cap M_c$ near c, appears to asymptotically approach the closed integral curve γ. As it does so, it continually intersects the plane M_c, giving a sequence of points $P(x), P^2(x), P^3(x), \ldots$, that appear to approach c. If we can show that P is a contraction on $U \cap M_c$, then we will know that indeed, $\lim_{k \to \infty} P^k(x) = c$ for any $x \in U \cap M_c$. The general results in Appendix B show that P will be a contraction if the eigenvalues of the matrix $P'(c)$ all have modulus less than 1.

Example 6.5 We return to the two-dimensional system discussed in Example 6.4. The polar coordinate version of the system:

$$r' = (r^2 - 1)r$$
$$\theta' = b$$

provides a quick way to understand and sketch the integral curves. First $\theta(t) = bt + \theta_0$ and so the angle $\theta(t)$ between the x-axis and the position vector to the point $(x(t), y(t))$ is positive and increases uniformly with t. The radial equation, $r' = (r^2 - 1)r$, says that r decreases at times when $r < 1$ and increases when $r > 1$. The circular integral curve $\gamma(t) = (\cos bt, \sin bt)$ corresponds to $r = 1$. Based on these observations, we can do a quick sketch of the phase portrait. See Figure 6.13.

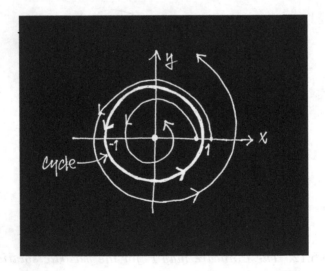

Figure 6.13: *A sketch of the phase portrait.*

To get the Poincaré map for γ, we will need an explicit formula for the flow $\phi_t(x, 0)$ starting at points x on the x-axis. For this we solve the radial equation

$$r' = (r^2 - 1)r,$$

by separation of variables

$$\frac{1}{(r^2 - 1)r} \, dr = dt,$$

or

$$\left(\frac{\frac{1}{2}}{r-1} + \frac{\frac{1}{2}}{r+1} - \frac{1}{r}\right) dr = dt$$

Integration gives

$$\frac{1}{2}\ln|r-1| + \frac{1}{2}\ln|r+1| - \ln|r| = t + c$$

or

$$\ln\left|\frac{r^2-1}{r^2}\right| = 2t + 2c.$$

Equivalently

$$\frac{|r^2-1|}{r^2} = ke^{2t},$$

where we have relabeled the arbitrary constant: $k - e^{2c}$. To solve this for r, assume first that $r(0) = r_0 > 1$, then for t near zero, we will have $r > 1$ (and $r^2 - 1 > 0$). Consequently the last equation is

$$\frac{r^2-1}{r^2} = ke^{2t}.$$

Taking $t = 0$ in the above equation gives $k = \frac{r_0^2-1}{r_0^2}$. Then, solving the above equation for r^2 yields

$$r^2 = \frac{1}{1-ke^{2t}} = \frac{1}{1 - \frac{r_0^2-1}{r_0^2}e^{2t}} = \frac{1}{1 - (1-r_0^{-2})e^{2t}}.$$

Since $r_0 > 1$, the denominator of the above expression can be zero. This occurs when

$$t = \frac{1}{2}\ln\left(\frac{r_0^2}{r_0^2-1}\right) \equiv T(r_0)$$

In summary, we have found that

$$r(t) = [1 + (r_0^{-2} - 1)e^{2t}]^{-1/2} \tag{6.39}$$

which has maximal interval of definition $t \in I_{r_0} = (-\infty, T(r_0))$. This is for the case $r_0 > 1$. In the case when $r_0 < 1$, calculations like those above will give the *same* formula for $r(t)$ but now the maximum interval of existence is $I_{r_0} = \mathbb{R}$.

The general integral curve for the Cartesian system has the form

$$\phi_t(x_0, y_0) = \left(r(t) \cos \theta(t),\ r(t) \sin \theta(t) \right),$$

and has a complicated dependence on (x_0, y_0) (i.e., $r_0 = (x_0^2 + y_0^2)^{1/2}$ in the formula for $r(t)$ and $\theta_0 = \tan^{-1}(y_0/x_0)$ in the formula for $\theta(t)$). However, for initial points on the positive x-axis $(y_0 = 0)$, we have $r_0 = x_0$ and $\theta_0 = 0$, and

$$\phi_t(x_0, 0) = \left([1 + (x_0^{-2} - 1)e^{2t}]^{-1/2} \cos bt,\ [1 + (x_0^{-2} - 1)e^{2t}]^{-1/2} \sin bt \right),$$

Then

$$\phi_{2\pi/b}(x_0, 0) = \left([1 + (x_0^{-2} - 1)e^{4\pi/b}]^{-1/2},\ 0 \right),$$

gives the point on the positive x-axis to which $(x_0, 0)$ first returns under the flow. Note: The τ in Proposition 6.9 is a constant function for this example: $\tau(x_0) = 2\pi/b$, for every x_0 that returns. It is important to note in the case $x_0 > 1$, that the condition $2\pi/b \in I_{x_0} = (-\infty, T(x_0))$ is required in order for there any return at all. This condition is

$$\frac{2\pi}{b} < \frac{1}{2} \ln \left(\frac{x_0^2}{x_0^2 - 1} \right).$$

Equivalently.

$$x_0 < [1 - e^{-4\pi/b}]^{-1/2} \equiv L_b.$$

In the case $x_0 < 1$, there is no condition on returns, as you can readily see from Figure 6.13. Consequently we can define the Poincaré map $P : (0, L_b) \to \mathbb{R}$ by

$$P(x) = [1 + (x^{-2} - 1)e^{4\pi/b}]^{-1/2}.$$

It is clear that $P(1) = 1$, so $x = 1$ is a fixed point of P. From the geometry of how we got P it is clear that the iterates $P^k(x),\ k = 1, 2, 3, \ldots$ get smaller if $x < 1$ and larger if $x > 1$. You should also note that the number of iterates is limited when $x > 1$. To quantify this, observe that for a given $x > 1$, the time $T(x)$ is when the radius goes to infinity $(\lim_{t \to T(x)} r(t) = \infty)$. The integral curve $t \mapsto \phi_t(x, 0)$, after wrapping around the origin a number of times, heads off to infinity, becoming asymptotic to the ray with polar equation: $\theta = \theta_x = bT(x)$. The angle θ_x is the limiting angle (in radians) and the number of returns (the number of wraps) is the greatest integer less than or equal to

$$\theta_x/(2\pi) = bT(x)/(2\pi) = \frac{b}{4\pi} \ln \left(\frac{x^2}{x^2 - 1} \right).$$

For example, suppose the parameter $b = 20$ in the system of equations. Then the right endpoint of the domain for P is $L_{20} \approx 1.464$. Taking $x = 1.01$ as an initial point gives $\theta_{1.01} \approx 39.269$ radians, and so the integral curve starting at $(1.01, 0)$ will wrap around the origin 6 times before heading off to infinity at angle $\theta_{1.01} \approx 1.57 \approx \pi/2$. This is shown in Figure 6.14. Note that the sequence of iterates $P^k(1.01), k = 0, 1, \ldots, 6$ is $1.01, 1.018994917, 1.036523531, 1.071971735, 1.149539859, 1.355769346, 2.623260528$, the last one being outside the domain of P.

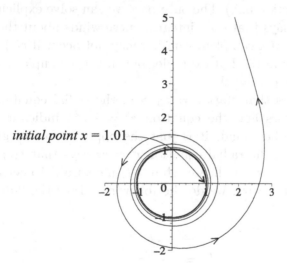

Figure 6.14: *An integral curve that eventually does not return, but rather becomes asymptotic to the ray $\theta \approx \pi/2$.*

The instability of the circular cycle γ is indicated by the Poincaré map P, and the behavior of the iterates of P near its fixed point $x = 1$ is controlled by $P'(1)$. From the above formula for P, we get

$$P'(x) = -\tfrac{1}{2}[1 + (x^{-2} - 1)e^{4\pi/b}]^{-3/2}(-2x^{-3})e^{4\pi/b}$$

Consequently $P'(1) = e^{4\pi/b}$. Thus, P is not a contraction since $P'(1)$ is not less than 1. Also note that $e^{4\pi/b}$, as seen in Example 6.4, is one of the eigenvalues of the characteristic matrix G for γ.

Example 6.6 For some systems the Poincaré map P for a cycle γ is not explicitly computable, but the geometrical action of P is easy to describe. This is the case for $\gamma(t) = (\cos t, \sin t, 0)$ and the system in Example 6.3:

$$\begin{aligned} x' &= -y + xz \\ y' &= x + yz \\ z' &= -z(x^2 + y^2) \end{aligned}$$

Transforming to cylindrical coordinates ($x = r\cos\theta$, $y = r\sin\theta$, $z = z$) gives

$$\begin{aligned} r' &= rz \\ z' &= -r^2 z \\ \theta' &= 1 \end{aligned}$$

(See Exercise 8 in Section 5.2.) The only part we can solve explicitly here is $\theta(t) = t + \theta_0$, which says that each integral curve winds about the z-axis in the positive direction. (A complete winding may not occur if $r(t) \to \infty$ in a finite time, as we saw in the last example, but in this example, we shall see that the windings are continual.)

For integral curves that start $z = z_0 > 0$, the radial equation $r' = rz$ says that $r(t)$ increases and the equation $z' = -r^2 z$ indicates that $z(t)$ decreases. On the other hand, it is just the opposite for integral curves that start $z = z_0 < 0$: the radial equation $r' = rz$ says that $r(t)$ decreases and the equation $z' = -r^2 z$ indicates that $z(t)$ increases. To get additional information how z an r vary, we do the following. Take the 2-dimensional system

$$\begin{aligned} r' &= rz \\ z' &= -r^2 z \end{aligned}$$

and multiply the 1st equation by r and then use the 2nd equation to get

$$rr' = -z', \quad \text{or} \quad \frac{d}{dt}\left(\tfrac{1}{2}r^2 + z\right) = 0.$$

Hence we see that each solution $(r(t), z(t))$ of the system must satisfy

$$\tfrac{1}{2}r^2(t) + z(t) = \tfrac{1}{2}r_0^2 + z_0,$$

for all t in the maximum interval of existence. Otherwise said, the curve $t \mapsto (r(t), z(t))$ lies on the graph of the parabola

$$z = -\tfrac{1}{2}r^2 + a_0,$$

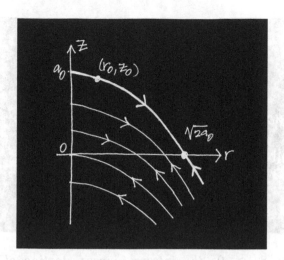

Figure 6.15: *The integral curve through (r_0, z_0) lies on the parabola $z = -\frac{1}{2}r^2 + a_0$ where $a_0 = \frac{1}{2}r_0^2 + z_0$.*

in the r-z plane, where $a_0 = \frac{1}{2}r_0^2 + z_0$. See Figure 6.15.

Because of the geometric significance of r, the graph is only for $r \geq 0$. Note that when $a_0 \geq 0$, the r-intercept of this (half) parabola is $(\sqrt{2a_0},\, 0)$. While for $a_0 < 0$, there is no r-intercept. Based on the above discussion about the way r and z vary, we have marked the directions of flow of the integral curves tracing out the (half) parabolas. Note that all points on the z-axis are fixed points and that each integral curve with $a_0 \leq 0$ flows toward the fixed point $(0, a_0)$. On the other hand, an integral curve with $a_0 > 0$ will flow toward the point $(\sqrt{2a_0}, 0)$.

In the three dimensional system, each integral curve

$$\alpha(t) = \Big(r(t)\cos(t + \theta_0),\, r(t)\sin(t + \theta_0),\, z(t) \Big),$$

of the system lies on a paraboloid of revolution

$$z = -\tfrac{1}{2}(x^2 + y^2) + a_0.$$

From our previous observations, this integral curve α, while remaining on this paraboloid, will wind clockwise about the z-axis as $z(t) \to 0$. The limiting radius depends on $a_0 = \frac{1}{2}r_0^2 + z_0$.

If $a_0 > 0$, then $r(t) \to \sqrt{2a_0}$ as $t \to \infty$. Thus, α asymptotically ap-

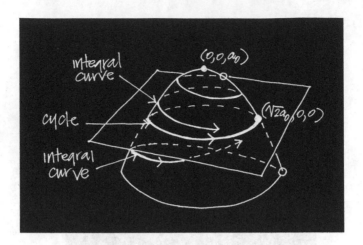

Figure 6.16: *For $a_0 > 0$, the integral curves starting at points on the paraboloid $z = -\frac{1}{2}(x^2 + y^2) + a_0$ spiral toward the cycle $x^2 + y^2 = 2a_0$ In the x-y plane.*

proaches the circle

$$\beta_{a_0}(t) = \left(\sqrt{2a_0} \cos t, \sqrt{2a_0} \sin t, 0 \right),$$

in the x-y-plane. See Figure 6.16.

On the other hand, if $a_0 \leq 0$, then $r(t) \to a_0$, a fixed pont on the z-axis, as $t \to \infty$.

While it is not possible to write an explicit formula for the Poincaré map P in this example, it is possible to understand the action of P geometrically. This is based on our qualitative understanding of the phase portrait as discussed above.

Now $\gamma(t) = (\cos bt, \sin bt, 0)$, $c = \gamma(0) = (1,0,0)$, and $X(c) = (0,1,0)$. Thus, M_c is the x-z plane, and for a ball $B(c, R)$ in $\mathbb{R}^3$, the set $W = B(c, R) \cap M_c$ is a disk of radius R in M_c. Integral curves that start at a point $p_0 = (x_0, 0, z_0) \in W$ (with $z_0 \neq 0$) will spiral around the z-axis while remaining on the paraboloid $z = -\frac{1}{2}(x^2 + y^2) + a_0$, where $a_0 = \frac{1}{2}r_0^2 + z_0$. Upon first return to the x-z plane this integral hits a point p_1 on the parabola $z = -\frac{1}{2}r^2 + a_0$. The point p_1 will be closer to the x-y plane that p_0 was. Continuing like this it is easy to see that the sequence of iterates $p_0, p_1, p_2, \ldots$ of the Poincaré map will approach $(\sqrt{2a_0}, 0)$ in the limit. See Figure 6.17

While P is not given explicitly in this example, we did find $G(p)$ exactly. We will see below that, in general, the Jacobian matrix $P'(c)$ coincides with

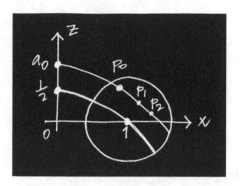

Figure 6.17: *A sequence of iterates $p_0, p_1 = P(p_0), p_2 = P(p_1), \ldots$ of the Poncaré map converges to the point $(\sqrt{2a_0}, 0)$.*

a certain sub-matrix of $G(p)$. Using that result in this example allows us to conclude that

$$P'(c) = \begin{bmatrix} 1 & 1 - e^{-2\pi} \\ 0 & e^{-2\pi} \end{bmatrix}.$$

We now turn to the general task of calculating the Jacobian matrix of P at c. We proceed in a somewhat indirect route. This is mainly necessitated by the fact that P is a map from $W \equiv U \cap M_c$ into M_c and we have only defined Jacobians of maps with domains that are open subsets of $\mathbb{R}^m$. We get around this technicality by identifying, via a translation and rotation, M_c with $\mathbb{R}^{n-1}$ and W with an open set of $\mathbb{R}^{n-1}$. (If you are familiar with the theory for derivatives (Jacobians) of functions between two differentiable manifolds you can use this to directly calculate $P'(c)$).

First, for convenience, we define the tensor product $a \otimes b$ of two vectors $a, b \in \mathbb{R}^n$ to be the operator: $a \otimes b : \mathbb{R}^n \to \mathbb{R}^n$ defined by

$$(a \otimes b)v \equiv (b \cdot v)\, a,$$

for all $v \in \mathbb{R}^n$. Identifying $a \otimes b$ with the $n \times n$ matrix that represents it with respect to the standard basis for $\mathbb{R}^n$, it is easy to see that the i-jth entry of this matrix is

$$(a \otimes b)_{ij} = a_i b_j \tag{6.40}$$

(exercise).

Next note that while the Poincaré map P is defined on $W = U \cap M_c$, the proposition above also gives a map $\widetilde{P} : U \to \mathbb{R}^n$ defined by

$$\widetilde{P}(x) = \phi_{\tau(x)}(x) = \phi(\tau(x), x),$$

for $x \in U$. It is an easy exercise, using the chain rule and the property $d(\phi_t(x))/dt = X(\phi_t(x))$ of the flow map, to show that

$$\widetilde{P}'(x) = X(\widetilde{P}(x)) \otimes \nabla\tau(x) + H(\tau(x), x), \tag{6.41}$$

for all $x \in U$. Thus, in particular, for $x = c$ we get

$$\widetilde{P}'(c) = X(c) \otimes \nabla\tau(c) + G(p). \tag{6.42}$$

Properly interpreted, this equation relates the characteristic matrix $G(p)$ to the Jacobian matrix of P at c and shows that, with the exception of $\mu = 1$, they have the same eigenvalues. Specifically:

Proposition 6.10 *With the prior notation, assume that $c = 0$ and that $X(0) = \varepsilon_n = (0, 0, \ldots, 0, 1)$. Then the characteristic matrix $G(p)$ has the form*

$$G(p) = \begin{bmatrix} b_{11} & \cdots & b_{1,n-1} & 0 \\ \vdots & & \vdots & \vdots \\ b_{n1} & \cdots & b_{n,n-1} & 0 \\ * & \cdots & * & 1 \end{bmatrix} \tag{6.43}$$

and the matrix $\widetilde{P}'(0)$ has the form

$$\widetilde{P}'(0) = \begin{bmatrix} b_{11} & \cdots & b_{1,n-1} & 0 \\ \vdots & & \vdots & \vdots \\ b_{n1} & \cdots & b_{n,n-1} & 0 \\ * & \cdots & * & * \end{bmatrix}. \tag{6.44}$$

We identify the hyperplane $M_0 = \{ (x_1, \ldots, x_{n-1}, 0) \mid x_i \in \mathbb{R} \}$ with $\mathbb{R}^{n-1}$, and identify $W = M_0 \cap U$ with an open set V in $\mathbb{R}^{n-1}$. Then the Poincaré map, considered as a map $P : V \to V$, has Jacobian at $0 \in V$ given by

$$P'(0) = \begin{bmatrix} b_{11} & \cdots & b_{1,n-1} \\ \vdots & & \vdots \\ b_{n1} & \cdots & b_{n,n-1} \end{bmatrix}. \tag{6.45}$$

Hence if $\mu_1 = 1, \mu_2, \ldots, \mu_n$, are the characteristic multipliers of the closed curve γ, then $\mu_2, \ldots, \mu_n$, are the eigenvalues of $P'(0)$.

Proof: Since $G(p)X(0) = X(0)$ and $X(0) = \varepsilon_n$, it follows that the nth column of $G(p)$ is ε_n. Thus, $G(p)$ has the form (6.43). From equation (6.40) and the fact that $X(0) = \varepsilon_n$, it is clear that $X(0) \otimes \tau(0)$ has the form

$$X(0) \otimes \tau(0) = \begin{bmatrix} 0 & \cdots & 0 & 0 \\ \vdots & & \vdots & \vdots \\ 0 & \cdots & 0 & 0 \\ * & \cdots & * & * \end{bmatrix}. \tag{6.46}$$

From equation (6.42), we know that $\widetilde{P}'(0) = X(0) \otimes \tau(0) + G(p)$ and thus $\widetilde{P}'(0)$ has the form shown in equation (6.45).

Let I_{n-1} denote the $(n-1) \times (n-1)$ identity matrix and let B be the $(n-1) \times (n-1)$ matrix given by $B = \{b_{ij}\}_{i,j=1}^{n-1}$. Then because of the way this square matrix enters as a submatrix of $G(p)$, it is clear that

$$\det(G(p) - \mu I_n) = (1 - \mu) \det(B - \mu I_{n-1}).$$

Thus, the characteristic multipliers are $\mu_1 = 1, \mu_2, \ldots, \mu_n$, where $\mu_2, \ldots, \mu_n$ are the eigenvalues of B.

All that remains is to show that $P'(0) = B$. To be explicit about the identifications we are making, let $e : \mathbb{R}^{n-1} \to \mathbb{R}^n$ be the standard embedding:

$$e(x_1, \ldots, x_{n-1}) = (x_1, \ldots, x_{n-1}, 0),$$

and let $\rho : \mathbb{R}^n \to \mathbb{R}^{n-1}$ be the projection

$$\rho(x_1, \ldots, x_{n-1}, x_n) = (x_1, \ldots, x_{n-1}).$$

Then the Poincaré map is given by (or identified with) the composite map

$$P(v) = \rho\left(\widetilde{P}(e(v))\right),$$

for $v \in V \equiv \rho(W) = \rho(M_0 \cap U)$. Hence by the chain rule we get

$$P'(0) = R\widetilde{P}'(0)E,$$

where $R = [I_{n-1}, 0]$ is the $(n-1) \times n$ matrix obtained by augmenting I_{n-1} with a column of zeros and $E = [I_{n-1}, 0]$ is the $n \times (n-1)$ matrix obtained by augmenting I_{n-1} with a row of zeros. Because of the forms of these matrices it is easy to see that

$$R\widetilde{P}'(0)E = B,$$

and this completes the proof. $\square$

Comments: While the assumptions on c and $X(c)$ in the proposition are rather special, it is easy to see that when these assumptions do not hold we can always translate and rotate the coordinate system so that the resulting vector field does satisfy the assumptions. Specifically, let $V_c = \{v \in \mathbb{R}^n | v \cdot X(c) = 0\}$ be the subspace orthogonal to $X(c)$, and choose an orthonormal basis $\{e_1, \ldots, e_{n-1}\}$ for V_c. Let

$$Q = [e_1, \ldots, e_{n-1}, X(c)/|X(c)|^2],$$

be the $n \times n$ matrix with rows $e_1, \ldots, e_{n-1}, X(c)/|X(c)|^2$. Then Q is an orthogonal matrix and we can assume that $\det(Q) = 1$ (otherwise permute and relabel the e_i's so that this is so). Thus, Q is a rotation and the map $f : \mathbb{R}^n \to \mathbb{R}^n$ defined by

$$f(x) = Q(x - c) \tag{6.47}$$

is a diffeomorphism such that the vector field $Y \equiv f_*(X)$ has the property: $Y(0) = \varepsilon_n$ (exercise). The proposition then applies to Y and the closed integral curve $f \circ \gamma$ of Y.

We now come to the main theorem which shows how the characteristic multipliers determine orbital asymptotic stability of closed integral curves. The proof given here is an elaboration, with many more details, of the proof given by Hartman (cf. [Ha 82, p. 254]).

Theorem 6.3 (Orbital Asymptotic Stability) *Suppose $\gamma : \mathbb{R} \to \mathbb{R}^n$ is a closed integral curve of the vector field $X : \mathcal{O} \subseteq \mathbb{R}^n \to \mathbb{R}^n$ and let p be the period of γ and $c = \gamma(0)$. Suppose that $n - 1$ of the characteristic multipliers $\mu_2, \ldots, \mu_n$ of γ have modulus less than 1. Then γ is orbitally asymptotically stable. In addition, suppose a is a number such that $|\mu_j| < a < 1$, for $j = 2, \ldots, n$. Then for each neighborhood Ω of $\Gamma = \{\gamma(t) | t \in \mathbb{R}\}$, there exists an $L > 0$ and neighborhood $\Omega_0 \subseteq \Omega$ of Γ, such that for every $x \in \Omega_0$, there is a $T \in \mathbb{R}$ such that*

$$|\phi_{t+T}(x) - \gamma(t)| \leq L\, a^{t/p}, \tag{6.48}$$

for all $t \geq p$. The number T depends on x and is called the asymptotic phase of x.

Proof: Assume that $c = 0 \in \mathbb{R}^n$ and $X(0) = \varepsilon_n = (0, \ldots, 0, 1)$. After proving the theorem for this special case, we will easily be able to obtain the general case.

Suppose $\Omega \subseteq \mathbb{R}^n$ is a neighborhood of $\Gamma = \{\gamma(t) \,|\, t \in \mathbb{R}\}$. We can assume, without loss of generality, that $\overline{\Omega}$ is compact.

Note that we can choose an $R > 0$ such that $[0, 2p] \times \overline{B}(0, R) \subseteq \phi^{-1}(\Omega)$. To see this observe that $(t, 0) \in \phi^{-1}(\Omega)$, for all $t \in \mathbb{R}$, and $\phi^{-1}(\Omega)$ is an open set. Thus, for each $t \in \mathbb{R}$, there are numbers $a_t < t < b_t$, $R_t > 0$, such that

$$(a_t, b_t) \times B(0, R_t) \subseteq \phi^{-1}(\Omega).$$

Now $\{(a_t, b_t)\}_{t \in [0,2p]}$ is an open cover of $[0, 2p]$, and so there exists a finite subcover $\{(a_{t_i}, b_{t_i})\}_{i=1}^N$ of $[0, 2p]$. Thus, choosing $R > 0$, such that $R < R_{t_i}$, for $i = 1, \ldots, N$, gives

$$(a_{t_i}, b_{t_i}) \times \overline{B}(0, R) \subseteq \phi^{-1}(\Omega),$$

for $i = 1, \ldots, N$. Hence $[0, 2p] \times \overline{B}(0, R) \subseteq \phi^{-1}(\Omega)$. This result says that $\phi_t(x) \in \Omega$ for all $x \in \overline{B}(0, R)$ and $t \in [0, 2p]$, i.e., the flow through each $x \in \overline{B}(0, R)$ remains in Ω for all $t \in [0, 2p]$.

Next, by Proposition 6.8, there is an open set $U \subseteq \mathcal{O}$ and a time of return map $\tau : U \to \mathbb{R}$, which is C^1, such that $\phi_{\tau(x)}(x)$ is in the hyperplane $M_0 = \{z \in \mathbb{R}^n \,|\, z_n = 0\}$, for all $x \in U$. We can assume that $\overline{B}(0, R) \subseteq U$ (otherwise choose a smaller R so that this is so). The crucial inequality we need is the contraction map inequality for the Poincaré map P. Thus, let $\widetilde{P} : U \to \mathbb{R}^n$ be the map $\widetilde{P}(x) = \phi_{\tau(x)}(x)$. We can assume that $U = B(0, R)$. With the notation used in the in the proof of Proposition 6.9, $P = \widetilde{P}|_{B(0,R) \cap M_0}$. Identifying P with the map $\rho \circ \widetilde{P} \circ e$ and using the results of that proposition together with the supposition here, we have that the eigenvalues of $P'(0)$ all have modulus less than a and $a < 1$. By Theorem B.3 in Appendix B (with $f = P$ and $c = 0$), there is a norm $\| \cdot \|_0$ on $\mathbb{R}^n$ and a $\delta > 0$ such that

$$\|P(w)\|_0 \le a\|w\|_0, \tag{6.49}$$

for all $w \in B(0, R) \cap M_0 \cap B(0, \delta)$. We can assume that $\delta < R$ and $\delta < 1$ (otherwise choose a smaller δ). Thus, let $W = B(0, \delta) \cap M_0$. Then because of the contraction property (6.49), we have $P(W) \subseteq W$. Consequently, all the iterates $P^k(w)$, $k = 1, 2, 3, \ldots$ exist for each $w \in W$. This is needed to prove orbital stability. The fact that these iterates tend to zero figures in the proof of orbital asymptotic stability.

Below, we will derive all the estimates we need using the norm $\| \cdot \|_0$ and then convert to the usual Euclidean norm $| \cdot |$ at the end of the proof. This we can do since all norms on $\mathbb{R}^n$ are equivalent and thus, in particular, there exist positive constants $K, \bar{K}$ such that

$$K|v| \leq \|x\|_0 \leq \bar{K}|v|, \tag{6.50}$$

for all $v \in \mathbb{R}^n$.

We first show orbital stability. Recall that $W = B(0, \delta) \cap M_0$ and introduce

$$\Omega_0 \equiv \{ \phi_t(w) \mid w \in W, \, t \in [0, \tau(w)] \}.$$

Since $\tau : W \to \mathbb{R}$ is continuous, we can assume (by choosing a smaller δ if necessary) that $|\tau(x) - p| \leq p/2$ for all $x \in W$, i.e., $\tau(x) \in [p/2, 3p/2]$, for all $x \in W$. In particular, $\tau(x) \leq 2p$. By this and the results in the third paragraph of the proof, it follows that $\Omega_0 \subseteq \Omega$.

Claim: $\phi_t(x) \in \Omega_0$, for all $x \in \Omega_0$ and $t \in [0, \infty]$.

Proving the claim will establish orbital stability of γ. Note that part of the claim is that $[0, \infty] \subseteq I_x$.

Thus, fix $x \in \Omega_0$. Then, by definition, $x = \phi_{t_0}(w_0)$ for some $w_0 \in B(0, \delta)$ and $t_0 \in [0, \tau(w_0)]$.

Using the time of return map and the Poincaré map we get sequences $\{t_k\}_{k=0}^{\infty}$ and $\{w_k\}_{k=0}^{\infty}$ of times and points when and where the flow through x strikes the hyperplane: $\phi_{t_k}(x) = w_k \in M_0$. Specifically, with t_0, w_0, already chosen, let

$$w_1 = P(w_0) \qquad \text{and} \qquad t_1 = \tau(w_0) - t_0,$$

Note that t_1 is the time for x to reach the hyperplane M_0, since $\phi_{t_1}(x) = \phi_{t_1}(\phi_{t_0}(w_0)) = \phi_{\tau(w_0)}(w_0) = w_1$. The ensuing choices for the points and times are:

$$w_k = P(w_{k-1}) \qquad \text{and} \qquad t_k = \tau(w_{k-1}) + t_{k-1},$$

for $k = 2, 3, 4, \ldots$. (Note the $+$ sign in the definition of t_k, $k \geq 2$. Only t_1 involves a $-$ sign.) From the semigroup property (Theorem 3.7), it is easy to show that $t_k \in I_x$ and

$$\phi_{t_k}(x) = w_k, \tag{6.51}$$

for all k (exercise).

From the definition of the t_is and the fact that $\tau(y) \geq p/2$ for all $y \in B(0, \delta)$, it is easy to see that

$$t_k = t_1 + \sum_{j=1}^{k} \tau(w_j) \geq t_1 + \sum_{j=1}^{k} \frac{p}{2} = t_1 + \frac{(k-1)p}{2},$$

for all $k \geq 2$. Hence, $\lim_{k \to \infty} t_k = \infty$. This shows that $[0, \infty) \subseteq I_x$, i.e., that $\phi_t(x)$ is defined for all $t \geq 0$.

We can now prove the Claim by noting that if $t \in [t_0, \infty]$, then there is a k such that $t_k \leq t < t_{k+1}$. Then

$$\phi_t(x) = \phi_{t-t_k}(\phi_{t_k}(x)) = \phi_{t-t_k}(w_k).$$

But this says that $\phi_t(x) \in \Omega_0$, since $w_k \in W$ and $t - t_k \in [0, \tau(w_k)]$. In the exceptional case that $t \in (0, t_0)$ we have either that (1) $t + t_0 \geq \tau(w_0)$ or (2) $t + t_0 < \tau(w_0)$. In Case (1), it follows that $t \geq \tau(w_0) - t_0 = t_1$ and this case is covered above. On the other hand, in Case (2) we have $\phi_t(x) = \phi_t(\phi_{t_0}(w_0)) = \phi_{t+t_0}(w_0)$. But this says that $\phi_t(x) \in \Omega_0$, since $w_0 \in W$ and $t - t_0 \in [0, \tau(w_0)]$. This proves the Claim.

Next, to prove orbital *asymptotic* stability, we continue with the above assumptions and constructions based on the chosen $x \in \Omega_0$. We need three estimates which are as follows.

First, since $\overline{\Omega}$ is compact and X is continuous, there is a constant $L_1 > 0$ such that

$$\|X(y)\|_0 \leq L_1, \tag{6.52}$$

for all $y \in \overline{\Omega}$.

Since the flow $\phi : \mathcal{D} \to \mathbb{R}^n$ for X is C^1, we can apply Proposition B.1 from Appendix B to get a constant $L_2 > 0$ such that

$$\|\phi_t(y) - \phi_t(z)\|_0 \leq L_2 \|y - z\|_0, \tag{6.53}$$

for all $y, z \in \overline{B}(0, R)$ and all $t \in [0, 2p]$.

Again applying Proposition B.1 from Appendix B to τ restricted to $\overline{B}(0, R)$ gives a constant $L_3 > 0$ such that

$$|\tau(y) - \tau(z)| \leq L_3 \|y - z\|_0, \tag{6.54}$$

for all $y, z \in \overline{B}(0, R)$. With the constants L_1, L_2, and L_3 thus selected, we define L by

$$L \equiv \left(L_2 + \frac{L_1 L_3}{(1-a)} \right) \frac{1}{aK}.$$

As we shall see below, this is the constant needed to get inequality (6.48)

Using the contraction inequality (6.49), we get

$$\|w_k\|_0 = \|P(w_{k-1})\|_0 \leq a\|w_{k-1}\|_0,$$

for all k and so by induction

$$\|w_k\|_0 \leq a^k \|w_0\|_0,$$

for all k. We can use this to get a number T, called the time phase shift for x, as follows.

Let $T_k = t_k - kp$, for $k = 0, 1, 2, \ldots$. Observe that for any k,

$$
\begin{aligned}
|T_{k+1} - T_k| &= |t_{k+1} - (k+1)p - (t_k - kp)| = |t_{k+1} - t_k - p| \\
&= |\tau(w_k) - p| = |\tau(w_k) - \tau(0)| \\
&\leq L_3 \|w_k\|_0 \leq L_3 a^k \|w_0\|_0
\end{aligned}
$$

Using this inequality successively, we get for all $0 \leq k < m$

$$
\begin{aligned}
|T_m - T_k| &\leq |T_m - T_{m-1}| + \cdots + |T_{k+1} - T_k| \\
&\leq L_3(a^{m-1} + \cdots + a^k)\|w_0\|_0 = L_3 a^k (1 - a^{m-k})\|w_0\|_0/(1-a) \\
&\leq \frac{L_3 \|w_0\|_0}{1-a} a^k \tag{6.55}
\end{aligned}
$$

Since $a < 1$, this shows that the sequence $\{T_k\}_{k=0}^{\infty}$ is a Cauchy sequence. Thus, there exists a real number T such that

$$\lim_{k \to \infty} T_k = T.$$

Now we prove Inequality (6.48. Suppose $t \geq p$ is given. Then there is a k such that $kp \leq t < (k+1)p$. Let

$$s \equiv t - kp.$$

Then $\phi_{s+t_k}(x) = \phi_s(\phi_{t_k}(x)) = \phi_s(w_k)$ and $\phi_s(0) = \gamma(s)$. Since $s \in [0, p]$ we can apply inequality (6.53) to get

$$
\begin{aligned}
\|\phi_{s+t_k}(x) - \gamma(s)\|_0 &= \|\phi_s(w_k) - \phi_s(0)\|_0 \\
&\leq L_2 \|w_k\|_0 \leq L_2 a^k. \tag{6.56}
\end{aligned}
$$

Furthermore, by the Mean Value Theorem, for any two times $s_1 \neq s_2$, there is a time s_* between s_1 and s_2, such that

$$\frac{\phi_{s_2}(x) - \phi_{s_1}(x)}{s_2 - s_1} = \frac{\partial \phi}{\partial t}(s_*, x) = X(\phi_{s_*}(x)).$$

But using inequality (6.52), we get from this that

$$\|\phi_{s_2}(x) - \phi_{s_1}(x)\|_0 \leq L_1 |s_2 - s_1|,$$

for all $s_1, s_2 \geq 0$. Hence in particular, for $s_2 = s + t_k$ and $s_1 = s + kp + T$, we have $s_2 - s_1 = T_k - T$, and so the last inequality reads

$$\|\phi_{s+t_k}(x) - \phi_{s+kp+T}(x)\|_0 \leq L_1 |T - T_k| \leq \frac{L_1 L_3}{1 - a} a^k.$$

Using this, inequality (6.56), and the Triangle Inequality gives

$$\|\phi_{s+kp+T}(x) - \gamma(s)\|_0 \leq \left(L_2 + \frac{L_1 L_3}{1 - a} \right) a^k = KLa^{k+1}.$$

Now rewrite this using the facts that $s + kp = t$ and $\gamma(s) = \gamma(s + kp) = \gamma(t)$. The result is

$$|\phi_{t+T}(x) - \gamma(t)| \leq La^{k+1} \leq La^{t/p}.$$

The last inequality here follows from $t/p < k + 1$ and $0 < a < 1$. Using inequality (6.50) we can convert this into inequality (6.48) and establishes orbital asymptotic stability of γ in the special case when $c = 0$ and $X(0) = \varepsilon_n$.

To prove the same thing for a general c and X, we use the translation and rotation map $f(x) = Q(x - c)$ discussed in the comments prior to the theorem. Letting $Y = f_*(X)$, we get that the theorem is true for Y and $f \circ \gamma$. Let ϕ^X and ϕ^Y denote the flows for X and Y, and for a given neighborhood $\widetilde{\Omega}$ of $\Gamma = \{\gamma(t) \mid t \in \mathbb{R}\}$, let $\Omega = f(\widetilde{\Omega})$ be the corresponding neighborhood of $f \circ \gamma$. Then with all the constructions and notation in the first part of the proof (relative to Y and $f \circ \gamma$), we get that $\widetilde{\Omega}_0 \equiv f^{-1}(\Omega_0)$ is the necessary neighborhood of γ to establish orbital stability.

To establish orbital asymptotic stability, note that the theorem applied to Y and $f \circ \gamma$ gives a $\delta > 0$ and an $L > 0$ such that for every $y \in B(0, \delta)$ there is a $T \in \mathbb{R}$ such that for all $t \geq p$

$$\begin{aligned} La^{t/p} &\geq \left| \phi_{t+T}^Y(y) - f \circ \gamma(t) \right| \\ &= \left| f \circ \phi_{t+T}^X \circ f^{-1}(y) - f \circ \gamma(t) \right| \\ &= \left| Q(\phi_{t+T}^X(f^{-1}(y)) - c) - Q(\gamma(t) - c) \right| \\ &= \left| Q\left(\phi_{t+T}^X(f^{-1}(y)) - \gamma(t) \right) \right| \\ &= \left| \phi_{t+T}^X(f^{-1}(y)) - \gamma(t) \right| \end{aligned}$$

Hence if we choose a $\widetilde{\delta} > 0$, such that $B(\widetilde{\delta}, c) \subseteq f^{-1}(B(0, \delta))$, then for all $x \in B(\widetilde{\delta}, c)$, there is a T such that $|\phi_{t+T}^X(x) - \gamma(t)| \leq La^{t/p}$, for all $t \geq p$. This proves orbital asymptotic stability. $\square$

Corollary 6.3 (Orbital Stability) *Suppose* $\gamma : \mathbb{R} \to \mathbb{R}^n$ *is a closed integral curve of the vector field* $X : \mathcal{O} \subseteq \mathbb{R}^n \to \mathbb{R}^n$. *Let* $c = \gamma(0)$ *and let* $P : U \cap M_c \to M_c$ *be the Poincaré map for* γ. *Suppose each neighborhood* $\widetilde{W}$ *of* c *in* $U \cap M_c$ *has a neighborhood* $W \subseteq \widetilde{W}$ *of* c *in* $U \cap M_c$ *such that*

$$P(W) \subseteq W.$$

Then γ *is orbitally stable.*

Proof: The proof of orbital stability in the first part of Theorem 6.3 only required what we are assuming in this corollary. There, the contraction property of P was used to ensure that $P(W) \subseteq W$. $\square$

Example 6.7 (Spherical Coordinates) We end the section with a 3-D system having cycles in its phase portrait which are orbitally stable. These cycles comprise the sphere of radius 1 at the origin. The system is related to a particularly simple one in spherical coordinates. For this, we use ρ as the standard radial distance from the origin: $\rho^2 = x^2 + y^2 + z^2$. Also $b > 0$ is a constant. The system in Cartesian coordinates is

$$
\begin{aligned}
x' &= (1 - \rho^2)x - by &&= x - by - x^3 - xy^2 - xz^2 \\
y' &= bx + (1 - \rho^2)y &&= bx + y - x^2 y - y^3 - yz^2 \\
z' &= (1 - \rho^2)z &&= z - x^2 z - y^2 z - z^3
\end{aligned}
$$

Transforming this to spherical coordinates ρ, θ, ϕ

$$x = \rho \sin \phi \cos \theta, \quad y = \rho \sin \phi \sin \theta, \quad z = \rho \cos \phi,$$

gives (see Proposition 5.1, Chapter 5)

$$
\begin{aligned}
\rho' &= (1 - \rho^2)\rho \\
\theta' &= b \\
\phi' &= 0
\end{aligned}
$$

This readily yields that $\theta(t) = bt + \theta_0$ and that $\phi(t) = \phi_0$, which say that each integral curve winds uniformly, and counterclockwise, about the z-axis while remaining on the cone $\phi = \phi_0$. The radial distance ρ from the origin is governed by $\rho' = (1 - \rho^2)\rho$, which we will analyze further in a minute.

The particular solution $\rho = 1$, $\theta = bt + \theta_0$, $\phi = \phi_0$ gives a circular integral curve on the unit sphere centered at the origin. This integral curve lies on the intersection of the sphere $\rho = 1$ with the cone $\phi = \phi_0$. See Figure 6.18. Each of these circles is an orbitally stable cycle of the system.

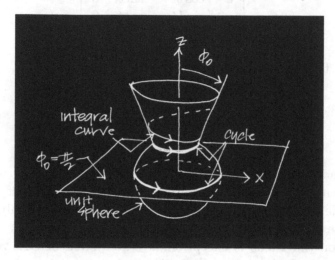

Figure 6.18: *A sketch of.*

We will examine this for the one that lies in the x-y plane ($\phi_0 = \pi/2$): $\gamma(t) = (\cos bt, \sin bt, 0)$.

The vector field for the system is

$$X(x, y, z) = \left(x - by - x^3 - xy^2 - xz^2, \, bx + y - x^2 y - y^3 - yz^2, \, z - x^2 z - y^2 z - z^3 \right),$$

and this has Jacobian matrix

$$X'(x, y, z) = \begin{bmatrix} 1 - 3x^2 - y^2 - z^2 & -b - 2xy & -2xz \\ b - 2xy & 1 - x^2 - 3y^2 - z^2 & -2yz \\ -2xz & -2yz & 1 - x^2 - y^2 - 3z^2 \end{bmatrix}.$$

Therefore the variational matrix for γ is

$$A(t) = \begin{bmatrix} -2\cos^2 bt & -b - 2\cos bt \sin bt & 0 \\ b - 2\cos bt \sin bt & -2\sin^2 bt & 0 \\ 0 & 0 & 0 \end{bmatrix}$$

$$= \begin{bmatrix} -1 - \cos 2bt & -b - \sin 2bt & 0 \\ b - \sin 2bt & -1 + \cos 2bt & 0 \\ 0 & 0 & 0 \end{bmatrix}$$

One can show that the fundamental matrix for A is

$$G(t) = \begin{bmatrix} e^{-2t}\cos bt & -\sin bt & 0 \\ e^{-2t}\sin bt & \cos bt & 0 \\ 0 & 0 & 1 \end{bmatrix}.$$

Consequently, the characteristic matrix for γ is

$$G(2\pi/b) = \begin{bmatrix} e^{-4\pi/b} & 0 & 0 \\ 0 & 1 & 0 \\ 0 & 0 & 1 \end{bmatrix},$$

and the characteristic multipliers are $\mu = e^{-4\pi/b}, 1, 1$. Thus, we cannot use the theorem to get orbital asymptotic stability of γ. In fact, γ is only orbitally stable. This is easy to conclude from a qualitative analysis of the system's phase portrait as follows.

The flow on the z-axis is exceptional. The system of DEs reduces to $x' = 0, y' = 0$, and $z' = (1 - z^2)z$. Thus, there are three fixed points: $(0,0,0), (0,0,1), (0,0,-1)$. The flow on the positive z-axis is toward $(0,0,1)$, while the flow on the negative z-axis is toward $(0,0,-1)$.

From the spherical coordinate equations, we have seen that the flow through an initial point (x_0, y_0, z_0) on the unit sphere $x^2 + y^2 + z^2 = 1$ will be a circular cycle of radius $R_0 = \sqrt{x_0^2 + y_0^2}$, center $(0,0,z_0)$ and period $p = 2\pi/b$.

For any other initial point (x_0, y_0, z_0) (not on the z-axis, not on the unit sphere), the flow through (x_0, y_0, z_0) will be a counterclockwise spiral which will asymptotically approach the cycle on the unit sphere with radius

$$R_0 = \frac{\sqrt{x_0^2 + y_0^2}}{\sqrt{x_0^2 + y_0^2 + z_0^2}},$$

while remaining on the conical surface $\phi = \phi_0 = \sin^{-1}(R_0)$. Figure 6.19 illustrates some of these features of the phase portrait.

It is clear from this qualitative analysis that the cycle γ is orbitally stable (or, for that matter, each of the cycles on the unit sphere is orbitally stable).

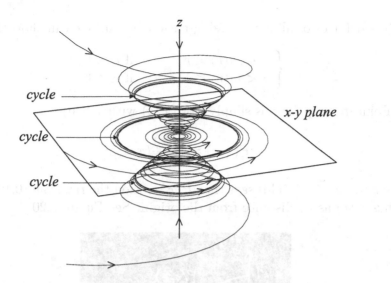

Figure 6.19: *Computer plots of six integral curves spiraling toward three different cycles on the unit sphere.*

We just have to restrict to a neighborhood of $c = (0, 1, 0)$ where all points flow toward cycles that are near γ. To see how this fits in the above corollary, we compute the Poincaré map.

First note that the solution of $\rho' = (1 - \rho^2)\rho$, with $\rho(0) = \rho_0 > 0$ is

$$\rho(t) = [1 + (\rho_0^{-2})e^{-2t}]^{-1/2},$$

defined for all $t \in [0, \infty)$. The plane M_c is, in this case, the x-z plane. The flow through a point $(x_0, 0, z_0) \in M_c$ (we assume $x_0 > 0$, $z_0 > 0$) is

$$\Phi_t(x_0, 0, z_0) = \left(\frac{\rho(t)x_0}{r_0} \cos bt, \; \frac{\rho(t)x_0}{r_0} \sin bt, \; \frac{\rho(t)z_0}{r_0} \right),$$

where $r_0 = \sqrt{x_0^2 + z_0^2}$. Then, the point of first return for $(x_0, 0, z_0)$ to M_c is

$$\Phi_{2\pi/b}(x_0, 0, z_0) = \left(\frac{m(r_0)x_0}{r_0}, \; 0, \; \frac{m(r_0)z_0}{r_0} \right),$$

where

$$m(r) = [1 + (r^{-2} - 1)e^{-4\pi/b}]^{-1/2} = \frac{re^{2\pi/b}}{[1 + (e^{4\pi/b} - 1)r^2]^{1/2}},$$

which is defined for all $r > 0$. Using basic algebra, one can show that

$$\begin{cases} 1 < m(r) < r & \text{if } r > 1 \\ r < m(r) < 1 & \text{if } r < 1 \end{cases}$$

The Poincaré map for γ is simply (for (x, z) with $x > 0$)

$$P(x, z) = \frac{m(r)}{r}(x, z),$$

where $r = \sqrt{x^2 + z^2}$. This says that $P(x, z)$ is on the ray from $(0, 0)$ to (x, z) and has $m(r)$ as its distance from the origin. See Figure 6.20.

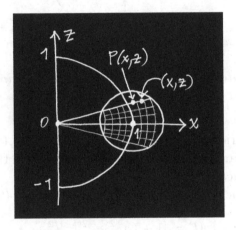

Figure 6.20: *Geometry of the Poincaré map P.*

Thus, in the x-z plane, each sector of an annulus:

$$W = \{(r, \theta) \,|\, r_1 < r < r_2, \theta_1 < \theta < \theta_2\},$$

($r_1 < 1 < r_2$, $-\pi/2 < \theta_1 < \theta_2 < \pi/2$) is invariant: $P(W) \subseteq W$ under P. Using this and Corollary 6.3, one can analytically prove the orbital stability of γ (even though it's rather obvious geometrically). It is also interesting to compute directly the iterations of the map $m : (0, \infty) \to (0, \infty)$ and show that the kth iterate is

$$m^k(r) = [1 + (r^{-2} - 1)e^{-4k\pi/b}]^{-1/2}.$$

Consequently $\lim_{k \to \infty} m^k(r) = 1$ for every $r > 0$. Of course, this also follows from properties of the flow map, since that's where m comes from.

Exercises 6.5

1. Prove that if H is the deformation matrix for the flow ϕ, then

$$\frac{\partial H}{\partial t}(t, x) = X'(\phi(t, x))H(t, x),$$

for all $(t, x) \in \mathcal{D}$ and that $H(0, x) = I$, for all $x \in \mathcal{O}$.

2. Show that equations (6.40)-(6.41) hold.

3. For each of the systems $x' = X(x)$ and closed curves γ shown below (with $c = \gamma(0)$ and $b > 0$), do the following:

   (i) Compute the variational matrix $A(t) = X'(\gamma(t))$ and find the corresponding characteristic matrix G (i.e., the fundamental matrix for A). Determine the characteristic multipliers for γ.

   (ii) Determine (where possible) an explicit formula for the Poincaré map P for γ. Calculate (where possible) $P'(c)$, the eigenvalues of this, and compare with the characteristic multipliers.

   iii) Sketch, by hand, the phase portrait of the system. Use this and the work in Parts (i)-(ii) to determine the orbital stability, orbital asymptotic stability, or instability of γ. Fully justify your answers.

Note: In these systems: $r^2 = x^2 + y^2$ and $\rho^2 = x^2 + y^2 + z^2$. Also, (1) transform to polar coordinates in Parts (a)-(c), (2) transform to cylindrical coordinates in Parts (d)-(f), and (3) transform to spherical coordinates in Part (h).

   (a) $\gamma(t) = (\cos bt, \sin bt)$

$$\begin{aligned} x' &= (1 - r^2)x - by \\ y' &= bx + (1 - r^2)y \end{aligned}$$

   (b) $\gamma(t) = (\cos bt, \sin bt)$

$$\begin{aligned} x' &= (1 - r)x - by \\ y' &= bx + (1 - r)y \end{aligned}$$

   (c) $\gamma(t) = (\cos bt, \sin bt)$, $\gamma(t) = (2\cos bt, 2\sin bt)$

$$\begin{aligned} x' &= (2 - r)(1 - r)x - by \\ y' &= bx + (2 - r)(1 - r)y \end{aligned}$$

   (d) $\gamma(t) = (\cos bt, \sin bt, 0)$.

$$\begin{aligned} x' &= -by + xz \\ y' &= bx + yz \\ z' &= -rz \end{aligned}$$

(e) $\gamma(t) = (\cos bt, \sin bt, 0)$, $\gamma(t) = (2\cos bt, 2\sin bt, 0)$.

$$\begin{aligned} x' &= -by + xz \\ y' &= bx + yz \\ z' &= -4r^2(1 - r^2)z \end{aligned}$$

(f) $\gamma(t) = (\cos bt, \sin bt, 0)$

$$\begin{aligned} x' &= (1 - r^2)x - by \\ y' &= bx + (1 - r^2)y \\ z' &= -z \end{aligned}$$

(g) $\gamma(t) = (\cos bt, \sin bt, 0)$ (This one is easy, since it's a canonical linear system.)

$$\begin{aligned} x' &= -by \\ y' &= bx \\ z' &= -z \end{aligned}$$

(h) $\gamma(t) = (\cos bt, \sin bt, 0)$, $\gamma(t) = (2\cos bt, 2\sin bt, 0)$.

$$\begin{aligned} x' &= (2 - \rho)(1 - \rho)x - by \\ y' &= bx + (2 - \rho)(1 - \rho)y \\ z' &= (2 - \rho)(1 - \rho)z \end{aligned}$$

5. As a generalization of Example 6.3 and Exercises (d) and (e) above, consider the following system where $V : [0, \infty) \to \mathbb{R}$ is a twice continuously differentiable function:

$$\begin{aligned} x' &= -by + xz \\ y' &= bx + yz \\ z' &= -V'(r)rz \end{aligned}$$

Transform to cylindrical coordinates and show that the integral curves in the r-z plane lie on the graph of $z = -V(r) + a_0$ for some constant a_0. Describe the directions of flow on these graphs. Describe how the integral curves of the original system lie on the surface of revolution obtained by revolving these graphs about the z-axis. How do the critical points of V play a role in the phase portrait. Select a V of your choice (one with at least one local maxima and one local minima). From your selection, sketch a phase portrait for the system.

6. Read the material in CDChapter 6 on the electronic component that pertains to periodic points of discrete dynamical systems and work the exercises there.

7. Prove that equation (6.51) holds for all k.

Chapter 7

Integrable Systems

In this chapter we consider a special class of autonomous systems, $x' = X(x)$, on open sets $\mathcal{O} \subseteq \mathbb{R}^n$, whose integral curves are completely "determined" by $n-1$ functions, $F^1, F^2, \ldots, F^{n-1} : U \subseteq \mathcal{O} \to \mathbb{R}$, defined on an open dense subset U of $\mathcal{O}$. These functions are called first integrals, or constants of the motion, and have, by definition, constant values along each integral curve of X. In addition, there are conditions on $F^1, F^2, \ldots, F^{n-1}$, so that the level sets $F^i(x) = k_i$, $i = 1, \ldots, n-1$, intersect to give 1-dimensional submanifolds or curves in $\mathbb{R}^n$ and these curves coincide, in a sense, with the integral curves of X. Such systems are called *integrable systems* and will be defined more precisely below.

Integrable systems are often called *completely integrable systems* in accordance with the terminology used in the more general subject of Pfaffian systems (see [BCG 91], [Sl 70], [Di 74]). However, in the study of Hamiltonian systems (Chapter 9), there is the well-accepted term of completely integrable *Hamiltonian* system, which is related to but quite distinct from the type of system studied here. Thus, we will use the terms "integrable" and "completely integrable" to distinguish between the two distinct types of the systems studied in this chapter and in Chapter 9, respectively. This naming convention was suggested by Olver [Olv 96, p. 70].

While there are many important examples of integrable systems, like Euler's equations for rigid-body motion in Chapter 8, these systems are rather exceptional. We study them here primarily to develop more geometric intuition about phase portraits in dimensions three and higher. Indeed, the phase portrait for an integrable system in $\mathbb{R}^3$ is visualized as the collection of curves that are the intersections of two families of level surfaces. In simple cases these can be drawn by hand and in more complicated cases there is some special Maple code on the electronic component to plot the curves of

D. Betounes, *Differential Equations: Theory and Applications*, DOI 10.1007/978-1-4419-1163-6_7,

intersection with a computer. Furthermore, even for dimensions larger than three it is possible to visualize the integral curve that is the intersection of the level sets $F^i(x) = k_i$, $i = 1, \ldots, n - 1$, as the image of a level curve $g_{k_1 \ldots k_{n-2}}(x_1, x_2) = k_{n-1}$ in the plane $\mathbb{R}^2$. Thus, the study of the phase portrait is reduced to the study of the level curves of a family $\{g_{k_1 \ldots k_{n-2}}\}$ of functions of two variables.

7.1   First Integrals (Constants of the Motion)

There are several equivalent ways of defining what a first integral, or constant of the motion, is for an autonomous system $x' = X(x)$. The following definition is the easiest to check when presented with a candidate for a first integral. The actual process of finding first integrals is considerably more difficult.

Definition 7.1 (First Integrals: Constants of the Motion) Suppose $X : \mathcal{O} \to \mathbb{R}^n$ is a vector field on an open set $\mathcal{O}$ in $\mathbb{R}^n$. A *first integral* (or *constant of the motion*) for the autonomous system $x' = X(x)$, is a differentiable function $F : U \to \mathbb{R}$, defined on an open, dense subset $U \subseteq \mathcal{O}$, such that

$$\nabla F(x) \cdot X(x) = 0,$$

for every $x \in U$. Geometrically, this says that ∇F is perpendicular to X at each point of U.

There are several alternative, and conceptually useful, ways of expressing the orthogonality property: $\nabla F \cdot X = 0$, for first integrals. All of these are based on the following elementary result.

Proposition 7.1 *Suppose $F : U \to \mathbb{R}$ is a differentiable function defined on an open, dense subset U of $\mathcal{O}$. Then F is a first integral if and only if for each integral curve $\alpha : I \to \mathbb{R}$ of X that lies in U, i.e., $\alpha(t) \in U$ for all $t \in I$, there is a constant k such that*

$$F(\alpha(t)) = k,$$

for all $t \in I$.

Proof: The proof is an elementary consequence of the chain rule. The details are left for an exercise. $\square$

The proposition says that a first integral has a constant value along each integral curve. Specifically, F has the same value at each point in the set $\alpha(I) = \{\alpha(t) \,|\, t \in I\}$, whenever α is an integral curve. This is why first integrals are also called constants of the motion. As we have seen for particular systems arising from physics, the function F also gives rise to a *conservation law*, usually expressed in the form $F(\alpha(t)) = k$ for all t, where α is any motion of the system. The motion α in this context contains the various positions and momenta of the objects constituting the system. Chapters 8 and 9 contain more detailed studies of conservation laws.

There is a slightly different, more geometric way of expressing what a first integral means for a system. If F is a first integral for $x' = X(x)$, then each integral curve of X lies on one of the level sets of F, i.e., there is a constant k, such that $\alpha(t) \in S_F^k$, for all $t \in I$. Here

$$S_F^k = \{\, x \in U \,|\, F(x) = k \,\} = F^{-1}(\{k\})$$

is the *level set* of F corresponding to k. In dimension $n = 2$ this says that each integral curve must lie on one of the level curves of F and thus, as we shall see in the next section, the plots of the level curves of F give a complete picture of the phase portrait for X. See Figure 7.1.

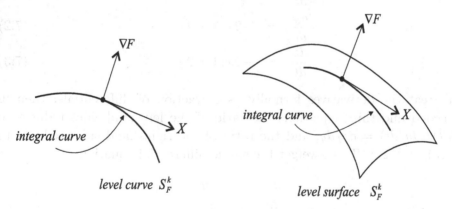

Figure 7.1: *If F is a first integral, each integral curve of X lies on a level set S_F^k of F. Geometrically ∇F is normal and X is tangential to the level set S_F^k at each of its points. The figure displays the cases $n = 2$ and $n = 3$ where the level sets are curves and surfaces, respectively.*

In dimension $n = 3$ the level sets of F are surfaces and so each integral curve of X is constrained to lie completely on one of these level surfaces.

Since phase portraits in $\mathbb{R}^3$ are difficult to visualize, knowing a first integral for the system often helps understand the behavior of the integral curves. See Figure 7.1.

In general, the Submanifold Theorem from Appendix A gives us that the level set S_F^k is an $n-1$ dimensional submanifold of $\mathbb{R}^n$ (assuming k is a value of F at a point other than a critical point). The gradient ∇F is normal to S_F^k at each of its points and if F is a first integral for X, then X is tangential to S_F^k.

Expressing the above proposition in terms of the flow for X, we get the following: F is a constant of the motion if an only if for each $c \in U$

$$F(\phi_t(c)) = F(c),$$

for all $t \in I_c$. This latter statement is equivalent to saying that each level set of F is an invariant set for the flow.

Example 7.1 Often it is possible to determine first integrals by the heuristic method illustrated in this example. Consider the system:

$$\frac{dx}{dt} = 4xz - 1 \tag{7.1}$$

$$\frac{dy}{dt} = -2x(1 + 2z) \tag{7.2}$$

$$\frac{dz}{dt} = -2x(1 + 2y). \tag{7.3}$$

If we treat the derivatives formally as a fraction of differentials, then in the second and third equations the ratio of the left-hand sides reduces to $(dz/dt)/(dy/dt) = dz/dy$, and the ratio of the right-hand sides reduces to $(1+2y)/(1+2z)$. Thus, we get the single differential equation

$$\frac{dz}{dy} = \frac{1 + 2y}{1 + 2z},$$

which is a separable DE. We solve this by formally separating the variables

$$(1 + 2y)dy - (1 + 2z)dz = 0,$$

and integrating to get the (implicit) solution:

$$y + y^2 - z - z^2 = k.$$

The claim is that this formal process actually gives us a constant of the motion F. This is the function

$$F(x, y, z) \equiv y + y^2 - z - z^2,$$

which is defined on all of $\mathbb{R}^3$. It is easy to see that

$$\nabla F(x, y, z) = \left(0, 1 + 2y, -(1 + 2z) \right)$$

is perpendicular to

$$X(x, y, z) = \left(4xz - 1, -2x(1 + 2z), -2x(1 + 2y) \right),$$

at each point (x, y, z), and thus F is a first integral.

The precise formulation and proof of the validity of the formal manipulations in the example are contained in the following proposition.

Proposition 7.2 *Suppose $X : \mathcal{O} \to \mathbb{R}^n$ is a vector field on $\mathcal{O} \subseteq \mathbb{R}^n$ and assume that for two indices $i \neq j$, the differential equation*

$$-X^j + X^i \frac{dx_j}{dx_i} = 0 \tag{7.4}$$

has an integrating factor. That is, assume that there is a function $\mu : U \subseteq \mathcal{O} \to \mathbb{R}$, defined on an open, dense subset of $\mathcal{O}$ such that $-\mu X^j, \mu X^i$ depend only on the variables x_i, x_j and such that

$$-\mu X^j + \mu X^i \frac{dx_j}{dx_i} = 0, \tag{7.5}$$

is an exact differential equation. Then there is a function $F : U \subseteq \mathcal{O} \to \mathbb{R}$, which depends only on x_i, x_j, and such that

$$\frac{\partial F}{\partial x_i} = -\mu X^j \tag{7.6}$$

$$\frac{\partial F}{\partial x_j} = \mu X^i, \tag{7.7}$$

on U. Thus, $\nabla F \cdot X = 0$ on U.

Proof: There is really not much to prove, once we agree upon what it means for a function $f : U \to \mathbb{R}$ to "depend only on the variables x_i, x_j." To define this, let $P_{ij} : \mathbb{R}^n \to \mathbb{R}^2$ be the projection: $P_{ij}(x) \equiv (x_i, x_j)$, and let $\widetilde{U} \subseteq \mathbb{R}^2$ be the open set: $\widetilde{U} \equiv P_{ij}(U)$. Then saying that f depends only on the variables x_i, x_j, means that $f = \widetilde{f} \circ P_{ij}$ for some function $\widetilde{f}$ on $\widetilde{U}$ (which is differentiable if f is).

Now by assumption there are functions $\widetilde{M} = \widetilde{\mu X^i}, \widetilde{N} = \widetilde{\mu X^j}$ on $\widetilde{U}$ such that the differential equation

$$\widetilde{M} + \widetilde{N}\frac{dx_j}{dx_i} = 0,$$

is exact. Thus, by definition of exactness, there is a differentiable function $\widetilde{F} : \widetilde{U} \to \mathbb{R}$, such that $\widetilde{F}_{x_i} = \widetilde{M}$ and $\widetilde{F}_{x_j} = \widetilde{N}$ on $\widetilde{U}$. Thus, we can let $F \equiv \widetilde{F} \circ P_{ij}$, to get a function on $\mathcal{O}$ for which equations (7.6)-(7.7) hold. But then

$$\nabla F = (0, \dots, -\mu X^j, \dots, \mu X^i, \dots, 0),$$

is clearly perpendicular to X at each point of $\mathcal{O}$. $\square$

The proposition just makes precise the heuristic process we usually use in solving a first order DE. Thus, phrasing the above example in terms of the proposition, we have that the associated DE (7.4) is

$$-2x(1 + 2y) + 2x(1 + 2z)\frac{dz}{dy} = 0,$$

and if we multiply this by $\mu = 1/2x$, we get the exact DE

$$-(1 + 2y) + (1 + 2z)\frac{dz}{dy} = 0.$$

In this example, $U = \{(x, y, z) \in \mathbb{R}^3 | x \neq 0\}$ and $\widetilde{U} = \{(x, y) \in \mathbb{R}^2 | x \neq 0\}$, are the domains used in the proposition.

Before defining and studying integrable systems in general, we discuss this concept for systems in dimensions two and three, where its geometric interpretation is visualized more easily. This is presented in the next two sections.

Exercises 7.1

1. Prove Proposition 7.1. You may use the fact that if a C^1 function on an interval has derivative that is identically zero, then the function is a constant function.

2. Use the method from Example 7.1 and Proposition 7.2 to find a first integral for each of the following systems. Be sure to specify the domain U for each first integral (make it as large as possible). Where possible find additional first integrals for the system, ones that are essentially "different" from the first one you found.

   (a)

   $$\begin{aligned} x' &= x^2 y \\ y' &= x(1+y) \end{aligned}$$

   (b)

   $$\begin{aligned} x' &= y(z-1) \\ y' &= -x(z+1) \\ z' &= -2xy \end{aligned}$$

   (c)

   $$\begin{aligned} x' &= y(z+1) \\ y' &= -x(z+1) \\ z' &= -2xy \end{aligned}$$

3. Show that if $F, G : U \subseteq \mathcal{O} \to \mathbb{R}$ are first integrals for a system $x' = X(x)$, then so are $F + G, FG$, and F/G (assuming G is never zero on U).

7.2   Integrable Systems in the Plane

Definition 7.2 Suppose $X : \mathcal{O} \subseteq \mathbb{R}^2 \to \mathbb{R}^2$ is a vector field on the plane. The system $(x', y') = X(x, y)$ is called an *integrable system* if it has a first integral (or constant of the motion) $F : U \to \mathbb{R}$, such that $\nabla F(x, y) \neq 0$, for all $(x, y) \in U$.

For integrable systems in the plane, any first integral gives a more or less complete description of the phase portrait for the system. Indeed, each integral curve in U must lie on one of the level curves of F. Thus, a plot of the level curves gives a picture of the phase portrait. The following examples exhibit what is meant by this.

Example 7.2 The linear system

$$\frac{dx}{dt} = -x - y$$

$$\frac{dy}{dt} = x - y,$$

has a phase portrait with integral curves that spiral toward the origin, which is the only fixed point of the system. This is the canonical focus studied in Chapter 5 and shown in Figure 4.25. To see that this system is integrable, we construct a first integral using the method of Example 7.1 and Proposition 7.2. Namely, taking the ratio of the second equation by the first gives the associated DE:

$$\frac{dy}{dx} = \frac{y - x}{x + y}.$$

This is a homogeneous (nonautonomous) differential equation and we can solve it by elementary methods. We find that the general solution y is defined implicitly as a function of x by the equation

$$\tfrac{1}{2}\ln(x^2 + y^2) + \tan^{-1}\left(\frac{y}{x}\right) = k.$$

Taking $U = \{(x, y) \in \mathbb{R}^2 | x \neq 0\}$ and defining $F : U \to \mathbb{R}$ by

$$F(x, y) = \tfrac{1}{2}\ln(x^2 + y^2) + \tan^{-1}\left(\frac{y}{x}\right)$$

gives a first integral for the linear system. Indeed,

$$\nabla F = \left(\frac{x - y}{x^2 + y^2}, \frac{x + y}{x^2 + y^2}\right),$$

and $X : \mathbb{R}^2 \to \mathbb{R}^2$ is

$$X = (-(x + y), x - y).$$

Thus, clearly $\nabla F \cdot X = 0$ on U and U is dense in $\mathbb{R}^2$. Also, $\nabla F \neq 0$ in U and thus the system is an integrable system. Figure 7.2 shows a plot of the graph of F and a plot of a number of its level curves.

Note that F approaches $\pm\infty$ as $x \to 0$ and thus its graph is hard to render near the line $x = 0$. Also, while each integral curve is a continuous spiral, the corresponding level curve on which it lies is divided into infinitely many pieces by the line $x = 0$ (exercise). Some of these can be discerned from the figure, but most are too small to show up.

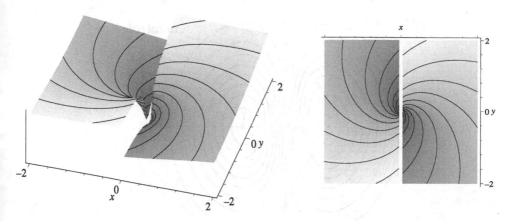

Figure 7.2: *Graph of the first integral* $F(x, y) = \ln(x^2 + y^2)/2 + \tan^{-1}(y/x)$, *for the linear system* $x' = -x - y$, $y' = x - y$.

It is easy to see that the integrability of the linear system in the above example generalizes as follows.

Proposition 7.3 *Any constant coefficient, linear system*

$$\frac{dx}{dt} = ax + by$$
$$\frac{dy}{dt} = px + qy.$$

in the plane is an integrable system.

Proof: The proof is left as an exercise. $\square$

The above example of a linear system illustrates why we cannot generally expect a first integral $F : U \to \mathbb{R}$ to be defined on all of $\mathcal{O}$. For this reason, in dimension two, the level curves of F may not contain all the information about the system. For instance, the origin $(0, 0)$ is a fixed point in the example, but is not in the domain U of F. However, in some examples such as the following, the fixed points come from the critical points of a first integral.

Example 7.3 Let $X : \mathbb{R}^2 \to \mathbb{R}^2$ be given by

$$X(x, y) = \left(2xy\, e^{-x^2 - y^2},\ (1 - 2x^2)e^{-x^2 - y^2} \right).$$

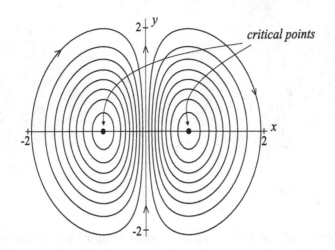

Figure 7.3: *A plot of a collection of level curves for the function $F(x,y) = xe^{-x^2-y^2}$. Note that the orientation of each curve is determined by evaluating X at one of its points.*

Then the function $F : \mathbb{R}^2 \to \mathbb{R}$ defined by

$$F(x,y) = xe^{-x^2-y^2}$$

has gradient

$$\nabla F(x,y) = \left((1 - 2x^2)e^{-x^2-y^2},\ -2xy\, e^{-x^2-y^2} \right),$$

and this is clearly perpendicular to X at each point. It is also easy to see that the critical points of F and the fixed points of X coincide and there are exactly two such points:

$$\{\, (1/\sqrt{2}, 0),\ (-1/\sqrt{2}, 0) \,\}.$$

A plot of a collection of level curves of F is shown in Figure 7.3 and by evaluating X at certain points on each level curve we get the directions as shown. By the theory, each integral curve of X lies on a level curve and so the figure represents the phase portrait for the system. Note that in this example $\mathcal{O} = \mathbb{R}^2$ and the first integral F is defined on all of $\mathbb{R}^2$. To meet the condition for integrability, we must restrict F to $U = \mathbb{R}^2 \setminus \{(\pm 1/\sqrt{2}, 0)\}$. However, as mentioned, the fixed points in this example arise from the critical points of the first integral.

In examining Example 7.3 above, you might discover a method of manufacturing integrable systems in the plane. Namely, start with a differentiable function $F : \mathcal{O} \to \mathbb{R}$ on an open set $\mathcal{O}$ in $\mathbb{R}^2$, take its gradient $\nabla F = (F_x, F_y)$, and then switch the partial derivatives, taking the negative of one of them, to get a vector field

$$X_F \equiv (-F_y, F_x),$$

on $\mathcal{O}$ that is perpendicular to ∇F and has fixed points that are the same as the critical points of F. Thus, F is a first integral for X_F. Furthermore, if the subset U, obtained by deleting the critical points of F from $\mathcal{O}$, is open and dense, then $(x', y') = X_F(x, y)$ is an integrable system.

The construction of X_F takes advantage of the normal operator on $\mathbb{R}^2$.

Definition 7.3 The *normal operator* on $\mathbb{R}^2$ is the linear map $N : \mathbb{R}^2 \to \mathbb{R}^2$ given by

$$N(x, y) = (-y, x). \tag{7.8}$$

This operator has the property that $N(v) \cdot v = 0$ for every vector $v \in \mathbb{R}^2$. More generally, it is easy to see that $N(v) \cdot w = \det(v, w)$, for every $v, w \in \mathbb{R}^2$.

As a linear operator, N is represented by the 2×2 matrix $-J$, where

$$J = \begin{bmatrix} 0 & 1 \\ -1 & 0 \end{bmatrix}.$$

That is, $N(v) = -Jv$, for every $v \in \mathbb{R}^2$. The matrix J is called the *canonical complex structure* on $\mathbb{R}^2$. There is such a structure on each *even* dimensional Euclidean space $\mathbb{R}^{2k}$. *Note*: We could have defined N slightly differently so as to have $N(v) = Jv$, but our choice was motivated by the wish that $\{v, N(v)\}$ be a right-handed frame for each nonzero vector v.

Using the normal operator, we can phrase the above construction of an integrable, planar system as

$$X_F = N(\nabla F),$$

where F is any differentiable function. The construction can be generalized slightly by multiplying by a differentiable function $\rho : \mathcal{O} \to \mathbb{R}$ to get an integrable system of the form

$$X_{\rho, F} = \rho N(\nabla F).$$

By considering the geometry of the situation as shown in Figure 7.4, it is not too hard to guess that every integrable system on $\mathcal{O}$ must be of this form. The following theorem gives the proof of this conjecture.

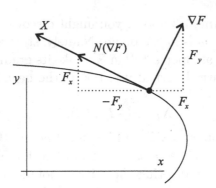

Figure 7.4: *If F is a first integral for a planar vector field X, then, at each point, X must be a multiple of $N(\nabla F)$.*

Theorem 7.1 *Suppose $X : \mathcal{O} \to \mathbb{R}^2$ is a vector field on the plane. Then the system $(x', y') = X(x, y)$ is integrable if and only if there is an open, dense subset $U \subseteq \mathcal{O}$, on which X has the form*

$$X = \rho N(\nabla F),$$

where $\rho, F : U \to \mathbb{R}$ are differentiable functions and $\nabla F \neq 0$ on U.

Proof: Assume the system is integrable, with first integral $F : U \to \mathbb{R}$, such that $\nabla F \neq 0$ on U. Let $e_1 = \nabla F$ and $e_2 = N(\nabla F)$. This gives an orthogonal, moving frame $\{e_1, e_2\}$ on U. The reciprocal frame $\{e^1, e^2\}$ (cf. Section 10.1) in this case is given simply by $e^1 = e_1/a, e^2 = e_2/a$, where $a = e_1 \cdot e_1 = F_x^2 + F_y^2 = e_2 \cdot e_2$. Since $e^1 \cdot X = a^{-1} e_1 \cdot X = 0$, we can write X in terms of this frame as

$$X = (e^1 \cdot X)e_1 + (e^2 \cdot X)e_2 = (e^2 \cdot X)e_2,$$

on U. Thus, we can take $\rho = e^2 \cdot X$ to get the desired form for X. $\square$

Exercises 7.2

1. Prove Proposition 7.3. Discuss all the cases depending on which of the coefficients a, b, p, q are zero or not.

2. For each of the following functions $F : \mathbb{R}^2 \to \mathbb{R}$, compute the vector field $X_F = N(\nabla F)$ on $\mathcal{O} = \mathbb{R}^2$. Use a computer to construct the phase portrait of the system of DEs from the level curves of F. Mark the directions of flow on the flow lines. Determine and classify all the fixed points for the system.

(a) $F(x,y) = (x^2 + y)e^{-x^2-y^2}$.

(b) $F(x,y) = (x^3 + y^3 - 4x^2y^2)e^{-x^2-y^2}$.

(c) $F(x,y) = (\frac{1}{3}x^4 + \frac{1}{2}y^4 - 4xy^2 - 2x^2 - 2y^2 + 3)e^{-x^2-y^2}$. Be sure to find all the fixed points. Hint: there are seven of them.

7.3 Integrable Systems in 3-D

For a system in $\mathbb{R}^3$, any first integral F^1 gives a family of surfaces $S_{F^1}^{k_1}$ (the level surfaces of F^1 in $\mathbb{R}^3$) on which the integral curves of the system must lie. If another first integral F^2 exists, then any integral curve of the system must lie on two level surfaces $S_{F^1}^{k_1}, S_{F^2}^{k_2}$, one for each first integral. Consequently, the integral curve coincides with part of the "curve" of intersection

$$S_{F^1 F^2}^{k_1 k_2} \equiv S_{F^1}^{k_1} \cap S_{F^2}^{k_2},$$

of these two level surfaces. See Figure 7.5.

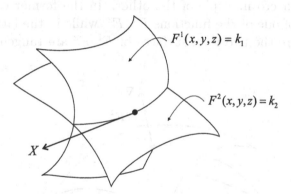

$F^1(x,y,z) = k_1$

$F^2(x,y,z) = k_2$

X

Figure 7.5: *If a system has two independent first integrals F^1, F^2, then each integral curve lies on the curve of intersection of two particular level surfaces $F^1 = k_1, F^2 = k_2$.*

Otherwise said, the integral curve is part of the solution set $S_{F^1 F^2}^{k_1 k_2}$ of the system of equations:

$$
\begin{aligned}
F^1(x,y,z) &= k_1 \\
F^2(x,y,z) &= k_2.
\end{aligned}
$$

Thus, in a sense, the two first integrals F^1, F^2 completely determine the phase portrait for a system $x' = X(x)$ in dimension three. This discussion and the drawing in Figure 7.5 assume that the functions F^1, F^2 are different, or independent, in some sense, so that the level surfaces intersect to give a 1-dimensional manifold. More specifically:

Definition 7.4 (Integrability in $\mathbb{R}^3$) Suppose $X : \mathcal{O} \to \mathbb{R}^3$ is a vector field on an open set $\mathcal{O}$ in $\mathbb{R}^3$.

(1) Two differentiable functions $F^1, F^2 : U \subseteq \mathcal{O} \to \mathbb{R}$ are called *functionally independent* on U, if $\nabla F^1(x, y, z)$ and $\nabla F^2(x, y, z)$ are linearly independent for each $(x, y, z) \in U$.

(2) The system $(x', y', z') = X(x, y, z)$ on $\mathcal{O} \subseteq \mathbb{R}^3$ is called *integrable* if it has two first integrals $F^1, F^2 : U \to \mathbb{R}$, which are functionally independent on U.

Note that linear independence of two vectors $\nabla F^1(x, y, z), \nabla F^2(x, y, z)$ means that neither is zero and they are not parallel. On the other hand linear dependence means that either one of these gradient vectors is zero or one is a nonzero multiple of the other. In the former case (x, y, z) is a critical point of one of the functions F^1, F^2, while in the latter case (x, y, z) is a point where the two level surfaces of F^1, F^2 are tangent to each other. See Figure 7.6.

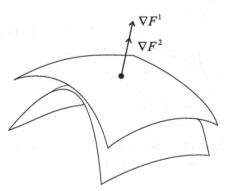

Figure 7.6: *The level surfaces of F^1, F^2 are tangent at a point (x, y, z) for which their gradient vectors $\nabla F^1(x, y, z), \nabla F^2(x, y, z)$ are parallel.*

The two first integrals F^1, F^2 for an integrable system in $\mathbb{R}^3$ satisfy

$$\nabla F^1 \cdot X = 0, \qquad \nabla F^2 \cdot X = 0,$$

on $\mathcal{O}$ and we can interpret this as saying that the vector fields $\nabla F^1, \nabla F^2$ give two directions that X must be perpendicular to at each point of U.

This assertion is illustrated in Figure 7.7 and geometrically you can see that it determines X up to a scalar multiple.

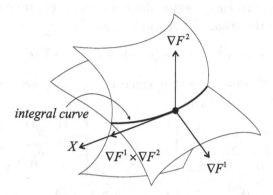

Figure 7.7: *For an integrable system $x' = X(x)$ in $\mathbb{R}^3$, the vector field X must be perpendicular to the gradient vector fields $\nabla F^1, \nabla F^2$ at each point of U.*

More precisely, we have the following theorem:

Theorem 7.2 *Suppose $X : \mathcal{O} \to \mathbb{R}^3$ is a vector field on an open set $\mathcal{O}$ in $\mathbb{R}^3$. Then the system $(x', y', z') = X(x, y, z)$ is integrable if and only if there is an open, dense subset $U \subseteq \mathcal{O}$ on which X has the form*

$$X = \rho(\nabla F^1 \times \nabla F^2), \tag{7.9}$$

where $\rho, F^1, F^2 : U \to \mathbb{R}^3$ are differentiable functions with F^1, F^2 functionally independent on U.

Proof: It is easy to see that if X has the form (7.9), then $(x', y', z') = X(x, y, z)$ is integrable (with F^1, F^2 as first integrals). Conversely, suppose the system is integrable and let $F^1, F^2 : U \to \mathbb{R}$, be two first integrals that are functionally independent on U. On U, define the vector fields

$$\begin{aligned}
e_1 &= \nabla F^1 \\
e_2 &= \nabla F^2 \\
e_3 &= \nabla F^1 \times \nabla F^2.
\end{aligned}$$

Then $\{e_1, e_2, e_3\}$ is an orthogonal, moving frame on U. The reciprocal frame $\{e^1, e^2, e^3\}$ is given by $e^i = \sum_{j=1}^{3} g^{ij} e_j$, where the g^{ij}'s are the entries of the inverse matrix G^{-1} of the metric matrix $G = \{g_{ij}\} = \{e_i \cdot e_j\}$. By construction $e^i \cdot e_j = 0$ for $i \neq j$, and $e^i \cdot e_i = 1$, for all i, j. This latter property allows one to easily write the coordinate expression of any vector field Y relative to the frame $\{e_1, e_2, e_3\}$. Namely,

$$Y = (e^1 \cdot Y)e_1 + (e^2 \cdot Y)e_2 + (e^3 \cdot Y)e_3.$$

To apply this to X, first note that since $e_3 \cdot e_i = 0$, for $i = 1, 2$, it follows that G and G^{-1} have the forms

$$G = \begin{bmatrix} g_{11} & g_{12} & 0 \\ g_{21} & g_{22} & 0 \\ 0 & 0 & g_{33} \end{bmatrix}, \qquad G^{-1} = \begin{bmatrix} g^{11} & g^{12} & 0 \\ g^{21} & g^{22} & 0 \\ 0 & 0 & g_{33}^{-1} \end{bmatrix}.$$

Consequently, $e^i = g^{i1}e_1 + g^{i2}e_2$, for $i = 1, 2$ and $e^3 = e_3/g_{33}$. In particular, we see from the first two relations that $e^1 \cdot X = 0 = e^2 \cdot X$. Thus,

$$
\begin{aligned}
X &= (e^1 \cdot X)e_1 + (e^2 \cdot X)e_2 + (e^3 \cdot X)e_3 \\
  &= (e^3 \cdot X)e_3 \\
  &= (e^3 \cdot X)(\nabla F^1 \times \nabla F^2),
\end{aligned}
$$

at all points of U. Thus, taking $\rho = e^3 \cdot X$ gives the result (7.9).

We now look at some examples of integrable systems in $\mathbb{R}^3$ and develop several techniques for sketching the curves that are the intersections of the respective level surfaces. The vector fields in these examples are manufactured by taking $X = \nabla F^1 \times \nabla F^2$, where F^1, F^2 are two functions for which the level surfaces $F^1 = k_1, F^2 = k_2$ are fairly simple.

Example 7.4 (Two Cylinders) A simple example, one that is easily sketched by hand, involves two families of cylinders. These cylinders are the level surfaces $F^1 = k_1$ and $F^2 = k_2$, of the functions

$$
\begin{aligned}
F^1(x, y, z) &\equiv \tfrac{1}{2}(x^2 + y^2) \\
F^2(x, y, z) &\equiv \tfrac{1}{2}(x^2 + z^2).
\end{aligned}
$$

Defining $X : \mathbb{R}^3 \to \mathbb{R}^3$ by $X = \nabla F^1 \times \nabla F^2$, one easily calculates

$$
\begin{aligned}
\nabla F^1 &= (x, y, 0) \\
\nabla F^2 &= (x, 0, z) \\
X = \nabla F^1 \times \nabla F^2 &= (yz, -xz, -xy).
\end{aligned}
$$

Thus, the corresponding system of differential equations is

$$x' = yz$$
$$y' = -xz$$
$$z' = -xy.$$

Each integral curve of the system coincides with the curve obtained by intersecting two cylinders $x^2 + y^2 = 2k_1, x^2 + z^2 = 2k_2$ (with $k_1 \geq 0, k_2 \geq 0$).

To sketch the phase portrait we use the following strategy. Choose a particular cylinder, say $x^2 + z^2 = 1$ (which has its axis along the y-axis). Now try to visualize the curves obtained by slicing this cylinder with the cylinders $x^2 + y^2 = 2k_1 = r^2$, with r varying from small to large. The z-axis is the axis for each of these cylinders. See Figure 7.8.

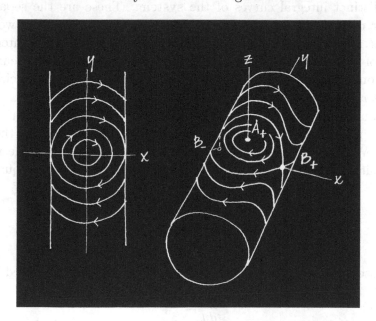

Figure 7.8: *Two views of the curves obtained by slicing the cylinder $x^2 + z^2 = 1$ with a family of cylinders $x^2 + y^2 = r^2$, for various values of r. On the left is a sketch of how these curves look when projected orthogonally on the x-y plane. On the right is a sketch of these curves in space.*

Attempt to sketch the curves on the surface of the fixed cylinder $x^2 + z^2 = 1$. Before doing this in space, try some sketches in two dimensions, i.e., look down along the z-axis at the projection of the fixed cylinder onto the x-y plane. In projection, the curves of intersection with $x^2 + y^2 = r^2$ are either circles (for $r \leq 1$) or a pair of arcs of circles (for $r > 1$). By evaluating $X(x, y, z) = (yz, -xz, -xy)$ at various points, we can determine the directions to mark on the curves in the planar projection.

The spatial view reveals four fixed points $A_\pm = (0, 0, \pm 1)$, $B_\pm = (\pm 1, 0, 0)$ on the surface of the fixed cylinder. $A_\pm$ are stable center points and $B_\pm$ are unstable saddle points. We see that when $r < 1$, the intersection of the cylinders consists of two closed ovals, one on the top half and the other on the bottom half of the fixed cylinder. These ovals are centered on the z-axis and are oppositely directed. Each is an integral curve of the system. For $r = 1$, the pair of ovals on the top and bottom of the fixed cylinder are pinched into cusp points at $B_\pm$. The intersection of the cylinders, while being one "curve," is actually a pair of circles (exercise) and is properly composed of four distinct integral curves of the system. These are the separatrices that flow toward or away from the saddle points $B_\pm$ (see the worksheet spcurves.mws on the electronic component). For $r > 1$, each intersection consists of a pair of closed curves on the fixed cylinder. These curves are centered on the y-axis, are oppositely directed, and become more circular in nature as r becomes larger.

We can find another first integral, in addition to F^1, F^2, that is also helpful in understanding the origin of the saddle points $B_\pm$ in the phase portrait. We use the method from Example 7.1, which was made rigorous in Proposition 7.2. For the example here, consider the last two equations:

$$\frac{dy}{dt} = -xz$$

$$\frac{dz}{dt} = -xy,$$

in the system. Taking the ratio of each side leads to the associated DE

$$\frac{dz}{dy} = \frac{y}{z},$$

which is a separable DE. Separating variable and integrating gives

$$z^2 - y^2 = k_3.$$

Thus, if we let

$$F^3(x, y, z) = z^2 - y^2,$$

then it is not hard to verify that F^3 is a first integral for the system and that F^2, F^3 are functionally independent. Thus, we can use the level surfaces

F^2, F^3 to describe the phase portrait. The level surfaces for F^3 are $z^2 - y^2 = m^2$ and these are surfaces generated by translating the hyperbolas $z^2 - y^2 = m^2$, in the y-z plane, along the x-axis. The exceptional case $m = 0$ is a pair of planes (generated by translating the asymptotes $z = \pm y$ along the x-axis). To visualize how these surfaces intersect the fixed cylinder $x^2 + z^2 = 1$, first take a two-dimensional view by looking along the x-axis. Figure 7.9 shows the resulting picture.

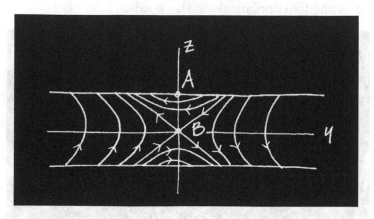

Figure 7.9: *Sketch of the curves of intersection of the surfaces $x^2 + z^2 = 1$ and $z^2 - y^2 = m^2$, for various values of m. The view is the orthogonal projection of these curves onto the y-z plane.*

Now this picture can aid in understanding the three-dimensional view of the integral curves in Figure 7.8. Note that the discussion here illustrates that the first integrals used in determining integrability of a system of DEs are not unique, and often some choices are more convenient than others.

To analyze the fixed points of X further, note that since $X = \nabla F^1 \times \nabla F^2$, the fixed points occur precisely at the critical points of F^1, or the critical points of F^2, or at the points of tangency of the level surfaces of F^1, F^2. In this example,

$$\begin{aligned} \nabla F^1 &= (x, y, 0) \\ \nabla F^2 &= (x, 0, z), \end{aligned}$$

and so the z-axis, comprising the critical points of F^1, is a line of fixed points which, as we have seen, are stable centers. The y-axis, comprising the critical points of F^2 is a line of fixed points that are stable centers. On the other hand the x-axis is comprised of points $(x, 0, 0)$, where $\nabla F^1 = \nabla F^2$

and these are points where two of the cylinders are tangent to each other.
This line of fixed points consists of unstable saddle points. The exceptional
point $(0, 0, 0)$ lies on all three of the lines of fixed points and is neither stable
nor unstable.

In summary, we sketch a number of integral curves on the surface of a
several different fixed cylinders, as above, and obtain a representation of the
phase portrait in $\mathbb{R}^3$. This is shown in Figure 7.10. Note the pairs of circles
intersecting at the fixed points along the x-axis.

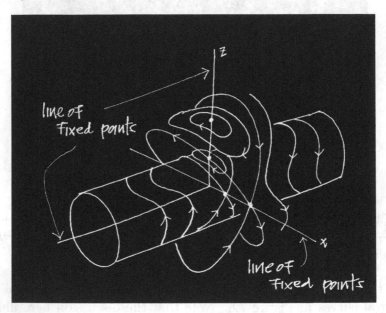

Figure 7.10: *Sketch of the phase portrait for the system $x' = yz, y' = -xz, z' = -xy$. The three coordinate axes are lines of fixed points. All other integral curves shown lie on the surface of one of the cylinders $x^2 + z^2 = b^2$ and are obtained by intersecting these cylinders with the cylinders $x^2 + y^2 = r^2$.*

The exercises in this section will give additional opportunities for you
to develop your ability to visualize and sketch the intersections of two families $F^1 = k_1, F^2 = k_2$ of level surfaces. Even for simple surfaces this can
be difficult and so use of a computer is often necessary. Using Maple's
`implicitplot3d` command to plot a pair of level surfaces will give an indication of what the curve of intersection looks like (see the worksheet
`spcurves.mws` on the electronic component). However, producing a clear

picture of how one family of level surfaces intersects a fixed level surface in the other family is usually difficult. For simple examples, like the one above, one can solve the equations $F^1 = k_1, F^2 = k_2$ explicitly, obtaining formulas for the curves of intersection, and then plot these curves with a computer. Alternatively, one can use the following method, which will be most useful in higher dimensions than three.

Consider the two equations for the level surfaces:

$$F^1(x, y, z) = k_1 \qquad (7.10)$$
$$F^2(x, y, z) = k_2. \qquad (7.11)$$

Solve the first equation for one of the variables x, y, z, say z, to get

$$z = h(x, y, k_1).$$

Generally this is possible only for restricted values of x and y, but where possible, the above equation is equivalent to equation (7.10) and is an equation for a surface that is the graph of the function h. Substituting $z = h(x, y, k_1)$ into equation (7.11) gives

$$F^2(x, y, h(x, y, k_1)) = k_2.$$

This is an equation for a curve in $\mathbb{R}^2$ and is in fact a level curve for the function:

$$g_{k_1}(x, y) \equiv F^2(x, y, h(x, y, k_1)).$$

The construction gives a function g_{k_1} of two variables (and one parameter k_1) whose level curves can be "lifted" from the x-y plane to curves on the level surface $F^1 = k_1$ (using h) and these will be the curves of intersection of the family $F^2 = k_2$ with the fixed surface $F^1 = k_1$ (see Figure 7.11).

More specifically, if (x, y) satisfies $g_{k_1}(x, y) = k_2$, i.e., lies on a level curve of g_{k_1}, and we let $z = h(x, y, k_1)$, then (x, y, z) satisfies equations (7.10)-(7.11), i.e., lies on the respective level surfaces of F^1 and F^2. This construction relies upon being able to solve the first equation, equation (7.10), for one of the variables. Theoretically this is always possible locally (except at fixed points) using the Implicit Function Theorem. This will be explained in general later in this chapter. In the examples here and in the exercises, solving the equation for one of the variables will be easy. Note that we could also do a similar construction by solving the second equation, equation (7.11), for one of the variables. In any particular example, one or the other way may be more advantageous.

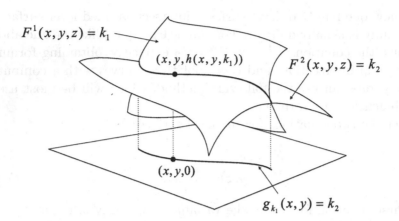

Figure 7.11: *The curve of intersection of $F^1 = k_1, F^2 = k_2$ projects onto a level curve of a function g_{k_1} of two variables. Conversely, each level curve of g_{k_1} can be "lifted" to a curve of intersection.*

Example 7.5 (Hyperbolic Paraboloids and Spheres) Here we consider the two functions $F^1, F^2 : \mathbb{R}^3 \to \mathbb{R}$ given by:

$$F^1(x, y, z) = z - \tfrac{1}{2}(x^2 - y^2)$$
$$F^2(x, y, z) = \tfrac{1}{2}(x^2 + y^2 + z^2).$$

The level surfaces for these are hyperbolic paraboloids and spheres, respectively. Figure 7.12 shows a computer plot of two of the level surfaces. Even though these are well-known surfaces, sketching the curves of intersection of a family of spheres with a fixed hyperbolic paraboloid: $z = (x^2 - y^2)/2$ may not be so easy for some students. However, applying the above method is easy and then a computer can be used to plot the curves.

The equations for the level surfaces are

$$z - \tfrac{1}{2}(x^2 - y^2) = k_1$$
$$\tfrac{1}{2}(x^2 + y^2 + z^2) = k_2.$$

Solving the first equation for z gives

$$z = \tfrac{1}{2}(x^2 - y^2) + k_1,$$

so that h is given by $h(x, y, k_1) = \tfrac{1}{2}(x^2 - y^2) + k_1$. Substituting this in the second equation gives the family of functions g_{k_1}:

$$g_{k_1}(x, y) = \tfrac{1}{2}x^2 + \tfrac{1}{2}y^2 + [\tfrac{1}{2}(x^2 - y^2) + k_1]^2.$$

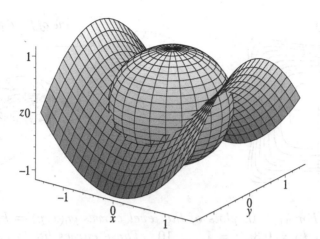

Figure 7.12: *Plots of the hyperbolic paraboloid $z = \frac{1}{2}(x^2 - y^2)$ and the sphere $x^2 + y^2 + z^2 = 1$.*

Taking a specific value for k_1, the level curves of g_{k_1} can be plotted on a computer, as shown in Figure 7.13, and these curves can then be "lifted" to curves on the surface of the hyperbolic paraboloid $z - \frac{1}{2}(x^2 - y^2) = k_1$.

Examining the pictures of the level curves and their lifts to the surface of the $k_1 = 0$ hyperbolic paraboloid, we can more readily understand how these are the lines of intersection with a series of spheres. The saddle point for the hyperbolic paraboloid is at the origin $(0, 0, 0)$ and near there the surface is relatively flat (coinciding with its tangent plane: $z = 0$). Thus, small spheres will intersect it in a series of nearly circular ovals, and this is what Figure 7.13 illustrates.

For the value $k_1 = 3$, the level curves of g_3 and their lifts to the hyperbolic paraboloid $z = \frac{1}{2}(x^2 - y^2) + 3$ are quite different. In this case the hyperbolic paraboloid has been shifted up three units to have its saddle point at $(0, 0, 3)$. Thus, small spheres about the origin will first begin to intersect the saddle on the stirrups' sides, giving a series of *pairs* of ovals. These eventually coalesce to form a figure eight with crossing at the saddle point. Figure 7.14 should help you visualize the curves of intersection.

The figures in both cases, $k_1 = 0, k_1 = 3$, show explicitly the integral curves of the system $x' = X(x)$ that lie on the surface $F^1 = k_1$ determined by the constant of the motion F^1. Included in these pictures are fixed points of the system as well. The spatial view in Figure 7.13 shows a stable center point $(0, 0, 0)$ on the surface of the $k_1 = 0$ hyperbolic paraboloid. On the

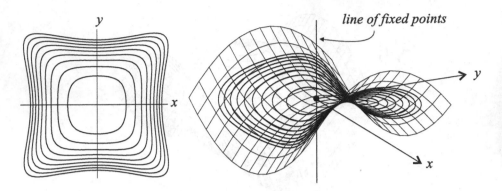

Figure 7.13: *For $k_1 = 0$, plots of the level curves $g_0(x, y) = k_2$ are shown on the left for $k_2 = 0.3i$, $i = 1, \ldots, 10$. These curves lift to curves on the hyperbolic paraboloid $z = \frac{1}{2}(x^2 - y^2)$, giving the curves of intersection of this surface with the family of spheres $x^2 + y^2 + z^2 = 2k_2$. This is shown on the right.*

other hand Figure 7.14 shows two stable center points and one unstable saddle point on the surface of the $k_1 = 3$ hyperbolic paraboloid. The figures also display lines that are lines of fixed points for the system. These come from looking at the explicit form for the vector field X. One easily calculates from $F^1 = z - \frac{1}{2}(x^2 - y^2)$ and $F^2 = \frac{1}{2}(x^2 + y^2 + z^2)$ that

$$
\begin{aligned}
\nabla F^1 &= (-x, y, 1) \\
\nabla F^2 &= (x, y, z) \\
X &= \Big(y(z - 1), x(z + 1), -2xy \Big).
\end{aligned}
$$

Thus, the fixed points constitute a set comprised of three straight lines

$$\{(x, 0, -1) | x \in \mathbb{R}\} \cup \{(0, y, 1) | y \in \mathbb{R}\} \cup \{(0, 0, z) | z \in \mathbb{R}\}.$$

These three lines of fixed points as well as a series of other integral curves for X are shown in Figure 7.15. The first two lines are comprised of points where the respective hyperbolic paraboloids and spheres are tangent to each other. The third line is comprised of the critical points for F^1, F^2. You can see from the figure that the lines $\{(x, 0, -1) | x \in \mathbb{R}\}$ and $\{(0, y, 1) | y \in \mathbb{R}\}$ are comprised of fixed points that are stable center points, and the line $\{(0, 0, z) | z \in \mathbb{R}\}$ consists of unstable saddle points.

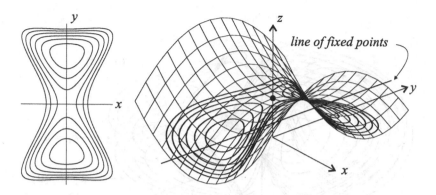

Figure 7.14: *For $k_1 = 3$, plots of the level curves $g_3(x, y) = k_2$ are shown on the left for $k_2 = 0.3i$, $i = 1, \ldots, 6$. These curves lift to curves on the hyperbolic paraboloid $z = \frac{1}{2}(x^2 - y^2) + 3$, giving the curves of intersection of this surface with the family of spheres $x^2 + y^2 + z^2 = 2k_2$. This is shown on the right.*

Exercises 7.3

The first group of exercises studies integrable systems in $\mathbb{R}^3$ with vector field $X_{F^1 F^2}$, where F^1, F^2 are two given functions. In each exercise you are to compute the vector field $X_{F^1 F^2}$ and write down the corresponding system of DEs. Determine all the fixed points and classify their types. For the latter you should be able to compute, by hand, the eigenvalues of $X'_{F^1 F^2}(x, y, z)$ at a general fixed point. Study the phase portrait of the system by drawing, by hand and by computer, a number of the curves of intersections of the two families of level surfaces $F^1 = k_1$, $F^2 = k_2$. Mark the direction of flow on the curves, label the lines of fixed points (if any), and generally annotate your drawings.

1. **[Cylinders and Hyperboloids]** The system here has integral curves that lie on the intersections of a family of cylinders and a family of hyperboloids. Since a hyperboloid of 1 sheet is almost as simple to visualize as a cylinder, this exercise is very similar to the families of intersecting cylinders in Example 7.6. The two functions are

$$
\begin{aligned}
F^1(x, y, z) &= \tfrac{1}{2}(x^2 + y^2 - z^2) \\
F^2(x, y, z) &= \tfrac{1}{2}(y^2 + z^2).
\end{aligned}
$$

Strategy: Sketch, by hand, the curves of intersection of a sequence of cylinders $F^2(x, y, z) = k$ (varying k) with the fixed hyperboloid $x^2 + y^2 - z^2 = 1$. Then try to visualize the curves on several different hyperboloids. Find another first integral F^3 by the method in Proposition 7.2 and then visualize the intersections of its level surfaces with the cylinder. Produce at least three good sketches (two planar and one spatial).

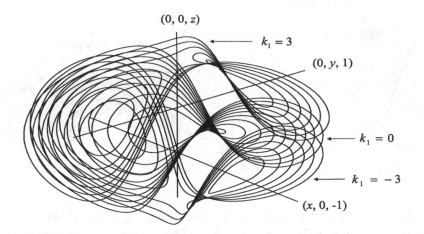

Figure 7.15: *Three lines of fixed points for the system* $x' = y(z - 1), y' = x(z + 1), z' - 2xy$. *The other integral curves shown lie on the hyperbolic paraboloids* $z = \frac{1}{2}(x^2 - y^2) + k_1$, *for* $k_1 = -3, 0, 3$.

2. **[Cylinders and Circular Paraboloids]** The system here is also similar in complexity to that in Example 7.6 and so you should be able to do the sketches by hand. The two functions are

$$F^1(x, y, z) = z - \tfrac{1}{2}(x^2 + y^2)$$
$$F^2(x, y, z) = \tfrac{1}{2}(x^2 + z^2).$$

Strategy: Sketch, by hand, the curves of intersection of the circular paraboloids $F^1(x, y, z) = k$ with the fixed cylinder $x^2 + z^2 = 1$. For this, you might find it helpful to find another first integral F^3 by the method in Proposition 7.2 and then visualize the intersections of its level surfaces with the cylinder. *Hint*: You will need to complete the square to recognize the level surfaces of F^3 as obtained by translating a certain hyperbola. Produce at least three good sketches (two planar and one spatial).

3. **[Hyperbolic Paraboloids]** Consider the system of DEs whose integral curves lie on the intersections of two families of hyperbolic paraboloids, say the level surfaces of the following two functions:

$$F^1(x, y, z) = z - \tfrac{1}{2}(x^2 - y^2)$$
$$F^2(x, y, z) = y - \tfrac{1}{2}(z^2 - x^2).$$

It is a real exercise in visualization to try to draw, by hand, the curve of intersection for two such hyperbolic paraboloids. Thus, you probably should use a computer and follow the work in Example 7.5. Also take advantage of the special Maple code on the electronic component for lifting level curves onto a given surface. This is in the worksheet `integrable3d.mws`.

4. **[Circular Paraboloids]** Consider the system of DEs whose integral curves lie on the intersections of two families of circular paraboloids, say the level surfaces of the following two functions:

$$
\begin{aligned}
F^1(x, y, z) &= z - \tfrac{1}{2}(x^2 + y^2) \\
F^2(x, y, z) &= y - \tfrac{1}{2}(x^2 + z^2).
\end{aligned}
$$

This one is tough to visualize too, but you should be able to do the work by hand if you use another first integral F^3. Complete the square in the expression for F^3 to recognize that its level surfaces are obtained by translating hyperbolas in the y-z plane along the x-axis. Visualize how these cut a fixed circular paraboloid $F^1 = k_1$. *Note*: In general, visualization is easier if one of the surfaces is a surface of translation. Then it can be viewed as a "cookie cutter" and the other surface as the "dough" when trying to determine the curve of intersection.

7.4 Integrable Systems in Higher Dimensions

For higher dimensions, the definitions and theorems follow the pattern that is exhibited for integrable systems in $\mathbb{R}^2$ and $\mathbb{R}^3$. We present this here in general for $\mathbb{R}^n$, along with some other results which were not mentioned for $n = 2, 3$. In particular, the generalization of the normal operator to an operator on $\mathbb{R}^n$ should be interesting to you.

Definition 7.5 (Integrable Systems in $\mathbb{R}^n$) Suppose $X : \mathcal{O} \to \mathbb{R}^n$ is a vector field on an open set in $\mathbb{R}^n$. The system $x' = X(x)$ is called an *integrable system* if it has $n - 1$ constants of the motion: $F^i : U \subseteq \mathcal{O} \to \mathbb{R}^n$, $i = 1, \dots, n - 1$, which are functionally independent on U. This latter requirement means that $\nabla F^1(x), \nabla F^2(x), \dots, \nabla F^{n-1}(x)$ are linearly independent for every $x \in U$.

To characterize the integrable systems, we need the generalization of the normal operator:

Definition 7.6 (The Normal Operator) We will use the notation

$$
(\mathbb{R}^n)^{n-1} = \mathbb{R}^n \times \mathbb{R}^n \times \cdots \times \mathbb{R}^n
$$

for the Cartesian product $\mathbb{R}^n$ with itself $n - 1$ times. The *normal operator* on $\mathbb{R}^n$ is the map $N : (\mathbb{R}^n)^{n-1} \to \mathbb{R}^n$ defined as follows. For vectors

$v_1, \ldots, v_{n-1} \in \mathbb{R}^n$, the vector $N(v_1, \ldots, v_{n-1})$ is the vector in $\mathbb{R}^n$ with components

$$N(v_1, \ldots, v_{n-1})_i = \det(v_1, \ldots, v_{n-1}, \varepsilon_i), \tag{7.12}$$

for $i = 1, \ldots, n$. Here ε_i is the ith standard unit vector in $\mathbb{R}^n$ and the determinant is of the matrix which has $v_1, \ldots, v_{n-1}, \varepsilon_i$ as its rows (or columns, if you prefer). We call $N(v_1, \ldots, v_{n-1})$ the *normal product* of $v_1, \ldots, v_{n-1}$, and use the following alternative notation to suggest a product structure:

$$v_1 \times \cdots \times v_{n-1} \equiv N(v_1, \ldots, v_{n-1}). \tag{7.13}$$

As the name suggests $v_1 \times \cdots \times v_{n-1}$ gives us a vector that is perpendicular to each of the vectors $v_1, \ldots, v_{n-1}$ and is the natural generalization the N we used in dimensions 2 and 3. It is easy to see that N is a multilinear map. This, as well as some of the more important properties of N listed in the next proposition, follow easily from properties of determinants.

Proposition 7.4 *Suppose $v_1, \ldots, v_{n-1} \in \mathbb{R}^n$.*

(1) *For every $w \in \mathbb{R}^n$,*

$$(v_1 \times \cdots \times v_{n-1}) \cdot w = \det(v_1, \ldots, v_{n-1}, w), \tag{7.14}$$

and consequently:

(2) *$v_1 \times \cdots \times v_{n-1}$ is perpendicular to $v_1, \ldots, v_{n-1}$.*

(3) *$v_1 \times \cdots \times v_{n-1} = 0$ if and only if $v_1, \ldots, v_{n-1}$ are linearly dependent.*

(4) *If P is an invertible $n \times n$ matrix, then*

$$P(v_1 \times \cdots \times v_{n-1}) = \det(P)[P^{-T}v_1 \times \cdots \times P^{-T}v_{n-1}]. \tag{7.15}$$

(5) *In dimension three $N(v_1, v_2) = v_1 \times v_2$, while in dimension two $N(v_1) = -Jv_1$ (where J is the canonical symplectic matrix in dimension two).*

Proof: We just prove (4) and leave the rest for an exercise. To prove it, we use property (1). Suppose w is any vector in $\mathbb{R}^n$. Then by (1) and properties of determinants and the dot product we get

$$
\begin{aligned}
[P(v_1 \times \cdots \times v_{n-1})] \cdot w &= [v_1 \times \cdots \times v_{n-1}] \cdot P^T w \\
&= \det(v_1, \ldots, v_{n-1}, P^T w)) \\
&= \det(P^T P^{-T} v_1, \ldots, P^T P^{-T} v_{n-1}, P^T w)) \\
&= \det(P^T) \det(P^{-T} v_1, \ldots, P^{-T} v_{n-1}, w)) \\
&= \det(P)[P^{-T} v_1 \times \cdots \times P^{-T} v_{n-1}] \cdot w.
\end{aligned}
$$

Since w was arbitrary, this gives equation (7.15). $\square$

In general, the computation of $v_1 \times \cdots \times v_{n-1}$ for any given $v_1, \ldots, v_{n-1}$ is just like the computation of the cross product in dimension 3, but of course involves more work. The examples and exercises will exhibit this.

We can now characterize the integrable systems in $\mathbb{R}^n$ in an identical way to what was done for $n = 2, 3$. This gives us a standard way of producing examples of integrable systems in any dimension.

Theorem 7.3 *Suppose $X : \mathcal{O} \to \mathbb{R}^n$ is a vector field on an open set $\mathcal{O}$ in $\mathbb{R}^n$. Then the system $x' = X(x)$ is integrable if and only if there is an open dense subset $U \subseteq \mathcal{O}$ on which X has the form*

$$X = \rho(\nabla F^1 \times \cdots \times \nabla F^{n-1}), \tag{7.16}$$

where $\rho, F^1, \ldots, F^{n-1} : U \to \mathbb{R}^3$ are differentiable functions with $F^1, \ldots, F^{n-1}$ functionally independent on U.

Proof: The proof is essentially identical to that of Theorem 7.2, except now we use properties of the normal product in $\mathbb{R}^n$.

It is easy to see that if X has the form (7.16), then $x' = X(x)$ is integrable (with $F^1, \ldots, F^{n-1}$ as first integrals). Conversely suppose the system is integrable and let $F^1, \ldots, F^{n-1}$ be first integrals which are functionally independent on an open, dense subset $U \subseteq \mathcal{O}$. On U, define the vector fields:

$$e_1 = \nabla F^1$$
$$\vdots$$
$$e_{n-1} = \nabla F^{n-1}$$
$$e_n = \nabla F^1 \times \cdots \times \nabla F^{n-1}.$$

Then $\{e_i\}_{i=1}^n$ is an orthogonal, moving frame on U. The reciprocal frame $\{e^i\}_{i=1}^n$ is given by $e^i = \sum_{j=1}^n g^{ij} e_j$, where the g^{ij}'s are the entries of the inverse matrix G^{-1} of the metric matrix $G = \{g_{ij}\} = \{e_i \cdot e_j\}$. One has, in general, that $e^i \cdot e_j = 0$, for $i \neq j$, and $e^i \cdot e_i = 1$, for all i. This property allows one to easily write the coordinate expression of any vector field relative to the frame $\{e_i\}_{i=1}^n$. In particular, to express X, we note that from $e_i \cdot X = 0$ for $i = 1, \ldots, n-1$, it follows that $e^i \cdot X = 0$, for $i = 1, \ldots, n-1$ (exercise).

Hence we get

$$
\begin{aligned}
X &= \sum_{i=1}^{n} (e^i \cdot X) e_i \\
&= (e^n \cdot X) e_n \\
&= (e^n \cdot X)(\nabla F^1 \times \cdots \times \nabla F^{n-1}),
\end{aligned}
$$

at all points of U. Thus, taking $\rho = e^n \cdot X$ gives the result (7.16). $\square$

As mentioned in the lower-dimensional cases, there is often additional information about the system contained in $\rho, F^1, \ldots, F^{n-1}$. In particular, if we start with differentiable functions $\rho, F^1, \ldots, F^{n-1} : \mathcal{O} \to \mathbb{R}$ and define $X_{\rho, F^1, \ldots, F^{n-1}} : \mathcal{O} \to \mathbb{R}^n$ by

$$
X_{\rho, F^1, \ldots, F^{n-1}} \equiv \rho(\nabla F^1 \times \cdots \times \nabla F^{n-1}), \tag{7.17}
$$

then we get an integrable system provided $F^1, \ldots, F^{n-1}$ are functionally independent on some open, dense subset of $\mathcal{O}$. The extra information contained in equation (7.17) is that the equation holds on all of $\mathcal{O}$. Thus, the fixed points of $X_{\rho, F^1, \ldots, F^{n-1}}$ coincide the zeros of ρ and the points where $\nabla F^1, \ldots, \nabla F^{n-1}$ are linearly dependent. From a property of the normal product , we know that $\nabla F^1(x), \ldots, \nabla F^{n-1}(x)$ are linearly dependent if and only if their normal product is zero

$$
\nabla F^1 x) \times \cdots \times \nabla F^{n-1}(x) = 0.
$$

This extra information is often helpful.

Another important property of the construction is that the normal product of gradient vector fields is always a divergence-free vector field:

Proposition 7.5 *Suppose* $F^1, \ldots, F^{n-1} : \mathcal{O} \to \mathbb{R}$ *are differentiable functions, and let*

$$
X_{F^1, \ldots, F^{n-1}} \equiv \nabla F^1 \times \cdots \times \nabla F^{n-1}. \tag{7.18}
$$

Then $X_{F^1, \ldots, F^{n-1}}$ *is a divergence-free vector field on* $\mathcal{O}$,

$$
\mathrm{div}(X_{F^1, \ldots, F^{n-1}}) = 0, \tag{7.19}
$$

identically on $\mathcal{O}$.

Proof: By definition, the ith component of $X_{F^1,\ldots,F^{n-1}}$ is

$$X^i_{F^1,\ldots,F^{n-1}} = \det(\nabla F^1,\ldots,\nabla F^{n-1},\varepsilon_i).$$

Using a property of derivatives of such determinential expressions, we get

$$
\begin{aligned}
\mathrm{div}(X_{F^1,\ldots,F^{n-1}}) &= \sum_{i=1}^{n} \frac{\partial X^i_{F^1,\ldots,F^{n-1}}}{\partial x_i} \\
&= \sum_{i=1}^{n} \frac{\partial}{\partial x_i}\left[\det(\nabla F^1,\ldots,\nabla F^{n-1},\varepsilon_i)\right] \\
&= \sum_{i=1}^{n}\sum_{j=1}^{n-1} \det\left(\nabla F^1,\ldots,\frac{\partial}{\partial x_i}\nabla F^j,\ldots,\nabla F^{n-1},\varepsilon_i\right) \\
&= \sum_{i=1}^{n}\sum_{j=1}^{n-1} \det\left(\nabla F^1,\ldots,\sum_{k=1}^{n}\frac{\partial^2 F^j}{\partial x_k \partial x_i}\varepsilon_k,\ldots,\nabla F^{n-1},\varepsilon_i\right) \\
&= \sum_{j=1}^{n-1}\left[\sum_{i,k=1}^{n}\frac{\partial^2 F^j}{\partial x_k \partial x_i}\det(\nabla F^1,\ldots,\varepsilon_k,\ldots,\nabla F^{n-1},\varepsilon_i)\right] \\
&= 0.
\end{aligned}
$$

The last equation comes from the observation that $\partial^2 F^j/\partial x_k \partial x_i$ is symmetric in i and k, while $\det(\nabla F^1,\ldots,\varepsilon_k,\ldots,\nabla F^{n-1},\varepsilon_i)$ is antisymmetric in i and k. $\square$

Example 7.6 (Paraboloids in $\mathbb{R}^4$) Consider the vector field X_{F^1,F^2,F^3} on $\mathbb{R}^4$ generated by the functions:

$$
\begin{aligned}
F^1(x,y,z,w) &= w - \tfrac{1}{2}(x^2+y^2+z^2) &\qquad (7.20)\\
F^2(x,y,z,w) &= z - \tfrac{1}{2}(x^2+w^2) &\qquad (7.21)\\
F^3(x,y,z,w) &= y - \tfrac{1}{2}(x^2+z^2). &\qquad (7.22)
\end{aligned}
$$

The level sets $F^i = k_i$ for these functions are "paraboloids" in $\mathbb{R}^4$. The calculation of X_{F^1,F^2,F^3} is fairly simple. First, the gradient vector fields are

$$
\begin{aligned}
\nabla F^1 &= (-x,-y,-z,1) &\qquad (7.23)\\
\nabla F^2 &= (-x,0,1,-w) &\qquad (7.24)\\
\nabla F^3 &= (-x,1,-z,0). &\qquad (7.25)
\end{aligned}
$$

The normal product of these is then computed as

$$X_{F^1,F^2,F^3}$$
$$= \nabla F^1 \times \nabla F^2 \times \nabla F^3$$

$$= \begin{vmatrix} -x & -y & -z & 1 \\ -x & 0 & 1 & -w \\ -x & 1 & -z & 0 \\ \varepsilon_1 & \varepsilon_2 & \varepsilon_3 & \varepsilon_4 \end{vmatrix}$$

$$= \left(-\begin{vmatrix} -y & -z & 1 \\ 0 & 1 & -w \\ 1 & -z & 0 \end{vmatrix}, \begin{vmatrix} -x & -z & 1 \\ -x & 1 & -w \\ -x & -z & 0 \end{vmatrix}, -\begin{vmatrix} -x & -y & 1 \\ -x & 0 & -w \\ -x & 1 & 0 \end{vmatrix}, \begin{vmatrix} -x & -y & -z \\ -x & 0 & 1 \\ -x & 1 & -z \end{vmatrix} \right)$$

$$= \Big(1 - zw(y+1),\ x(z+1),\ x[w(y+1)+1],\ x(y+1)(z+1) \Big).$$

By the theorems, the vector field X_{F^1,F^2,F^3} is divergence-free, which is easy to check directly, and gives an integrable system of DEs:

$$x' = 1 - zw(y+1) \tag{7.26}$$
$$y' = x(z+1) \tag{7.27}$$
$$z' = x[w(y+1)+1] \tag{7.28}$$
$$w' = x(y+1)(z+1). \tag{7.29}$$

The integrability follows from the construction, once we verify functional independence of F^1, F^2, F^3 on a open, dense subset U of $\mathbb{R}^4$. As mentioned above, for $u \equiv (x,y,z,w) \in U$, the vectors $\nabla F^1(u), \nabla F^2(u), \nabla F^3(u)$ are linearly dependent if and only if

$$X_{F^1,F^2,F^3}(u) \equiv \nabla F^1(u) \times \nabla F^2(u) \times \nabla F^3(u) = 0.$$

Thus, F^1, F^2, F^3 are functionally independent on the set U consisting of $\mathbb{R}^4$ minus the fixed points of X_{F^1,F^2,F^3}. The fixed points are easy to determine from the above system of DEs. They are

$$\{ (0, (zw)^{-1} - 1, z, w) \,|\, z, w \in \mathbb{R} \} \cup \{ (x, -w^{-1} - 1, -1, w) \,|\, x, w \in \mathbb{R} \}.$$

Each of the sets in the union is a 2-dimensional submanifold of $\mathbb{R}^4$ and a closed set. Thus, U is an open, dense set, as required.

The integral curves of the system lie on the curves of intersections of the three level sets:

$$w = \tfrac{1}{2}(x^2 + y^2 + z^2) + k_1 \tag{7.30}$$
$$z = \tfrac{1}{2}(x^2 + w^2) + k_2 \tag{7.31}$$
$$y = \tfrac{1}{2}(x^2 + z^2) + k_3. \tag{7.32}$$

To visualize these curves, we use the technique explained prior to Example 7.5 and employed in that example. If we substitute the w from equation (7.30) above into equation (7.31), the system reduces to

$$z = \tfrac{1}{2}x^2 + \tfrac{1}{2}[\tfrac{1}{2}(x^2 + y^2 + z^2) + k_1]^2 + k_2 \qquad (7.33)$$
$$y = \tfrac{1}{2}(x^2 + z^2) + k_3. \qquad (7.34)$$

Now, in this reduced system of equations, substitute the y from the second equation into the first equation to get

$$z = \tfrac{1}{2}x^2 + \tfrac{1}{2}\left[\tfrac{1}{2}x^2 + \tfrac{1}{2}\left(\tfrac{1}{2}x^2 + \tfrac{1}{2}z^2 + k_3\right)^2 + \tfrac{1}{2}z^2 + k_1\right]^2 + k_2. \qquad (7.35)$$

The reasoning here is that if (x, y, z, w) is a solution of the system (7.30)-(7.32), i.e., is a point on the intersection of the level sets, then (x, z) is a solution of the equation (7.35), i.e., a point on the level curve of the function g_{k_1,k_3} defined by

$$g_{k_1,k_3}(x, z) = z - \tfrac{1}{2}x^2 - \tfrac{1}{2}\left[\tfrac{1}{2}x^2 + \tfrac{1}{2}\left(\tfrac{1}{2}x^2 + \tfrac{1}{2}z^2 + k_3\right)^2 + \tfrac{1}{2}z^2 + k_1\right]^2. \qquad (7.36)$$

Conversely, if (x, z) satisfies $g_{k_1,k_3}(x, z) = k_2$, then defining y by equation (7.34) and then w by equation (7.30) gives a point (x, y, z, w) that satisfies the system (7.30)-(7.32). Geometrically, this means that the level curves of g_{k_1,k_3} lift first to curves on the surface of the paraboloid

$$y = \tfrac{1}{2}(x^2 + z^2) + k_3$$

in $\mathbb{R}^3$, and then from there are lifted to integral curves on the surface of the paraboloid

$$w = \tfrac{1}{2}(x^2 + y^2 + z^2) + k_1,$$

in $\mathbb{R}^4$. While we cannot plot the latter curves, we can certainly plot the level curves of g_{k_1,k_3} for various values of k_1, k_3 and visualize their lifts to the corresponding paraboloids in $\mathbb{R}^3$.

Figure 7.16 shows plots of the level curves of $g_{0,0}$ and $g_{0,-4}$. The figure indicates that $g_{0,0}$ has a single critical point at $x = 0, z = 1.01595$ (approximately). Letting $y = (x^2 + z^2)/2 + k_3 = .5160$ and then $w = (x^2 + y^2 + z^2)/2 + k_1 = .659245$ (approximately), we get a point (x, y, z, w) on an integral curve in $\mathbb{R}^4$. Indeed this point is a fixed point, since $(zw)^{-1} - 1 = y$ (approximately). The picture of the level curves of $g_{0,0}$ lend credence to the

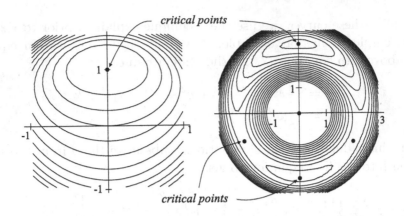

Figure 7.16: *Plots of the level curves of the function $g_{0,0}$ (on the left) and the function $g_{0,-4}$ (on the right).*

conjecture that this fixed point is a stable center. In a similar way the picture of the level curves of $g_{0,-4}$ reveals five critical points given approximately by

$$(0, -.0417), (0, 2.4957), (0, -2.4005), (\pm 2.10595, -1).$$

These correspond to five fixed points in phase space, the first three appearing to be stable centers and the latter two appearing to be unstable saddle points.

Figure 7.17 shows the lifts of the level curves of $g_{0,-4}$ to curves on the surface of the circular paraboloid $y = (x^2 + z^2)/2 + k_3$. One would conjecture from this analysis that the fixed points of the form $(0, (zw)^{-1} - 1, z, w)$ are stable centers, while those of the form $(x, -w^{-1} - 1, -1, w)$ are unstable saddles, except where these two forms coincide. This is left as an exercise.

While these plots of the level curves of g_{k_1, k_3}, for two sets of choices for k_1, k_3 enable us to understand *some* of the behavior of the system of DEs, it does not enable us to understand totally all the features that may be present. Further investigation is needed (exercise).

The technique used in the above and previous examples allows us to discern information about the integral curves of an integrable system by analyzing the level curves of a family $\{g_{k_1 \cdots k_{n-2}}\}$, of functions of two variables. The technique applies in general, but with some restriction on the domain of $g_{k_1 \cdots k_{n-2}}$, which is necessary unless we have more particular information about the system. In the following theorem we use the notation

$$S_{F^1 \ldots F^r}^{k_1 \cdots k_r} \equiv S_{F^1}^{k_1} \cap S_{F^2}^{k_2} \cap \cdots \cap S_{F^r}^{k_r},$$

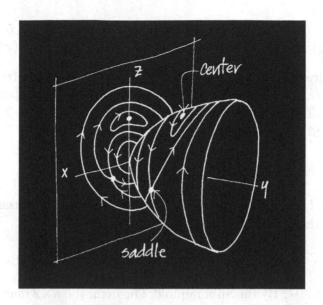

Figure 7.17: *Lifts of the level curves of* $g_{0,-4}$ *to the paraboloid* $y = (x^2+z^2)/2$. *This is for clarity. The paraboloid should be translated by* $k_3 = -4$ *to give the true lifts.*

for the intersection of the level sets (level hypersurfaces)

$$S_{F^i}^{k_i} \equiv \{\, x \in \mathcal{O} \,|\, F^i(x) = k_i \,\}.$$

Theorem 7.4 *Suppose that* $F^1, \ldots, F^{n-1} : \mathcal{O} \to \mathbb{R}$ *are differentiable functions on an open set in* $\mathbb{R}^n$ *and that* $c \in \mathcal{O}$ *is a point such that*

$$\nabla F^1(c), \ldots, \nabla F^{n-2}(c)$$

are linearly independent. Let $k_i = F^i(c)$, $i = 1, \ldots, n-2$. *Then there is an open set* $U \subseteq \mathbb{R}^2$, *a neighborhood* $V \subseteq \mathcal{O}$ *of* c, *and differentiable functions* $g_{k_1 \cdots k_{n-2}} : U \to \mathbb{R}$, *and* $h : U :\to \mathbb{R}^n$, *such that*

(1) $h : U \to S_{F^1 \ldots F^{n-2}}^{k_1 \cdots k_{n-2}} \cap V$ *is 1-1 and onto.*

(2) h *maps each level curve*

$$g_{k_1 \cdots k_{n-2}}(x_1, x_2) = k_{n-1},$$

    onto the curve

$$S_{F^1 \ldots F^{n-2} F^{n-1}}^{k_1 \cdots k_{n-2} k_{n-1}} \cap V.$$

(3) *A point $(x_1, x_2) \in U$ is a critical point of $g_{k_1 \cdots k_{n-2}}$ if and only if $h(x_1, x_2)$ is a fixed point for the system $x' = X_{F^1 \cdots F^{n-1}}(x)$.*

Proof: A direct application of the Submanifold Theorem (see Appendix A) gives the U, V and the differentiable map h with the property stated in assertion (1). This just says that h is a parametrization of the 2-dimensional submanifold $S^{k_1 \cdots k_{n-2}}_{F^1 \cdots F^{n-2}}$. Now define $g_{k_1 \cdots k_{n-2}}$ by

$$g_{k_1 \cdots k_{n-2}}(x_1, x_2) = F^{n-1}\Big(h(x_1, x_2) \Big),$$

for $(x_1, x_2) \in U$. Then it is easy to verify assertion (2) (exercise). Finally, note that by the chain rule

$$\nabla g_{k_1 \cdots k_{n-2}}(x_1, x_2) = \nabla F^{n-1}\Big(h(x_1, x_2) \Big) h'(x_1, x_2),$$

for all $(x_1, x_2) \in U$. By the Submanifold Theorem, the $n \times 2$ matrix $h'(x_1, x_2)$ consists of two linearly independent columns, each of which is a vector in $\mathbb{R}^n$ that is tangent to $S^{k_1 \cdots k_{n-2}}_{F^1 \cdots F^{n-2}}$ at $h(x_1, x_2)$. assertion (3) now follows (exercise).
$\square$

The concept of integrability is preserved under diffeomorphisms, i.e., if two systems are differentiably equivalent and one of the systems is integrable, then the other system is also integrable.

Theorem 7.5 *Suppose $F^1, \ldots, F^{n-1}, \rho : \mathcal{O} \to \mathbb{R}^3$ are differentiable functions on an open set $\mathcal{O}$ in $\mathbb{R}^n$ and let $X_{\rho, F^1, \ldots, F^{n-1}} : \mathcal{O} \to \mathbb{R}^n$ be the vector field*

$$X_{\rho, F^1, \ldots, F^{n-1}} \equiv \rho(\nabla F^1 \times \cdots \times \nabla F^{n-1}).$$

If $h : \mathcal{O} \to \tilde{\mathcal{O}}$ is any diffeomorphism onto an open set $\tilde{\mathcal{O}}$ of $\mathbb{R}^n$, then

$$h_*(X_{\rho, F^1, \ldots, F^{n-1}}) = X_{\mu, G^1, \ldots, G^{n-1}}, \tag{7.37}$$

where

$$\mu = (\rho \det(h')) \circ h^{-1} \tag{7.38}$$
$$G^i = F^i \circ h^{-1} \qquad \text{(for } i = 1, \ldots, n-1\text{)}. \tag{7.39}$$

Consequently, integrability is preserved under transformations induced by diffeomorphisms h and (7.39) gives the relationship between the respective constants of the motion.

Proof: To prove identity (7.37), fix $y \in \tilde{\mathcal{O}}$ and let $P = h'(x)$, where $x = h^{-1}(y)$. Using the definition of the transformation of a vector field by a diffeomorphism and identity (7.15), we get

$$
\begin{aligned}
h_*(X)(y) &= h'(x)X(x) \\
&= \rho(x)P(\nabla F^1(x) \times \cdots \times \nabla F^{n-1}(x)) \\
&= \rho(x)\det(P)(P^{-T}\nabla F^1(x) \times \cdots \times P^{-T}\nabla F^{n-1}(x)).
\end{aligned}
$$

Next note that we can consider the expression $P^{-T}\nabla F^i(x)$ as a product of matrices if the gradient vector $\nabla F^i(x)$ is viewed as a $n \times 1$ column matrix, i.e., $\nabla F^i(x) = ((F^i)')^T$. Then

$$
\begin{aligned}
P^{-T}\nabla F^i(x) &= h'(x)^{-T}(F^i)'(x)^T \\
&= \left((F^i)'(x)h'(x)^{-1}\right)^T \\
&= \left((F^i \circ h^{-1})'(y)\right)^T \\
&= \nabla(F^i \circ h^{-1})(y).
\end{aligned}
$$

Note that we have used the identity: $(h^{-1})'(y) = h'(h^{-1}(y))^{-1}$, which comes from the Inverse Function Theorem (see Appendix A). This shows that $h_*(X)(y)$ has the stated form. The rest of the assertions in the theorem follow from the main relation (7.37). $\square$

An immediate corollary of the theorem is that locally, near any regular point, the concept of integrability is not very special.

Corollary 7.1 (Local Integrability) *If $c \in \mathcal{O}$ is not a fixed point of X, then there is a neighborhood W of c in $\mathcal{O}$ such that X restricted to W gives an integrable system on W.*

Proof: By the Flow Box Theorem in Chapter 6, there is a neighborhood W of c and a diffeomorphism $h : W \to \widetilde{W}$, such that $h_*(X) = Y$, is the constant vector field: $Y(y) = \varepsilon_n$ for all $y \in \widetilde{W}$. Now the flow ψ for Y is uniform flow in the last coordinate direction:

$$
\psi_t(y) = (y_1, y_2, \ldots, y_n + t).
$$

Thus, it is clear that the system $y' = Y(y)$ is integrable because the first $n - 1$ coordinate projections: $G^i(y) = y_i$, for $i = 1, \ldots, n - 1$, are constants of the motion. Hence by the theorem, X restricted to W is integrable. $\square$

Exercises 7.4

1. Prove Properties (1), (2), and (3) for the normal product in Proposition 7.4.

2. In the proof of Theorem 7.2, we used the identities $e^i \cdot X = 0$, for $i = 1, \ldots, n-1$. Show that these follow from the identities $e_i \cdot X = 0$, for $i = 1, \ldots, n-1$. *Hint*: first show that $g^{in} = 0 = g^{ni}$, for $i = 1, \ldots, n-1$.

3. Continue the analysis of the system in Example 7.6. Study additional graphs of the level curves of $g_{k_1 k_3}$ for other values of k_1, k_3. Do any new features appear? The fixed points c are of two general types (see the discussion), but it is somewhat tedious to compute the eigenvalues of $X'_{F^1 F^2 F^3}(c)$ by hand and apply the Linearization Theorem. Try this and use Maple if necessary.

4. Complete the proof of Theorem 7.4.

5. In the following $F^1, F^2, F^3 : \mathbb{R}^4 \to \mathbb{R}$ are given functions on $\mathbb{R}^4$ and you are to study the system of DEs with vector field X_{F^1, F^2, F^3}. Calculate this vector field explicitly and write down the corresponding system of differential equations. Find all the fixed points c, calculate the eigenvalues of the Jacobian matrix $X'_{F^1, F^2, F^3}(c)$ at a general fixed point, and classify its type. Use the technique from Example 7.6 to study the integral curves of the system and to help classify the fixed points.

(a)

$$
\begin{aligned}
F^1(x, y, z, w) &= w - \tfrac{1}{2}(x^2 + y^2 - z^2) \\
F^2(x, y, z, w) &= z - \tfrac{1}{2}(x^2 - w^2) \\
F^3(x, y, z, w) &= y - \tfrac{1}{2}(x^2 - z^2)
\end{aligned}
$$

(b)

$$
\begin{aligned}
F^1(x, y, z, w) &= w - \tfrac{1}{2}(x^2 + y^2 + z^2) \\
F^2(x, y, z, w) &= z - \tfrac{1}{2}(x^2 - w^2) \\
F^3(x, y, z, w) &= y - \tfrac{1}{2}(x^2 - z^2)
\end{aligned}
$$

Chapter 8

Newtonian Mechanics

In this chapter we discuss a few aspects of Newtonian mechanics for systems of discrete particles. Such systems are *the* primary classical examples of systems of differential equations and serve to illustrate many of the concepts that have been developed for the study of systems of DEs.

The case for $N = 2$ particles (the two-body problem) was completely solved, for forces of mutual attraction, in Chapter 2, and here we consider the general N-particle case. The N-body problem has been studied in depth for many centuries, and an enormous body of important and deep results has accrued (cf. [Th 79, p. 176], [W 47, p. 233], and [Wh 65, p. 339]). You should realize however that the general problem (even when the forces are an inverse square law of attraction) is not solvable in closed form, as the $N = 2$ case and certain special cases of the $N = 3$ problem are. However, numerical methods are always available and we will see below that they are quite effective in studying the complex motions of a system when the number of particles is not too large.

It will be convenient, and customary as well, to denote the derivative with respect to time by a dot rather than a prime. Thus,

$$\dot{\mathbf{r}} = \frac{d\mathbf{r}}{dt}, \qquad \ddot{\mathbf{r}} = \frac{d^2\mathbf{r}}{dt^2}.$$

The prime will otherwise be used as before, e.g., $X'(c)$ denotes the Jacobian matrix at c for a vector field $X : \mathbb{R}^n \to \mathbb{R}^n$.

Philosophically, the discussion and development here is from a mathematical point of view. Namely, we assume no physics background or prior knowledge of such concepts as energy, momentum, etc., and show how these concepts arise from the mathematical model for the motion of a system of particles. Indeed, when Newton formulated his physical laws, the current

D. Betounes, *Differential Equations: Theory and Applications*, DOI 10.1007/978-1-4419-1163-6_8,   371
© Springer Science + Business Media, LLC 2010

ideas about (and names for) kinetic and potential energy, linear and angular momentum, and total energy were still quite primitive. However, once Newton's second law, $F = ma$, is viewed as a system of differential equations, then all the physical concepts can be derived from an analysis of this system.

Indeed, Newton's second law is in essence just the general form for a 2nd-order systems of DEs in normal form:

$$\ddot{\mathbf{r}} = G(t, \mathbf{r}, \dot{\mathbf{r}}).$$

This is so *provided* we ignore the requirements on the specific nature of the forces, the way the masses enter as parameters, and the invariance under Galilean transformations. The beauty of Newton's conception is that when these particulars are added, this system of DEs has, for three centuries, successfully explained the mechanics of a vast range of phenomena.

After discussing a few aspects of the general problem and the Euler numerical scheme for approximate solutions, we consider the special case of systems undergoing rigid-body motions. While this will only give you a modest introduction to the subject of Newtonian mechanics, it should nevertheless make the previous abstract concepts more concrete. For treatments of mechanics see [AM 78], [Ar 78b], [Go 59], [MT 95], [Th 78], [W 47], and [Wh 65].

8.1   The N-Body Problem

Newton's 2nd law is the quite reasonable postulate that the acceleration of the motion of a body is directly due to whatever forces are acting on it. Thus, in the absence of any force, there is no change in the velocity of the body. It is also reasonable to assume that the acceleration produced by a force is in the direction of the force and is inversely proportional to the mass of the body (bodies with large mass are accelerated less by a given force than bodies with small mass). These assumptions then explicitly give Newton's famous 2nd law for the motion of a single body:

$$m\mathbf{a} = F.$$

For the motion of N bodies (ideally point particles) with masses $m_i, i = 1, \ldots, N$, and positions $\mathbf{r}_i = (x_i, y_i, z_i), i = 1, \ldots, N$, the 2nd law applies to each of the bodies and gives the following system of differential equations:

Newton's Second Law:

$$m_1\ddot{\mathbf{r}}_1 = F_1(t, \mathbf{r}_1, \ldots, \mathbf{r}_N, \dot{\mathbf{r}}_1, \ldots, \dot{\mathbf{r}}_N)$$
$$m_2\ddot{\mathbf{r}}_2 = F_2(t, \mathbf{r}_1, \ldots, \mathbf{r}_N, \dot{\mathbf{r}}_1, \ldots, \dot{\mathbf{r}}_N)$$
$$\vdots$$
$$m_N\ddot{\mathbf{r}}_N = F_N(t, \mathbf{r}_1, \ldots, \mathbf{r}_N, \dot{\mathbf{r}}_1, \ldots, \dot{\mathbf{r}}_N).$$

Here each $\mathbf{r}_i$ is a curve in $\mathbb{R}^3$, with $\mathbf{r}_i(t)$ representing the position of the ith particle at time t:

$$\mathbf{r}_i(t) = (x_i(t), y_i(t), z_i(t)),$$

for $i = 1, \ldots, N$ and t in some interval I. See Figure 8.1. The force $F_i(t, \mathbf{r}_1, \ldots, \mathbf{r}_N, \dot{\mathbf{r}}_1, \ldots, \dot{\mathbf{r}}_N)$, depends in general on the positions and velocities of all the bodies in the system as well as the time t.

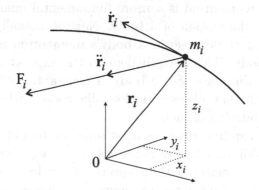

Figure 8.1: *Motion of the ith body in the system based on Newton's second law.*

Mathematically, a solution of this system of DEs is a curve in the space $\mathbb{R}^3 \times \mathbb{R}^3 \times \cdots \times \mathbb{R}^3 = (\mathbb{R}^3)^N \cong \mathbb{R}^{3N}$ and for convenience we will denote such a curve by

$$\mathbf{r} = (\mathbf{r}_1, \ldots, \mathbf{r}_N).$$

Thus, $\mathbf{r}(t)$ gives the configuration of the system at time t and

$$\dot{\mathbf{r}} = (\dot{\mathbf{r}}_1, \ldots, \dot{\mathbf{r}}_N)$$

gives the configuration of velocities for the system. With this notation we will often write the system as:

$$m_i\ddot{\mathbf{r}}_i = F_i(t, \mathbf{r}, \dot{\mathbf{r}}), \qquad i = 1, \ldots, N. \tag{8.1}$$

These equations are also called the *equations of motion*. The system is determined by the masses m_i and the given forces F_i. Generally each F_i : $J \times \mathcal{O} \to \mathbb{R}^3$ is a time-dependent function on an open set $\mathcal{O} \subseteq \mathbb{R}^{3N} \times \mathbb{R}^{3N}$ and represents the force exerted on the ith particle. This force, as the notation indicates, can in general depend on the positions and velocities of all the particles, i.e., on the *state* $(\mathbf{r}, \dot{\mathbf{r}})$ of the system at time t. Often the phase space has the form $\mathcal{O} = \mathcal{U} \times \mathbb{R}^{3N}$, when the nature of the forces preclude certain configurations $\mathbf{r} = (\mathbf{r}_1, \ldots, \mathbf{r}_N)$ of the system. This was the case with the two-body problem in Chapter 1 and occurs in the N-body problem with gravitational (and other) force systems as well.

The *momentum* (or *linear momentum*) of the ith body is, by definition,

$$\mathbf{p}_i = m_i \dot{\mathbf{r}}_i.$$

In some respects the momentum is a more fundamental quantity than the velocity in regards to the motion of a body. Newton actually phrased the 2nd law as saying that the change in a body's momentum is equal to the force acting on the body. This is equivalent to the expression for this law given above *provided* the mass of the body is constant. Otherwise it is a more general statement that allows situations like rockets which continually lose mass due to the burning of fuel.

Newton's 2nd law can be written as a 1st-order system in the customary way by introducing additional functions: $\mathbf{v} = (\mathbf{v}_1, \ldots, \mathbf{v}_N)$, with $\mathbf{v}_i \equiv \dot{\mathbf{r}}_i$, for $i = 1, \ldots, N$, being the velocities of the respective particles. For convenience, we let $F : J \times \mathcal{O} \to \mathbb{R}^{3N}$ be the vector-valued function whose components are the forces on the individual bodies:

$$F = (F_1, \ldots, F_N),$$

and let μ be the following $3N \times 3N$ diagonal matrix (called the *mass matrix*):

$$\mu = \begin{bmatrix} m_1 I_3 & & & \\ & m_2 I_3 & & \\ & & \ddots & \\ & & & m_N I_3 \end{bmatrix}. \tag{8.2}$$

Here I_3 denotes the 3×3 identity matrix. Then Newton's law, as a 2nd-order system, is

$$\ddot{\mathbf{r}} = \mu^{-1} F(t, \mathbf{r}, \dot{\mathbf{r}}),$$

and the corresponding 1st-order system is

$$\dot{\mathbf{r}} = \mathbf{v}$$
$$\dot{\mathbf{v}} = \mu^{-1}F(t, \mathbf{r}, \mathbf{v}).$$

The vector field $X_F : J \times \mathcal{O} \to \mathbb{R}^{3N} \times \mathbb{R}^{3N}$ for this 1st-order system is then given by

$$X_F(t, \mathbf{r}, \mathbf{v}) = \left(\mathbf{v}, \mu^{-1}F(t, \mathbf{r}, \mathbf{v}) \right).$$

8.1.1 Fixed Points

When the forces do not depend on the time, the fixed points of the vector field X_F are points $(\mathbf{r}, \mathbf{v}) \in \mathcal{O}$ such that

$$\mathbf{v} = 0$$
$$F(\mathbf{r}, 0) = 0,$$

that is, $\mathbf{v}_i = 0$, $F_i(\mathbf{r}, 0) = 0$, for $i = 1, \dots, N$. In real space $\mathbb{R}^3$ this amounts to a configuration $\mathbf{r} = (\mathbf{r}_1, \dots, \mathbf{r}_N)$ of N positions where the bodies could be located at rest and experience no forces acting on them. Because of this interpretation, a fixed point is traditionally referred to as an *equilibrium point*.

The stability of a fixed point $(\mathbf{r}, 0)$ can be analyzed by using the Linearization Theorem and Liapunov functions. Since X_F is quite general (it is the vector field associated with a general 2nd-order system in normal form), there is little we can conclude about the stability. It is important to note, however, that the derivative (Jacobian matrix) for X_F has the special form

$$X'_F = \begin{bmatrix} 0 & I \\ \mu^{-1}\frac{\partial F}{\partial \mathbf{r}} & \mu^{-1}\frac{\partial F}{\partial \mathbf{v}} \end{bmatrix}.$$

Here 0 and I denote the $3N \times 3N$ zero and identity matrices, respectively, and

$$\frac{\partial F}{\partial \mathbf{r}} = \left\{ \frac{\partial F^i}{\partial \mathbf{r}_j} \right\}_{i,j=1\cdots N}, \quad \frac{\partial F}{\partial \mathbf{v}} = \left\{ \frac{\partial F^i}{\partial \mathbf{v}_j} \right\}_{i,j=1\cdots N},$$

are $3N \times 3N$ matrices as well. The notation here is very natural, but at first can be confusing. To be explicit, the above matrices are actually in block form, with each block being a 3×3 matrix. For example, if F^i has component form $F^i = (f^i, g^i, h^i)$ then

$$\frac{\partial F^i}{\partial \mathbf{r}_j} = \begin{bmatrix} f^i_{x_j} & f^i_{y_j} & f^i_{z_j} \\ g^i_{x_j} & g^i_{y_j} & g^i_{z_j} \\ h^i_{x_j} & h^i_{y_j} & h^i_{z_j}, \end{bmatrix}, \qquad \frac{\partial F^i}{\partial \mathbf{v}_j} = \begin{bmatrix} f^i_{u_j} & f^i_{v_j} & f^i_{w_j} \\ g^i_{u_j} & g^i_{v_j} & g^i_{w_j} \\ h^i_{u_j} & h^i_{v_j} & h^i_{w_j} \end{bmatrix},$$

where the subscripts denote partial derivative with respect to x_j, y_j and z_j and u_j, v_j and w_j, respectively. With additional information on the nature of the forces, we see below that conclusions can be made about the eigenvalues of $X'_F(\mathbf{r}, 0)$ and the stability of fixed points.

8.1.2   Initial Conditions

If we assume that the partial derivatives $\partial F^i / \partial \mathbf{r}_j, \partial F^i / \partial \mathbf{v}_j$, for $i, j = 1, \ldots, N$ are continuous on $\mathcal{O}$, then the general Existence and Uniqueness Theorem from Chapter 3 applies. In the N-body setting, this theorem says that if the initial positions and initial velocities of all the bodies are known, then this initial data uniquely determines the motion of the system for all time (or at least for all the times in the maximum interval of existence).

Specifically, given an initial configuration of positions $\mathbf{a} = (\mathbf{a}_1, \ldots, \mathbf{a}_N)$ and an initial configuration of velocities $\mathbf{b} = (\mathbf{b}_1, \ldots, \mathbf{b}_N)$, with $(\mathbf{a}, \mathbf{b}) \in \mathcal{O}$, and an initial time $t_0 \in J$, there is a unique solution $\mathbf{r} = (\mathbf{r}_1, \ldots, \mathbf{r}_N) : I \to \mathbb{R}^{3N}$ of the equations of motion:

$$m_i \ddot{\mathbf{r}}_i = F_i(t, \mathbf{r}, \dot{\mathbf{r}}), \qquad i = 1, \ldots, N,$$

defined on a maximal interval $I = I_{(t_0, \mathbf{a}, \mathbf{b})}$, and satisfying the initial conditions:

$$\begin{aligned} \mathbf{r}_i(t_0) &= \mathbf{a}_i \\ \dot{\mathbf{r}}_i(t_0) &= \mathbf{b}_i, \end{aligned}$$

for $i = 1, \ldots, N$. For each time $t \in I$, the state of the system $\mathbf{r}(t) = (\mathbf{r}_1(t), \ldots, \mathbf{r}_N(t))$, is known completely and so the system is called *deterministic*. The evolution of the system over time, from state to state, is given by the flow: $t \mapsto \phi_t^{t_0}(\mathbf{a}, \mathbf{b})$, which also explicitly exhibits the dependence of the evolution on the initial state $(\mathbf{a}, \mathbf{b})$ of the system.

Definition 8.1 (System of Particles) For convenience of expression in the ensuing discussions, the term *system* (or *system of particles*) will refer to a particular solution $\mathbf{r} = (\mathbf{r}_1, \ldots, \mathbf{r}_N) : I \to \mathcal{O}$ of the equations of motion with given initial state $(\mathbf{a}, \mathbf{b})$. This is, of course, to be distinguished from the system comprising the differential equations that model the motion.

8.1.3 Conservation Laws

Except for the structure of the configuration space $\mathcal{O}$ as a subset of $\mathbb{R}^{3N} \times \mathbb{R}^{3N}$ and the appearance of the masses m_i, the 2nd-order system of DEs:

$$m_i \ddot{\mathbf{r}}_i = F_i(t, \mathbf{r}, \dot{\mathbf{r}}), \qquad i = 1, \ldots, N,$$

comprising Newton's 2nd law, is quite general. Despite this generality, the form of the system of differential equations does give us Galileo's principle:

Newton's First Law: *If there is no force on the ith body of the system* $(F_i = 0)$*, then the ith body will either remain at rest or move in a straight line with constant velocity.*

In terms of differential equations this just says that the solution of $\ddot{\mathbf{r}}_i = 0$ is $\mathbf{r}_i(t) = \mathbf{v}_i(0)t + \mathbf{r}_i(0)$.

Other than this, there is little we can conclude from Newton's 2nd law without specifying the nature of the forces acting on the system. All of the conservation laws we derived for the two-body problem (for gravitational attraction) extend to the N-body case, with more general systems of forces. As with Newton's 1st law, these conservation laws and other physical principles follow directly from the form of the system of differential equations. Before seeing how this is so, we introduce some standard terminology.

Definition 8.2 (Total Momenta and Center of Mass) For a system of particles $\mathbf{r} = (\mathbf{r}_1, \ldots, \mathbf{r}_N) : I \to \mathcal{O}$, the *total mass* of the system is

$$M = \sum_{i=1}^{N} m_i.$$

Furthermore, let

$$\mathbf{P}(t) = \sum_{i=1}^{N} m_i \dot{\mathbf{r}}_i(t) \tag{8.3}$$

$$\mathbf{L}(t) = \sum_{i=1}^{N} m_i [\mathbf{r}_i(t) \times \dot{\mathbf{r}}_i(t)] \tag{8.4}$$

$$\mathbf{R}(t) \;=\; \sum_{i=1}^{N} \frac{m_i}{M}\, \mathbf{r}_i(t), \tag{8.5}$$

denote respectively, the *total linear momentum*, the *total angular momen-tum* (about the origin), and *center of mass* of the system at time $t \in I$. Additionally, let

$$\mathbf{F}(t) \;=\; \sum_{i=1}^{N} F_i\Big(t, \mathbf{r}(t), \dot{\mathbf{r}}(t)\Big) \tag{8.6}$$

$$\mathbf{T}(t) \;=\; \sum_{i=1}^{N} \Big[\mathbf{r}_i(t) \times F_i\Big(t, \mathbf{r}(t), \dot{\mathbf{r}}(t)\Big)\Big], \tag{8.7}$$

denote the *total force* and *total torque* (about the origin) acting on the system at time t.

Note: The total force $\mathbf{F}$ should not to be confused with the vector-valued function $F = (F_1, \ldots, F_N)$, comprised of the forces on the respective bodies.

In general, the linear and angular momenta of the system, $\mathbf{P}$ and $\mathbf{L}$, are *not* conserved, i.e., not constant in time, but rather evolve according to the following relations:

$$\dot{\mathbf{P}}(t) = \mathbf{F}(t) \tag{8.8}$$
$$\dot{\mathbf{L}}(t) = \mathbf{T}(t) \tag{8.9}$$
$$M\ddot{\mathbf{R}}(t) = \mathbf{F}(t), \tag{8.10}$$

(exercise). The first equation is interpreted as saying that the change in the total linear momentum of the system is equal to the total force acting on the system. Similarly, the second equation says that the change in the total angular momentum, about the origin, is equal to the total torque, about the origin, applied to the system. The last equation says that the center of mass moves as if it were a body of mass M subject to the total force acting on the system.

From these equations (8.8)-(8.10), it is easy to show (exercise) that the following conservation laws hold

N-Body Conservation Laws:

1. *If the total force on the system is zero,* $\mathbf{F}(t) = 0$, *for all* t, *then*

    (a) *the total linear momentum is constant in time, and*

    (b) *the center of mass moves in a straight line with constant velocity.*

2. *If the total torque on the system is zero,* $\mathbf{T}(t) = 0$, *for all* t, *then the total angular momentum is constant in time.*

In terms of equations, these conservation laws, are expressed as follows. The laws apply to each solution $\mathbf{r} : I \to \mathbb{R}^{3N}$ of the equations of motion.

1. **(Conservation of Linear Momentum):** If $\mathbf{F}(t) = 0$, for all t, then

$$\text{(a)} \qquad \sum_{i=1}^{N} m_i \dot{\mathbf{r}}_i(t) = \mathbf{P}_0$$

$$\text{(b)} \qquad \sum_{i=1}^{N} \frac{m_i}{M} \mathbf{r}_i(t) = \frac{t}{M} \mathbf{P}_0 + \mathbf{R}_0,$$

for all $t \in I$.

2. **(Conservation of Angular Momentum):** If $\mathbf{T}(t) = 0$, for all t, then

$$\sum_{i=1}^{N} m_i [\mathbf{r}_i(t) \times \dot{\mathbf{r}}_i(t)] = \mathbf{L}_0,$$

for all $t \in I$.

Here $\mathbf{P}_0 = \sum_{i=1}^{N} m_i \dot{\mathbf{r}}_i(0)$ is the initial total linear momentum of the system, $\mathbf{R}_0 = \sum_{i=1}^{N} \frac{m_i}{M} \mathbf{r}_i(0)$ is the initial position of the center of mass, and $\mathbf{L}_0 = \sum_{i=1}^{N} m_i [\mathbf{r}_i(0) \times \dot{\mathbf{r}}_i(0)]$ is the initial total angular momentum of the system about the origin. Strictly speaking only 1(a) and 2 are conservation laws, while 1(b) is a physical principle since the quantity $\sum_{i=1}^{N} m_i \mathbf{r}_i(t)/M$ does not remain constant in time. Generally, conservation laws, also called *constants of the motion*, are expressions involving the positions and velocities that do not change over time.

Another, more geometrical, way to express conservation laws 1(a) and 2 is to say that the curve $t \mapsto (\mathbf{r}(t), \dot{\mathbf{r}}(t))$ lies on each of the two submanifolds in phase space given by:

$$S_1 = \left\{ (\mathbf{r}, \mathbf{v}) \in \mathcal{O} \,\Big|\, \sum_{i=1}^{N} m_i \mathbf{v}_i = \mathbf{P}_0 \right\}$$

$$S_2 = \left\{ (\mathbf{r}, \mathbf{v}) \in \mathcal{O} \,\Big|\, \sum_{i=1}^{N} m_i [\mathbf{r}_i \times \mathbf{v}_i] = \mathbf{L}_0 \right\}$$

Here $(\mathbf{r}, \mathbf{v}) = (\mathbf{r}_1, \ldots, \mathbf{r}_N, \mathbf{v}_1, \ldots, \mathbf{v}_N)$ denotes a general point in phase space $\mathcal{O} \subseteq \mathbb{R}^{3N} \times \mathbb{R}^{3N}$. In the description of S_1, the system of equations is a linear system of algebraic equations, so S_1 is a *flat* submanifold of $\mathbb{R}^{6N}$. In S_2, the system of equations is nonlinear and the submanifold S_2 is "curved." The intersection $S_1 \cap S_2$, of these two submanifolds gives a lower-dimensional submanifold in $\mathbb{R}^{6N}$. Each additional conservation law, which is independent in some sense of the above two, will give yet another submanifold on which the curve $t \mapsto (\mathbf{r}(t), \dot{\mathbf{r}}(t))$ must lie. In this view, the combination of a number of conservation laws will constrain the motion of the state $(\mathbf{r}(t), \dot{\mathbf{r}}(t))$ of the system to the intersection $S_1 \cap S_2 \cap \cdots \cap S_k$, of a number of submanifolds in phase space.

Many systems of forces have their total force $\mathbf{F}(t) = 0$ and total torque $\mathbf{T}(t) = 0$, for all time t. A standard, and important, class of such a systems are those where the forces arise from the *interactions* of each pair of bodies and these forces of interaction obey Newton's 3rd law. Mathematically, this is described as follows.

Definition 8.3 (Forces of Interaction) Suppose J is an interval of times, $\mathcal{O}$ is an open set in $\mathbb{R}^{3N} \times \mathbb{R}^{3N}$, and $F_{ij} : J \times \mathcal{O} \to \mathbb{R}^3$, for $i, j \in \{1, \ldots, N\}$, $i \neq j$, are given functions. For each $i \in \{1, \ldots, N\}$ define $F_i : J \times \mathcal{O} \to \mathbb{R}^3$ by

$$F_i = \sum_{j \neq i} F_{ij}. \tag{8.11}$$

Here the sum is over all indices j different than i, that is, $j \in \{1, \ldots, N\} \backslash \{i\}$. The function F_{ij} is the *force of interaction* between the ith and jth bodies. More specifically,

$$F_{ij} = \text{the force exerted on the } i\text{th body by the } j\text{th body.}$$

Thus, F_i is the sum of the forces of exerted on the ith body by all the other bodies. Note that no body exerts a force on itself (i.e., there is no F_{ii} in the above sum). A system of forces $\{F_1, \ldots, F_N\}$ is said to *arise from forces of interaction* when each F_i has the form (8.11). Such a system is said to obey Newton's 3rd law if

Newton's 3rd Law: *Each body exerts a force on every other body that is equal in magnitude and opposite in direction to the force that the other body exerts on it. That is, $F_{ji} = -F_{ij}$, for all $i \neq j$.*

The system of forces is said to obey the *strong form of Newton's 3rd law* if $F_{ji} = -F_{ij}$ and $F_{ij}(t, \mathbf{r}, \mathbf{v})$ is parallel to $\mathbf{r}_i - \mathbf{r}_j$, for all $i \neq j$, and all $(t, \mathbf{r}, \mathbf{v}) \in J \times \mathcal{O}$.

It is easy to construct quite arbitrary systems of forces arising from forces of interactions that obey Newton's 3rd law; just choose F_{ij}, for $i < j$, arbitrarily and then define $F_{ji} \equiv -F_{ij}$, for $i < j$. Similarly, to construct a system that obeys the strong form of Newton's 3rd law, choose any functions $f_{ij} : J \times \mathcal{O} \rightarrow \mathbb{R}$, for $i < j$, take $f_{ji} \equiv f_{ij}$, for $i < j$, and define $F_{ij} : J \times \mathcal{O} \rightarrow \mathbb{R}^3$ by

$$F_{ij}(t, \mathbf{r}, \mathbf{v}) = f_{ij}(t, \mathbf{r}, \mathbf{v})(\mathbf{r}_i - \mathbf{r}_j),$$

for any i, j and $(t, \mathbf{r}, \mathbf{v}) \in J \times \mathcal{O}$. An example of this latter construction is the following:

Newton's Law of Universal Gravitation: *Every particle of mass in the universe is attracted to every other particle of mass with a force directed along the line joining the particles and with magnitude that is proportional to the product of the masses and reciprocally proportional to the square of their distance apart. The proportionally constant is the gravitational constant G.*

For the N-body problem, with gravitational forces of attraction, the forces of interaction are

$$F_{ij}(t, \mathbf{r}, \mathbf{v}) = \frac{Gm_i m_j}{r_{ij}^3}(\mathbf{r}_j - \mathbf{r}_i), \tag{8.12}$$

where $r_{ij} \equiv |\mathbf{r}_i - \mathbf{r}_j|$.

The following proposition is an elementary consequence of the above definition.

Proposition 8.1 *If a system of forces $\{F_1, \ldots, F_N\}$ arises from forces of interaction and obeys Newton's 3rd law, then for any solution of the equations of motion, the total force on the system is zero for all time. If in addition the system obeys the strong form of Newton's 3rd law, then the total torque on the system is also zero for all time.*

Proof: Exercise.

A final, and perhaps the most important, conservation law for the N-body system is the conservation of energy law. This requires a special system of forces, called a *conservative* system of forces. The mathematical motivation for how this law arises from the form of the system of DEs is as follows. Suppose that $\mathbf{r} : I \to \mathbb{R}^{3N}$ is a solution of the equations of motion:

$$m_i \ddot{\mathbf{r}}_i = F_i(t, \mathbf{r}, \dot{\mathbf{r}}), \qquad i = 1, \ldots, N.$$

Taking the dot product of the ith equation with $\dot{\mathbf{r}}_i$ and then summing on i gives

$$\sum_{i=1}^{N} m_i \ddot{\mathbf{r}}_i \cdot \dot{\mathbf{r}}_i = \sum_{i=1}^{N} F_i(t, \mathbf{r}, \dot{\mathbf{r}}) \cdot \dot{\mathbf{r}}_i.$$

Now recognize that $\frac{d}{dt}(\frac{1}{2}|\dot{\mathbf{r}}_i|^2) = \ddot{\mathbf{r}}_i \cdot \dot{\mathbf{r}}_i$, for each i, and so the last equation is

$$\frac{d}{dt}\left[\sum_{i=1}^{N} \tfrac{1}{2} m_i |\dot{\mathbf{r}}_i|^2 \right] = \sum_{i=1}^{N} F_i(t, \mathbf{r}, \dot{\mathbf{r}}) \cdot \dot{\mathbf{r}}_i.$$

The quantity in the square brackets here is known as the *kinetic energy* of the system and the integral of the quantity on the right side of the equation is called the *work* done by the system. Thus, the equation, when integrated between two times, says that the change in kinetic energy is equal to the work done on the system. We formalize this in the following definition.

Definition 8.4 (Work, Potentials, and Energy)

(a) Suppose $\mathbf{r} : I \to \mathbb{R}^{3N}$ is a solution of the equations of motion and $t_1 < t_2 \in I$ are two times. Then the *work done by the system* in going from state $(\mathbf{r}(t_1), \dot{\mathbf{r}}(t_1))$ to state $(\mathbf{r}(t_2), \dot{\mathbf{r}}(t_2))$ is defined as

$$W(t_1, t_2) = \sum_{i=1}^{N} \int_{t_1}^{t_2} F_i\left(t, \mathbf{r}(t), \dot{\mathbf{r}}(t) \right) \cdot \dot{\mathbf{r}}_i(t) \, dt. \qquad (8.13)$$

(b) The *kinetic energy* of the system at time t is

$$T(t) = \sum_{i=1}^{N} \tfrac{1}{2} m_i |\dot{\mathbf{r}}_i(t)|^2.$$

(c) The system of forces F_i, $i = 1, \ldots, N$ is said to be *conservative* if (i) F_i depends only on $\mathbf{r}$, (ii) the forces have a common domain $F_i : \mathcal{U} \subseteq \mathbb{R}^{3N} \rightarrow \mathbb{R}^3$, $i = 1, \ldots, N$, and (iii) there is a differentiable function $V : \mathcal{U} :\rightarrow \mathbb{R}$ such that

$$F_i(\mathbf{r}) = -\frac{\partial V}{\partial r_i}(\mathbf{r}) \tag{8.14}$$

$$\equiv -\left(\frac{\partial V}{\partial x_i}(\mathbf{r}), \frac{\partial V}{\partial y_i}(\mathbf{r}), \frac{\partial V}{\partial z_i}(\mathbf{r}),\right), \tag{8.15}$$

for all $\mathbf{r} \in \mathcal{U}$. The function V is called a *potential* for the system for forces. *Note*: The partial derivative in equation (8.14) is symbolic, or just notation, for what is given in (8.15).

(d) For a conservative system of forces, with potential V, the *total energy* for a motion of the system is

$$E = \sum_{i=1}^{N} \tfrac{1}{2} m_i |\dot{\mathbf{r}}_i(t)|^2 + V(\mathbf{r}(t)) \tag{8.16}$$

$$= T(t) + V(\mathbf{r}(t)). \tag{8.17}$$

The second term $V(\mathbf{r}(t))$ is called the *potential energy* of the system at time t. Thus, the total energy E of the system is the sum of the kinetic and potential energies, each of which will vary in time, in general. However, as the notation above indicates the total energy is constant in time. This is the content of the energy conservation law.

N-Body Conservation Law (Conservative Systems):

3. *If the system of forces is conservative, then the total energy of the system, for any choice of potential, is constant in time.*

It is an easy exercise to prove this.

Since the total energy is constant throughout the motion, any increase in the kinetic energy of the system must be compensated for by a decrease in the potential energy, and vice-versa. While the total energy is constant in time, it *does* depend on the solution of the equations of motion. Different solutions can have different total energies. The value of the total energy for a solution $\mathbf{r} : I \rightarrow \mathbb{R}^{3N}$ is determined from the initial conditions, i.e., the initial positions and velocities of the bodies in the system:

$$E = E_0 = \sum_{i=1}^{N} \tfrac{1}{2} m_i |\dot{\mathbf{r}}_i(t_0)|^2 + V(\mathbf{r}(t_0)),$$

where t_0 is the initial time.

Geometrically the conservation of energy means that each curve $t \mapsto (\mathbf{r}(t), \dot{\mathbf{r}}(t))$, describing the evolution of a state of the system, lies on an *energy hypersurface*:

$$S_3 = \left\{ (\mathbf{r}, \mathbf{v}) \in \mathcal{O} \,\middle|\, \sum_{i=1}^{N} \tfrac{1}{2} m_i |\mathbf{v}_i|^2 + V(\mathbf{r}) = E_0 \right\},$$

in phase space.

Example 8.1 (The 3-Body Problem) Suppose $h_{12}, h_{13}, h_{23} : (0, \infty) \to \mathbb{R}$, are differentiable functions. The general version of the classical 3-body problem that satisfies the strong form of Newton's 3rd Law is

$$m_1 \ddot{\mathbf{r}}_1 = \frac{h_{12}(r_{12})}{r_{12}} (\mathbf{r}_2 - \mathbf{r}_1) + \frac{h_{13}(r_{13})}{r_{13}} (\mathbf{r}_3 - \mathbf{r}_1) \tag{8.18}$$

$$m_2 \ddot{\mathbf{r}}_2 = \frac{h_{12}(r_{12})}{r_{12}} (\mathbf{r}_1 - \mathbf{r}_2) + \frac{h_{23}(r_{23})}{r_{23}} (\mathbf{r}_3 - \mathbf{r}_2) \tag{8.19}$$

$$m_3 \ddot{\mathbf{r}}_3 = \frac{h_{13}(r_{13})}{r_{13}} (\mathbf{r}_1 - \mathbf{r}_3) + \frac{h_{23}(r_{23})}{r_{23}} (\mathbf{r}_2 - \mathbf{r}_3). \tag{8.20}$$

The *classical* 3-body problem is for an inverse square law of attraction, and in this case the h_{ij}'s are given by

$$h_{ij}(r) = \frac{G m_i m_j}{r^2}. \tag{8.21}$$

In the classical as well as the general case it is easy to see that the system of forces arises from forces of interaction and obeys the strong form of Newton's 3rd Law. Thus, the linear and angular momenta of the system are conserved. It is also true that the system of forces is conservative. For the classical case, with the h_{ij}'s given by equation (8.21), it is easy to verify that a potential for the system of forces is

$$V(\mathbf{r}) = V(\mathbf{r}_1, \mathbf{r}_2, \mathbf{r}_3) = -\frac{G m_1 m_2}{r_{12}} - \frac{G m_1 m_3}{r_{13}} - \frac{G m_2 m_3}{r_{23}}. \tag{8.22}$$

See the exercises for the form of the potential for the general h_{ij} case (and for arbitrary N).

8.1.4   Stability of Conservative Systems

A potential $V : \mathcal{U} \subseteq \mathbb{R}^{3N} \to \mathbb{R}$ for a conservative system of forces is also intimately connected with the stability of the motion. Indeed, the critical points of V correspond to the fixed points of the N-body system and V provides the essential part of an expression for a Liapunov function Λ. Thus, define $\Lambda : \mathcal{U} \times \mathbb{R}^{3N} \to \mathbb{R}$ by

$$\Lambda(\mathbf{r}, \mathbf{v}) = \sum_{i=1}^{N} \tfrac{1}{2} m_i |\mathbf{v}_i|^2 + V(\mathbf{r}). \tag{8.23}$$

This is also known as the *total energy function* on phase space (it's level sets are the energy hypersurfaces). Recall that to be a Liapunov function for a fixed point $c = (\mathbf{r}_*, 0)$, there must be a neighborhood W of c on which the covariant derivative is nonpositive: $\nabla_{X_F} \Lambda \le 0$, and c must be the absolute minimum value of Λ in W. In the situation at hand:

$$\nabla\Lambda = \left(\frac{\partial V}{\partial \mathbf{r}_1}(\mathbf{r}), \ldots, \frac{\partial V}{\partial \mathbf{r}_N}(\mathbf{r}), m_1\mathbf{v}_1, \ldots, m_N\mathbf{v}_N \right), \tag{8.24}$$

and

$$X_F = \left(\mathbf{v}_1, \ldots, \mathbf{v}_N, -\frac{1}{m_1}\frac{\partial V}{\partial \mathbf{r}_1}(\mathbf{r}), \ldots, -\frac{1}{m_N}\frac{\partial V}{\partial \mathbf{r}_N}(\mathbf{r}) \right), \tag{8.25}$$

and so clearly

$$\nabla_{X_F}\Lambda = \nabla\Lambda \cdot X_F = 0,$$

on $\mathcal{O} = \mathcal{U} \times \mathbb{R}^{3N}$. That is the first requirement for Λ to be a Liapunov function. For the second requirement, note that $c = (\mathbf{r}_*, 0)$ is a local minimum for Λ if and only if $\mathbf{r}_*$ is a local minimum for V (exercise). Thus, *points $\mathbf{r}_*$ where the potential has local minima give stable fixed points $(\mathbf{r}_*, 0)$ of the system.* This is part of the content of the next theorem.

Before stating this theorem, it is important to observe that from (8.24) and (8.25) it follows that $(\mathbf{r}_*, 0)$ is a fixed point of X_F if and only if $\mathbf{r}_*$ is a critical point of the potential V, if and only if $(\mathbf{r}_*, 0)$ is a critical point of Λ. The usual analysis of the extrema of V employs the second derivative test, involving the *Hessian matrix*:

$$\mathcal{H}_V = \begin{bmatrix} \dfrac{\partial^2 V}{\partial \mathbf{r}_1^2} & \cdots & \dfrac{\partial^2 V}{\partial \mathbf{r}_1 \partial \mathbf{r}_N} \\ \vdots & & \vdots \\ \dfrac{\partial^2 V}{\partial \mathbf{r}_N \partial \mathbf{r}_1} & \cdots & \dfrac{\partial^2 V}{\partial \mathbf{r}_N^2} \end{bmatrix}, \tag{8.26}$$

of V. This is the $3N \times 3N$ symmetric matrix consisting of all the 2nd-order partial derivatives of V. Note that the matrix shown above is in block form, each block being a 3×3 matrix. Specifically, the i-jth block is

$$\frac{\partial^2 V}{\partial \mathbf{r}_i \partial \mathbf{r}_j} = \begin{bmatrix} \dfrac{\partial^2 V}{\partial x_i \partial x_j} & \dfrac{\partial^2 V}{\partial x_i \partial y_j} & \dfrac{\partial^2 V}{\partial x_i \partial z_j} \\[2mm] \dfrac{\partial^2 V}{\partial y_i \partial x_j} & \dfrac{\partial^2 V}{\partial y_i \partial y_j} & \dfrac{\partial^2 V}{\partial y_i \partial z_j} \\[2mm] \dfrac{\partial^2 V}{\partial z_i \partial x_j} & \dfrac{\partial^2 V}{\partial z_i \partial y_j} & \dfrac{\partial^2 V}{\partial z_i \partial z_j} \end{bmatrix}. \tag{8.27}$$

The second derivative test says that if $\mathcal{H}_V(\mathbf{r}_*)$ is positive definite at a critical point $\mathbf{r}_*$ of V, then V has a local minimum at $\mathbf{r}_*$ (see Appendix A).

Theorem 8.1 (Stability of Conservative Systems) *Suppose V is a potential for the system of forces in the N-body problem: $(\dot{\mathbf{r}}, \dot{\mathbf{v}}) = X_F(\mathbf{r}, \mathbf{v})$.*

(1) *If V has a local minimum at $\mathbf{r}_*$, then $(\mathbf{r}_*, 0)$ is a stable fixed point for the system.*

(2) *The Jacobian matrix X_F' has the form*

$$X_F'(\mathbf{r}, \mathbf{v}) = \begin{bmatrix} 0 & I \\ -\mu^{-1}\mathcal{H}_V(\mathbf{r}) & 0 \end{bmatrix}, \tag{8.28}$$

where $0, I$ are the $3N \times 3N$ zero and identity matrices, $\mathcal{H}_V$ is the Hessian of V given by (8.26), and μ is the $3N \times 3N$ diagonal, mass matrix, given in equation (8.2). Hence the eigenvalues of $X_F'(\mathbf{r}, \mathbf{v})$ are

$$\pm\sqrt{\lambda_1}, \ldots, \pm\sqrt{\lambda_{3N}},$$

where $\lambda_1, \ldots, \lambda_{3N}$ are the eigenvalues of $-\mu^{-1}\mathcal{H}(\mathbf{r})$. Therefore the eigenvalues of $X_F'(\mathbf{r}, \mathbf{v})$ are either purely imaginary, zero, or real, each occuring in $\pm$ pairs.

(3) *If $(\mathbf{r}_*, 0)$ is a fixed point of X_F and if all the eigenvalues of $X_F'(\mathbf{r}_*, 0)$ are purely imaginary, then the Hessian $\mathcal{H}_V(\mathbf{r}_*)$ is positive definite, the potential V has a local minimum at $\mathbf{r}_*$, and the fixed point $(\mathbf{r}_*, 0)$ is a stable center for the N-body system.*

(4) *If $(\mathbf{r}_*, 0)$ is a simple fixed point (i.e., $\det(X_F'(\mathbf{r}_*, 0)) \neq 0$), and if at least one of the eigenvalues of $X_F'(\mathbf{r}_*, 0)$ is real, then $(\mathbf{r}_*, 0)$ is an unstable fixed point of the system.*

Proof: The proof of (1) follows from the discussion before the theorem. It is clear from the form of X_F in (8.25) that its Jacobian matrix has the form given in (8.28). It is an elementary exercise (see Chapter 4) to show that if a matrix A has block form

$$A = \begin{bmatrix} 0 & I \\ B & 0 \end{bmatrix},$$

with B a $k \times k$ matrix and I the $k \times k$ identity matrix, then the eigenvalues of A are $\pm\sqrt{\mu_1}, \ldots, \pm\sqrt{\mu_k}$, where $\mu_1, \ldots, \mu_k$ are the eigenvalues of B. This gives the result (2) about the eigenvalues of $X_F'(\mathbf{r}, \mathbf{v})$.

   To prove (3) suppose $(\mathbf{r}_*, 0)$ is a fixed point of X_F. If all the eigenvalues of $X_F'(\mathbf{r}_*, 0)$ are purely imaginary, then by part (2), all the eigenvalues of $-\mu^{-1}\mathcal{H}_V(\mathbf{r}_*)$ must be negative, say: $\lambda_1 \leq \cdots \leq \lambda_{3N} < 0$. (Note that $-\mu^{-1}\mathcal{H}_V(\mathbf{r}_*)$ is a real, symmetric matrix and so its eigenvalues must be real). Then by the Principal Axes Theorem (see Appendix C), the function

$$f(\mathbf{r}) \equiv -\frac{\mathcal{H}_V(\mathbf{r}_*)\mathbf{r} \cdot \mathbf{r}}{\mu\mathbf{r} \cdot \mathbf{r}},$$

which is the ratio of the indicated quadratic forms on $\mathbb{R}^{3N}$, has maximum value $\lambda_{3N} < 0$. Thus, $f(\mathbf{r}) < 0$, for all $\mathbf{r} \neq 0$, and so $\mathcal{H}_V(\mathbf{r}_*)\mathbf{r} \cdot \mathbf{r} > 0$, for all $\mathbf{r} \neq 0$. Hence $\mathcal{H}_V(\mathbf{r}_*)$ is positive definite and thus V has a local minimum value at $\mathbf{r}_*$ (see Appendix A). By the discussion prior to the theorem this makes the energy function Λ defined by (8.23) a Liapunov function for the fixed point $(\mathbf{r}_*, 0)$ and hence it is a stable fixed point.

   To prove (4), note that none of the eigenvalues of $X_F'(\mathbf{r}_*, 0)$ can be zero (since by assumption the fixed point is simple) and so if one of its eigenvalues λ is real it must be either positive or negative. But $-\lambda$ is also an eigenvalue (by (2)) and so $X_F'(\mathbf{r}_*, 0)$ has a negative eigenvalue. This makes $(\mathbf{r}_*, 0)$ an unstable fixed point. (See Corollary 6.2, Exercise 4 in Section 6.2, and Exercise 1 in Section 6.3.) $\square$

   The theorem tells us what types of fixed points (equilibria) we can expect to have for conservative systems. However, in practice it is often quite difficult to determine if a potential V has a local minimum at a known critical point. Using the second derivative test may not be easy since computing the Hessian matrix $\mathcal{H}_V$ can be very tedious and then determining if this matrix is positive definite can be equally challenging.

Example 8.2 (Hooke's Law) Consider a two-body problem with forces of interaction of the following form:

$$m_1 \ddot{\mathbf{r}}_1 = \frac{f(r_{12})}{r_{12}}(\mathbf{r}_2 - \mathbf{r}_1) \tag{8.29}$$

$$m_2 \ddot{\mathbf{r}}_2 = \frac{f(r_{12})}{r_{12}}(\mathbf{r}_1 - \mathbf{r}_2). \tag{8.30}$$

Here $f : (0, \infty) \to \mathbb{R}$ is a given function. It is easy to see that the fixed points occur for positions $\mathbf{r}_1, \mathbf{r}_2$ of the two bodies such that their separation $r_{12} = |\mathbf{r}_1 - \mathbf{r}_2|$ is a zero of f, i.e., $f(r_{12}) = 0$. For gravitational forces, f has no zeros and so there are no fixed points. However, many other cases do admit fixed points and a standard, relatively reasonable example is when f has the form

$$f(r) = k(r - L),$$

where k and L are two positive constants, called *the spring constant* and *the unstretched length*, respectively. Physically, we view the two masses as attached to each other with a spring, which pushes them apart when compressed and draws them together when extended. The model, which assumes this linear relation between the change in length of the spring and the resulting force, is known as *Hooke's Law*.

Clearly, for Hooke's Law, any two positions $\mathbf{r}_1, \mathbf{r}_2$ that do not require a change in the length of the spring, $r_{12} = L$, result in an equilibrium point for the system when the two bodies are released with no velocity. There are thus infinitely many fixed points, $\{(\mathbf{r}_1, \mathbf{r}_2) \in \mathbb{R}^6 \,|\, |\mathbf{r}_1 - \mathbf{r}_2| = L\}$, for this system.

The system of forces is conservative, since it can easily be verified that

$$V(\mathbf{r}_1, \mathbf{r}_2) = \frac{k}{2}(r_{12} - L)^2,$$

is a potential for the system. This also enables us to easily see that any fixed point is stable. This is so since the form of V readily reveals that

$$V(\mathbf{r}_1, \mathbf{r}_2) \geq 0,$$

for all $\mathbf{r}_1, \mathbf{r}_2 \in \mathbb{R}^3$ and $V(\mathbf{r}_1, \mathbf{r}_2) = 0$ at a fixed point. Hence each fixed point corresponds to an absolute minimum of V.

This example is related to the coupled springs example in Chapter 4, except now the masses are not constrained to lie in a straight line and are not attached by springs to the wall. The exercises in the this and the next section contain some generalizations and connections to the Chapter 4 example.

Exercises 8.1

1. Show that equations (8.8)-(8.10), for the rates of change of the total linear and angular momenta and position of the center of mass, hold. Use these to show that the Conservation Laws 1(a), 1(b), and 2 hold.

2. (a) Suppose $\mathbf{r} = (\mathbf{r}_1, \ldots, \mathbf{r}_N) : I \to \mathbb{R}^{3N}$ is a system of particles and $\mathbf{a}$ is a point in $\mathbb{R}^3$. Let $\mathbf{r}_i^a(t) = \mathbf{r}_i(t) - \mathbf{a}$, denote the position of the ith particle (body) at time t *relative to* $\mathbf{a}$. The definitions of the total, linear and angular momenta *relative to* $\mathbf{a}$ are

$$\mathbf{P_a}(t) = \sum_{i=1}^{N} m_i \dot{\mathbf{r}}_i^a(t),$$

$$\mathbf{L_a}(t) = \sum_{i=1}^{N} m_i [\mathbf{r}_i^a(t) \times \dot{\mathbf{r}}_i^a(t)].$$

Show that

$$\mathbf{P}_0 = \mathbf{P_a},$$

that is, the total linear momentum of the system relative the origin is the same as that relative to any point $\mathbf{a}$. Also show that

$$\mathbf{L}_0 = \mathbf{L_a} + \mathbf{a} \times \mathbf{P_a}.$$

This relates the total angular momentum about the origin and to that about $\mathbf{a}$ and shows that these are different in general. Under what conditions are they the same ? Describe this geometrically.

(b) Show that there are functions $\mathcal{P}, \mathcal{L}, \mathcal{R}, \mathcal{F}, \mathcal{T} : J \times \mathcal{O} \to \mathbb{R}^3$ such that for any solution $\mathbf{r} : I \to \mathbb{R}^{3N}$ of the equations of motion, the total linear and angular momenta, center of mass, and total force and torque are given by $\mathbf{P}(t) = \mathcal{P}(\mathbf{r}(t), \dot{\mathbf{r}}(t)), \mathbf{L}(t) = \mathcal{L}(\mathbf{r}(t), \dot{\mathbf{r}}(t)), \mathbf{R}(t) = \mathcal{R}(\mathbf{r}(t), \dot{\mathbf{r}}(t)), \mathbf{F}(t) = \mathcal{F}(t, \mathbf{r}(t), \dot{\mathbf{r}}(t))$, and $\mathbf{T}(t) = \mathcal{T}(t, \mathbf{r}(t), \dot{\mathbf{r}}(t))$ respectively. The text uses $\mathbf{P}, \mathbf{L}, \mathbf{R}, \mathbf{F}, \mathbf{T}$ for these respective quantities, which is the usual practice in the physics literature, but $\mathcal{P}, \mathcal{L}, \mathcal{R}, \mathcal{F}, \mathcal{T}$ are sometimes more useful mathematically and help display more clearly how the various quantities depend on the particular motion $\mathbf{r} : I \to \mathbb{R}^{3N}$ of the system.

3. Prove Proposition 8.1.

4. Give an example of a system of forces arising from forces of interaction that satisfies Newton's 3rd law but *not* the strong form of Newton's 3rd law. Compute the total torque for the system.

5. Suppose $\mathbf{r} : I \to \mathbb{R}^{3N}$ is a solution of the equations of motion and $t_1 < t_2 \in I$ are two times. Prove that the work done by the system in going from state

$(\mathbf{r}(t_1), \dot{\mathbf{r}}(t_1))$ to state $(\mathbf{r}(t_2), \dot{\mathbf{r}}(t_2))$ is equal to the difference in the kinetic energy of the system at times t_1 and t_2. Suppose $\mathbf{r}(t_1) = \mathbf{r}(t_2)$, i.e., the system returns to a state where all the positions of the bodies are the same. What can be said about $W(t_1, t_2)$, the work done? With the same assumption, show that if the system of forces is conservative, then the work done is zero.

6. Prove the conservation of energy law for conservative systems. *Note*: As mentioned in the introduction, our approach here is mathematically oriented. Thus, you are not really being asked to prove a physical law, but rather to prove that the mathematical model consisting of the differential equations $m_i \ddot{\mathbf{r}}_i = F_i(t, \mathbf{r}, \dot{\mathbf{r}})$, $i = 1, \ldots, N$, has solutions whose total energy is conserved when the forces are derived from a potential.

7. **(Inverse Power Laws of Attraction)** The usual law of gravitational attraction is an inverse square law, but we could consider an inverse power law for any power p. Thus, suppose that p is any real number and define $F_{ij} : \mathcal{U} \to \mathbb{R}^3$, for $i \neq j$ by

$$F_{ij}(\mathbf{r}) = \frac{Gm_i m_j}{r_{ij}^{p+1}} (\mathbf{r}_j - \mathbf{r}_i), \tag{8.31}$$

where $\mathbf{r} = (\mathbf{r}_1, \ldots, \mathbf{r}_N)$ and $r_{ij} \equiv |\mathbf{r}_i - \mathbf{r}_j|$. Here the domain $\mathcal{U}$ is

$$\mathcal{U} = \{(\mathbf{r}_1, \ldots, \mathbf{r}_N) \in \mathbb{R}^{3N} \mid \mathbf{r}_i \neq \mathbf{r}_j, \forall i \neq j\},$$

if $p > -1$, and

$$\mathcal{U} = \mathbb{R}^{3N},$$

if $p \leq -1$. This gives as large a domain as possible for the forces of interaction. The inverse square law is $p = 2$. The corresponding system of forces for the N-body problem is

$$F_i(\mathbf{r}) = \sum_{j \neq i} \frac{Gm_i m_j}{r_{ij}^{p+1}} (\mathbf{r}_j - \mathbf{r}_i). \tag{8.32}$$

Clearly this system of forces obeys the strong form of Newton's 3rd law and so the total force and torque on the system are zero. In this exercise you are to show, among other things, that this system of forces is conservative (for any value of p). Specifically,

(a) Show that for any $i \neq j$ and any $q \neq 0$,

$$\frac{\partial}{\partial \mathbf{r}_i} (r_{ij}^{-q}) = q\, r_{ij}^{-(q+2)} (\mathbf{r}_j - \mathbf{r}_i).$$

(b) Let $V : \mathcal{U} \to \mathbb{R}$ be the function defined by

$$V(\mathbf{r}) = -\sum_{i<j} \frac{Gm_i m_j}{(p-1) r_{ij}^{p-1}}, \tag{8.33}$$

for $p \neq 1$, and

$$V(\mathbf{r}) = \sum_{i<j} Gm_i m_j \ln(r_{ij}), \qquad (8.34)$$

for $p = 1$. *Note*: The sum here is over all ordered pairs (i,j) with $i < j$. Show that V is a potential for the system of forces, i.e. that

$$\frac{\partial V}{\partial \mathbf{r}_k} = -F_k,$$

for all k, and thus the system of forces is conservative.

8. **(Electrical Attraction and Repulsion Laws)** The model for the force system on charged particles, say protons with positive charge and electrons with negative charge, is entirely similar to that for gravitational forces of attraction. Indeed, now each particle has a charge e_i (which is either positive or negative), in addition to a mass m_i (which is always positive), and because of this there is an electrical interaction between each pair of particles, as well the gravitational interaction. Particles with like charges (either both positive or both negative) repel each other, while particles with opposite charges (one positive, the other negative) attract each other. Coulomb established by direct measurement that the magnitude of the repulsion or attraction is inversely proportional to the square of the distance between the particles. Thus, the *electrical force* E_{ij} exerted on the ith particle by the jth particle is

$$E_{ij}(\mathbf{r}) = -\frac{e_i e_j}{r_{ij}^3}(\mathbf{r}_j - \mathbf{r}_i), \qquad (8.35)$$

and the electrical force on the i particle, due to all the other particles is

$$E_i(\mathbf{r}) = -\sum_{j \neq i} \frac{e_i e_j}{r_{ij}^3}(\mathbf{r}_j - \mathbf{r}_i). \qquad (8.36)$$

If we combine the electrical forces E_i with the gravitational forces F_i from equation (8.12), the equations of motion for a system of charged, massive particles is

$$m_i \ddot{\mathbf{r}}_i = \sum_{j \neq i} \frac{Gm_i m_j - e_i e_j}{r_{ij}^3}(\mathbf{r}_j - \mathbf{r}_i), \qquad (8.37)$$

for $i = 1, \ldots, N$. For this model, do the following:

(a) Explain why the definition of E_{ij} in equation (8.35) says that *like charges repel each other, while unlike charges attract each other.*

(b) Show that the system of forces $\{E_1, \ldots, E_N\}$ is conservative, and therefore when combined with the system of gravitational forces gives a conservative system of forces in the equations of motion (8.37).

(c) The gravitational force is very weak in comparison with the electrical force between two particles. For example, if m is the mass of a proton and e the charge on the proton, then the two constants in the forces of interaction compare approximately as

$$e \cdot e \approx 10^{36} G m \cdot m.$$

This being the case, explain, in a qualitative way, why gravity has such a predominant effect in everyday life. *Hint*: since some of the products $e_i e_j$ in the electrical force can be negative as well as positive, there can be cancelation of electrical forces in the sum over all $j \neq i$ in (8.37).

9. **(3-Body, Hooke)** Suppose in the 3-body equations (8.18)-(8.20), the functions $h_{ij} : (0, \infty) \to \mathbb{R}$ are

$$h_{ij}(\mathbf{r}) = k_{ij}(r_{ij} - L_{ij}),$$

where $k_{ij} \geq 0$ and $L_{ij} > 0$ are constants and, as usual, $r_{ij} = |\mathbf{r}_i - \mathbf{r}_j|$. Also assume the symmetry conditions $k_{ji} = k_{ij}, L_{ji} = L_{ij}$, for all i, j. Then the equations of motion, written out explicitly, are

$$m_1 \ddot{\mathbf{r}}_1 = \frac{k_{12}(r_{12} - L_{12})}{r_{12}}(\mathbf{r}_2 - \mathbf{r}_1) + \frac{k_{13}(r_{13} - L_{13})}{r_{13}}(\mathbf{r}_3 - \mathbf{r}_1) \quad (8.38)$$

$$m_2 \ddot{\mathbf{r}}_2 = \frac{k_{12}(r_{12} - L_{12})}{r_{12}}(\mathbf{r}_1 - \mathbf{r}_2) + \frac{k_{23}(r_{23} - L_{23})}{r_{23}}(\mathbf{r}_3 - \mathbf{r}_2) \quad (8.39)$$

$$m_3 \ddot{\mathbf{r}}_3 = \frac{k_{13}(r_{13} - L_{13})}{r_{13}}(\mathbf{r}_1 - \mathbf{r}_3) + \frac{k_{23}(r_{23} - L_{23})}{r_{23}}(\mathbf{r}_2 - \mathbf{r}_3) \quad (8.40)$$

and the system of forces is easily seen to satisfy the strong form of Newton's 3rd Law. The model here is for three bodies coupled together with three springs having spring constants k_{12}, k_{13}, k_{23} and unstretched lengths L_{12}, L_{13}, L_{23}. See Figure 8.2.

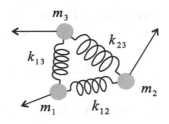

Figure 8.2: *Three bodies coupled together with three springs.*

For this model do the following:

(a) Show that the system of forces is conservative.

(b) Find all the equilibrium points. For this you may assume that $k_{ij} = 1$, for all i, j. Here are some suggestion for working the exercise:

(i) Simplify the work by introducing the following notation:

$$a = \frac{r_{12} - L_{12}}{r_{12}}$$

$$b = \frac{r_{13} - L_{13}}{r_{13}}$$

$$c = \frac{r_{23} - L_{23}}{r_{23}}.$$

Then $(\mathbf{r}_1, \mathbf{r}_2, \mathbf{r}_3)$ is an equilibrium point if

$$a(\mathbf{r}_2 - \mathbf{r}_1) + b(\mathbf{r}_3 - \mathbf{r}_1) = 0$$
$$a(\mathbf{r}_1 - \mathbf{r}_2) + c(\mathbf{r}_3 - \mathbf{r}_2) = 0$$
$$b(\mathbf{r}_1 - \mathbf{r}_3) + c(\mathbf{r}_2 - \mathbf{r}_3) = 0.$$

To simplify this further, assume (without loss of generality) that $\mathbf{r}_2, \mathbf{r}_3$ have the form

$$\mathbf{r}_2 = \mathbf{u}_2 + \mathbf{r}_1$$
$$\mathbf{r}_3 = \mathbf{u}_3 + \mathbf{r}_1,$$

and rewrite the last system of equations in terms of the new unknowns $\mathbf{u}_2, \mathbf{u}_3$. These are known as heliocentric coordinates (see Exercise 6 in Section 2 below).

(ii) Divide into cases depending on whether all, some, or none of a, b, c are zero. The case when $a = 0, b = 0, c = 0$ gives an obvious type of fixed point for the system. Show that the case when $a \neq 0, b \neq 0, c \neq 0$, is the only other case for which there can be fixed points. Further show that in this case, either $\mathbf{u}_2 = 0, \mathbf{u}_3 = 0$, or a, b, c must satisfy

$$ab + ac + bc = 0.$$

Of course, $\mathbf{u}_2 = 0, \mathbf{u}_3 = 0$ is not possible (why ?). So this leaves the above condition on a, b, c and this determines c in terms of a and b. Now find two other equations that a, b must satisfy and solve these to get the values of a, b. You should be led to the conditions

$$2L_{12} + L_{23} > L_{13}$$
$$2L_{13} + L_{23} > L_{12},$$

as necessary and sufficient conditions for existence of fixed points in this case. Interpret this and comment on how the fixed points in this case make sense physically.

10. Suppose $V : \mathcal{U} :\to \mathbb{R}$ is any function on an open set $\mathcal{U} \subseteq \mathbb{R}^{3N}$ and define $\Lambda : \mathcal{U} \times \mathbb{R}^{3N} \to \mathbb{R}$ by

$$\Lambda(\mathbf{r}, \mathbf{v}) = \sum_{i=1}^{N} \tfrac{1}{2} m_i |\mathbf{v}_i|^2 + V(\mathbf{r}).$$

Show that V has a local minimum at $r_* \in \mathcal{U}$ if and only if Λ has a local minimum at $c = (r_*, 0) \in \mathcal{U} \times \mathbb{R}^{3N}$.

11. For $i, j \in \{1, \dots, N\}$, with $i < j$, suppose $h_{ij} : (0, \infty) \to \mathbb{R}$ is a given continuous function. Define $h_{ji} = h_{ij}$. Let

$$\mathcal{U} = \{ (\mathbf{r}_1, \dots, \mathbf{r}_N) \in \mathbb{R}^{3N} \,|\, \mathbf{r}_i \neq \mathbf{r}_j, \forall\, i \neq j \},$$

For each i define $F_i : \mathcal{U} \to \mathbb{R}^3$ by

$$F_i(\mathbf{r}_1, \dots, \mathbf{r}_N) = \sum_{j \neq i} \frac{h_{ij}(r_{ij})}{r_{ij}} (\mathbf{r}_j - \mathbf{r}_i).$$

It is clear that the system of forces $\{F_1, \dots, F_N\}$ arises from forces of inter-action and that it satisfies the strong form of Newton's 3rd Law. Show that this system of forces is conservative.

12. Show that the matrix $\frac{\partial^2 V}{\partial \mathbf{r}_i \partial \mathbf{r}_j}$ defined by equation (8.27) is, in general, for $i \neq j$, not symmetric and

$$\frac{\partial^2 V}{\partial \mathbf{r}_i \partial \mathbf{r}_j} \neq \frac{\partial^2 V}{\partial \mathbf{r}_j \partial \mathbf{r}_i}.$$

Show that, nevertheless, the $\mathcal{H}_V$, defined in terms of these matrices, is indeed a symmetric matrix.

8.2   Euler's Method and the N-body Problem

As a rudimentary first approach to obtaining numerical solutions of the N-body problem, we discuss here Euler's numerical method. This discrete scheme directly exhibits, in its algorithm, how the forces alter the rectilinear motion of each body and the graphical displays of the numerical solutions enable us to understand more about this system of DEs, which is not ana-lytically solvable.

For simplicity, we assume the forces do not depend on the time. The discussion below can easily be altered to include time-dependent forces in the numerical algorithm (exercise).

The Euler method was discussed in general for 1st-order systems in Chapter 2 and here we want to apply it to a 2nd-order system. Thus, we introduce the velocities $\mathbf{v}_i = \dot{\mathbf{r}}_i$, $i = 1, \ldots, N$ and write the system (8.1) as

$$m_i \dot{\mathbf{v}}_i = F_i(\mathbf{r}, \mathbf{v}) \tag{8.41}$$

$$\dot{\mathbf{r}}_i = \mathbf{v}_i. \tag{8.42}$$

To replace this system by a discrete, finite difference system, we divide the time interval $[0, T]$ into K equal subintervals, each of length $h = T/K$. Thus, h is the magnitude of the *time step* and its smallness is critical to the goodness of the approximate solution.

The Euler scheme generates a discrete sequence

$$\mathbf{v}_i^k = (u_i^k, v_i^k, w_i^k)$$
$$\mathbf{r}_i^k = (x_i^k, y_i^k, z_i^k),$$

of approximate positions $\mathbf{r}_i^1, \mathbf{r}_i^2, \ldots, \mathbf{r}_i^K$ and velocities $\mathbf{v}_i^1, \mathbf{v}_i^2, \ldots, \mathbf{v}_i^K$ for the ith body at the times $0, h, 2h, \ldots, Kh = T$. This is accomplished by replacing the derivatives $\dot{\mathbf{v}}_i, \dot{\mathbf{r}}_i$ at time kh by the approximating finite differences $(\mathbf{v}_i^{k+1} - \mathbf{v}_i^k)/h$, $(\mathbf{r}_i^{k+1} - \mathbf{r}_i^k)/h$. Then the discrete analog of the system of DEs (8.41)-(8.42) is the following system of difference equations:

$$m_i \left(\frac{\mathbf{v}_i^{k+1} - \mathbf{v}_i^k}{h} \right) = F_i(\mathbf{r}^k, \mathbf{v}^k) \tag{8.43}$$

$$\frac{\mathbf{r}_i^{k+1} - \mathbf{r}_i^k}{h} = \mathbf{v}_i^{k+1}. \tag{8.44}$$

Here $\mathbf{r}^k = (\mathbf{r}_1^k, \ldots, \mathbf{r}_N^k)$ is the vector of the approximate positions of the N bodies at time kh and $\mathbf{v}^k = (\mathbf{v}_1^k, \ldots, \mathbf{v}_N^k)$ is the vector of approximate velocities.

A *solution* of the finite difference system is a sequence $\{(\mathbf{r}^k, \mathbf{v}^k)\}_{k=1\cdots K}$ of position and velocity vectors that satisfies equations (8.43)-(8.44). Given the initial position and velocity vectors $(\mathbf{r}^0, \mathbf{v}^0)$ for all the bodies, it is easy to "solve" the system (8.43)-(8.44) of difference equations. That is, we can easily manufacture a solution directly from the given initial data. This is so because the system (8.43)-(8.44) can be rewritten in the following more convenient form:

N-Body Euler Algorithm:

$$\mathbf{v}_i^{k+1} = \mathbf{v}_i^k + \frac{h}{m_i} F_i(\mathbf{r}^k, \mathbf{v}^k) \tag{8.45}$$

$$\mathbf{r}_i^{k+1} = \mathbf{r}_i^k + h\mathbf{v}_i^{k+1}, \tag{8.46}$$

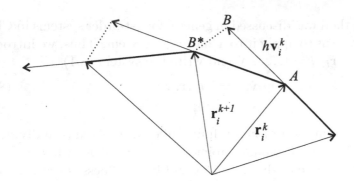

Figure 8.3: *Motion of the ith body governed by the discrete law of motion* (8.47).

for $i = 1, \ldots, N$. This explicitly gives the algorithm for determining the approximate positions and velocities at each of the discrete time steps. Knowing $\mathbf{v}^k, \mathbf{r}^k$ at the kth time step, equations (8.45)-(8.46) allow us to compute $\mathbf{v}^{k+1}, \mathbf{r}^{k+1}$ at the next time step. Note that $\mathbf{v}^{k+1}$ must be computed first in (8.45), so that its value can be used in computing $\mathbf{r}^{k+1}$ in (8.46). Thus, from the initial positions and velocities, we can generate a solution $\{(\mathbf{r}^k, \mathbf{v}^k)\}_{k=1\cdots K}$ of the finite difference system.

It is instructive to note that the numerical algorithm (8.45)-(8.46) can help us understand some of the physics behind Newton's law of motion. This is easy to see if we substitute $\mathbf{v}_i^{k+1}$ from equation (8.45) into equation (8.46) to get

$$\mathbf{r}_i^{k+1} = \mathbf{r}_i^k + h\mathbf{v}_i^k + \frac{h^2}{m_i}F_i(\mathbf{r}^k, \mathbf{v}^k). \qquad (8.47)$$

This gives the ith body's next position in terms of its present position, velocity, and force acting on it. Figure 8.3 shows an interpretation of the motion based on this discrete law of motion. As shown, the ith body is at point A at time k (the kth time step) and, in the absence of any force acting, would move in a straight line in the direction of its present velocity $\mathbf{v}_i^k$ to reach point B at time $k + 1$. However, the small increment of acceleration $\frac{h^2}{m_i}F_i(\mathbf{r}^k, \mathbf{v}^k)$, arising from the force, changes the body's path, causing it to end up at point B^* at time $k + 1$. This continual alteration, by the force, of the straight-line motion of the body is what causes the body's curved trajectory.

Using Maple (or other CASs), it is easy to program the above numerical scheme and produce a graphical display of the positions of the bodies at the discrete time steps. You can find the code for this on the electronic component, which has various Maple worksheets organized according to the type of force system. For example, `gravity2.mws` contains the code for two bodies attracting each other mutually with an inverse power law and with motion in the x-y plane. The worksheet `hooke2.mws` is for a similar model, but with forces arising from Hooke's Law. You will note that there is little difference between the worksheets `gravity2.mws` and `hooke2.mws`, except the nature of the forces. We could have combined these worksheets, along with the worksheets `gravity3.mws` and `hooke3.mws` for three bodies, into one worksheet which could be altered as needed. Or we could have written one interactive program for handling all cases based on the input parameters. However, we have not done this because duplicating the code is no trouble and keeping the cases separate seems to promote clarity and easy of use.

The following examples employ this code, which you will also use in working some of the exercises.

Example 8.3 (2-Body, Inverse Square Law) Figure 8.4 shows the trajectories of two bodies of equal mass under an inverse square law of mutual attraction.

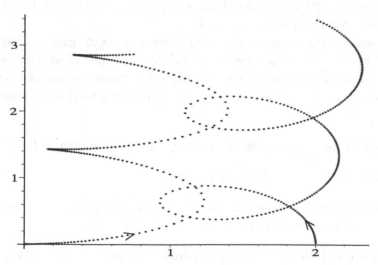

Figure 8.4: *The trajectories of two bodies of equal masses, $m_1 = 1 = m_2$, under an inverse square law of attraction. The initial positions are $\mathbf{r}_1 = (0, 0, 0)$, $\mathbf{r}_2 = (0, 2, 0)$ and the initial velocities are $\mathbf{v}_1 = (0.05, 0, 0)$, $\mathbf{v}_2 = (0.01, 0.5, 0)$. The Euler algorithm was used with $K = 250$ times steps and step size $h = 0.05$*

The analytical, exact solution of the 2-body problem was given in Chapter 2 and it was shown that, with the right initial conditions (like those in the figure), the relative motion of each body is an elliptical orbit about the other. This is not completely evident from Figure 8.4; however, the figure does show the absolute trajectories of both bodies in a fixed coordinate frame. On the other hand, you can perhaps visualize how the relative, elliptical motion given by the analytic solution from Chapter 2, when combined with the uniform motion of the center of mass, will give the absolute trajectories shown in Figure 8.4.

Note also that the initial linear momentum of the system is

$$\mathbf{P}(0) = m_1\mathbf{v}_1 + m_2\mathbf{v}_2 = (0.06, 0.5, 0),$$

while the initial center of mass is $\mathbf{R}(0) = (1, 0, 0)$. Since the system of forces satisfies Newton's 3rd Law, the conservation laws for the total linear momentum: $\mathbf{P}(t) = \mathbf{P}(0)$, and uniform motion of the center of mass: $\mathbf{R}(t) = \mathbf{R}(0) + t\mathbf{P}(0)/M$, hold for the system of DEs. We will see below that the system of *difference* equations also has some of the same conservation laws. In particular, the center of mass in this example moves uniformly along the line through $(1, 0, 0)$ in the direction of $\mathbf{P}(0)/M = (0.03, 0.25, 0)$. This can perhaps be discerned from Figure 8.4.

You can use the code on the worksheet gravity2.mws to look at the details of this motion and view several different types of animations of it. You can also alter this worksheet to study other two-body problems with an inverse power law of attraction. For such studies the worksheet enables you to

- plot the static picture of the motion (like that shown in Figure 8.4),

- produce a movie of the actual motion,

- produce a movie showing the motion of line joining the two bodies along with the motion of the center of mass, and

- produce a static and moving picture of the motion of one body relative to another and display the coordinates versus time and separation versus time of this relative motion.

The code for the Euler algorithm has a sensitivity to numerical error that is most easily discerned from the movie for the relative motion of the situation depicted in Figure 8.4. If the step size is not small enough, the relative

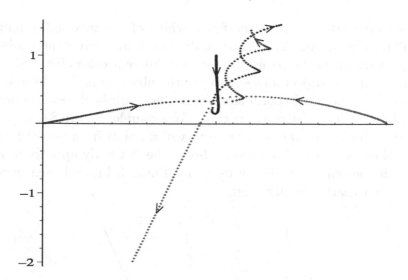

Figure 8.5: *The trajectories of three bodies of equal mass under an inverse square law of attraction. The initial positions are* $\mathbf{r}_1 = (0,0,0), \mathbf{r}_2 = (0,2,0), \mathbf{r}_3 = (1,1,0)$ *and the initial velocities are* $\mathbf{v}_1 = (0.05,0,0), \mathbf{v}_2 = (0.01,0.1,0), \mathbf{v}_3 = (0,-0.2,0)$. *The Euler algorithm was used with* $K = 180$ *time steps and step size* $h = 0.01$.

motion does not appear to be elliptical; indeed, the orbit is not even closed (exercise). However, the theory says that for an inverse square law, and these initial conditions, the relative motion must be elliptical. Thus, a careful choice of step sizes is crucial when using this code.

Example 8.4 (3-body, Inverse Square Law) Unlike the two-body problem, the trajectories of three bodies under an inverse square law of attraction generally will require all of $\mathbb{R}^3$ and are not be confined to a single plane. However, with the right initial conditions, the motion of each body can be made to take place in a single plane, say the x-y plane. This gives what is known as the *planar three-body problem* and even this limited version of the general problem is not solvable analytically. Nevertheless we can use the Maple code on the worksheet `gravity3.mws` to study this planar motion. Figure 8.5 shows the motions (trajectories) of the three bodies for a particular set of initial conditions. As you can see, the three bodies move toward a central region where they interact more strongly and then move off to infinity. Two of the bodies (bodies 1 and 3) orbit each other as they move off to infin-

ity. It is hard to discern from the figure which of the two bodies undergo
this orbital motion, but this can be easily ascertained from the worksheet
`gravity3.mws`, where the paths are assigned different colors for clarity.

While the three trajectories shown are the absolute paths of the bodies
with respect to the fixed coordinate frame, the worksheet also studies the
motion of each body relative to another. For example, Figure 8.6 shows the
trajectory traced out by the vector $\mathbf{r}_3 - \mathbf{r}_1$, which points from the 1st body to
the 3rd. Thus, as viewed from the 1st body, the 3rd body appears to move
in toward it, become captured by its gravitational field, and then begin to
orbit it with a nearly circular orbit.

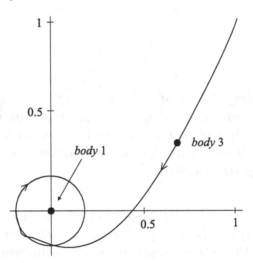

Figure 8.6: *The trajectory of the 3rd body relative to the 1st body.*

The numerical calculations of the relative motion $\mathbf{r}_3 - \mathbf{r}_1$ in the x-y plane
also allow us to view how the x and y coordinates of the relative motion, as
well as the separation $r = \sqrt{x^2 + y^2}$, between the bodies, vary in time. This
is shown in Figure 8.7. The figure actually gives the plots of $x = x^k, y = y^k$,
and $r = r^k$, $k = 1, \ldots, K$, as discrete functions of the time step k. These
approximate the actual continuous graphs at the times $t = kh$, where h is
the step size. Thus, from the figure, we see that the orbital motion begins
at approximately $t = 126 \times 0.01 = 1.26$ time units after the initial time.
The figure also shows the periodic variation in the x and y coordinates
of the relative motion, which always accompanies an orbital motion. The
period of the orbit can be estimated from the figure to be approximately
$T = 35 \times 0.01 = 0.35$ time units.

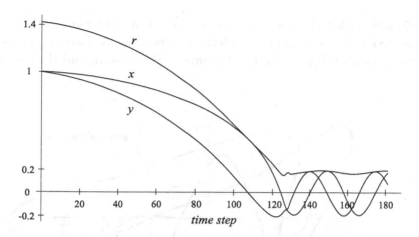

Figure 8.7: *Time variation of the x and y coordinates in the relative motion from Figure 8.6. Also shown is the time variation of the separation $r = \sqrt{x^2 + y^2}$, between the bodies.*

Example 8.5 (2-Body, Modified Hooke Law) A slight generalization of the model for two bodies coupled by a spring is the model with the following equations of motion:

$$m_1 \ddot{\mathbf{r}}_1 = \frac{k(r_{12} - L)}{r_{12}}(\mathbf{r}_2 - \mathbf{r}_1) - a\mathbf{r}_1 \qquad (8.48)$$

$$m_2 \ddot{\mathbf{r}}_2 = \frac{k(r_{12} - L)}{r_{12}}(\mathbf{r}_1 - \mathbf{r}_2) - b\mathbf{r}_2. \qquad (8.49)$$

When the constants a, b are zero, the system is the spring model from Example 8.2. Otherwise, in addition to the spring force, each body experiences a force directed toward the origin and proportional to its distance from the origin. This model will serve to illustrate several mathematical topics, but should not be construed to be a physically relevant model for any system.

In general, except for certain values of a, b, the linear momentum of the system is not conserved and the center of mass does not move in a straight line. Indeed, one can prove that for a, b positive and such that

$$\frac{a}{m_1} = \frac{b}{m_2},$$

the center of mass traces out an elliptical path (exercise). This is shown in Figure 8.8, which was drawn using the Maple code on the worksheet `hooke2.mws`. Other choices of a, b give center-of-mass motions that are more complicated and impossible to determine analytically (see the exercises). The exercises also discuss the fixed points of this system and their stability.

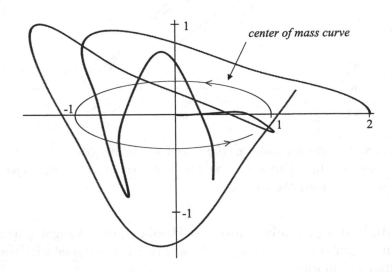

Figure 8.8: *The trajectories of two bodies of equal masses, $m_1 = 1 = m_2$, attached together by a spring and also attracted to the origin. The initial positions are $\mathbf{r}_1 = (0, 0, 0)$, $\mathbf{r}_2 = (0, 2, 0)$ and the initial velocities are $\mathbf{v}_1 = (0.05, 0, 0)$, $\mathbf{v}_2 = (0.01, 0.5, 0)$.*

The worksheet `hooke2.mws` contains an animation of the movement of the line joining the two bodies along with the path traced out by the center of mass. Since the line joining the bodies coincides physically with the spring between the bodies, this movie gives the best understanding of the motion of the system. The spring follows the center of mass, rotating while expanding and contracting during the motion.

The motion of the 2nd body relative to the 1st body is also rather interesting, as shown in Figure 8.9. The figure lends experimental evidence for the conjecture that the relative motion is a closed curve, but this needs to be verified theoretically. Often, as we shall see in the next section, it can be difficult to determine when the relative motion describes a closed a curve.

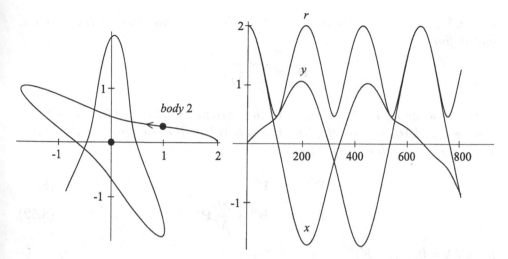

Figure 8.9: **Left:** *Motion of the 2nd body relative to the 1st body.* **Right:** *Plots of the coordinates and the separation in the relative motion as functions of the time step* $k = 1, \ldots, 800$.

8.2.1 Discrete Conservation Laws

In general, there is no reason to expect that the conservation laws for a system of differential equations will also hold, or be present in, a discrete, finite difference version of the system. Indeed, there are usually several ways to build the numerical model out of finite differences and it is often the case that one way preserves more of the physics, i.e., inherits more conservation laws from the original system.

The following theorem shows that the numerical scheme (8.43)-(8.44) for the N-body problem will inherit the laws for conservation of linear and angular momentum as well as the law for the uniform rectilinear motion of the center of mass.

Theorem 8.2 *Suppose that in the N-body system*

$$m_i \ddot{\mathbf{r}}_i = F_i(\mathbf{r}, \dot{\mathbf{r}}), \qquad i = 1, \ldots, N. \tag{8.50}$$

the total force $\mathbf{F} \equiv \sum_{i=1}^{N} F_i$ *is zero. Let* $\mathbf{r}_i^k, \mathbf{v}_i^k, i = 1, \ldots, N, k = 0, 1, \ldots, K$ *be a solution of the system of finite differences (8.43)-(8.44) discussed above. Define*

$$\mathbf{P}^k = \sum_{i=1}^{N} m_i \mathbf{v}_i^k,$$

$k = 0, 1, \ldots, K$, *to be the total, discrete, linear momentum at time step* k, *and define*

$$\mathbf{R}^k = \sum_{i=1}^{N} \frac{m_i}{M} \mathbf{r}_i^k,$$

to be the discrete, center of mass of the system at time step k. *Then* $\mathbf{P}^k$ *is constant and* $\mathbf{R}^k$ *lies on the line through* $\mathbf{R}^0$ *in the direction of* $\mathbf{P}^0$. *More specifically, the following hold:*

$$\mathbf{P}^k = \mathbf{P}^0 \tag{8.51}$$

$$\mathbf{R}^k = \mathbf{R}^0 + \frac{kh}{M}\mathbf{P}^0, \tag{8.52}$$

for all $k = 0, 1, \ldots, K$.

Proof: Since the finite difference solution satisfies equation (8.43), summing both sides if this equation as $i = 1, \ldots, N$ gives

$$\sum_{i=1}^{N} m_i \left(\frac{\mathbf{v}_i^{k+1} - \mathbf{v}_i^k}{h} \right) = \sum_{i=1}^{N} F_i(\mathbf{r}^k, \mathbf{v}^k) = 0.$$

This says, after multiplying through by h and rearranging, that $\mathbf{P}^{k+1} = \mathbf{P}^k$, for all $k = 0, 1, \ldots, K - 1$. This leads directly to identity (8.51).

Next, if we multiply both sides of equation (8.44) by m_i, then sum as $i = 1, \ldots, N$, and use the result just derived, we get

$$\sum_{i=1}^{N} m_i \left(\frac{\mathbf{r}_i^{k+1} - \mathbf{r}_i^k}{h} \right) = \sum_{i=1}^{N} m_i \mathbf{v}_i^{k+1} = \mathbf{P}^{k+1} = \mathbf{P}^0,$$

for each $k = 0, 1, \ldots, K - 1$. Dividing by M, multiplying by h, and rearranging this gives

$$\mathbf{R}^{k+1} = \mathbf{R}^k + \frac{h}{M}\mathbf{P}^0,$$

for $k = 0, 1, \ldots, K - 1$. Using this last identity successively, we get

$$\mathbf{R}^{k+1} = \mathbf{R}^0 + \frac{(k+1)h}{M}\mathbf{P}^0,$$

for $k = 0, 1, \ldots, K - 1$. This proves (8.52). $\square$

Exercises 8.2

1. **(2-Body, Gravity)** Study the motion of the two mutually attracting bodies in Example 8.3 in more detail as follows. Execute the worksheet `gravity2.mws` with the given values and view all the plots and animations. Note that for the given step size $h = 0.05$, the motion of the 2nd body relative to the 1st body is not a closed curve, in particular not an ellipse as the theory predicts it should be. Choose a smaller step size $h = 0.01$ and re-execute the appropriate parts of the code in order to do the following:

   (a) Plot the motion of the 2nd body relative to the 1st body, confirming that it is approximately an ellipse.

   (b) Find (approximately from the plot in part (a)) the lengths of the major and minor axes and the coordinates of the center. Determine the pericenter and apocenter (points of closest and furthest approach of the bodies) and the approximate time at which these occur. Find a good approximation to the period of the relative elliptical motion.

   (c) At what time (or times) does the relative position of the 2nd body make a 45 degree angle with the x-axis.

2. **(2-Body, Gravity)** Consider the two-body problem in Example 8.3 and on worksheet `gravity2.mws`, with the same initial data and parameters, but now with $v_2 = (-0.3, 0.5)$. Do a complete study of the new motion of the system, using the discussion in Example 8.3 and worksheet `gravity2.mws` as a guide. Also extend the study as in Exercise 1, determining the extra information about the relative motion. Be sure to include plots of (1) the absolute trajectories of the bodies, (2) the path traced out by the center of mass, (3) the path described by the motion of the 2nd body relative to the 1st body, and (4) the x and y coordinates and separation of the bodies in the relative motion as functions of the time step.

3. **(2-Body, Gravity)** This problem is designed to develop some experience with how the initial conditions and the masses affect the motion of two bodies under an inverse square law of attraction. As in Example 8.3 and on worksheet `gravity2.mws`, take $G = 1$, the initial positions $r_1 = (0, 0, 0), r_2 = (2, 0, 0)$, and the initial velocity of the 1st body to be $v_1 = (0.05, 0, 0)$. The rest of the data is as follows:

   (a) Take the masses to be $m_1 = 1, m_2 = 1$ and the initial velocity of the 2nd body to be $v_2 = (0.01, b, 0)$, with b varying between $b = 1.3$ and $b = 2$. The value used in Example 8.3, was $b = 0.5$ and clearly gives trajectories where the bodies orbit each other. You should find that the bodies still orbit each other for $b = 1.3$, but that for $b = 2$ they do not appear to do so. Try to determine experimentally the critical value b_0 such that if $b \leq b_0$, the bodies orbit each other, and for $b > b_0$, the bodies

do not orbit each other. Also using the theory that was discussed in
Chapter 2, determine the *exact* value of b_0. Produce some static plots of
the absolute trajectories and relative motions to document your studies.
CAUTION: You might have to use a large number of time steps to see
how the motion goes, so beware of memory overload if you want to look
at animations of the motion. You can do the whole exercise without
looking at the animations, but viewing them can add to your experience
with how bodies behave under the attraction of gravity.

(b) Take the masses to be $m_1 = 4, m_2 = 1$, and the initial velocity of the
2nd body to be $\mathbf{v}_2 = (0.01, b, 0)$. Do a study like that in part (a). *Note*:
Use the default step size $h = 0.05$ on worksheet gravity2.mws for the
first plot. This will give trajectories that are *not* accurate. So use a
smaller step size, say $h = 0.01$. Compare the results here with those in
part (a) and comment on how the larger value for m_1 affects the results.

4. **(3-Body, Gravity)** Consider the 3-body problem with inverse square law
of attraction. Assume $G = 1$, all the masses are the same, say $m_1 = m_2 = m_3 = 1$, and the initial positions of the three bodies are the vertices of an
equilateral triangle, say $\mathbf{r}_1 = (0,0,0), \mathbf{r}_2 = (2,0,0)$, and $\mathbf{r}_3 = (1, \sqrt{3}, 0)$.
Suppose the initial velocities have the form

$$\begin{aligned}
\mathbf{v}_1 &= (b,0,0) \\
\mathbf{v}_2 &= (0,b,0) \\
\mathbf{v}_3 &= (-b,0,0).
\end{aligned}$$

Do a numerical study of the motion of this system for the following special
values of b.

(a) For $b = 0$, all three bodies collide after a finite amount of time T.
Verify this numerically and approximate the time of collision T. For this
use the code on the worksheet gravity3.mws, making the appropriate
alterations of the initial conditions. You will have to experiment to
determine good values to use for the step size h and number of time
steps nt. Do the bodies seem to "survive" the collision in the numerical
simulation, i.e., emerge and speed away from the collision point? How
can you explain this? Does this happen theoretically? For extra credit
prove, theoretically, that the bodies collide after a finite amount of time.

(b) For $b = 0.1$ and $b = 0.2$, show, numerically, that the bodies do not
collide, but after "near" misses, each body runs off to infinity. Estimate
the time of interaction, that is, the time before they are all separated
by more than, say, two units.

(c) For $b = 0.5$, show numerically that all three bodies initially orbit one
another. For this take the step size to be $h = 0.1$ and the number of
time steps to be $nt = 1000$. Next, for $nt = 2000$ and $nt = 2200$ and step

size $h = 0.005$ (which give the same and slightly longer time intervals, but greater accuracy) determine if two or more of the bodies appear to collide. Numerically show that is *not* the case by using $h = 0.001$ and $nt = 10,000$ and then $nt = 11,000$. CAUTION: This requires a decent computer. Show that two bodies leave the interaction orbiting each other while the third body goes off to infinity. Plot the relative motions.

5. **(3-Body, Gravity)** Consider the 3-body problem with inverse square law of attraction. Assume $G = 1$, and the initial positions of the bodies are $r_1 = (0,0,0), r_2 = (2,0,0), r_3 = (1,1,0)$. This exercise studies the nature of several motions when two of the bodies have equal mass, say $m_2 = 1 = m_3$, and the mass of the other body is relatively large, say $m_1 = 500$. This is a model for two small planets orbiting a large star, or sun. The theory (see Exercise 6 below) says that we can almost treat the three-body problem as if it decomposed into a couple of central force problems, one for each planet orbiting the sun (which serves as the central of force). The numerical studies should lend credence to this idea. The theory depends on how large the magnitude of m_1 is. So the exercise looks at two values, $m_1 = 50$ and $m_1 = 500$.

(a) Suppose $m_1 = 50$ and the initial velocities are $v_1 = (0,0,0), v_2 = (0,-5,0)$, and $v_3 = (-5,0,0)$. Use a step size $h = 0.001$ and number of time steps $nt = 2800$ (or more if you wish) to do a numerical study of the motions of the three bodies.

    (i) Plot the trajectories of the three bodies, all in the same figure. The motion of the sun should be a small, wiggly (but almost straight) line, while the motion of each planet should appear to be on a "precessing" ellipse with one focus on the moving sun.

    (ii) Plot the motion of each planet relative to the sun. Each of these should appear to be an ellipse. Print each out (be sure to use the 1-1 scale, so that angles are not distorted) and locate approximately the center, the major axis, the minor axis, and the angle δ that the major axis makes with the x-axis. Measure the appropriate quantities and write an equation for the ellipse (in polar coordinates). Approximate the closest and farthest approach of each planet to the sun. Determine the approximate period of each planet (the time for one orbit about the sun. Document your work with printouts of various plots and graphs, all of which should be properly annotated.

(b) Suppose $m_1 = 500$ and the initial velocities are $v_1 = (0,0,0), v_2 = (0,-20,0)$, and $v_3 = (-20,0,0)$. Use a step size $h = 0.001$ and number of time steps $nt = 2800$ (or more if you wish) to do a numerical study like that in part (a) above.

(c) Compare and contrast the results in parts (a) and (b).

6. (Heliocentric Coordinates) Consider a general type of N-body problem of the form

$$m_i\ddot{\mathbf{r}}_i = \sum_{j \neq i} h_{ij}(\mathbf{r}_j - \mathbf{r}_i) \qquad (i = 1, \ldots, N), \qquad (8.53)$$

where $h_{ij} = h_{ij}(r_{ij})$ is a function that depends only on the distance $r_{ij} = |\mathbf{r}_i - \mathbf{r}_j|$, between the ith and jth bodies. Let $\mathbf{u}_2, \ldots, \mathbf{u}_N$ denote the positions of bodies $2, \ldots, N$ relative to body 1, i.e., let

$$\mathbf{u}_i \equiv \mathbf{r}_i - \mathbf{r}_1 \qquad (i = 2, \ldots, N).$$

Considering $\mathbf{r}_1, \mathbf{u}_2, \ldots, \mathbf{u}_N$ as the basic unknowns for the problem, show that the equations of motion (8.53) can be rewritten in the following form:

Heliocentric Equations of Motion:

$$\ddot{\mathbf{r}}_1 = \sum_{j=2}^{N} \frac{h_{1j}}{m_1} \mathbf{u}_j \qquad (8.54)$$

$$\ddot{\mathbf{u}}_i = -\left(\frac{m_1 + m_i}{m_1 m_i}\right) h_{1i} \mathbf{u}_i + \sum_{j \in \{2,\ldots,N\}\setminus\{i\}} \left[\frac{h_{ij}}{m_i}(\mathbf{u}_j - \mathbf{u}_i) - \frac{h_{1j}}{m_1}\mathbf{u}_j\right]$$

for $i = 2, \ldots, N$. Here $h_{1j} = h_{1j}(u_j)$ and $h_{ij} = h_{ij}(u_{ij})$, for $i \neq j, j \neq 1$, are functions of the distances $u_j = |\mathbf{u}_j|$ and $u_{ij} = |\mathbf{u}_i - \mathbf{u}_j|$. These equations are often called the equations of motion in *heliocentric coordinates*. The name comes from our solar system where the 1st body (with position $\mathbf{r}_1$) represents the sun. In general, equations (8.54) are the most convenient form of the equations of motion when the mass of the 1st body is overwhelmingly larger than the masses of all the other bodies: $m_1 \gg m_i$, for $i = 2, \ldots, N$. To see this do the following:

(a) Write out equations (8.54) explicitly, simplifying where possible, for the case of gravity $h_{ij} = Gm_im_j/r_{ij}^3$.

(b) Show that for $m_i = 0, i = 2, \ldots, N$, the heliocentric equations of motion reduce to the equations

$$\ddot{\mathbf{r}}_1 = 0$$

$$\ddot{\mathbf{u}}_i = -\frac{Gm_1}{u_i^3} \mathbf{u}_i,$$

for $i = 2, \ldots, N$. This system is the idealized limit of a system (like our solar system) where the masses $m_2, \ldots, m_N$ are very small relative to m_1. Discuss how the idealized system can be completely solved and what types of solutions it has.

7. **(2-Body, Modified Hooke)** This exercise studies the system

$$m_1 \ddot{\mathbf{r}}_1 = \frac{k(r_{12} - L)}{r_{12}}(\mathbf{r}_2 - \mathbf{r}_1) - a\mathbf{r}_1$$

$$m_2 \ddot{\mathbf{r}}_2 = \frac{k(r_{12} - L)}{r_{12}}(\mathbf{r}_1 - \mathbf{r}_2) - b\mathbf{r}_2.$$

from Example 8.5.

(a) Determine all the equilibrium points of the system. *Suggestions*: Supposing that $(\mathbf{r}_1, \mathbf{r}_2)$ is an equilibrium point, let

$$c = \frac{k(r_{12} - L)}{r_{12}}.$$

Assuming that $a + b \neq 0$, show first that

$$c = -\frac{ab}{a + b}$$

and then use this to show that

$$r_{12} = \frac{kL(a + b)}{ab + k(a + b)}.$$

Now use this to construct the equilibrium points. *Also*: Comment on the configuration of the masses that gives the equilibrium points (is the spring extended or compressed?) and argue that such a configuration seems physically reasonable.

(b) Show that the system of forces in this model is conservative. Are there choices of a, b for which the total force is zero ? Does the system of forces, for some choices of a, b satisfy Newton's 3rd Law (either weak or strong form)?

(c) Suppose a, b are such that

$$\frac{a}{m_1} = \frac{b}{m_2}.$$

Call this ratio q. If q is positive let $q = \omega^2$, and otherwise let $q = -\omega^2$. Show that the position vector $\mathbf{R}$ for the center of mass has the form

$$\mathbf{R} = \mathbf{A}\cos\omega t + \mathbf{B}\sin\omega t,$$

when q is positive and otherwise it has the form

$$\mathbf{R} = \mathbf{A}\cosh\omega t + \mathbf{B}\sinh\omega t,$$

when q is negative. Here $\mathbf{A}, \mathbf{B}$ are constant vectors. Relate these vectors to the initial data for the center of mass and then describe the curve (trajectory) for the center of mass. Next let $\mathbf{r} = \mathbf{r}_2 - \mathbf{r}_1$, be the position vector of the second mass relative to the first. Show that the relative motion is governed by the equation

$$\ddot{\mathbf{r}} = -\left[q + \frac{sk}{r}(r - L)\right]\mathbf{r},$$

where $s = (m_1 + m_2)/m_1 m_2$ and $r = |\mathbf{r}|$.

(d) Use the Maple code on the worksheet hooke2.mws to study the motion of the masses and their center of mass in each of the following choices of the parameters: $(a, b) = (0.5, 0.5), (a, b) = (0.1, 0.5), (a, b) = (0.1, -0.5), (a, b) = (-0.1, 0.5)$, and $(a, b) = (-0.5, 0.1)$. In each case take the masses to be $m_1 = 1 = m_2$, the spring constant $k = 2$, and the initial conditions: $\mathbf{r}_1(0) = (0, 0), \mathbf{r}_2(0) = (2, 0), \dot{\mathbf{r}}_1(0) = (0.01, 0), \dot{\mathbf{r}}_2(0) = (0.01, 0.5)$.

8. **(3-Body, Modified Hooke)** As in the last exercise, modify the model for the three masses coupled together with springs by adding attractive forces toward the origin to each equation. The resulting model is

$$m_1\ddot{\mathbf{r}}_1 = \frac{k(r_{12} - L)}{r_{12}}(\mathbf{r}_2 - \mathbf{r}_1) + \frac{k(r_{13} - L)}{r_{13}}(\mathbf{r}_3 - \mathbf{r}_1) - a\mathbf{r}_1$$

$$m_2\ddot{\mathbf{r}}_2 = \frac{k(r_{12} - L)}{r_{12}}(\mathbf{r}_1 - \mathbf{r}_2) + \frac{k(r_{23} - L)}{r_{23}}(\mathbf{r}_3 - \mathbf{r}_2) - b\mathbf{r}_2$$

$$m_3\ddot{\mathbf{r}}_3 = \frac{k(r_{13} - L)}{r_{13}}(\mathbf{r}_1 - \mathbf{r}_3) + \frac{k(r_{23} - L)}{r_{23}}(\mathbf{r}_2 - \mathbf{r}_3) - c\mathbf{r}_3$$

Study this system as you did in Exercise 7. In particular, find conditions on m_1, m_2, m_3, a, b, c, so that you can exactly determine the position vector $\mathbf{R}$ for the center of mass. Determining the fixed points is considerably more difficult here and so this is left open-ended (do as much on it as you can). Likewise, you can choose your own initial data for whatever numerical studies you wish to do. The worksheet hooke3.mws can be used for this.

9. **(Two Coupled Masses)** Consider the system of two masses, coupled to each other by a spring and also having each mass attached to a given point by a spring. See Figure 8.10. The equations of motion for this system are

$$m_1\ddot{\mathbf{r}}_1 = \frac{k_0(r_{01} - L)}{r_{01}}(\mathbf{r}_1 - \mathbf{r}_0) + \frac{k_1(r_{12} - L)}{r_{12}}(\mathbf{r}_2 - \mathbf{r}_1)$$

$$m_2\ddot{\mathbf{r}}_2 = \frac{k_1(r_{12} - L)}{r_{12}}(\mathbf{r}_1 - \mathbf{r}_2) + \frac{k_2(r_{23} - L)}{r_{23}}(\mathbf{r}_3 - \mathbf{r}_2).$$

Here $\mathbf{r}_0, \mathbf{r}_3$ are the position vectors of the two given points to which the masses m_1, m_2 are attached, respectively. For this system do the following:

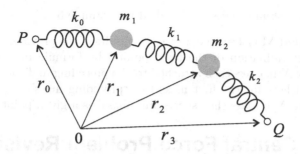

Figure 8.10: *A system of two masses attached to each other by a spring and also attached to given points P, Q with springs.*

(a) Show that the system of forces is conservative but does not satisfy Newton's 3rd Law. *Note:* Be sure to remember that $\mathbf{r}_0, \mathbf{r}_3$ are given vectors (constants) and are not variable.

(b) Do a study of the fixed points (equilibrium points) and their stability. For this, first show that, without loss of generality, one can assume that $\mathbf{r}_0 = 0$ (*Hint:* Let $\mathbf{r}_i = \mathbf{u}_i + \mathbf{r}_0$, $i = 1, 2$). Also, if you wish, you can restrict yourself to the special case when all the spring constants are the same.

(c) Suppose $\mathbf{r}_0 = 0$ and $\mathbf{r}_3 = (3L, 0, 0)$ and that the initial conditions are such that the motion takes place along the x-axis, i.e., $\mathbf{r}_1 = (x_1, 0, 0)$, $\mathbf{r}_2 = (x_2, 0, 0)$, for all time. See Figure 8.11.

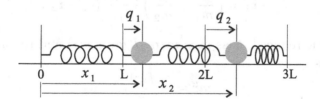

Figure 8.11: *A motion of the masses along the x-axis.*

Let $q_1 \equiv x_1 - L$, $q_2 \equiv x_2 - 2L$, denote the deviations from the equilibrium positions. Write out explicitly the system of DEs for q_1, q_2. Under the assumption that the deviations are small:

$$|q_1(t)| < L, \qquad |q_2(t)| < L,$$

for all t, show the deviations also satisfy the following system of DEs:

$$m_1\ddot{q}_1 = -k_0 q_1 + k_1(q_2 - q_1) \qquad (8.55)$$
$$m_2\ddot{q}_2 = -k_1(q_2 - q_1) - k_2 q_2. \qquad (8.56)$$

This system of DEs was studied in Example 5.7.

10. **(N Coupled Masses)** Generalize the results of the previous exercise (except for the determination of the fixed points) by formulating the equations of motion for N masses, each attached to the preceding and succeeding mass by a spring, where for the first mass the "preceding mass" is a given point $\mathbf{r}_0$ and for the Nth mass the "succeeding mass" is a given point $\mathbf{r}_{N+1}$.

8.3   The Central Force Problem Revisited

We return to the two-body problem

$$m_1\ddot{\mathbf{r}}_1 = \frac{f(r_{12})}{r_{12}}(\mathbf{r}_2 - \mathbf{r}_1) \tag{8.57}$$

$$m_2\ddot{\mathbf{r}}_2 = \frac{f(r_{12})}{r_{12}}(\mathbf{r}_1 - \mathbf{r}_2), \tag{8.58}$$

which we studied in Chapter 2 for an inverse square law of attraction $f(r) = k/r^2$, with some extensions and generalizations mentioned in the exercises. In essence the solution of this system of DEs reduces to the solution of the system for one body when we introduce the vector $\mathbf{r} = \mathbf{r}_2 - \mathbf{r}_1$ for the position of the 2nd body relative to the 1st body. Dividing equations (8.57)-(8.58) by m_1, m_2, respectively, and then subtracting, we get

$$\ddot{\mathbf{r}}_2 - \ddot{\mathbf{r}}_1 = \left(\frac{1}{m_1} + \frac{1}{m_2}\right)\frac{f(r_{12})}{r_{12}}(\mathbf{r}_1 - \mathbf{r}_2).$$

Introducing the reduced mass $m = m_1 m_2/(m_1 + m_2)$ and rewriting the last equation gives

$$m\ddot{\mathbf{r}} = -\frac{f(r)}{r}\mathbf{r},$$

which models the motion of a single body of mass m, attracted (or repelled) from the origin by a central force of magnitude $|f(r)|$.

The interest in the two-body problem, and its resolution via its reduction to a central force problem, is based on several factors. First, it is the only case of the N-body problem that we can completely "solve" by explicit computation of certain integrals. Second, the techniques involved in the resolution of the two-body problem help us understand the complexity of the problem for $N > 2$ bodies. For example, suppose the masses $m_2, m_3, \ldots, m_N$ are all quite small relative to the mass m_1 of the 1st body. By introducing heliocentric coordinates: $\mathbf{u}_i = \mathbf{r}_i - \mathbf{r}_1$, $i = 2, \ldots, N$, rewriting the N-body equations for, say, the inverse square law, and taking the limit $m_i \to 0$, $i = 2, \ldots, N$, we get the equations

$$\ddot{\mathbf{r}}_1 = 0$$

$$\ddot{\mathbf{u}}_i = -\frac{Gm_1}{u_i^3}\mathbf{u}_i \qquad (i = 2, \ldots, N)$$

(see Exercise 6 in the last section). Thus, the N-body problem is approximated by this system consisting of $N - 1$ separate central force problems. For this, and other reasons as well, it will be valuable to examine the central force problem again, but now in more detail.

As we saw in Chapter 2, the form of the central force equation

$$m\ddot{\mathbf{r}} = -\frac{f(r)}{r}\mathbf{r}, \qquad (8.59)$$

leads to the result that the angular momentum $\mathbf{L} = \mathbf{r} \times m\dot{\mathbf{r}}$ is constant, and thus the motion takes place in the plane perpendicular to $\mathbf{L}$. We assume, without loss of generality, that this plane is the x-y plane (exercise). Then the central force equation (8.59) for $\mathbf{r} = (x, y)$ is the system

$$m\ddot{x} = -f(r)x/r$$
$$m\ddot{y} = -f(r)y/r,$$

with $r = (x^2 + y^2)^{1/2}$. Transforming to polar coordinates: $x = r\cos\theta$, $y = r\sin\theta$, gives the central force equation in polar coordinates:

Central Force Equations:

$$m\ddot{r} = mr\dot{\theta}^2 - f(r) \qquad (8.60)$$
$$2\dot{r}\dot{\theta} + r\ddot{\theta} = 0. \qquad (8.61)$$

The 2nd equation here leads to Kepler's Second Law: $r^2\dot{\theta} = c$, where c is a constant. Because of the assumption about the motion being in the x-y plane, it is easy to see that $\mathbf{L} = (0, 0, mc)$, which gives the interpretation of mc as the third component of the angular momentum.

Now we take a slightly different route than in Chapter 2 for the analysis of this polar coordinate system. In essence we "decouple" the system of DEs to get a separate DE for r that does not involve θ. For this we replace the 2nd equation (8.61) by Kepler's law, $\dot{\theta} = c/r^2$, which is its consequence, and also use Kepler's law to eliminate $\dot{\theta}$ in the 1st equation (8.60). Thus, solution of the central force problem reduces to the solution of the system

Decoupled Central Force Equations:

$$m\ddot{r} = \frac{mc^2}{r^3} - f(r) \tag{8.62}$$

$$\dot{\theta} = \frac{c}{r^2}. \tag{8.63}$$

It is important to note, before getting lost in the details below, that the central force equations (8.62)-(8.63) involve the angular velocity c as a parameter. Thus, all the ensuing results implicitly involve c. Further, the appearance of c makes the system (8.62)-(8.63), different than the system (8.60)-(8.61), which does not involve c. The initial conditions $r(0) = r_0, \dot{r}(0) = \dot{r}_0, \theta(0) = \theta_0$, and $\dot{\theta}(0) = \dot{\theta}_0$, determine a unique solution $t \mapsto (r(t), \theta(t))$, of the system (8.60)-(8.61), and this is also a solution of the system (8.62)-(8.63), *provided* we take $c = r_0^2 \dot{\theta}_0$.

Also bear in mind that a solution $r = r(t)$ of equation (8.62), gives the radial variation, or distance from the origin, as a function of time, which is shown in Figure 8.12. The differential equation (8.63) for the angular deviation $\theta = \theta(t)$ from the x-axis, says that this angle is increasing if $c > 0$ (the orbit is counterclockwise) and is decreasing if $c < 0$ (the orbit is clockwise).

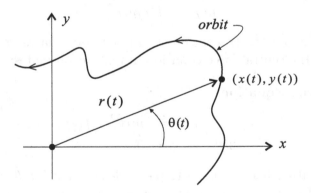

Figure 8.12: *Geometric meaning of the time variation of the radial distance r and angular deviation θ.*

Since the first equation, equation (8.62), in the central force equations does not involve θ, its solutions can be analyzed separately. Each solution $r = r(t)$, as we shall see below, can be computed formally by integrating twice and then inverting the equation found from the second integral. The first integral used in constructing the solution is a special case of the more general concept of *first integrals*, or conservation laws for differential equations (see Chapter 7). In general, two integrals are needed to construct a solution of a 2nd-order DE and this process historically came to be known as constructing solutions by *quadratures*.

Knowing a solution r of the first equation allows us to determine θ from the second equation simply by integration:

$$\theta = \int \frac{c}{r(t)^2} \, dt.$$

This integral and the ones needed to find r explicitly are, in general, not able to be calculated in closed form. However, some particular choices of f, such as an inverse square or inverse cube, give integrals that are computable using various integration techniques. Nevertheless, we are still able to draw a number of important conclusions about the general case when f is arbitrary. The remainder of the section is devoted to this, with particular cases left to the exercises.

8.3.1 Effective Potentials

In the system (8.62)-(8.63), we rewrite the first equation,

$$m\ddot{r} = \frac{mc^2}{r^3} - f(r),$$

so that the right-hand side is an exact derivative. We can do this since $f : (0, \infty) \to \mathbb{R}$ is C^1, and so it has an antiderivative g,

$$g' = f,$$

on $(0, \infty)$, for some C^2 function $g : (0, \infty) \to \mathbb{R}$. Using this and the standard antiderivative of mc^2/r^3, we get a function $V : (0, \infty) \to \mathbb{R}$, defined by

$$V(r) = \frac{mc^2}{2r^2} + g(r), \tag{8.64}$$

called an *effective potential* for the central force problem. The reason for this designation is that, by definition,

$$-V'(r) = \frac{mc^2}{r^3} - f(r),$$

and so equation (8.62) can be written as

$$m\ddot{r} = -V'(r). \tag{8.65}$$

Note that V implicitly involves the parameter c (as mentioned in the prior discussion). V is not a "true" potential for the central force since it involves the term $mc^2/(2r^2)$, which corresponds to an additional centrifugal force. However, equation (8.65) can be viewed as a differential equation for a one-dimensional conservative system. Then, as its name suggests, V effectively gives us a conservation law (or first integral) that each solution $r : I \to \mathbb{R}$ of equation (8.65) must satisfy. As in the customary conservation of energy law, this law is derived by multiplying each side of equation (8.65) by $\dot{r}$ to get

$$m\dot{r}\ddot{r} = -V'(r)\dot{r},$$

or

$$\frac{d}{dt}\left(\tfrac{1}{2}m\dot{r}^2\right) = \frac{d}{dt}\left(-V(r)\right).$$

Since this holds for $t \in I$, we get that there exists a constant E such that

$$\tfrac{1}{2}m\dot{r}^2 = E - V(r), \tag{8.66}$$

for all $t \in I$. This conservation law provides two avenues of approach to further analysis of the solution r of the radial differential equation. The first avenue is a qualitative analysis of the integral curves, examination of fixed points, and stability. We do this next and then pursue the second avenue which consists of integrating the DE (8.66) one more time to obtain an implicit relation between r and t.

8.3.2 Qualitative Analysis

We consider here the qualitative information about the solutions that arises because the DE $m\ddot{r} = -V'(r)$ has the conservation law (8.66). Indeed, we show how to sketch, by hand, the phase portrait for the corresponding 1st-order system directly from a sketch of the effective potential V.

The corresponding 1st-order system is

$$\dot{r} = p/m \tag{8.67}$$
$$\dot{p} = -V'(r), \tag{8.68}$$

and because of the conservation law (8.66), the integral curves $t \mapsto (r(t), p(t))$ of the 1st-order system lie on one of the curves in the r-p plane with the equation

$$\frac{p^2}{2m} + V(r) = E,$$

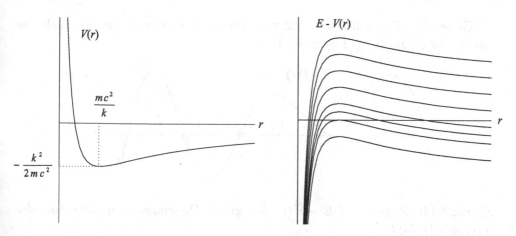

Figure 8.13: *Graph of an effective potential V and the graphs of $E - V(r)$ for various values of E.*

for some value of E. These are called the (*effective*) *energy curves* for the system and a plot of a number of them for various values of E will give the phase portrait for the system. While the plot can be done by a computer, a quick, rough sketch can always be done by hand using only a plot of the graph of V and some additional information that comes from the form of the equation for the energy curves.

For example, suppose the graph of a typical effective potential is as shown in Figure 8.13. The figure is for the effective potential $V(r) = mc^2/2r^2 - k/r$ corresponding to an inverse square law. Many other laws of attraction have effective potentials with graphs similar to this (see the exercises). Also shown in the figure are the graphs of $E - V(r)$, for various values of E. These latter graphs can be used to obtain the corresponding energy curves. Just note that each energy curve splits into the graphs of two functions of r,

$$p = \pm\sqrt{2m}\sqrt{E - V(r)},$$

one for each choice of $\pm$. The domains $\mathcal{D}_E = \{ r \mid E - V(r) \geq 0 \}$, for each of these functions depends on the value of E and can be discerned from the graphs of $E - V(r)$. For example, from Figure 8.13, we see that the domain $\mathcal{D}_E$ (for the type of potential shown) is either a bounded interval $[r_1, r_2]$ or an unbounded interval $[r_1, \infty)$. The degenerate cases are $\mathcal{D}_E = \{mc^2/k\}$, a single point, when $E = -k^2/(2mc^2)$, and $\mathcal{D}_E = \emptyset$, the empty set, when $E < -k^2/(2mc^2)$.

Generally, $\mathcal{D}_E$ will consist of the union of finitely many intervals, as shown, for example, in Figure 8.14.

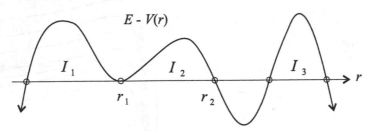

Figure 8.14: *A graph of $E - V(r)$ for which $\mathcal{D}_E$ consists of three bounded intervals I_1, I_2, I_3.*

The graph of $\sqrt{2m}\sqrt{E - V(r)}$ on its domain is similar to the graph of $E - V(r)$. Just delete the parts of the graph of $E - V(r)$ where $E - V(r)$ is negative and on the intervals $[r_1, r_2]$ where $E - V(r) \geq 0$, adjust the shape of the graph slightly to get the graph of $\sqrt{2m}\sqrt{E - V(r)}$. The adjustment is partly due to that fact that the square root decreases or increases the magnitude of a number depending on whether it is larger or smaller than 1. Additional adjustment is a *possible* rounding of the corners of the graph of $E - V(r)$ at one or the other of the endpoints r_1, r_2 of the interval. The rounding occurs when $\sqrt{2m}\sqrt{E - V(r)}$ has vertical tangents at r_1 or r_2. To see this note that since $E - V(r) = 0$ for $r = r_1, r_2$ and

$$\frac{d}{dr}\sqrt{2m}\sqrt{E - V(r)} = \frac{-\sqrt{2m}\,V'(r)}{2\sqrt{E - V(r)}},$$

a vertical tangent occurs at r_1 only if $V'(r_1) \neq 0$, or at r_2, only if $V'(r_2) \neq 0$, that is, only if $(r_1, 0)$ or $(r_2, 0)$ is *not* a fixed point of the system. The endpoints where $\sqrt{2m}\sqrt{E - V(r)}$ has a vertical tangents are called *turning points* because of their physical significance in the central force motion (as we shall see).

Having the graph of $\sqrt{2m}\sqrt{E - V(r)}$, the graph of $-\sqrt{2m}\sqrt{E - V(r)}$ is, of course, obtained by reflection about the x-axis. Using these observations, we can easily sketch the energy curves. For the effective potential in Figure 8.13, the energy curves obtained by this method are shown in Figure 8.15. The sketch gives a picture of the phase portrait for the system $\dot{r} = p$, $\dot{p} = -V'(r)$, since each integral curve of the system lies on one of the energy curves shown. In this example, we see that there is a fixed point $(r_*, 0)$

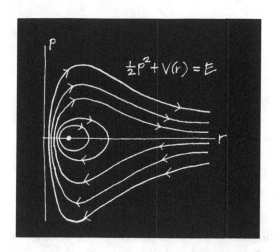

Figure 8.15: *A sketch of the energy curves $p^2/2m + V(r) = E$, for various values of E. Here V has graph as shown in Figure 8.13 and $m = 1$.*

corresponding to the critical point r_* of the effective potential V. The critical point is where the potential energy is a *minimum* and the corresponding fixed point is a center.

Figure 8.15 shows that some of the energy curves are closed (and bounded), while others are not (and are unbounded). In terms of central force motion, this indicates that if the initial radial distance r_0 and radial velocity $p_0 = \dot{r}_0$ are such that $E = m\dot{r}_0^2/2 + V(r_0)$ corresponds to one of the closed energy curves, then $r = r(t)$ will periodically oscillate between a minimum distance r_1 and a maximum distance r_2 from the origin. (We will prove that this is indeed the case in the theorem below.) Note also the direction marked on the energy curve. Thus, the corresponding orbit has the form shown in Figure 8.16.

The figure is for initial data $r_0 = r_1, \dot{r}_0 = 0$, so that the radius increases from an initial minimum value r_1, and for $c > 0$ (equivalently $\dot{\theta}_0 > 0$), so the orbit is counterclockwise. The figure indicates the orbit does not close on itself when $\theta = 2\pi$, but does not rule out the possibility that the orbit will eventually close up. Below we will derive a condition that tells us precisely when the orbit is closed or not. Note also that for an inverse square law, the orbits are conic sections (as shown in Chapter 2) and so the ones corresponding to closed energy curves must be ellipses (and close upon themselves after one revolution).

The situation discussed here is typical in general (assuming the potential

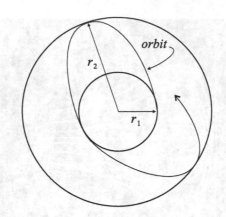

Figure 8.16: *The actual orbit corresponding to a solution $r = r(t)$ of (8.62), when $(r(t), m\dot{r}(t))$ lies on a closed energy curve.*

V is not too wild). The plots of the effective energy curves give most of the essential, qualitative information about the orbits in the central force problem. The angular variation θ along an orbit is determined by equation (8.63), which says that θ is either increasing or decreasing depending on the sign of c. However, to make these observations and assertions rigorous, we need to look at some of the analytical details. For this we begin with a discussion of how linearization and stability analyses from Chapters 5 and 6 apply to the radial equation (8.62) in the central force problem.

8.3.3 Linearization and Stability

We assume that $V : (0, \infty) \to \mathbb{R}$ is defined on $(0, \infty)$, is twice continuously differentiable, and has only finitely many critical points. Let $X : (0, \infty) \times \mathbb{R} \to \mathbb{R}^2$ be the vector field

$$X(r, p) = (p/m, -V'(r)),$$

for the system $\dot{r} = p/m$, $\dot{p} = -V'(r)$. Clearly the fixed points for X are the points $(r_*, 0)$, where r_* is a critical point of V. The Jacobian matrix of X at any (r, p) is easily seen to be

$$X'(r, p) = \begin{bmatrix} 0 & 1/m \\ -V''(r) & 0 \end{bmatrix},$$

and has characteristic equation: $\lambda^2 + V''(r)/m = 0$. Thus, its eigenvalues are

$$\lambda = \pm\sqrt{-V''(r)/m}.$$

Now it is possible to have fixed points that are not simple (which only happens when $V''(r_*) = 0$), but many physical systems have only simple fixed points. In that case the above shows that $(r_*, 0)$ is a saddle point when $V''(r_*) < 0$ (that is, when V has a local maximum at r_*) and is a possible center when $V''(r_*) > 0$ (i.e., when V has a local minimum at r_*). In the latter case, we are assured that the fixed point is *actually* a center, because the energy function

$$\Lambda(r, p) \equiv \frac{p^2}{2m} + V(r)$$

is in fact a Liapunov function for the system. To see this, note that

$$\nabla\Lambda(r, p) = (V'(r), p/m),$$

and so $\nabla\Lambda(r, p) \cdot X(r, p) = 0$, for all (r, p). Thus, the first condition for a Liapunov function is satisfied. The second condition on Λ requires that the fixed point be a local minimum of Λ. This can be checked by looking at its Hessian, which in this case is the 2×2 matrix

$$\mathcal{H}_\Lambda(r, p) = \begin{bmatrix} V''(r) & 0 \\ 0 & 1/m \end{bmatrix}.$$

Clearly this matrix is positive definite when $V''(r) > 0$, and thus at a possible center $(r_*, 0)$, the Liapunov function Λ has a local minimum. Hence the possible center is an actual center.

Note: The above is a standard application of the fixed point and stability analysis to the radial equation (8.62) when written in 1st-order form. You should realize, however, within the setting of the central force problem, the fixed points of the radial equation correspond to *circular* orbits of a body (or particle) about the center of force. This is discussed in the next subsection.

8.3.4   Circular Orbits

The central force equations (8.62)-(8.63), written in terms of the effective potential V, are

$$m\ddot{r} = -V'(r)$$
$$\dot{\theta} = \frac{c}{r^2}.$$

The system is in polar coordinates and so solutions $r, \theta : I \to \mathbb{R}$ of it are interpreted as a radial distance r from the origin and an angular displacement θ from the x-axis. As we have seen in the last section, a critical point r_* of V gives a fixed-point solution of the first equation (with initial conditions $r(0) = r_*$, $\dot{r}(0) = 0$). Then the second equation has solution $\theta(t) = ct/r_*^2 + \theta_0 = \dot{\theta}_0 t + \theta_0$ (note that $c = r_*^2 \dot{\theta}_0$). Thus, the corresponding solution of the central force equation in Cartesian coordinates lies on a circle of radius r_*, centered at the origin. Specifically, this solution is

$$\alpha(t) = \left(r_* \cos(\dot{\theta}_0 t + \theta_0), \; r_* \sin(\dot{\theta}_0 t + \theta_0) \right),$$

for $t \in \mathbb{R}$. *Note*: We are assuming that $\dot{\theta}_0 \neq 0$, so that $c = r_*^2 \dot{\theta}_0 \neq 0$. To launch the body on such a circular orbit, we must place it at a distance $r_0 = r_*$ from the center of force, give it no initial radial velocity: $\dot{r}_0 = 0$, give it an initial angular displacement θ_0, say $\theta_0 = 0$, and give it just the right initial angular velocity. This initial angular velocity is $\dot{\theta}_0 = c/r_*^2$.

The analysis of the stability of the fixed point $(r_*, 0)$ in the previous section was phrased in terms of the effective potential V. But since

$$V(r) = \frac{mc^2}{2r^2} + g(r),$$

where $g' = f$, we can make the discussion more specific to the central force problem by expressing the derivatives of V in terms of the given function f:

$$V'(r) = -\frac{mc^2}{r^3} + f(r)$$

$$V''(r) = \frac{3mc^2}{r^4} + f'(r).$$

Then in summary, we have the following:

Proposition 8.2 (Circular Orbits) *A central force problem has circular orbits if and only if the equation*

$$\frac{mc^2}{r^3} - f(r) = 0 \tag{8.69}$$

has real roots r. In addition, if r is a real root of this equation, then the corresponding circular orbit is stable if r satisfies

$$\frac{3mc^2}{r^4} + f'(r) > 0. \tag{8.70}$$

One can apply these results, for example, to inverse power laws $f(r) = kr^{-q}$, for $q \in \mathbb{R}$, to find that these central forces always admit circular orbits and these orbits are stable if and only if $q < 3$ (exercise).

8.3.5 Analytical Solution

Here we consider, in detail, the analytical solution of the radial equation

$$m\ddot{r} = -V'(r).$$

We will show that the maximal interval of existence for any initial conditions $r_0, \dot{r}_0$ is always $\mathbb{R}$ and that, depending on the value of the energy

$$E = \tfrac{1}{2}m\dot{r}_0^2 + V(r_0),$$

the maximal solution $r : \mathbb{R} \to \mathbb{R}$ is either periodic or has limit ∞ as $t \to \pm\infty$. Since $t \mapsto (r(t), \dot{r}(t))$ lies on an energy curve, as for instance shown in Figure 8.15, we would expect this type of behavior.

The rigorous argument is based on the following heuristic argument. As we have seen above, any solution $r : I \to \mathbb{R}$ satisfies the conservation of energy law:

$$\tfrac{1}{2}m\dot{r}^2 = E - V(r).$$

View this as a 1st-order differential equation that r must satisfy. To put it in normal form, take square roots of each side of the equation to get the separable DE:

$$\frac{dr}{dt} = \pm\sqrt{2m}\sqrt{E - V(r)}. \tag{8.71}$$

Heuristically this DE has solution

$$t = \pm \int \frac{1}{\sqrt{2m}\sqrt{E - V(r)}}\, dr. \tag{8.72}$$

This gives t as a function of r, but inverting will give r as a function of t and thus a solution of the original separable differential equation.

To make this pedagogical argument precise, suppose $V : (0, \infty) \to \mathbb{R}$ is twice continuously differentiable and has finitely many local extrema and E is a real number. Theorem 8.3 below describes the precise nature of the integral curve (or curves) that lie on the energy curve $p^2/2m + V(r) = E$. But, first we need the following lemma.

Lemma 8.1 *Suppose $[r_1, r_2]$ is a subinterval of $(0, \infty)$ for which $E - V(r) >$ 0 for $r \in (r_1, r_2)$ and $V(r_1) = E = V(r_2)$. Suppose $H : (0, \infty) \to$ is continuously differentiable. Then the improper integral*

$$\int_{r_1}^{r_2} \frac{H'(u)}{\sqrt{2m}\sqrt{E - V(u)}} \, du, \qquad (8.73)$$

converges if $V'(r_1) \neq 0$ and $V'(r_2) \neq 0$.

Proof: Suppose first that $V'(r_1) \neq 0$. Note that necessarily $-V'(r_1) > 0$, since $r \mapsto E - V(r)$ is a positive function on (r_1, r_2) and is zero at r_1. We show that the improper integral

$$\int_{r_1}^{r_0} \frac{H'(u)}{\sqrt{2m}\sqrt{E - V(u)}} \, du, \qquad (8.74)$$

converges, for any $r_0 \in (r_1, r_2)$. To see this, let $\alpha : I \to \mathbb{R}$, be a solution of the initial value problem:

$$
\begin{aligned}
m\ddot{\alpha}(t) &= -V'(\alpha(t)) \\
\alpha(0) &= r_1 \\
\dot{\alpha}(0) &= 0,
\end{aligned}
$$

for all $t \in I$. Then, as we have seen above, α satisfies the conservation law $m[\dot{\alpha}(t)]^2/2 = E_* - V(\alpha(t))$, for all $t \in I$ and some constant E_*. But from the initial conditions that α satisfies, it follows that

$$E_* = \tfrac{1}{2}m[\dot{\alpha}(0)]^2 + V(\alpha(0)) = V(r_1) = E.$$

Using this in the conservation law and taking square roots, we get that

$$\dot{\alpha}(t) = \pm\sqrt{2m}\sqrt{E - V(\alpha(t))}, \qquad (8.75)$$

for all $t \in I$. The $\pm$ sign must in fact be a $+$ sign. To see this recall that $\dot{\alpha}(0) = 0$ and $\ddot{\alpha}(0) = -V'(\alpha(0)) = -V'(r_1) > 0$. Thus, α has a local minimum at $t = 0$ (see Figure 8.17).

As indicated in the figure, there is an interval $[0, \delta]$ on which α is strictly increasing. Now without loss of generality in proving that the integral (8.74) converges, we can assume that $r_0 \in \alpha((-a, a))$ (as shown in Figure 8.17). Now, in the integral make the change of variables

$$u = \alpha(t),$$

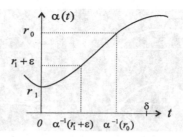

Figure 8.17: *Graph of the function α near t = 0.*

so that the change in the differentials is

$$du = \dot{\alpha}(t)\, dt = \sqrt{2m}\sqrt{E - V(\alpha(t))}\, dt.$$

Using this gives

$$\lim_{\varepsilon \to 0} \int_{r_1+\varepsilon}^{r_0} \frac{H'(u)}{\sqrt{2m}\sqrt{E-V(u)}}\, du = \lim_{\varepsilon \to 0} \int_{\alpha^{-1}(r_1+\varepsilon)}^{\alpha^{-1}(r_0)} H'(\alpha(t))\, dt$$

$$= \lim_{\varepsilon \to 0} \Big(H(r_0) - H(r_1 + \varepsilon) \Big)$$

$$= H(r_0) - H(r_1).$$

Thus, the improper integral converges.

An entirely similar argument shows that if $V'(r_2) \neq 0$, then the improper integral

$$\int_{r_0}^{r_2} \frac{H'(u)}{\sqrt{2m}\sqrt{E - V(u)}}\, du,$$

converges for any $r_0 \in (r_1, r_2)$. $\Box$

The determination of the periodic and bounded nature (or aperiodic and unbounded nature) of solutions is based on the behavior of the function $r \mapsto E - V(r)$ at the endpoints of the intervals where this function is positive.

There are four possibilities, as shown in Figure 8.18, which correspond to whether $V'(r_1)$, $V'(r_2)$ are zero or not.

Theorem 8.3 *Suppose $[r_1, r_2]$ is a subinterval for which: $E - V(r) > 0$, for $r \in (r_1, r_2)$, and $V(r_1) = E = V(r_2)$. Then there is a twice continuously*

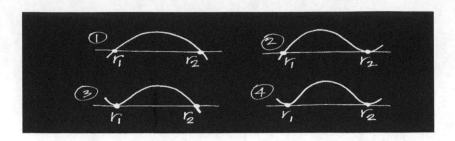

Figure 8.18: *Four possibilities for the behavior of the function $E - V(r)$ at the endpoints of an interval where it is positive.*

differentiable function $\rho : \mathbb{R} \to [r_1, r_2]$, which satisfies

$$m\ddot{\rho}(t) = -V'(\rho(t)),$$

for all $t \in \mathbb{R}$ and has one (and only one) of the following sets of properties depending on the values of V' at r_1, r_2.

(1) *If $V'(r_1) \neq 0$ and $V'(r_2) \neq 0$, then ρ is periodic with period $2T$, where*

$$T = \int_{r_1}^{r_2} \frac{1}{\sqrt{2m}\sqrt{E - V(u)}}\, du, \tag{8.76}$$

Furthermore, $\rho(0) = r_1$, and $\rho(T) = r_2$.

(2) *If $V'(r_1) \neq 0$ and $V'(r_2) = 0$, then $r_1 \leq \rho(t) < r_2$ for all $t \in \mathbb{R}$, $\rho(0) = r_1$, and*

$$\lim_{t \to \pm\infty} \rho(t) = r_2.$$

(3) *If $V'(r_1) = 0$ and $V'(r_2) \neq 0$, then $r_1 < \rho(t) \leq r_2$ for all $t \in \mathbb{R}$, $\rho(0) = r_2$, and*

$$\lim_{t \to \pm\infty} \rho(t) = r_1.$$

(4) *If $V'(r_1) = 0$ and $V'(r_2) = 0$, then $r_1 < \rho(t) < r_2$ for all $t \in \mathbb{R}$, and*

$$\lim_{t \to \infty} \rho(t) = r_2, \qquad \lim_{t \to -\infty} \rho(t) = r_1.$$

Proof: Suppose that $V'(r_1) \neq 0$ and $V'(r_2) \neq 0$. Define a time function $\tau : [r_1, r_2] :\to \mathbb{R}$ by

$$\tau(r) = \int_{r_1}^{r} \frac{1}{\sqrt{2m}\sqrt{E - V(u)}}\, du, \tag{8.77}$$

for $r \in [r_1, r_2]$. By Lemma 8.1 this is well-defined: $\tau(r_1) = 0$ and $\tau(r_2)$ is finite. Let $T = \tau(r_2)$. According to this theorem, T is the time it takes for the radius to increase from r_1 to r_2. By the Fundamental Theorem of Calculus: $\tau'(r) = 1/(\sqrt{2m}\sqrt{E - V(r)})$. Thus, $\tau'(r) > 0$, for $r \in (r_1, r_2)$, and $\tau'(r_i) = \infty$, for $i = 1, 2$. Hence τ is strictly increasing on $[r_1, r_2]$ and has vertical tangent lines at the endpoints of the interval. See Figure 8.19.

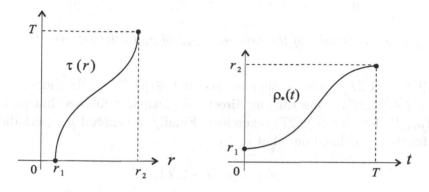

Figure 8.19: *Graphs of the functions τ and $\rho_* = \tau^{-1}$.*

By the Inverse Function Theorem, τ has an inverse $\rho_* \equiv \tau^{-1} : [0, T] \rightarrow [r_1, r_2]$. This function, as shown in Figure 8.19 is strictly increasing on $[0, T]$, has horizontal tangent lines at the end points, and $\rho_*(0) = r_1$, $\rho_*(T) = r_2$. Also by the formula for differentiating an inverse function, we get

$$\begin{aligned} \dot{\rho}_*(t) &= (\tau^{-1})'(t) \\ &= \frac{1}{\tau'(\rho_*(t))} \\ &= \sqrt{2m}\sqrt{E - V(\rho_*(t))}, \end{aligned} \tag{8.78}$$

for all $t \in [0, T]$. From this it is easy to show that $m\ddot{\rho}_*(t) = -V(\rho_*(t))$, for all $t \in [0, T]$ (exercise). Next we extend ρ_* to a function $\rho_{**} : [0, 2T] :\rightarrow [r_1, r_2]$ by

$$\rho_{**}(t) = \begin{cases} \rho_*(t) & \text{if } t \in [0, T] \\ \rho_*(2T - t) & \text{if } t \in [T, 2T] \end{cases}$$

Figure 8.20 shows the graph of this extension of ρ_*. From the figure, or by direct calculation from the above defining formula, one can see that

$$\dot{\rho}_{**}(t) = \pm\sqrt{2m}\sqrt{E - V(\rho_{**}(t))},$$

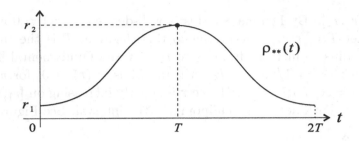

Figure 8.20: *Graphs of the extension ρ_{**} of ρ_* to the interval $[0, 2T]$.*

for all $t \in [0, 2T]$. The $+$ sign is used if $t \in [0, T]$, while the $-$ sign is used if $t \in [T, 2T]$. From this, by direct calculation, it follows that $\ddot{\rho}_{**}(t) = -V'(\rho_{**}(t))$ for all $t \in [0, 2T]$ (exercise). Finally, we extend ρ_{**} periodically to a function ρ defined on all of $\mathbb{R}$, by

$$\rho(t) = \rho_{**}(t - 2iT),$$

if $t \in [2iT, 2(i+1)T]$, for $i = 0, \pm 1, \pm 2, \dots$. Then clearly ρ satisfies $\ddot{\rho}(t) = -V'(\rho(t))$, for all $t \in \mathbb{R}$. Also ρ is periodic with period $2T$, and $\rho(0) = r_1$, $\rho(T) = r_2$. This completes the proof of Case (1).

Next assume that $V'(r_1) \neq 0$ and $V'(r_2) = 0$. Define $\tau : [r_1, r_2) \to \mathbb{R}$ by formula (8.77) above. Now however, we claim, τ is *not* defined at r_2, i.e., the improper integral $\tau(r_2)$ is divergent. If not, then $\tau(r_2)$ is finite, say $\tau(r_2) = T$. Then, as in the first part of the proof, we can construct a differentiable, periodic function $\rho : \mathbb{R} \to [r_1, r_2]$, such that $\ddot{\rho}(t) = -V'(\rho(t))$, for all $t \in \mathbb{R}$ and $\rho(0) = r_1$, $\rho(T) = r_2$. But the constant function $\beta : \mathbb{R} \to \mathbb{R}$, defined by $\beta(t) \equiv r_2$, for all $t \in \mathbb{R}$, also satisfies this same initial value problem. Thus, $\rho = \beta$, which is contradiction since $r_1 \neq r_2$. With this established, it follows that τ has a vertical asymptote at r_2 as shown in Figure 8.21. As in the first part of the proof, we get that τ has an inverse $\rho_* \equiv \tau^{-1} : [0, \infty) \to \mathbb{R}$, which satisfies $\ddot{\rho}_*(t) = -V'(\rho_*(t))$, for all $t \in [0, \infty)$, and $\rho_*(0) = r_1$, while $\lim_{t \to \infty} \rho_*(t) = r_2$ (see Figure 8.21). Extending ρ_* to a function $\rho : \mathbb{R} \to \mathbb{R}$ by

$$\rho(t) = \begin{cases} \rho_*(t) & \text{if } t \in [0, \infty) \\ \rho_*(-t) & \text{if } t \in (-\infty, 0] \end{cases}$$

gives the function with the desired properties for Case (2) (exercise).

The proof of Cases (3) is similar to Case (2), except that the definition

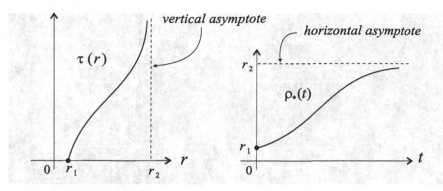

Figure 8.21: *Graphs of the functions τ and $\rho_* = \tau^{-1}$.*

of τ is

$$\tau(r) = \int_r^{r_2} \frac{1}{\sqrt{2m}\sqrt{E - V(u)}}\, du,$$

for $r \in (r_1, r_2]$ (exercise). Case (5) can be proved in an analogous manner (exercise). $\square$

Corollary 8.1 *With the same assumptions as in the theorem, it follows that the improper integral*

$$\int_{r_1}^{r_2} \frac{1}{\sqrt{2m}\sqrt{E - V(u)}}\, du, \tag{8.79}$$

converges if and only if $V'(r_1) \neq 0$ and $V'(r_2) \neq 0$.

Proof: Lemma 8.1 proved the "if" part. The "only if" part follows from the proof of Case (2) of the theorem. $\square$

Figure 8.22 shows the energy curves $p^2/2m + V(r) = E$ corresponding to each of the four graphs of $E - V(r)$ shown in Figure 8.18.

To summarize the results of the above theorem:

- **Case (1):** Neither of the end points r_1, r_2 corresponds to a fixed point of the system and the energy curve consists of one closed integral curve that passes periodically through r_1, r_2 at times $2iT, (2i + 1)T, \ i = 0, \pm 1, \pm 2, \ldots$, respectively.

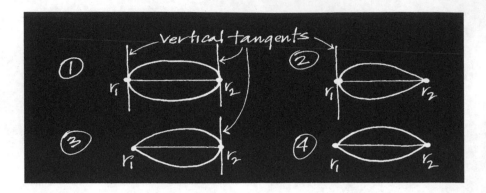

Figure 8.22: *Four energy curves corresponding to the graphs in Figure 8.18.*

- **Case (2) and Case (3):** One of the endpoints r_1, r_2, corresponds to a fixed point of the system, while the other does not. The energy curves consists of one, nonclosed curve that "starts" at the fixed point at time $t = -\infty$ and "returns" again to the fixed point at time $t = \infty$.

- **Case (4):** Both of the endpoints r_1, r_2, correspond to fixed points of the system. The energy curve consists of two integral curves, each of which passes from one fixed point to the other as t runs from $-\infty$ to ∞.

In Case (1), it is important to note that even though the radial solution $\rho : \mathbb{R} \to \mathbb{R}$ of $\ddot{r} = -V'(r)$ with initial conditions $\rho(0) = r_1, \dot{\rho}(0) = 0$, is periodic, the corresponding solution $t \mapsto (\rho(t) \cos \theta(t), \rho(t) \sin \theta(t))$, of the central force problem may not be a closed orbit about the center of force. Figure 8.23 shows an example of a central force motion with a periodic radial function $\rho : \mathbb{R} \to [r_1, r_2]$.

Because ρ oscillates periodically between its maximum value r_2 and minimum value r_1, the orbit oscillates between two circles with radii r_2 and r_1.

It is clear that the trajectory will close on itself (be a closed orbit) if and only if there is a time t_* such that $\rho(t_*) = \rho(0)$ and $\theta(t_*) = \theta(0) + 2k\pi$, for some positive integer k. If such a t_* exists, then since ρ has period $2T$, it follows that $t_* = 2nT$ for some positive integer n. Recalling that the angular deviation θ is the solution of

$$\dot{\theta}(t) = \frac{c}{\rho(t)^2},$$

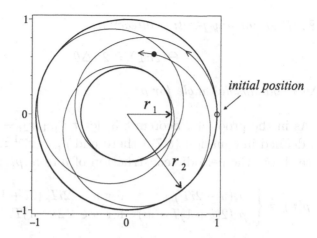

Figure 8.23: *An example of a central force motion with periodic radial function ρ .*

that is

$$\theta(t) = \theta(0) + \int_0^t \frac{c}{\rho(u)^2} \, du,$$

for $t \in \mathbb{R}$, we see that the condition for a closed orbit is the existence of two positive integers n, k such that

$$\int_0^{2nT} \frac{c}{\rho(u)^2} \, du = 2k\pi.$$

Determining whether such integers k, n exist might, at first, seem to be a difficult problem, but the next proposition shows that the problem is reduced to the computation of a certain integral defining the *apsidal angle*.

Definition 8.5 (Apsidal Angle) With the notation and assumptions in Theorem 8.3, Case (1), let $\Delta\theta$ be the positive number defined by the integral

$$\Delta\theta = \int_{r_1}^{r_2} \frac{cr^{-2}}{\sqrt{2m}\sqrt{E - V(r)}} \, dr. \tag{8.80}$$

By Lemma 8.1 this improper integral converges and so $\Delta\theta$, called the *apsidal angle*, is well defined.

Proposition 8.3 *Let $\rho : \mathbb{R} \to \mathbb{R}$ be the periodic, radial function with period $2T$ constructed in Case (1) in Theorem 8.3 above. Define $\Theta : \mathbb{R} \to \mathbb{R}$, by*

$$\Theta(t) = \int_0^t \frac{c}{\rho(u)^2} \, du, \tag{8.81}$$

for $t \in \mathbb{R}$. Then for any positive integer n,

$$\Theta(2nT) = 2n\Delta\theta, \tag{8.82}$$

where $\Delta\theta$ is the apsidal angle for ρ.

Proof: As in the proof of Theorem 8.3, let $\tau : [r_1, r_2] \to [0, T]$ be the time function defined in equation (8.77) there and $\rho_* = \tau^{-1} : [0, T] \to [r_1, r_2]$ be its inverse. From the periodic construction of ρ from ρ_*, it follows that

$$\rho(u) = \begin{cases} \rho_*(u - 2iT) & \text{for } u \in [2iT, (2i+1)T] \\ \rho_*(2(i+1)T - u) & \text{for } u \in [(2i+1)T, 2(i+1)T] \end{cases}$$

for $i = 0, \ldots, n-1$. Thus, by a simple change of variables we have

$$\int_{2iT}^{(2i+1)T} \frac{c}{\rho(u)^2} \, du = \int_{2iT}^{(2i+1)T} \frac{c}{\rho_*(u - 2iT)^2} \, du$$

$$= \int_0^T \frac{c}{\rho_*(u)^2} \, du,$$

and

$$\int_{(2i+1)T}^{(2(i+1)T} \frac{c}{\rho(u)^2} \, du = \int_{(2i+1)T}^{2(i+1)T} \frac{c}{\rho_*(2(i+1)T - u)^2} \, du,$$

$$= \int_0^T \frac{c}{\rho_*(u)^2} \, du,$$

for $i = 0, \ldots, n-1$. Consequently,

$$\begin{aligned} \Theta(2nT) &= \int_0^{2nT} \frac{c}{\rho(u)^2} \, du \\ &= \sum_{i=0}^{n-1} \left(\int_{2iT}^{(2i+1)T} \frac{c}{\rho(u)^2} \, du + \int_{(2i+1)T}^{(2(i+1)T} \frac{c}{\rho(u)^2} \, du \right) \\ &= 2n \int_0^T \frac{c}{\rho_*(u)^2} \, du. \end{aligned}$$

Thus, all we have to show is that the latter integral above is the same as the integral (8.80) for the apsidal angle. But this is easily done by making the change of variables:

$$u = \tau(r),$$

with the corresponding change of differentials:

$$du = \tau'(r)\, dr = \frac{1}{\sqrt{2m}\sqrt{E - V(r)}}\, dr.$$

This establishes the result. $\square$

Corollary 8.2 *With the assumptions in the above discussion, the trajectory*

$$t \mapsto \Big(\rho(t)\cos\theta(t),\ \rho(t)\sin\theta(t) \Big),$$

is a closed orbit if and only if the apsidal angle $\Delta\theta$ is a rational multiple of π:

$$\Delta\theta = \frac{k}{n}\,\pi,$$

for two positive integers k, n.

Exercises 8.3

1. Suppose V is the effective potential for which $E - V(r)$ has the graph shown in Figure 8.14, for a particular value of E. Based on this figure draw, by hand, the corresponding energy curves $p^2/2m + V(r) = E^*$ in the r-p plane, for various values of E^*. This latter drawing represents the phase portrait for the system: $\dot{r} = p/m$, $\dot{p} = -V'(r)$. Thus, indicate the direction of flow, the fixed points, and their type on the drawing.

2. For each of the following functions f compute the corresponding effective potential V, plot (by hand or by computer) the graphs of $E - V(r)$ for various values of E, and use these to draw, by hand, the corresponding energy curves $p^2/2m + V(r) = E$ in the r-p plane. This latter drawing represents the phase portrait for the system: $\dot{r} = p/m$, $\dot{p} = -V'(r)$. Thus, indicate the direction of flow, the fixed points, and their type on the drawing.

   (a) $f(r) = r^{-2.5}$.

   (b) $f(r) = r^{-3}$.

   (c) $f(r) = r^{-0.5}$.

   (d) $f(r) = r^2$.

   (e) $f(r) = 2(r - 1)$.

   Comment on the various types of integral curves that occur and what type of motion they represent in the central force problem.

3. Suppose the central force is attractive with a magnitude that is an inverse power law, i.e., $f(r) = kr^{-q}$, for $q \in \mathbb{R}$. Show that these central forces always admit circular orbits and these orbits are stable if and only if $q < 3$. Show that a central force obeying Hooke's Law: $f(r) = k(r-L)$, with $k > 0$, admits stable, circular orbits

4. Add graphical evidence to what you proved in Exercise 3. Specifically, let V_q denote an effective potential for inverse power law $f(r) = kr^{-q}$. Take $m = 1 = k$, and $c = 0.5$. Consider the following groups of values of q:

   (i) $q = -2, -1, 0.5, 0.8, 1$,

   (ii) $q = 1.3, 1.5, 2, 2.2, 2.5$,

   (iii) $q = 3, 3.2, 3.5, 4, 4.5$.

   For each group, plot the graphs of the effective potential V_q, all in the same figure. Annotate the three figures appropriately, by identifying the specific values of r that give circular orbits, specifying which are stable/unstable, and describing why, based on the graph, the orbit is stable or unstable. Also describe why the value of r that gives a circular orbit varies with q in the way it does.

5. **(Rigid-Body Motions)** The theory discussed in the section above showed how to solve the general 2-body problem:

$$m_1 \ddot{\mathbf{r}}_1 = \frac{f(r_{12})}{r_{12}}(\mathbf{r}_2 - \mathbf{r}_1) \tag{8.83}$$

$$m_2 \ddot{\mathbf{r}}_2 = \frac{f(r_{12})}{r_{12}}(\mathbf{r}_1 - \mathbf{r}_2), \tag{8.84}$$

   in terms of the relative position vector $\mathbf{r} = \mathbf{r}_2 - \mathbf{r}_1$ and the position vector $\mathbf{R}$ for the center of mass. In particular, Proposition 8.2 gives the conditions under which there is a solution of these equations such that $r = |\mathbf{r}|$ is constant. Thus, the two bodies move so that their distance apart is always the same. This is known as a *rigid-body motion* for the system and the next section studies such motions for N bodies in detail. Here you are to study this phenomenon for two bodies and for the following choices of f.

   (i) **(Hooke's Law)** $f(r) = k(r - L)$.

   (ii) **(Gravity)** $f(r) = Gm_1 m_2/r^2$

   (iii) **(Inverse Cube)** $f(r) = Gm_1 m_2/r^3$

   For each of these, which are assigned to you, do the following:

   (a) Determine the values r_* that are the roots of equation (8.69). These correspond to a circular orbit of the 2nd body around the 1st body. Note that r_* will involve the $c = r_0^2 \dot{\theta}_0^2$, as well as the other constants in the problem.

(b) Validate the theory by doing the following (numerical) experiments, using the worksheets hooke2.mws and/or gravity2.mws. Assume the masses are $m_1 = 1 = m_2$ and for (i) $k = 1, L = 1$, while for (ii)-(iii) $G = 1$. Let the initial positions be $\mathbf{r}_1 = (0,0,0), \mathbf{r}_2 = (r_*,0,0)$ and the initial velocities be $\mathbf{v}_1 = (0,0,0), \mathbf{v}_2 = (0,\nu,0)$, where r_* is the value you found in part (a), and ν is chosen so that the relative motion is circular. Plot the relative motion to verify that it is circular. View the animation of the motion of the line joining the two bodies and verify that this line has length r_* at all times. Plot the absolute motion, i.e., the curves traced out by each body. (For extra credit determine the formula for each of these curves.) Explain, in your own words, why the bodies always remain the same distance apart even though they are drawn toward each other by the given force (with magnitude $f(r_*)$).

(c) Study the stability or instability of the circular motion of one body around the other by varying, by small amounts, the initial conditions, specifically the value of ν used in part (b). Produce plots of the relative and absolute motions. Include any other graphics that you think will help to display the stability/instability.

6. Using equation (8.78), show that $m\ddot{\rho}_*(t) = -V(\rho_*(t))$, for all $t \in [0,T]$.

7. Show that $m\ddot{\rho}_{**}(t) = -V'(\rho_{**}(t))$ for all $t \in [0,2T]$, where ρ_{**} is the function defined in the proof of Theorem 8.3.

8. Show that the function ρ defined by equation (8.3.5) has the properties required in Case (2) of Theorem 8.3.

9. Prove Cases (3) and (4) of Theorem 8.3.

10. (Apsidal Angle) For an inverse square law $f(r) = k/r^2$, show by direct calculation of the definite integral in (8.80) that the apsidal angle is

$$\Delta\theta = \pi.$$

Recall that E is a value such that $\{r | E - V(r) > 0\}$ is a bounded interval with endpoints $r_1 < r_2$. This means (see Figure 8.13) that

$$-\frac{k^2}{2mc^2} < E < 0,$$

and that $E - V(r) = 0$ for $r = r_1, r_2$. Hint: write out the integral in equation (8.80) specifically for the inverse square law case and make the change of variables $u = 1/r$ to get

$$\Delta\theta = \frac{c}{\sqrt{2m}} \int_{r_2^{-1}}^{r_1^{-1}} \frac{du}{\sqrt{E - c^2u^2/2 + ku}}.$$

Note that the radicand in this expression vanishes at $u = r_1^{-1}, r_2^{-1}$. Complete the square on this radicand and make another change of variables to show that the above integral can be written as

$$\Delta\theta = \int_{-a}^{a} \frac{dw}{\sqrt{a^2 - w^2}},$$

where a is a suitable constant positive constant.

8.4   Rigid-Body Motions

In this section we consider a system of particles which undergoes a special type of motion known as a rigid-body motion. Such a motion preserves the distance between any pair of particles in the system and is typical of the motion of a system of particles comprising a continuum or solid body that is "rigid" or not deformable. The constraints that this type of motion puts on the system of DEs reduces it to a smaller system, indeed, one where the number of unknowns is independent of the number of particles, and leads to important physical and geometrical concepts such as angular velocity and moments of inertia.

We use the general setup from the first section and assume there are N particles and $F_i : J \times \mathcal{O} \to \mathbb{R}^3$, $i = 1, \ldots, N$, is a general system of position, velocity, and time-dependent forces.

Definition 8.6 (Rigid-Body Motion) A curve in $\mathbb{R}^{3N}$

$$\mathbf{r} = (\mathbf{r}_1, \ldots, \mathbf{r}_N) : I \to \mathbb{R}^{3N},$$

is said to be a *rigid-body motion* if

$$|\mathbf{r}_i(t) - \mathbf{r}_j(t)| = |\mathbf{r}_i(0) - \mathbf{r}_j(0)|, \tag{8.85}$$

for all $i, j = 1, \ldots, N$ and all $t \in I$.

Usually we will only be concerned with rigid-body motions that satisfy the equations of motion $m_i \ddot{\mathbf{r}}_i = F_i(t, \mathbf{r}, \dot{\mathbf{r}})$. Thus, in addition to satisfying the equations of motion, the $\mathbf{r}_i$'s must maintain their respective distances apart (the same as their initial distances). See Figure 8.24. Clearly not all systems of forces allow for rigid-body motions, but for those that do, the motion $\mathbf{r} = (\mathbf{r}_1, \ldots, \mathbf{r}_N)$ has a very particular form, as the following theorem shows.

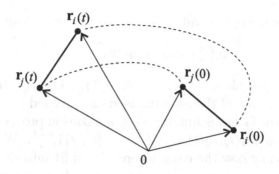

Figure 8.24: *The distances between the bodies in the system remain constant in a rigid-body motion.*

The theorem is of fundamental importance to the theory (as we will see) and while its proof is somewhat long, the concepts involved (such as orthogonal matrices, the Gram-Schmidt orthogonalization process, properties of determinants) are interesting and elementary. In the theorem, $\mathcal{M}_3$ denotes the set of all 3×3 real matrices.

Theorem 8.4 (Rigid-Body Motions) *Suppose* $\mathbf{r} = (\mathbf{r}_1, \ldots, \mathbf{r}_N) : I \to \mathbb{R}^{3N}$ *is a rigid-body motion. Then there is* (1) *a differentiable, matrix-valued function* $Q : I \to \mathcal{M}_3$, *with* $Q(t)$ *an orthogonal matrix:*

$$Q(t)^T Q(t) = I = Q(t)Q(t)^T, \qquad (8.86)$$

for each $t \in I$, *with* $Q(0) = I$, *and* (2) *vectors* $\mathbf{u}_i \in \mathbb{R}^3$, $i = 1, \ldots, N$, *with* $\sum_{i=1}^{N} m_i \mathbf{u}_i = 0$, *such that*

$$\mathbf{r}_i(t) = Q(t)\mathbf{u}_i + \mathbf{R}(t), \qquad (8.87)$$

for $i = 1, \ldots, N$ *and for all* $t \in I$. *Here* $\mathbf{R}(t)$ *is the position of the center of mass at time* t.

Proof: Let $w_i = \mathbf{r}_{i+1} - \mathbf{r}_1$, $i = 1, \ldots, N-1$. Then it is easy to see that

$$|w_i(t) - w_j(t)| = |w_i(0) - w_j(0)|$$
$$|w_i(t))| = |w_i(0)|$$

for every $t \in I$ and all $i, j \in \{1, \ldots, N-1\}$. The first equation gives $|w_i(t) - w_j(t)|^2 = |w_i(0) - w_j(0)|^2$, and expanding this in terms of the dot product gives

$$|w_i(t)|^2 - 2w_i(t) \cdot w_j(t) + |w_j(t)|^2 = |w_i(0)|^2 - 2w_i(0) \cdot w_j(0) + |w_j(0)|^2.$$

Reducing this by using the 2nd of the above identities leads to

$$w_i(t) \cdot w_j(t) = w_i(0) \cdot w_j(0), \tag{8.88}$$

for every $t \in I$ and all $i, j \in \{1, \ldots, N - 1\}$. The constancy of the dot products $w_i \cdot w_j$ is one of the essential elements needed for the proof.

We next use the Gram-Schmidt orthogonalization process to construct an orthonormal basis $\{e_1(t), e_2(t), e_3(t)\}$ from $\{w_i(t)\}_{i=1}^{N-1}$. We assume $N \geq 4$ and point out below how the cases where $N < 4$ fit into the proof.

Let $k \leq 3$ be the number of linearly independent vectors in $\{w_i(0)\}_{i=1}^{N-1}$ and relabel these vectors so that the linearly independent ones come first. Let

$$B_1 = w_1 \cdot w_1$$

$$B_2 = \begin{bmatrix} w_1 \cdot w_1 & w_1 \cdot w_2 \\ w_2 \cdot w_1 & w_2 \cdot w_2 \end{bmatrix}$$

$$B_3 = \begin{bmatrix} w_1 \cdot w_1 & w_1 \cdot w_2 & w_1 \cdot w_3 \\ w_2 \cdot w_1 & w_2 \cdot w_2 & w_2 \cdot w_3 \\ w_3 \cdot w_1 & w_3 \cdot w_2 & w_3 \cdot w_3 \end{bmatrix}.$$

Then by equation (8.88), the B_n's are constant matrices

$$B_n(t) = B_n(0),$$

for all $t \in I$. From linear algebra, we know that $w_1(t), w_2(t)$ are linearly independent if and only if $\det(B_2(t)) \neq 0$, and $w_1(t), w_2(t), w_3(t)$ are linearly independent if and only if $\det(B_3(t)) \neq 0$ (exercise). But since $B_n(t) = B_n(0)$, for any $t \in I$, it follows that the number of linearly independent vectors in $\{w_i(t)\}_{i=1}^{N-1}$ is the same as that for $t = 0$, namely k.

The orthogonalization part of the Gram-Schmidt algorithm amounts to introducing the vectors

$$\tilde{e}_1 = w_1 \tag{8.89}$$

$$\tilde{e}_2 = -(w_1 \cdot w_2)w_1 + (w_1 \cdot w_1)w_2 \tag{8.90}$$

$$\tilde{e}_3 = a_1 w_1 - a_2 w_2 + a_3 w_3, \tag{8.91}$$

where the coefficients in the last expression are given by

$$a_1 = \begin{vmatrix} w_1 \cdot w_2 & w_1 \cdot w_3 \\ w_2 \cdot w_2 & w_2 \cdot w_3 \end{vmatrix} \tag{8.92}$$

$$a_2 = \begin{vmatrix} w_1 \cdot w_1 & w_1 \cdot w_3 \\ w_2 \cdot w_1 & w_2 \cdot w_3 \end{vmatrix} \tag{8.93}$$

$$a_3 = \begin{vmatrix} w_1 \cdot w_1 & w_1 \cdot w_2 \\ w_2 \cdot w_1 & w_2 \cdot w_2 \end{vmatrix}. \tag{8.94}$$

It is important to note that while the vectors $\tilde{e}_i$ vary with t, the coefficients in the linear combinations on the left sides of their defining equations do not depend on t. Using the well-known Laplace expansion of a determinant about its last row, we get that, for any vector v

$$\tilde{e}_2 \cdot v = \begin{vmatrix} w_1 \cdot w_1 & w_1 \cdot w_2 \\ w_1 \cdot v & w_2 \cdot v \end{vmatrix}$$

$$\tilde{e}_3 \cdot v = \begin{vmatrix} w_1 \cdot w_1 & w_1 \cdot w_2 & w_1 \cdot w_3 \\ w_2 \cdot w_1 & w_2 \cdot w_2 & w_2 \cdot w_3 \\ w_1 \cdot v & w_2 \cdot v & w_3 \cdot v \end{vmatrix}.$$

In particular, since a determinant is zero when two rows are the same, we find from the above that

$$\tilde{e}_2 \cdot w_1 = 0, \quad \tilde{e}_3 \cdot w_1 = 0, \quad \tilde{e}_3 \cdot w_2 = 0.$$

Using this, it is easy to see that the $\tilde{e}_i$'s are mutually orthogonal

$$\tilde{e}_2 \cdot \tilde{e}_1 = 0, \quad \tilde{e}_3 \cdot \tilde{e}_1 = 0, \quad \tilde{e}_3 \cdot \tilde{e}_2 = 0.$$

Also, we get that

$$\tilde{e}_2 \cdot \tilde{e}_2 = (w_1 \cdot w_1) \det(B_2), \quad \tilde{e}_3 \cdot \tilde{e}_3 = \det(B_2) \det(B_3) \tag{8.95}$$

(exercise). We now divide into cases depending on the number of linearly independent vectors.

If $k = 3$, then $\tilde{e}_1, \tilde{e}_2, \tilde{e}_3$ are nonzero for each t, so we can normalize them

$$e_i \equiv \tilde{e}_i / |\tilde{e}_i|, \qquad i = 1, 2, 3, \tag{8.96}$$

and thus, get an orthonormal set $\{e_1(t), e_2(t), e_3(t)\}$ for each $t \in I$. Let $E(t) = [e_1(t), e_2(t), e_3(t)]$ be the 3×3 matrix with the $e_i(t)$'s as its columns and similarly let $W(t) = [w_1(t), w_2(t), w_3(t)]$ be the 3×3 matrix formed from the $w_i(t)$'s as columns. Then $E(t)$ is an orthogonal matrix for all $t \in I$. In terms of this, let

$$Q(t) \equiv E(t)E(0)^T.$$

Thus, $Q(t)$ is an orthogonal matrix as well (exercise).

With this notation, equations (8.89)-(8.91) can be rewritten, using (8.96), to get in matrix form $E(t) = W(t)A$, for a certain invertible matrix A, not depending on t (exercise). Inverting the relation gives

$$W(t) = E(t)P,$$

for all $t \in I$, where $P = A^{-1}$. Next note that for $i \in \{1, \ldots, N-1\}$ and $t \in I$

$$
\begin{aligned}
E(t)^T w_i(t) &= A^T W(t)^T w_i(t) \\
&= A^T W(0)^T w_i(0) \\
&= E(0)^T w_i(0).
\end{aligned}
$$

This follows from the identities (8.88). Consequently, if we expand $w_i(t)$ in terms of the orthonormal basis $\{e_1(t), e_2(t), e_3(t)\}$ and use some matrix algebra, we get

$$
\begin{aligned}
w_i(t) &= \sum_{j=1}^{3} \left(w_i(t) \cdot e_j(t) \right) e_j(t) \\
&= E(t) \begin{bmatrix} w_i(t) \cdot e_1(t) \\ w_i(t) \cdot e_2(t) \\ w_i(t) \cdot e_3(t) \end{bmatrix} \\
&= E(t) E(t)^T w_i(t) \\
&= E(t) E(0)^T w_i(0) \\
&= Q(t) w_i(0),
\end{aligned}
$$

for $i = 1, \ldots, N-1$. This is the key equation in the proof. We need to also show that this holds for the cases $k = 2$ and $k = 1$.

Suppose $k = 2$. Then $w_1(t), w_2(t)$ are linearly independent for each $t \in I$, but $w_1(t), w_2(t), w_3(t)$ are not, i.e., $\det(B_2(t)) \neq 0$ and $\det(B_3(t)) = 0$. Consequently, from equation (8.95), we see that $\tilde{e}_3 = 0$ and so from equation (8.91) we get

$$w_3 = b_3^1 w_1 + b_3^2 w_2,$$

where $b_3^1 = -a_1/a_3$ and $b_3^2 = a_2/a_3$, with a_1, a_2, a_3 given by equations (8.92)-(8.94) (note that $a_3 = \det(B_2) \neq 0$). Thus, b_3^1, b_3^2 do not depend on t. Replacing w_3 by any other w_i, $i = 4, \ldots, N-1$, gives, in a similar fashion that

$$w_i = b_i^1 w_1 + b_i^2 w_2,$$

where b_i^1, b_i^2 do not depend on t. *Note:* Since $w_i = r_{i+1} - r_1$, the $k = 2$ case occurs only when the bodies are initially coplanar (and therefore co-planar for all time)(Exercise). Now let $W(t) = [w_1(t), w_2(t)]$ be the 3×2 matrix with $w_1(t), w_2(t)$ as its columns and let $E(t) = [e_1(t), e_2(t), e_3(t)]$ be the 3×3 orthogonal matrix with columns formed from the time-dependent vectors:

$$e_1 = \tilde{e}_1/|\tilde{e}_1| \qquad e_2 = \tilde{e}_2/|\tilde{e}_2| \qquad e_3 = e_1 \times e_2.$$

Then it is easy to see that there is a 3×2 constant matrix P such that

$$W(t) = E(t)P, \tag{8.97}$$

for all $t \in I$ (exercise). Thus, in particular $W(0) = E(0)P$ and so $P = E(0)^T W(0)$. Using this to rewrite the above equation gives

$$W(t) = E(t)E(0)^T W(0).$$

Letting $Q(t) = E(t)E(0)^T$ in the last equation and interpreting it in terms of the columns of the respective matrices, we get

$$w_i(t) = Q(t)w_i(0),$$

for $i = 1, 2$, and all $t \in I$. From this we get the extension of this identity to $i > 2$:

$$
\begin{aligned}
w_i(t) &= b_1^i w_1(t) + b_2^i w_2(t) \\
&= b_1^i Q(t)w_1(0) + b_2^i Q(t)w_2(0) \\
&= Q(t)\left(b_1^i w_1(0) + b_2^i w_2(0) \right) \\
&= Q(t)w_i(0).
\end{aligned}
$$

With this established for $k = 2$ and $k = 3$, we turn to the remaining possibility, $k = 1$.

In the case $k = 1$, we have $\det(B_2) = 0$ and so (cf. equation (8.95) $\tilde{e}_2 = 0$. Consequently, from equation (8.90), we see that $w_2 = b_2 w_1$, where $b_2 = (w_1 \cdot w_2)/(w_1 \cdot w_1)$. Replacing w_2 by w_i, we get, in general,

$$w_i = b_i w_1,$$

where $b_i = (w_1 \cdot w_i)/(w_1 \cdot w_1)$ is a constant not depending on t. *Note:* Since $w_i = r_{i+1} - r_1$, the $k = 1$ case occurs only when the bodies are initially collinear (and therefore collinear for all time). Now let

$$e_1 = w_1/|w_1|$$
$$e_2 = \dot{e}_1$$
$$e_3 = e_1 \times e_2.$$

Then $\{e_1(t), e_2(t), e_3(t)\}$ is an orthonormal set for each $t \in I$ and consequently the matrix $E(t) \equiv [e_1(t), e_2(t), e_3(t)]$, formed with these vectors as columns, is an orthogonal matrix for each $t \in I$. If we let P be the 3×1 matrix $P = [|w_1|, 0, 0]$, then clearly $w_1(t) = E(t)P$ for each $t \in I$. This gives, in particular, that $P = E(0)^T w_1(0)$. Thus, if we let $Q(t) \equiv E(t)E(0)^T$, then we have

$$w_1(t) = Q(t)w_1(0),$$

for all t. From this and the result $w_i = b_i w_1$ from above, we get that

$$w_i(t) = Q(t)w_i(0),$$

for $i = 1, \ldots, N-1$ and for all t.

With all the cases established, we can now easily finish the proof. So far we have established the existence of an matrix-valued map $Q : I \to \mathcal{M}_3$, such that $Q(t)$ is an orthogonal matrix, $Q(0) = I$, and

$$w_i(t) = Q(t)w_i(0),$$

for all $t \in I$, and $i = 1, \ldots, N-1$.

Now change the notation slightly by letting

$$\tilde{\mathbf{u}}_i = w_{i-1}(0) = \mathbf{r}_i(0) - \mathbf{r}_1(0),$$

for $i = 1, \ldots, N$. With this notation, the equation $w_{i-1}(t) = Q(t)w_{i-1}(0)$ becomes, after rearrangement, the following equation

$$\mathbf{r}_i(t) = Q(t)\tilde{\mathbf{u}}_i + \mathbf{r}_1(t). \tag{8.98}$$

Computing $\mathbf{R}(t) = \sum_{i=1}^{N} \frac{m_i}{M} \mathbf{r}_i(t)$ from this and letting $\tilde{\mathbf{U}} = \sum_{i=1}^{N} \frac{m_i}{M} \tilde{\mathbf{u}}_i$ gives

$$\mathbf{R}(t) = Q(t)\left(\sum_{i=1}^{N} \frac{m_i}{M} \tilde{\mathbf{u}}_i \right) + \mathbf{r}_1(t)$$
$$= Q(t)\tilde{\mathbf{U}} + \mathbf{r}_1(t).$$

Subtracting the left and right sides of the last equation from the left and right sides of equation (8.98), gives

$$\mathbf{r}_i(t) - \mathbf{R}(t) = Q(t)(\tilde{\mathbf{u}}_i - \tilde{\mathbf{U}}).$$

Since $Q(t)$ is an orthogonal matrix, this last equation leads directly to the result that each body remains at the same distance from the center of mass: $|\mathbf{r}_i(t) - \mathbf{R}(t)| = |\tilde{\mathbf{u}}_i - \tilde{\mathbf{U}}|$ for all t. Finally, by introducing $\mathbf{u}_i \equiv \tilde{\mathbf{u}}_i - \tilde{\mathbf{U}}$, the above equation becomes

$$\mathbf{r}_i(t) = Q(t)\mathbf{u}_i + \mathbf{R}(t),$$

for all $i = 1, \ldots, N$ and $t \in I$. Note that with this notation, it follows that

$$\mathbf{U} \equiv \sum_{i=1}^{N} \frac{m_i}{M} \mathbf{u}_i = 0.$$

This completes the proof. □

The theorem says that, in a rigid-body motion, each particle of the body follows the motion of the center of mass while rotating about this center of mass. This is the content of the equations

$$\mathbf{r}_i = Q\mathbf{u}_i + \mathbf{R}, \qquad (8.99)$$

for $i = 1, \ldots, N$, which express the position vectors of the particles in terms of the position vector $\mathbf{R} = \mathbf{R}(t)$ of the center of mass, the rotation matrix $Q = Q(t)$, and the (constant) vectors $\mathbf{u}_i$, which are the initial positions of the bodies *relative* to the initial center of mass (since $\mathbf{r}_i(0) = \mathbf{u}_i + \mathbf{R}(0)$). See Figure 8.25.

8.4.1 The Rigid-Body Differential Equations

Theorem 8.4 shows that any rigid-body motion $\mathbf{r} : I \to \mathbb{R}^{3N}$ of N bodies is completely determined by the motion of its center of mass $\mathbf{R} : I \to \mathbb{R}^3$ and a certain a family $\{Q(t)\}_{t \in I}$ of orthogonal matrices. If we now require that the rigid-body motion also satisfy the equations of motion (Newton's 2nd Law), then this forces $\mathbf{R}$ and Q to satisfy a system of DEs, called the *rigid-body differential equations*. In this section we derive these equations and, along the way, explain the geometric and dynamical significance of Q.

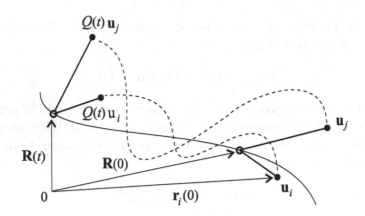

Figure 8.25: *Positions of the ith and jth bodies at times* 0 *and t, showing a general translational motion following the center of mass combined with a rotational motion about the center of mass.*

This derivation will take several pages because we need to introduce some important concepts, like angular velocity and the moment of inertia operator.

The individual forces $F_i = F_i(t, \mathbf{r}_1, \ldots, \mathbf{r}_N, \dot{\mathbf{r}}_1, \ldots, \dot{\mathbf{r}}_N)$, can be rewritten in terms of $\mathbf{R}, Q$ and their derivatives by using

$$\mathbf{r}_i = Q\mathbf{u}_i + \mathbf{R},$$

and its consequence

$$\dot{\mathbf{r}}_i = \dot{Q}\mathbf{u}_i + \dot{\mathbf{R}}. \tag{8.100}$$

Having done this, the total force $\mathbf{F}$ and torque $\mathbf{T}$ become functions of $\mathbf{R}, Q$ and their derivatives as well. Then since, as we have already seen, the position $\mathbf{R}$ of the center of mass satisfies

$$M\ddot{\mathbf{R}} = \mathbf{F},$$

we can interpret this equation as one part of the system of DEs for $\mathbf{R}$ and Q that we are seeking.

The differential equations for Q involve its derivative $\dot{Q}$ and since $\{Q(t)\}_{t \in I}$ is a 1-parameter family of rotations, it is naturally compounded from a 1-parameter family $\{\Omega(t)\}_{t \in I}$ of skew symmetric matrices. The next proposition, which is a special case of the more general result in Exercise 5 below, makes this precise.

Proposition 8.4 *Suppose* $Q : J \to O(n)$ *is a (smooth) 1-parameter family of* $n \times n$ *orthogonal matrices. Then there exists a (smooth) 1-parameter family* $\Omega : J \to o(n)$ *of* $n \times n$ *skew symmetric matrices such that* Q *satisfies the matrix differential equation:*

$$\dot{Q} = Q\Omega.$$

Furthermore, if $Q(0) = I$, *then* $\det(Q(t)) = 1$, *i.e.,* $Q(t)$ *is a rotation matrix, for each* $t \in J$.

Conversely if $\Omega : J \to o(n)$ *is a given, 1-parameter family of skew symmetric matrices, then any solution* Q *of the initial value problem*

$$\dot{Q} = Q\Omega$$
$$Q(0) = I,$$

is a 1-parameter family of rotation matrices.

Proof: To prove the first assertion, suppose $\{Q(t)\}_{t \in J}$ is a smooth 1-parameter family of orthogonal matrices. Define Ω by

$$\Omega \equiv -\dot{Q}^T Q.$$

Since $Q(t)$ is an orthogonal matrix, $Q(t)^T Q(t) = I$, for all t, and so differentiating this we get

$$0 = \frac{d}{dt}(Q^T Q) = \dot{Q}^T Q + Q^T \dot{Q}.$$

Rearranging gives

$$\dot{Q} = -Q\dot{Q}^T Q = Q\Omega.$$

To show that $\Omega(t)$ is skew symmetric for each $t \in J$, differentiate the identity $QQ^T = I$ and use $\dot{Q} = Q\Omega$ to get

$$\begin{aligned} 0 &= \frac{d}{dt}(QQ^T) = \dot{Q}Q^T + Q\dot{Q}^T \\ &= Q\Omega Q^T + Q\Omega^T Q^T \\ &= Q(\Omega + \Omega^T)Q^T. \end{aligned}$$

Hence $\Omega + \Omega^T = 0$. This shows skew symmetry.

Next, suppose that in addition $Q(0) = I$. To show that $\det(Q(t)) = 0$, for all $\in J$, we use Liouville's Formula from Chapter 4. For this, apply the matrix transpose to both sides of the equation $\dot{Q} = Q\Omega$ and use $\Omega^T = -\Omega$ to get

$$\dot{Q}^T = -\Omega Q^T.$$

Also note that $Q^T(0) = I$. Then it is clear from the development in Chapter 4 (see Theorem 4.2) that Q^T is the fundamental matrix for the linear system of DEs: $\dot{x} = -\Omega(t)x$. Hence by Liouville's Formula (Proposition 4.1) and the fact that the trace of a skew symmetric matrix is 0, we get

$$\det(Q(t)) = \det(Q(t)^T) = e^{\int_0^t \mathrm{tr}(-\Omega(s))ds} = e^0 = 1,$$

for all $t \in J$. Hence each $Q(t)$ is a rotation matrix. The proof of the rest of the proposition is left as an exercise. $\square$

Returning to the rigid-body mechanics, we note that the above result allows us to consider Q and Ω as interchangeable in terms of determining the motion. Knowing one is equivalent to knowing the other. On the other hand because of the skew symmetry condition, Ω is completely determined by three functions: $\omega_1, \omega_2, \omega_3 : J \to \mathbb{R}$. That is, Ω has the form

$$\Omega = \begin{bmatrix} 0 & -\omega_3 & \omega_2 \\ \omega_3 & 0 & -\omega_1 \\ -\omega_2 & \omega_1 & 0 \end{bmatrix}. \tag{8.101}$$

There is a reason for expressing Ω in terms of $\omega_1, \omega_2, \omega_3$ in this very particular way. Namely, if we let ω be the vector

$$\omega \equiv (\omega_1, \omega_2, \omega_3), \tag{8.102}$$

then for any vector $v = (v_1, v_2, v_3)$, we have

$$\Omega v = \omega \times v.$$

Thus, the action of the skew symmetric matrix Ω on vectors v is the same as the cross-product action of ω on v. This relationship between Ω and ω is true in general: *any 3×3, skew symmetric B matrix has its action $v \mapsto Bv$ represented by a cross-product action $v \mapsto b \times v$, with the fixed vector $b = (-B_{23}, B_{13}, -B_{12})$* (exercise). For mechanics, the cross product representation of Ω is important for the geometric interpretation of the motion.

Definition 8.7 (Angular Velocity) For a rigid-body motion, the (time dependent) vector ω in (8.102) is called the *angular velocity vector* and the corresponding (time-dependent) skew symmetric matrix Ω in (8.101) is called the *angular velocity operator*.

The point of view now is that ω is one of the fundamental unknowns in the rigid-body motion and we seek the DEs that will determine it. With ω determined, we get Ω and from this we get the rotations Q.

We can easily write the velocity of the ith particle in terms of the angular velocity as follows. From

$$\mathbf{r}_i = Q\mathbf{u}_i + \mathbf{R},$$

we get

$$
\begin{aligned}
\dot{\mathbf{r}}_i &= \dot{Q}\mathbf{u}_i + \dot{\mathbf{R}} \\
&= Q\Omega\mathbf{u}_i + \dot{\mathbf{R}} \\
&= Q(\omega \times \mathbf{u}_i) + \dot{\mathbf{R}}
\end{aligned}
$$

This expresses the velocity of the ith body as the sum of a translational velocity $\dot{\mathbf{R}}$ and a rotational velocity $Q(\omega \times \mathbf{u}_i) = Q\omega \times Q\mathbf{u}_i$. See Figure 8.26.

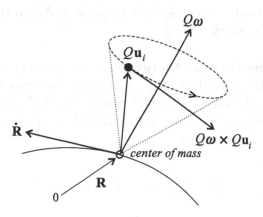

Figure 8.26: *Illustration of the instantaneous translational and rotational velocities, $\dot{\mathbf{R}}$ and $Q\omega \times Q\mathbf{u}_i$, of each particle in the rigid-body system.*

The former is viewed as an instantaneous translation, in the direction $\dot{\mathbf{R}}$, of all the bodies in the system, while the latter is viewed as an instantaneous rotation of all the bodies in the system about the axis through $Q\omega$.

Using the above expressions, we can also rewrite the total angular momentum $\mathbf{L}$ of the system. *Note*: A major simplification in $\mathbf{L}$ occurs because $\sum_{i=1}^{N} m_i \mathbf{u}_i = 0$. Thus,

$$
\begin{aligned}
\mathbf{L} &= \sum_{i=1}^{N} m_i \mathbf{r}_i \times \dot{\mathbf{r}}_i \\
&= \sum_{i=1}^{N} m_i \Big[Q\mathbf{u}_i + \mathbf{R} \Big] \times \Big[Q(\omega \times \mathbf{u}_i) + \dot{\mathbf{R}} \Big] \\
&= Q\left(\sum_{i=1}^{N} m_i \mathbf{u}_i \times (\omega \times \mathbf{u}_i) \right) + M\mathbf{R} \times \dot{\mathbf{R}}
\end{aligned}
$$

Here we have used the fact that rotation matrices distribute over cross products: $Q(v \times w) = Qv \times Qw$ (exercise). If we denote by $A\omega$ the expression in the large parentheses in the last line above, we get the following expression for the total angular momentum of the system:

Rigid-Body Angular Momentum:

$$
\mathbf{L} = QA\omega + M\mathbf{R} \times \dot{\mathbf{R}}. \tag{8.103}
$$

The notation $A\omega$ introduces an important operator A into the theory. This operator is defined by the following:

Definition 8.8 (The Inertia Operator A) Suppose $m_1, \ldots, m_N$ are given positive numbers (the masses) and $\mathbf{u}_1, \ldots, \mathbf{u}_N$ are given vectors (the particle positions relative to the initial center of mass) with $\sum_{i=1}^{N} m_i \mathbf{u}_i = 0$. These data comprise a "rigid body." Associated with this rigid body is the linear operator $A : \mathbb{R}^3 \to \mathbb{R}^3$ defined by

$$
Av \equiv \sum_{i=1}^{N} m_i \mathbf{u}_i \times (v \times \mathbf{u}_i), \tag{8.104}
$$

for $v \in \mathbb{R}^3$. This operator is called the *inertia operator* or *inertia tensor* for the rigid body.

The operator A is more than just convenient notation. It captures the geometrical characteristics of the rigid body that are important for the nature of its motion (as we shall see). Note that A does *not* depend on the time, but rather depends only on the initial (relative) positions $\mathbf{u}_1, \ldots, \mathbf{u}_N$ and masses of the particles.

The central differential equation for ω comes from rewriting the angular momentum equation $\dot{\mathbf{L}} = \mathbf{T}$ in terms of the above notation. First note that differentiating both sides of equation (8.103) gives

$$
\begin{aligned}
\dot{\mathbf{L}} &= QA\dot{\omega} + \dot{Q}A\omega + M\mathbf{R} \times \ddot{\mathbf{R}} \\
&= Q(A\dot{\omega} + \omega \times A\omega) + M\mathbf{R} \times \ddot{\mathbf{R}}.
\end{aligned}
\tag{8.105}
$$

Then calculating the total torque gives

$$
\begin{aligned}
\mathbf{T} &= \sum_{i=1}^{N} \mathbf{r}_i \times F_i \\
&= \sum_{i=1}^{N} (Q\mathbf{u}_i + \mathbf{R}) \times F_i \\
&= \left(\sum_{i=1}^{N} Q\mathbf{u}_i \times F_i \right) + \mathbf{R} \times \mathbf{F}.
\end{aligned}
\tag{8.106}
$$

Substituting these expressions into $\dot{\mathbf{L}} = \mathbf{T}$ and reducing each side using the fact that $m\ddot{\mathbf{R}} = \mathbf{F}$, gives

$$
Q(A\dot{\omega} + \omega \times A\omega) = \sum_{i=1}^{N} Q\mathbf{u}_i \times F_i,
$$

i.e.,

$$
Q(A\dot{\omega} + \omega \times A\omega) = Q\left(\sum_{i=1}^{N} \mathbf{u}_i \times Q^T F_i \right).
$$

Here the factorization of the Q on the left-hand side follows from the general identity: $Qv \times Qw = Q(v \times w)$. Canceling the Q in the last equation above gives us the desired differential equation for ω. Putting all of the above together we arrive at the following:

The Rigid-Body Equations of Motion:

$$
M\ddot{\mathbf{R}} = \mathbf{F}
\tag{8.107}
$$

$$
A\dot{\omega} + \omega \times A\omega = \sum_{i=1}^{N} \mathbf{u}_i \times Q^T F_i
\tag{8.108}
$$

$$
\dot{Q} = Q\Omega.
\tag{8.109}
$$

The unknowns here are $\mathbf{R}, \omega$, and Q (recall that Ω is an expression involving ω as in (8.101)). Thus, the above is a system of 15 scalar DEs for the 15 unknowns R_i, ω_i, Q_{ij}, $i, j = 1, 2, 3$. The initial conditions are (in terms of the original unknowns)

$$\mathbf{R}(0) = \sum_{i=0}^{N} \frac{m_i}{M} \mathbf{r}_i(0)$$

$$\dot{\mathbf{R}}(0) = \sum_{i=0}^{N} \frac{m_i}{M} \dot{\mathbf{r}}_i(0)$$

$$Q(0) = I$$

$$A\omega(0) = \mathbf{L}(0) - M\mathbf{R}(0) \times \dot{\mathbf{R}}(0).$$

The last initial condition only determines the initial value of ω when the inertia operator A is invertible. The case when A is not invertible occurs, as we shall see, only when all the bodies initially lie on a straight line. This is called the *degenerate case*.

The first differential equation (8.107) determines the motion of the center of mass, while the DE (8.108) determines the angular velocity vector ω, which in turn gives Ω and thus, via the last DE (8.109), the rotation Q of the particles about the center of mass.

To be more specific about the actual motion, we will look at some special cases and examples below. Before doing so we discuss a few additional results for rigid-body motion in general.

8.4.2 Kinetic Energy and Moments of Inertia

The general rigid-body motion under consideration is characterized by the initial configuration of the particles comprising the system, i.e., by the initial positions $\mathbf{u}_1, \ldots, \mathbf{u}_N$ (relative to the initial center of mass $\mathbf{R}(0)$) and by the masses $m_1, \ldots, m_N$, with $\sum_{i=1}^{N} m_i \mathbf{u}_i = 0$ and $\sum_{i=1}^{N} m_i = M$. This system *is* the "rigid body" and moves as a unit following the center of mass while rotating about it. While the system is a discrete system of finitely many (though possibly large number of) particles, the discussion and concepts extend to a continuum of infinitely many particles, which is what we typically think of as a rigid body. This will be treated in the exercises.

The influence of the "shape" of the rigid body on the motion is encoded in the inertia operator A and the next proposition describes the nature of A and shows how it enters into the expression for the kinetic energy of the system.

Proposition 8.5 *Let A be the inertia operator defined by*

$$Av = \sum_{i=1}^{N} m_i \mathbf{u}_i \times (v \times \mathbf{u}_i),$$

for $v \in \mathbb{R}^3$. Then

$$Av \cdot w = \sum_{i=1}^{N} m_i (w \times \mathbf{u}_i) \cdot (v \times \mathbf{u}_i), \tag{8.110}$$

for all $v, w \in \mathbb{R}^3$. In particular,

$$Av \cdot v = \sum_{i=1}^{N} m_i |v \times \mathbf{u}_i|^2 \geq 0, \tag{8.111}$$

for every $v \in \mathbb{R}^3$.

Hence A is a symmetric, positive semidefinite matrix, and thus its eigenvalues I_1, I_2, I_3 are nonnegative:

$$0 \leq I_1 \leq I_2 \leq I_3,$$

and three corresponding eigenvectors

$$Ae_i = I_i e_i, \qquad i = 1, 2, 3,$$

can be chosen so that they are orthogonal $e_i \cdot e_j = 0$, for $i \neq j$.
If $K = \frac{1}{2} \sum_{i=1}^{N} m_i |\dot{\mathbf{r}}|^2$ is the total kinetic energy of the system, then

$$K = \tfrac{1}{2} A\omega \cdot \omega + \tfrac{1}{2} M |\dot{\mathbf{R}}|^2. \tag{8.112}$$

Thus, the kinetic energy splits into two parts. The first part, called the **rotational kinetic energy***, is due to the instantaneous rotation, or spin, about the axis determined by the angular velocity vector ω. The second part, called the* **translational kinetic energy***, is equivalent to the kinetic energy of a particle of mass M located at and moving with the center of mass of the system.*

Proof: The identity (8.110) follows easily from the property

$$(a \times b) \cdot c = (c \times a) \cdot b,$$

for the cross product. Thus,

$$Av \cdot w = \sum_{i=1}^{N} m_i \Big[(\mathbf{u}_i \times (v \times \mathbf{u}_i) \Big] \cdot w$$

$$= \sum_{i=1}^{N} m_i (w \times \mathbf{u}_i) \cdot (v \times \mathbf{u}_i).$$

Next, since the expression on the right-hand side of identity (8.110) does not change if we interchange v and w, it follows that

$$Av \cdot w = Aw \cdot v,$$

for all $v, w \in \mathbb{R}^3$. Hence A is a symmetric matrix. If we take $w = v$ in identity (8.110), we get identity (8.111). Hence A is a positive definite matrix. Identity (8.111) also shows that if $Av = 0$, then $v \times \mathbf{u}_i = 0$, for $i = 1, \ldots, N$. Hence, either $v = 0$ or there are constants c_i such that $\mathbf{u}_i = c_i v$, for $i = 1, \ldots, N$. This shows that if 0 is an eigenvalue of A, then necessarily the $\mathbf{u}_i$'s lie on the same line through the origin. The converse of this assertion is easy to show as well (exercise).

The other assertions about the eigenvalues/vectors of A are standard results from linear algebra about symmetric, positive semidefinite matrices.

The calculation of the kinetic energy uses $\sum_{i=1}^{N} m_i \mathbf{u}_i = 0$, $|Qv| = |v|$, and the identity $|a \times b|^2 = (a \times b) \cdot (a \times b) = [(b \times (a \times b)] \cdot a$. Thus, we find

$$K = \tfrac{1}{2} \sum_{i=1}^{N} m_i |\dot{\mathbf{r}}_i|^2 = \tfrac{1}{2} \sum_{i=1}^{N} m_i |Q(\omega \times \mathbf{u}_i) + \dot{\mathbf{R}}|^2$$

$$= \tfrac{1}{2} \sum_{i}^{N} m_i \Big[|\omega \times \mathbf{u}_i|^2 + 2\dot{\mathbf{R}} \cdot (\omega \times \mathbf{u}_i) + |\dot{\mathbf{R}}|^2 \Big]$$

$$= \tfrac{1}{2} \sum_{i=1}^{N} m_i [\mathbf{u}_i \times (\omega \times \mathbf{u}_i)] \cdot \omega + \tfrac{1}{2} M |\dot{\mathbf{R}}|^2$$

$$= \tfrac{1}{2} A\omega \cdot \omega + \tfrac{1}{2} M |\dot{\mathbf{R}}|^2$$

This completes the proof. $\square$

As in the proposition, we will always label and order the eigenvalues of A as $I_1 \leq I_2 \leq I_3$ and we will select of three corresponding eigenvectors e_1, e_2, e_3, that are orthogonal, of unit length $|e_i| = 1$, $i = 1, 2, 3$, and such that $\{e_1, e_2, e_3\}$ is positively oriented (a right-hand frame). When the eigenvalues are distinct this selection is unique. When there are only two distinct eigenvalues, say $I_1 = I_2 \neq I_3$, the selection of two orthonormal vectors e_1, e_2 from the eigenspace E_{I_1} can be done in infinitely many ways, but having made one choice, then e_3 is uniquely determined. When all the eigenvalues are the same, the eigenspace E_{I_1} is all of $\mathbb{R}^3$ and so any positively oriented, orthonormal basis $\{e_1, e_2, e_3\}$ for $\mathbb{R}^3$ can be selected.

As is customary, we identify an operator on $\mathbb{R}^n$ with the $n \times n$ matrix which represents it with respect to the standard unit vector basis for $\mathbb{R}^n$. In the present case $n = 3$ and the standard unit vector basis $\{\varepsilon_1, \varepsilon_2, \varepsilon_3\}$ for $\mathbb{R}^3$ is

$$\varepsilon_1 = (1, 0, 0), \quad \varepsilon_2 = (0, 1, 0), \quad \varepsilon_3 = (0, 0, 1).$$

Thus, the inertia operator $A : \mathbb{R}^3 \to \mathbb{R}^3$ is identified with the 3×3 matrix $A = \{A_{ij}\}_{i,j=1,2,3}$, where the i-jth entry of A is

$$A_{ij} \equiv A\varepsilon_j \cdot \varepsilon_i.$$

Using the definition of A, it is easy to derive the following explicit representation of A.

Proposition 8.6 *The inertia operator* $A = \{A_{ij}\}_{i,j=1,2,3}$ *has entries*

$$A_{ij} = \sum_{k=1}^{N} m_k \left[|\mathbf{u}_k|^2 \delta_{ij} - u_{ki} u_{kj} \right], \tag{8.113}$$

where δ_{ij} *is the Kronecker delta symbol (i.e.,* $\delta_{ij} = 0$*, for* $i \neq j$*,* $\delta_{ii} = 1$*), and*

$$\mathbf{u}_k = (u_{k1}, u_{k2}, u_{k3}),$$

for $k = 1, \ldots N$*. Alternatively, without the use of the Kronecker delta symbol this can be written as*

$$A_{ii} = \sum_{k=1}^{N} m_k \left[|\mathbf{u}_k|^2 - u_{ki}^2 \right] \tag{8.114}$$

$$A_{ij} = -\sum_{k=1}^{N} m_k u_{ki} u_{kj} \qquad \text{(for } i \neq j\text{).} \tag{8.115}$$

Proof: Exercise.

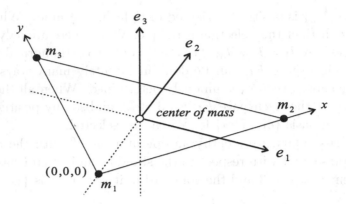

Figure 8.27: *A rigid body consisting of three particles with equal masses*
$m_k = 1$, $k = 1, 2, 3$, *and the three principal axes of inertia* e_1, e_2, e_3.

Example 8.6 Suppose there are three particles with equal masses, say
$m_k = 1$, $k = 1, 2, 3$ comprising the rigid body, and that the initial positions of these particles are

$$\begin{aligned}
\mathbf{r}_1(0) &= (0, 0, 0) \\
\mathbf{r}_2(0) &= (1, 0, 0) \\
\mathbf{r}_3(0) &= (0, 1, 0).
\end{aligned}$$

See Figure 8.27. To calculate the inertia tensor, we must first calculate the
relative positions $\mathbf{u}_k$, $k = 1, 2, 3$. For this note that the initial center of mass
is

$$\mathbf{R}(0) = \tfrac{1}{3}(0, 0, 0) + \tfrac{1}{3}(1, 0, 0) + \tfrac{1}{3}(0, 1, 0) = (\tfrac{1}{3}, \tfrac{1}{3}, 0).$$

Then we get

$$\begin{aligned}
\mathbf{u}_1 &= \mathbf{r}_1(0) - \mathbf{R}(0) = (-\tfrac{1}{3}, -\tfrac{1}{3}, 0) \\
\mathbf{u}_2 &= \mathbf{r}_2(0) - \mathbf{R}(0) = (\tfrac{2}{3}, -\tfrac{1}{3}, 0) \\
\mathbf{u}_3 &= \mathbf{r}_3(0) - \mathbf{R}(0) = (-\tfrac{1}{3}, \tfrac{2}{3}, 0).
\end{aligned}$$

To compute the inertia operator A, we use formulas (8.114)-(8.115). Note
that formula (8.114) says that $A_{11} = \sum_{k=1}^{N} m_k(u_{k2}^2 + u_{k3}^2)$ and so in this
example

$$A_{11} = \tfrac{1}{9} + \tfrac{1}{9} + \tfrac{4}{9} = \tfrac{2}{3}.$$

Similarly A_{22} involves the sum of the squares of the 1st and 3rd components of $\mathbf{u}_k$, while A_{33} involves the sum of the squares of the 1st and 2nd

components. Thus, we get

$$A_{22} = \tfrac{1}{9} + \tfrac{4}{9} + \tfrac{1}{9} = \tfrac{2}{3}$$
$$A_{33} = \tfrac{1}{9} + \tfrac{1}{9} + \tfrac{4}{9} + \tfrac{1}{9} + \tfrac{1}{9} + \tfrac{4}{9} = \tfrac{4}{3}.$$

To calculate A_{ij} for $i \neq j$, note that for this example $u_{k3} = 0$ for every k and so

$$A_{13} = 0 = A_{23}.$$

For A_{12} we find

$$A_{12} = -[u_{11}u_{12} + u_{21}u_{22} + u_{31}u_{32}] = -[\tfrac{1}{9} - \tfrac{2}{9} - \tfrac{2}{9}] = \tfrac{1}{3}.$$

Since the matrix A is symmetric, these are the only calculations we need. Thus,

$$A = \begin{bmatrix} \tfrac{2}{3} & \tfrac{1}{3} & 0 \\ \tfrac{1}{3} & \tfrac{2}{3} & 0 \\ 0 & 0 & \tfrac{4}{3} \end{bmatrix}.$$

From this it is easy to compute the following eigenvalues and eigenvectors of A:

$$I_1 = \tfrac{1}{3}, \quad e_1 = \tfrac{1}{\sqrt{2}}(-1, 1, 0)$$
$$I_2 = 1, \quad e_2 = \tfrac{1}{\sqrt{2}}(1, 1, 0)$$
$$I_3 = \tfrac{4}{3}, \quad e_3 = (0, 0, 1).$$

Figure 8.27 shows the three eigenvectors plotted with their initial points at the center of mass.

To interpret further the geometric significance of the eigenvalues and eigenvectors of the inertia operator, we look at the concept of moments of inertia for a rigid body.

As shown in Appendix A, each symmetric matrix B gives rise to a function $f(v) \equiv (Bv \cdot v)/(v \cdot v)$ whose extreme values are respectively the largest and smallest eigenvalues of B and these values are assumed at respective eigenvectors v of B. In the present setting, the function f is known as the moment of inertia function.

Definition 8.9 (Moment of Inertia) The function $I : \mathbb{R}^3 \setminus \{0\} \to \mathbb{R}$ defined by

$$I(v) = \frac{Av \cdot v}{v \cdot v},$$

is called the *moment of inertia function*. The number $I(v)$ is called the *moment of inertia of the system about the axis through the center of mass and in the direction of v*. Note that $I(cv) = I(v)$ for all nonzero scalars c and so I has the same value for all nonzero vectors in the line (axis) determined by v. The eigenvalues $I_1 \leq I_2 \leq I_3$ of A are called the *principal moments of inertia*.

It follows from the general theory that the principal moments of inertia are values of I at the eigenvectors of A:

$$I_i = I(e_i),$$

$i = 1, 2, 3$, and that

$$I_1 \leq I(v) \leq I_3,$$

for every $v \neq 0$ in $\mathbb{R}^3$. Thus, I_1 is the minimum moment of inertia of the system and it occurs as the system spins about the axis through e_1. Likewise, I_3 is the maximum moment of inertia of the system and it occurs as the system spins about the axis through e_3.

This is illustrated by the elementary configuration in Example 8.6. Spinning the configuration about the axis through e_1 (see Figure 8.27) gives the least moment of inertia $I_1 = 1/3$, while a spin about the axis through e_2 results in a larger moment of inertia $I_2 = 1$. This corresponds to the fact that the bodies are closer to the axis of revolution in the 1st case than in the 2nd case (and all the masses are the same). The greatest moment of inertia $I_3 = 4/3$ occurs for a revolution about the axis through e_3, where the bodies are furthest removed from the axis of revolution.

An alternative expression for the moment of inertia $I(v)$ helps clarify the idea alluded to in the last paragraph. This expression arises from the observation that $|v \times \mathbf{u}_i|$ is the area of the parallelogram determined by v and $\mathbf{u}_i$. This area is $|v| \, d_i(v)$, where $d_i(v)$ denotes the distance from the tip of $\mathbf{u}_i$ to the line through v. See Figure 8.28.

Thus, we find

$$Av \cdot v = \sum_{i=1}^{N} m_i |v \times \mathbf{u}_i|^2$$

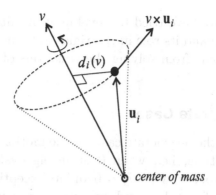

Figure 8.28: *The moment of inertia of the system of particles about the axis determined by v is the sum, over all particles $\mathbf{u}_i$, of the mass of the particle times the square of the particle's distance $d_i(v)$ from the axis of revolution.*

$$= |v|^2 \sum_{i=1}^{N} m_i \, d_i(v)^2.$$

This gives that the moment of inertia function is also expressed by

$$I(v) = \sum_{i=1}^{N} m_i \, d_i(v)^2. \tag{8.116}$$

This is the traditional definition of the moment of inertia of the system about an axis v. It is *the sum, over all particles, of the mass of the particle times the square of the particle's distance from the axis of revolution.* Note: In the setup here, $\mathbf{u}_i$, $i = 1, \ldots, N$, are the positions of the particles relative to the initial center of mass and so the moment of inertia above is about an axis through the center of mass.

The name for the moment of inertia arises from the way this quantity enters into the expression for the kinetic energy of the system. Rewriting the kinetic energy K from equation (8.112) in terms of I gives the following

Rigid-Body Kinetic Energy

$$K = \tfrac{1}{2}I(\omega)|\omega|^2 + \tfrac{1}{2}M|\dot{\mathbf{R}}|^2. \tag{8.117}$$

Thus, with respect to the angular velocity ω of the body, $I(\omega)$ plays the same role as mass does with respect to linear velocity. The greater the value of $I(\omega)$, the harder it is to stop (bring to rest) the spinning of the body about the axis through ω.

Up to this point we have tried to develop an intuitive understanding of the angular velocity ω and its role in the rigid-body motion. A more precise understanding will come from solving the equations of motion in a number of special cases.

8.4.3 The Degenerate Case

As mentioned above, the degenerate case for the motion of the system occurs when A is not invertible, i.e., when 0 is an eigenvalue of A. With our conventions, this case occurs when $I_1 = 0$ and its exceptional nature warrants a separate treatment from the nondegenerate case where all the principal moments of inertia $I_1 \leq I_2 \leq I_3$ are positive.

The case of a zero eigenvalue for A occurs only when the initial relative positions of the bodies are collinear, i.e., $\mathbf{u}_i, i = 1, \ldots, N$, lie on the same line through the origin. To see this note that, in the previous notation, the eigenvector e_1 corresponding to $I_1 = 0$ necessarily is in the null space of A, i.e., $Ae_1 = 0$. But then by identity (8.111), it follows that

$$0 = Ae_1 \cdot e_1 = \sum_{i=1}^{N} m_i |e_1 \times \mathbf{u}_i|^2.$$

Consequently, $e_1 \times \mathbf{u}_i = 0$ for each $i = 1, \ldots, N$, and so there exist constants $c_1, \ldots, c_N$, such that

$$\mathbf{u}_i = c_i e_1,$$

for $i = 1, \ldots, N$ (exercise). This says that all the bodies initially lie the line through the center of mass in the direction of e_1 (the 1st principal axis). See Figure 8.29.

Note also that $\sum_{i=1}^{N} m_i c_i = 0$. If we use this in the definition of the inertia operator, we find that in the degenerate case it has the form

$$Av = c[e_1 \times (v \times e_1)] = c[v - (e_1 \cdot v)e_1],$$

where $c = \sum_{i=1}^{N} m_i c_i^2$. This says that $A = cP$, where P is the orthogonal projection on the plane through the origin which is perpendicular to e_1.

Not only are the bodies collinear initially in the degenerate case, but they also remain collinear throughout the motion. This is so since

$$\mathbf{r}_i(t) = c_i Q(t)e_1 + \mathbf{R}(t), \tag{8.118}$$

for any $t \in I$, and so all the bodies lie on the line through $\mathbf{R}(t)$ in the direction of $Q(t)e_1$.

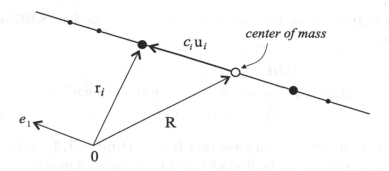

Figure 8.29: *In the degenerate case all the bodies initially lie on a line through the center of mass.*

In discussing the initial conditions for the rigid-body differential equations, we mentioned that the initial value $\omega(0)$ for the angular velocity vector ω must be determined from the equation

$$A\omega(0) = \mathbf{L}(0) - M\mathbf{R}(0) \times \dot{\mathbf{R}}(0),$$

where $\mathbf{L}(0), \mathbf{R}(0), \dot{\mathbf{R}}(0)$ are known. In the degenerate case A is not invertible and so it is not clear immediately how to choose $\omega(0)$ so that this equation holds. However, note that without loss of generality we can assume that the initial center of mass is at the origin: $\mathbf{R}(0) = 0$. Thus, from equation (8.118), we get that $\mathbf{r}_i(0) = c_i e_1$, for $i = 1, \ldots, N$. In addition, the equation for $\omega(0)$ reduces to

$$A\omega(0) = \mathbf{L}(0).$$

Furthermore, if we let $h = \sum_{i=1}^{N} m_i c_i \dot{r}_i(0)$, which is a known vector, then the initial angular momentum can be written as

$$\mathbf{L}(0) = \sum_{i=1}^{N} m_i c_i e_1 \times \dot{r}_i(0) = e_1 \times \left(\sum_{i=1}^{N} m_i c_i \dot{r}(0) \right) = e_1 \times h.$$

Hence taking

$$\omega(0) = c^{-1} e_1 \times h,$$

gives a solution of $A\omega(0) = \mathbf{L}(0)$.

8.4.4 Euler's Equation

Consider the case where the total force and torque on the system vanish: $\mathbf{F} = 0$ and $\mathbf{T} = 0$. From $\mathbf{T} = 0$, one gets that $\sum_{i=1}^{N} Q \mathbf{u}_i \times F_i = 0$ (see equation (8.106). Thus, the rigid-body equations of motion (8.107)-(8.109) reduce to

$$M\ddot{\mathbf{R}} = 0 \tag{8.119}$$
$$A\dot{\omega} + \omega \times A\omega = 0 \qquad \text{(Euler's Equation)} \tag{8.120}$$
$$\dot{Q} = Q\Omega. \tag{8.121}$$

As expected the 1st equation gives that $\mathbf{R}(t) = \dot{\mathbf{R}}(0)t + \mathbf{R}(0)$, so the center of mass moves in a straight line with constant speed. Equation (8.120) is known as *Euler's equation* and it's complete solution and interpretation is given as follows.

Euler's equation is a 1st-order, nonlinear system of three equations for three unknowns, the components of ω. As usual, the first part of the analysis of a system of DEs consists of finding and classifying fixed points.

Fixed Points: The fixed points of Euler's equation are *constant* angular velocity vectors ω and they must satisfy the algebraic equation:

$$\omega \times A\omega = 0.$$

This is equivalent to saying that ω and $A\omega$ lie on the same line, i.e., that ω is an eigenvector of the inertia operator A:

$$A\omega = \lambda\omega,$$

where $\lambda = I_1, I_2$, or I_3. Hence if the principal moments of inertia $I_i, i = 1, 2, 3$ are all distinct, then the fixed points of Euler's equation are precisely the points on the three principal axes: $\text{span}\{e_i\}$, $i = 1, 2, 3$. If two of the moments of inertia (i.e., eigenvalues) are the same, say $I_1 = I_2$, then the corresponding eigenspace $E_{I_1} = \text{span}\{e_1, e_2\}$ constitutes a plane of fixed points and the remaining principal axis $E_{I_3} = \text{span}\{e_3\}$ is a line of fixed points perpendicular to this plane. Similarly, if $I_2 = I_3$. If all the principal moments of inertia are the same $I_1 = I_2 = I_3$, then the body is called *perfectly symmetric*. In this case every point in $\mathbb{R}^3$ is a fixed point of Euler's equation and thus fixed points are the only types of solutions. See Figure 8.30.

We will determine the stability of the fixed points later, but first we examine what type of rigid-body motion corresponds to a fixed point of Euler's equation. When $\omega = (\omega_1, \omega_2, \omega_3)$ is a fixed point of the Euler equations (eigenvector of A), the corresponding rigid-body motion is particularly

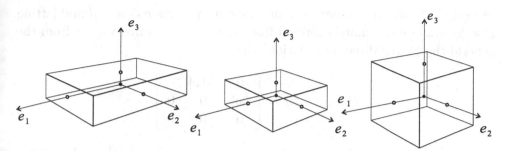

Figure 8.30: *Three examples illustrating rigid bodies with* (a) $I_1 < I_2 < I_3$, (b) $I_1 = I_2 < I_3$, *and* (c) $I_1 = I_2 = I_3$, *respectively. In case* (c) *the body is called perfectly symmetric.*

simple. To see this, observe that since w is a constant, the angular velocity matrix

$$\Omega = \begin{bmatrix} 0 & -\omega_3 & \omega_2 \\ \omega_3 & 0 & -\omega_1 \\ -\omega_2 & \omega_1 & 0 \end{bmatrix} \tag{8.122}$$

is constant as well. Hence the solution of $\dot{Q} = Q\Omega$ is simply

$$Q(t) = e^{\Omega t}.$$

Using the techniques from Chapter 4 on linear systems, we can demonstrate that, for each t, $Q(t)$ is a rotation about the line through w and that the speed of rotation (angular speed) is $|w|$. The details of this are as follows and, of course, the discussion is valid for any 3×3, skew symmetric matrix Ω.

We first determine the Jordan canonical form for Ω and use this to compute $e^{\Omega t}$.

With Ω expressed in the form (8.122), it is easy to calculate that

$$\det(\Omega - \lambda I) = \lambda(\lambda^2 + |w|^2),$$

and hence the eigenvalues of Ω are $\lambda = 0, \pm|w|i$ (exercise). Further, since $\Omega w = w \times w = 0$, it follows that w is an eigenvector corresponding to eigenvalue 0. One can verify that a complex eigenvector v corresponding to eigenvalue $|w|i$ is

$$v = \begin{bmatrix} \omega_1\omega_3 + |w|\omega_2 i \\ \omega_2\omega_3 - |w|\omega_1 i \\ -(\omega_1^2 + \omega_2^2) \end{bmatrix}$$

(exercise). Separating v into real and imaginary parts $v = u + qi$ and letting $P = [q, u, \omega]$ be the matrix formed from q, u, ω as columns, we get from the general theory (or direct calculation) that

$$P^{-1}\Omega P = J = \begin{bmatrix} 0 & -|\omega| & 0 \\ |\omega| & 0 & 0 \\ 0 & 0 & 0 \end{bmatrix}$$

is the Jordan canonical form for Ω. Thus, for a point $c \in \mathbb{R}^3$, let $b = P^{-1}c$. Then

$$c = Pb = b_1 q + b_2 u + b_3 \omega,$$

where $b = (b_1, b_2, b_3)$. Hence, as in Chapter 4, we find that

$$\begin{aligned} Q(t)c &= e^{\Omega t} c = e^{\Omega t} Pb = Pe^{Jt} P^{-1} Pb \\ &= (b_1 q + b_2 u) \cos|\omega|t + (-b_2 q + b_1 u) \sin|\omega|t + b_3 \omega. \quad (8.123) \end{aligned}$$

The motion of c described by equation (8.123) is a circular motion about the axis: $\text{span}\{\omega\}$, with constant angular speed $|\omega|$. See Figure 8.31.

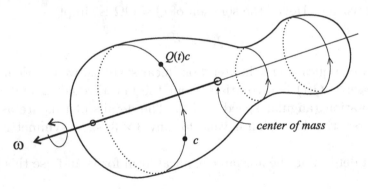

Figure 8.31: *A fixed point ω of Euler's equation corresponds to a rigid-body motion where each point c in the body undergoes uniform circular motion $t \mapsto Q(t)c$ about the axis through ω.*

Summary (Motion Corresponding to a Fixed Point): *For a fixed point ω of Euler's equation, the corresponding motion of the rigid body is one where each of its particles undergoes a uniform rotation, with constant angular speed $|\omega|$, about the axis: $\text{span}\{\omega\}$. When all three moments of inertia are distinct, then a constant spinning motion about one of the three* *principal axes* e_1, e_2, e_3 *is a possible movement of the body subject to no total force or torque. When two of the moments of the inertia are the same, say* $I_1 = I_2$, *then the body can spin with constant speed about the axis* e_3 *as well as about any axis lying in the plane spanned by* e_1, e_2. *In the case when all three principal moments of inertia are the same, i.e., when the body is perfectly symmetric, then the only possible motion of the body, subject to no external force and torque, is a constant spin about some fixed axis in space.*

8.4.5 The General Solution of Euler's Equation

The general solution to Euler's equation is easier to describe if we write this equation in terms of the axes for the principal moments of inertia.

Note: For convenience of notation we *now* use w_1, w_2, w_3 to denote the components of w relative to the eigenbasis for A:

$$w(t) = w_1(t)e_1 + w_2(t)e_2 + w_3(t)e_3. \tag{8.124}$$

Hopefully this will not cause confusion with what we did above, where the w_i's were the components with respect to the standard basis for $\mathbb{R}^3$ (and were constants). Now since $Ae_i = I_i e_i$, $i = 1, 2, 3$, it follows that

$$Aw = I_1 w_1 e_1 + I_2 w_2 e_2 + I_3 w_3 e_3 \tag{8.125}$$

and

$$A\dot{w} = I_1 \dot{w}_1 e_1 + I_2 \dot{w}_2 e_2 + I_3 \dot{w}_3 e_3. \tag{8.126}$$

Using these and a direct computation of $w \times Aw$ in terms of the e_i's, one can easily compute Euler's equation:

$$A\dot{w} + w \times Aw = 0,$$

in terms of the above expressions and get the following:

Euler's Equation in Principal Axes Coordinates:

$$\dot{w}_1 = \left(\frac{I_2 - I_3}{I_1}\right) w_2 w_3 \tag{8.127}$$

$$\dot{w}_2 = \left(\frac{I_3 - I_1}{I_2}\right) w_3 w_1 \tag{8.128}$$

$$\dot{w}_3 = \left(\frac{I_1 - I_2}{I_3}\right) w_1 w_2 \tag{8.129}$$

Note that, in each equation, the expression in the large parentheses is a constant since the principal moments of inertia I_1, I_2, I_3 are constants. We are also assuming that $I_k \neq 0$, $k = 1, 2, 3$.

This system has a particularly nice form and, indeed, one can readily discern two conservation laws that follow directly from it. These two laws completely determine the solutions of the system, as we shall see.

Suppose, then, that $\omega : I \rightarrow \mathbb{R}^3$ is a solution (integral curve) of Euler's equation. One conservation law comes about simply by multiplying the ith equation in the system by $I_i \omega_i$ (for $i = 1, 2, 3$) and then adding all three equations together. Thus,

$$I_1 \omega_1 \dot{\omega}_1 + I_2 \omega_2 \dot{\omega}_2 + I_3 \omega_3 \dot{\omega}_3 =$$
$$(I_2 - I_3)\omega_1\omega_2\omega_3 + (I_3 - I_1)\omega_1\omega_2\omega_3 + (I_1 - I_2)\omega_1\omega_2\omega_3 = 0.$$

Thus, there exists a constant k, which depends on the integral curve ω, such that:

Conservation Law I (Rotational Kinetic Energy):

$$I_1 \omega_1^2 + I_2 \omega_2^2 + I_3 \omega_3^2 = 2k, \tag{8.130}$$

for all $t \in I$.

A second conservation law results, somewhat less transparently, from multiplying the ith equation in the system by $I_i^2 \omega_i$ (for $i = 1, 2, 3$) and then adding all three equations together. One now gets the slightly different expression

$$I_1^2 \omega_1 \dot{\omega}_1 + I_2^2 \omega_2 \dot{\omega}_2 + I_3^2 \omega_3 \dot{\omega}_3 =$$
$$\left[(I_2 - I_3)I_1 + (I_3 - I_1)I_2 + (I_1 - I_2)I_3 \right] \omega_1\omega_2\omega_3 = 0.$$

Thus, there exists a constant $a > 0$, which depends on the integral curve ω, such that:

Conservation Law II (Angular Momentum):

$$I_1^2 \omega_1^2 + I_2^2 \omega_2^2 + I_3^2 \omega_3^2 = a^2, \tag{8.131}$$

for all $t \in I$.

It is easy to see from equations (8.124)-(8.125) that if $\dot{\mathbf{R}}(0) = 0$, then the constant $k = A\omega \cdot \omega$ is the rotational kinetic energy of the system. If we also assume that $\mathbf{R}(0) = 0$ (which is no loss of generality) then the constant $a = |A\omega| = |\mathbf{L}|$ is the magnitude of the angular momentum of the system. Thus, the two equations (8.130)-(8.131) are the laws for the *conservation of rotational kinetic energy and angular momentum* of the rigid body.

Equations (8.130)-(8.131) can be considered as the equations for two ellipsoids in $\mathbb{R}^3$ relative to the principal axes e_1, e_2, e_3. Even though ω_1, ω_2, and ω_3 in these equations are time-dependent functions, we will, when there is no danger of confusion, use $\omega_1, \omega_2, \omega_3$ to denote the coordinates of points in $\mathbb{R}^3$. Note that the three axes of each ellipsoid coincide with these principal axes of inertia. These two ellipsoids are called the *kinetic energy ellipsoid* and the *angular momentum ellipsoid*, respectively. Historically, the kinetic energy ellipsoid is also called the *Poinsot ellipsoid*.

The content of the two conservation laws is that any integral curve $\omega :$ $I \to \mathbb{R}^3$ of the Euler equation must lie on the surface of each ellipsoid and hence on the curve of intersection of these two surfaces. Thus, the curves of intersection are, except for the explicit time dependence, the same as the integral curves themselves.

Visualizing the curves of intersection of these ellipsoids, and hence the integral curves of Euler's equation, is difficult to do by hand. However, there are several ways of rewriting the conservation laws that lead to easier visualization. One way is to multiply equation (8.130) by $a^2/2k$ and then subtract equation (8.131) from it to get

$$I_1\left(\frac{a^2}{2k} - I_1\right)\omega_1^2 + I_2\left(\frac{a^2}{2k} - I_2\right)\omega_2^2 + I_3\left(\frac{a^2}{2k} - I_3\right)\omega_3^2 = 0. \qquad (8.132)$$

This equation describes a double cone in $\mathbb{R}^3$ or a degenerate case depending on the constants I_1, I_2, I_3, k, a. The reasoning leading to this gives us that each integral curve lies on this cone as well as the two ellipsoids. Since the intersection of two of these surfaces suffices to determine the integral curves, we will do the visualization with the cone C and kinetic energy (or Poinsot) ellipsoid E. Describing the intersection $C \cap E$ is considerably easier. Figure 8.32 shows a plot of one of the kinetic energy ellipsoids for a specific value of k and its intersection with one of the cones (8.132). As the picture shows, the axes of the ellipsoid coincide with the principal axes of inertia and the longest axis, of length $\sqrt{2k/I_1}$ corresponds to the smallest moment of inertia. We fix this ellipsoid and consider how the cone in (8.132), for various values of a, intersects this ellipsoid.

If we let $b_i = I_i(\frac{a^2}{2k} - I_i)$, for $i = 1, 2, 3$, then each of these is negative when $\frac{a^2}{2k} < I_1$ and each is positive when $\frac{a^2}{2k} > I_3$. Hence the equation for the

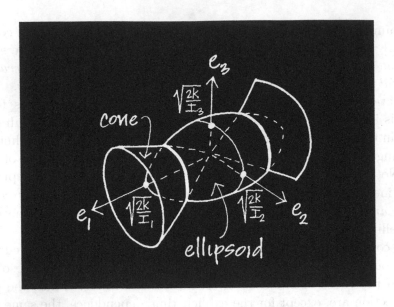

Figure 8.32: *A kinetic energy ellipsoid E and the curves obtained by intersecting it with a cone C.*

cone

$$b_1\omega_1^2 + b_2\omega_2^2 + b_3\omega_3^2 = 0$$

has no solution in either of these cases, and so there is no cone. Another way of saying this is the following: the magnitudes of angular momentum and rotational kinetic energy, a, k of any motion corresponding to a solution of the Euler equation must satisfy

$$I_1 \leq \frac{a^2}{2k} \leq I_3. \tag{8.133}$$

Thus, with k fixed, we restrict a so that $\frac{a^2}{2k}$ lies in the interval $[I_1, I_3]$ and look at the resulting family of cones $C = C_a$ as a varies over this interval. We divide into cases, since the description is different depending upon whether the moments of inertia are all distinct or not.

(1) **(Three Distinct Moments of Inertia)** Suppose $0 < I_1 < I_2 < I_3$. When a is the least that it can be, $b_1 = 0$, i.e., $\frac{a^2}{2k} = I_1$. Then the equation for the cone reduces to

$$b_2\omega_2^2 + b_3\omega_3^2 = 0,$$

with $b_2 > 0, b_3 > 0$ and, of course, this means that $\omega_2 = 0, \omega_3 = 0$ and the "cone" C reduces to a line along e_1. Thus, C is the first principal axis. Then $C \cap E = \{(\pm\sqrt{2k/I_1}, 0, 0)\}$ is a pair of points that are fixed points of the Euler equation.

Next, for a such that $I_1 < \frac{a^2}{2k} < I_2$, we have $b_1 > 0$, while $b_2 < 0, b_3 < 0$, and so the equation for C can be written as

$$\omega_1^2 = \frac{-b_2}{b_1}\omega_2^2 + \frac{-b_3}{b_1}\omega_3^2.$$

Thus, C is an elliptic cone with its axis coinciding with the first principal axis. The intersection $C \cap E$ is fairly easy to visualize—it consists of a pair of circular-like curves centered about the first principal axis. Figure 8.33 shows several of these pairs of curves.

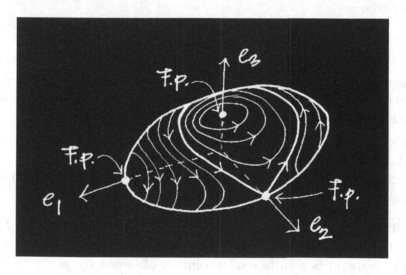

Figure 8.33: *Typical integral curves of Euler's equation when $I_1 < I_2 < I_3$. The integral curves lie on a specific kinetic energy ellipsoid and are separated into two sets. Each set consists of circular-like, or oval, curves, with one set centered about the principal axis e_1 and the other set centered about the principal axis e_3.*

These pairs of curves increase in size as a increases, but then when a reaches the value for which $\frac{a^2}{2k} = I_2$, the nature of $C \cap E$ changes. For this value of a, the coefficients $b_1 > 0, b_2 = 0, b_3 < 0$ and so the equation for C can be written as

$$\omega_1 = \pm\sqrt{\frac{-b_3}{b_1}}\,\omega_3.$$

Thus, C is a pair of planes, each of which is perpendicular to the plane span$\{e_1, e_3\}$. Intersecting this pair of planes with the ellipsoid E is easy to visualize and gives a crossing pair of curves as shown in Figure 8.33. This pair of curves separates the sets of other curves on the ellipsoid from each other. The pair actually consists of four integral curves, each of which runs from one of the fixed points on the e_2 axis to the other fixed point on this axis.

For a slightly larger value of a, so that $I_2 < \frac{a^2}{2k} < I_3$, the coefficients $b_1 > 0, b_2 > 0$, while $b_3 < 0$. So the equation for the cone can be written as

$$\omega_3^2 = \frac{b_1}{b_3}\omega_1^2 + \frac{b_2}{b_3}\omega_2^2.$$

Thus, the cone C is an elliptic cone with its axis along the third principal axis. As before, the intersection of $C \cap E$ is a pair of circular-like closed curves, but now centered about the third principal axis. See Figure 8.33.

Finally, when a is as large as possible, $a^2/2k = I_3$ and the cone C degenerates into a straight line corresponding with the third principal axis. Then $C \cap E$ is a pair of points that are fixed points for the Euler equation.

(2) **(Two Distinct Moments of Inertia).** Assume $I_1 = I_2 < I_3$. The other possibility: $I_1 < I_2 = I_3$, has a similar analysis. First note that the ellipsoid is a surface of revolution obtained by, say, revolving an ellipse in the e_2-e_3 plane about the e_3 axis. The analysis of how the cone C_a intersects this ellipsoid is as follows. For a the least it can be, we have $a^2/2k = I_1 = I_2$, and so the "cone" C degenerates into the plane $w_3 = 0$. In fact, for this value of a,

$$C = \text{span}\{e_1, e_2\} = E_{I_1}$$

is the two-dimensional eigenspace corresponding to the eigenvalue $I_1 = I_2$. Hence the intersection $C \cap E$ is a circle and each point on this circle is a fixed point of the Euler equation. See Figure 8.34.

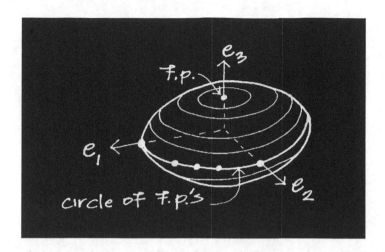

Figure 8.34: *Typical integral curves of the Euler equation when $I_1 = I_2 < I_3$.*
The integral curves lie on a specific kinetic energy ellipsoid and, except for
the fixed points, are circles with centers on the e_3 axis. There are two stable,
fixed points on the e_3 axis and a whole circle of unstable, fixed points lying
in the e_1-e_2 plane.

When a has a slightly larger value so that $I_2 < a^2/2k < I_3$, the cone C
is an circular cone whose axis coincides with the third principal axis.
Thus, the intersection $C \cap E$ is a pair of circular-like closed curves
centered on the third principal axis. See Figure 8.34.

Finally, when a is such that $a^2/2k = I_3$, the cone C_a degenerates into a
straight line that coincides with the third principal axis. Hence $C \cap E$ is
a pair of points and these points are fixed points of the Euler equation.
See Figure 8.34.

As the figure indicates, each fixed point from the plane $E_{I_1} = \text{span}\{e_1, e_2\}$
of fixed points is unstable, while each fixed point from the line $E_{I_1} =$
$\text{span}\{e_3\}$ of fixed points is stable.

(3) **(Only One Moment of Inertia)** In the case where all three moments
of inertia coincide: $I_1 = I_2 = I_3$, it is clear that there is only one value
of a for which C is not empty, i.e., for $a = \sqrt{2kI_3}$, and that for this
value, $C = \mathbb{R}^3$. In this case the entire space $\mathbb{R}^3$ consists of (stable)
fixed points of the Euler equation. Then too, $C \cap E = E$ is a sphere of
fixed points as well. See Figure 8.35. This is to be expected, since, as
we have mentioned, the rigid body is called perfectly symmetric when

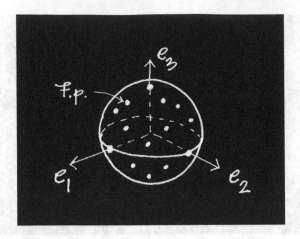

Figure 8.35: *Typical integral curves of the Euler equation when $I_1 = I_2 = I_3$, i.e., when the body is perfectly symmetric. In this case all of $\mathbb{R}^3$ consists of fixed points and so $C \cap E$ is a sphere of fixed points.*

all its principal moments of inertia are the same. The moments of inertia about *any* axis are the same and the only possible motions of the body (with total force and torque equal to zero) are spins about such axes with uniform angular velocity.

Stability of the Euler Fixed Points: The stability of the fixed points of the Euler equation is easily discerned from Figures 8.33-8.35 and depends on the number of distinct principal moments of inertia.

When there are three distinct moments of inertia as in Figure 8.33, then the fixed points on the axes e_3, e_3, corresponding to the least and greatest moments of inertia I_1 and I_3 are stable. The fixed points on the axis e_2 are unstable saddle points.

When there are two distinct principal moments of inertia as in Figure 8.34, then the fixed points on the axis e_3, corresponding to the greatest moment of inertia, are stable. Each of the fixed points on the circle in the e_1-e_2 plane is unstable.

When there is only one distinct principal moment of inertia, the fixed points, which comprise all of $\mathbb{R}^3$, are all stable.

Exercises 8.4

1. Suppose $v_1, \ldots, v_p \in \mathbb{R}^n$ and let $B = \{b_{ij}\}_{i,j=1}^p$ be the $p \times p$ matrix formed from the dot products of these vectors: $b_{ij} \equiv v_i \cdot v_j$. Show that $v_1, \ldots, v_p$ are linearly dependent if and only if $\det(B) = 0$.

2. Prove the assertions marked "exercise" in the proof of Theorem 8.4.

3. Prove the remaining part of Proposition 8.4.

4. Use Proposition 8.4 from Chapter 8 to show that rotation matrices distribute over cross products: $Q(v \times w) = Qv \times Qw$.

5. **(Matrix Lie Groups and Algebras)** Suppose that $G = \{g_{ij}\}_{i,j=1,\ldots,n}$ is a symmetric, positive definite matrix. Consider the following two sets of $n \times n$ matrices:

$$L_G = \{Q \in M_n \,|\, Q^T G Q = G\} \tag{8.134}$$
$$\mathcal{L}_G = \{\Omega \in M_n \,|\, \Omega G + \Omega^T G = 0\}. \tag{8.135}$$

In many applications G is the matrix of components for a metric. The sets $L_G, \mathcal{L}_G$ are then identified with the Lie group of isometries and Lie algebra of infinitesimal isometries of the metric. In this exercise do the following.

(a) Show that L_G is a group, i.e., (i) if $Q_1, Q_2 \in L_G$, then $Q_1 Q_2 \in L_G$, and (ii) if $Q \in L_G$, then Q is invertible and $Q^{-1} \in L_G$. Also show that $\det(Q) = \pm 1$, for every $Q \in L_G$.

(b) Show that $\mathcal{L}_G$ is an algebra with respect to the usual addition of matrices and with respect to the Lie product (commutator product) of matrices:

$$[\Omega_1, \Omega_2] \equiv \Omega_1 \Omega_2 - \Omega_2 \Omega_1.$$

Recall, that to be an algebra of matrices $\mathcal{L}_G$ must have the properties: (i) if $\Omega_1, \Omega_2 \in \mathcal{L}_G$, then $\Omega_1 + \Omega_2$ and $[\Omega_1, \Omega_2]$ are in $\mathcal{L}_G$, and (ii) if $\Omega \in \mathcal{L}_G$, then $c\Omega \in \mathcal{L}_G$, for every $c \in \mathbb{R}$.

(c) Define an inner product $\langle \cdot, \cdot \rangle_G$ on $\mathbb{R}^n$ by

$$\langle v, w \rangle_G = Gv \cdot w,$$

for $v, w \in \mathbb{R}^n$. Show that each $Q \in L_G$ preserves this inner product, i.e.,

$$\langle Qv, Qw \rangle_G = \langle v, w \rangle_G,$$

for every $v, w \in \mathbb{R}^n$. Let $|\cdot|_G$ be the corresponding norm on $\mathbb{R}^n$, i.e., for $v \in \mathbb{R}^n$ define $|v|_G = \langle v, v \rangle_G^{1/2}$. Show that each $Q \in L_G$ is an isometry of this norm, i.e.,

$$|Qv|_G = |v|_G,$$

for every $v \in \mathbb{R}^n$.

(d) Generalize Proposition 8.4 by proving the following:

   (i) Suppose $Q : I \to L_G$ is a differentiable matrix-valued map on an
        interval I, with $Q(t) \in L_G$ for every $t \in I$. Then there exists a
        continuous, matrix-valued map $\Omega : I \to \mathcal{L}_G$, such that

$$\dot{Q} = QG^{-1}\Omega G,$$

        for every $t \in I$. *Hint*: Write the identity $Q^T GQ = G$ as $G = Q^{-T}GQ^{-1}$ and then differentiate both sides using the identity:
        $\frac{d}{dt}(Q^{-1}) = -Q^{-1}\dot{Q}Q^{-1}$. (Prove this latter identity too.)

   (ii) Suppose $\Omega : I \to \mathcal{L}_G$ is a continuous. Let $Q : I \to \mathcal{M}_n$ be a
        solution of the matrix differential equation

$$\dot{Q} = QG^{-1}\Omega G.$$

        Then $Q(t) \in L_G$ for every $t \in I$.

      Comment: In the special case when the given matrix G is the identity
      matrix: $G = I$, then the $L_I = O(n)$ is the group of orthogonal ma-
      trices and $\mathcal{L}_I = o(n)$ is the Lie algebra of skew symmetric matrices.
      Another important choice for G is the 4×4 diagonal matrix with di-
      agonal entries $1, 1, 1, -1$, respectively. Then L_G is the group of Lorentz
      transformations, which is important in special relativity.

6. Suppose B is a 3×3, skew symmetric B. Show that there exists a vector
    $b \in \mathbb{R}^3$ such that $Bv = b \times v$, for all $v \in \mathbb{R}^3$. Conversely, show that if $b \in \mathbb{R}^3$
    is any vector, then there is a 3×3, skew symmetric matrix B such that
    $Bv = b \times v$, for all $v \in \mathbb{R}^3$.

7. Prove the property

$$u \times (v \times w) = (u \cdot w)v - (u \cdot v)w,$$

    for cross products of vectors $u, v, w \in \mathbb{R}^3$. Use this to prove that the inertia
    operator A for masses and relative positions $m_k, \mathbf{u}_k$, $k = 1, \ldots N$, can be
    expressed by

$$Av = \sum_{k=1}^{N} m_k \left[|\mathbf{u}_k|^2 v - (\mathbf{u}_k \cdot v)\mathbf{u}_k \right], \qquad (8.136)$$

    for all $v \in \mathbb{R}^3$. Use this to show how formula (8.113) for the entries A_{ij} of
    the matrix A is derived.

8. **(Animations of Rigid-Body Motions)** By using the Euler numerical
    method to discretize the rigid-body equations of motion, one can obtain fairly
    good animations of a number of particular motions that are solutions of these
    DEs. Some Maple code for this is on the worksheet `rigidbody.mws` on the

electronic component. To understand more fully how the angular velocity vector ω behaves during a motion and how the rotations Q are compounded out of the angular velocity matrix Ω, study the material on this worksheet and work the exercises listed there. The worksheet inertia.mws contains some code for calculating the moment of inertia operator, principal moments of inertia, and the principal axes. To understand more about how these influence rigid-body motions, read the material in CDChapter 8 on the electronic component and work the corresponding exercises.

9. Show that if $v \times w = 0$, for vectors $v, w \in \mathbb{R}^3$, then there exists a constant c such that $w = cv$.

10. **(The Inertia Operator for a Continuum)** It is important to note that the inertia operator A does not depend on the number of particles comprising the rigid body. While the discussion in the text was for a finite number N of particles, the treatment of a rigid body made up of a continuum of infinitely many particles is entirely analogous. Roughly speaking, we replace the summation over finitely many particles by an integration over the region in $\mathbb{R}^3$ that the rigid body occupies. More precisely, for a suitable region $B \subseteq \mathbb{R}^3$ and mass density function $\rho : B \to (0, \infty)$, we make the *assumption* that

$$\int_B \rho(\mathbf{u})\mathbf{u}\, d\mathbf{u} = 0.$$

This means that the center of mass of B is located at the origin. Then the inertia operator $A : \mathbb{R}^3 \to \mathbb{R}^3$ is defined by

$$
\begin{aligned}
Av &\equiv \int_B \rho(\mathbf{u})\left[\mathbf{u} \times (v \times \mathbf{u})\right] d\mathbf{u} \\
&= \int_B \rho(\mathbf{u})\left[|\mathbf{u}|^2 v - (\mathbf{u} \cdot v)\mathbf{u}\right] d\mathbf{u},
\end{aligned}
$$

for $v \in \mathbb{R}^3$. Here $d\mathbf{u} = du_1 du_2 du_3$ and the integral is the Lebesgue integral (triple integral) applied to each of the components of the vector expression. You should verify that the second equation above follows from the first. Also show that the entries A_{ij} of the matrix A are given by the following formulas:

$$A_{ii} = \int_B \rho(\mathbf{u})\left[|\mathbf{u}|^2 - u_i^2\right] d\mathbf{u} \qquad\qquad (8.137)$$

$$A_{ij} = -\int_B \rho(\mathbf{u})u_i u_j\, d\mathbf{u} \qquad (\text{for } i \neq j). \qquad (8.138)$$

The definitions of the principal moments of inertia and principal axes are the same as before (namely the eigenvalues and corresponding eigenvectors of A). Use formulas (8.137)-(8.138) to do the following.

(a) Suppose $B = [-a/2, a/2] \times [-b/2, b/2] \times [-c/2, c/2]$ is the box with its center at the origin and sides of lengths a, b, c. Assume the mass density ρ is constant. Show that the three principal moments of inertia (not in any order) and corresponding principal axes are

$$I_1 = \frac{\mu}{12}(b^2 + c^2), \qquad e_1 = (1, 0, 0)$$

$$I_2 = \frac{\mu}{12}(a^2 + c^2), \qquad e_2 = (0, 1, 0)$$

$$I_3 = \frac{\mu}{12}(a^2 + b^2), \qquad e_3 = (0, 0, 1),$$

where $\mu = \rho abc$ is the total mass of B.

(b) Suppose $B = \{(x, y, z) \mid x^2 + y^2 \leq R^2, |z| \leq h/2\}$ is the cylinder of height h and circular base of radius R (and centered at the origin). Show that the three principal moments of inertia (not in any order) and corresponding principal axes are

$$I_1 = I_2 = \mu\left(\frac{R^2}{4} + \frac{h^2}{12}\right), \qquad e_1 = (1, 0, 0), \; e_2 = (0, 1, 0)$$

$$I_3 = \mu\frac{R^2}{2}, \qquad e_3 = (0, 0, 1),$$

where $\mu = \rho\pi R^2 h$ is the total mass of B.

11. Suppose Ω is the skew symmetric matrix

$$\Omega = \begin{bmatrix} 0 & -\omega_3 & \omega_2 \\ \omega_3 & 0 & -\omega_1 \\ -\omega_2 & \omega_1 & 0 \end{bmatrix},$$

and let $\omega = (\omega_1, \omega_2, \omega_3)$. Show that

$$\det(\Omega - \lambda I) = \lambda(\lambda^2 + |\omega|^2).$$

12. Show that $a = |\mathbf{L}|$, where a is the constant in equation (8.131).

Chapter 9

Hamiltonian Systems

A great many of the dynamical systems: $x' = X(x)$ that arise in applications are *Hamiltonian systems*, and are important because of their special structure, as well as the fact that they are related to the dynamics of motion in classical systems (through Newton's second law). All of the previous theory and techniques apply to Hamiltonian systems, but now there are many additional features of the system, like conservation laws, a symplectic structure, and Poisson brackets, that enable us to study such systems in more detail.

This Chapter is independent of the chapters dealing with Newtonian mechanics (Chapter 8) and with integrable systems of DEs (Chapter 7). As a consequence there will be a certain amount of repetition here of ideas and concepts from those chapters. If you have read those chapters, then the material here should add to your understanding of the concepts. After all, Hamilton's approach is an alternative, in a sense equivalent, formulation of Newton's laws of motion for conservative systems.

Definition 9.1 (Hamiltonians) Suppose $\mathcal{O}$ is an open subset of $\mathbb{R}^n \times \mathbb{R}^n \cong \mathbb{R}^{2n}$. For historical reasons we denote a point in $\mathcal{O}$ by:

$$x = (q, p) = (q_1, \ldots, q_n, p_1, \ldots, p_n),$$

(1) A *Hamiltonian* on $\mathcal{O}$ is a differentiable function $H : \mathcal{O} \to \mathbb{R}$. Usually, physical considerations lead us to select particular types of functions H on $\mathcal{O}$, and these are distinguished from the rest with the designation "Hamiltonian." Most of the results we discuss later will require H to be a least a C^2 function.

(2) If H is a Hamiltonian, then the corresponding *Hamiltonian vector field*

D. Betounes, *Differential Equations: Theory and Applications*, DOI 10.1007/978-1-4419-1163-6_9,
© Springer Science + Business Media, LLC 2010

on $\mathcal{O}$ is the vector field X_H defined by:

$$X_H(q,p) = \left(\frac{\partial H}{\partial p_1}(q,p), \ldots, \frac{\partial H}{\partial p_n}(q,p), -\frac{\partial H}{\partial q_1}(q,p), \ldots, -\frac{\partial H}{\partial q_n}(q,p) \right)$$

$$= \left(\frac{\partial H}{\partial p}(q,p), -\frac{\partial H}{\partial q}(q,p) \right).$$

The last line above uses a special notation that we will often find convenient. Essentially $\partial H/\partial q$ and $\partial H/\partial p$ stand for the gradients of H with respect to the q and p variables, respectively.

(3) If X_H is a Hamiltonian vector field, the associated system of DEs is called a *Hamiltonian system*. Thus,

$$(\dot{q}, \dot{p}) = X_H(q,p),$$

is the form for a typical Hamiltonian system. In terms of the definition of X_H, this system is

$$\dot{q} = \frac{\partial H}{\partial p}(q,p) \tag{9.1}$$

$$\dot{p} = -\frac{\partial H}{\partial q}(q,p). \tag{9.2}$$

This is the form we will usually employ for a Hamiltonian system. It is a system of $2n$ differential equations for $2n$ unknown functions. Written out completely in component form it is

$$\dot{q}_1 = \frac{\partial H}{\partial p_1}(q,p)$$

$$\vdots$$

$$\dot{q}_n = \frac{\partial H}{\partial p_n}(q,p)$$

$$\dot{p}_1 = -\frac{\partial H}{\partial q_1}(q,p)$$

$$\vdots$$

$$\dot{p}_n = -\frac{\partial H}{\partial q_n}(q,p).$$

The above exhibits clearly the special structure that Hamiltonian systems have, but does little to motivate how such systems arise in physics and what important additional features these systems possess by virtue of this structure. Understanding of these latter things will come as we proceed.

Example 9.1 Suppose $n = 2$, so that $q = (q_1, q_2)$ and $p = (p_1, p_2)$. Let m, k be positive constants, and let H be the Hamiltonian defined by

$$
\begin{aligned}
H(q, p) &= \frac{|p|^2}{2m} - \frac{k}{|q|} \\
&= \frac{p_1^2 + p_2^2}{2m} - \frac{k}{(q_1^2 + q_2^2)^{1/2}}.
\end{aligned}
$$

The domain of H is $\mathcal{O} = (\mathbb{R}^2 \setminus \{(0,0)\}) \times \mathbb{R}^2$, and one easily computes:

$$
\begin{aligned}
\frac{\partial H}{\partial p}(q, p) &= \frac{p}{m} \\
\frac{\partial H}{\partial q}(q, p) &= \frac{kq}{|q|^3}
\end{aligned}
$$

Thus, the Hamiltonian system for this Hamiltonian is

$$
\begin{aligned}
\dot{q} &= \frac{p}{m} \\
\dot{p} &= -\frac{kq}{|q|^3}.
\end{aligned}
\tag{9.3}
$$

Using the first equation, we find that $m\ddot{q} = \dot{p}$, and so from the second equation we get the second-order system:

$$
m\ddot{q} = -\frac{kq}{|q|^3}.
\tag{9.4}
$$

This is Newton's second law for the motion of a particle of mass m attracted toward the origin with a force of magnitude reciprocally proportional to the square of the distance. The system (9.3) is Hamilton's version of the equations of motion and is (in a precise sense) equivalent to Newton's 2nd order system (9.4) (see the exercises).

As the above example and definitions show, Hamiltonian systems

$$
\begin{aligned}
\dot{q} &= \frac{\partial H}{\partial p}(q, p) \\
\dot{p} &= -\frac{\partial H}{\partial q}(q, p),
\end{aligned}
$$

are special is several respects. There are $2n$ equations for $2n$ unknown functions: $(q_1, \ldots, q_n, p_1, \ldots, p_n)$, and so Hamiltonian systems occur only in *even* dimensions. In addition, a Hamiltonian system derives its particular form from a Hamiltonian function: $H : \mathcal{O} \subseteq \mathbb{R}^{2n} \to \mathbb{R}$, and the phase space $\mathcal{O}$ consists of points (q, p) with $q = (q_1, \ldots, q_n) \in \mathbb{R}^n$ and $p = (p_1, \ldots, p_n) \in \mathbb{R}^n$. As we shall see, in many important examples the q's and p's represent the positions and momenta of a system of particles, while H is related to the energy of the system. To explain this and also to motivate the techniques to be used, we begin our study with 1-dimensional Hamiltonian systems (or systems with 1-degree of freedom).

9.1   1-Dimensional Hamiltonian Systems

In the case when $n = 1$, the Hamiltonian $H = H(q, p)$ is a function of two variables $(q, p) \in \mathbb{R}^2$ and the phase portrait for the corresponding Hamiltonian system is completely determined from the level curves of H in $\mathbb{R}^2$ (as we will see). In most physical examples, the integral curves, $t \mapsto (q(t), p(t))$, represent a quantity connected with the motion of something. For example $q = q(t)$ can represent the position of a particle moving along a straight line, or more generally a particle constrained to move on a given curve like the ball in a hoop (pendulum) discussed in Chapter 1. For a ball in a hoop, $q(t)$ is the angular displacement of the ball from equilibrium at time t and, even though the motion is in three dimensional space, the condition of being restrained to the hoop reduces the description of the motion to the single quantity $q(t)$. Thus, the motion is said to have only *one degree of freedom*. The quantity $p = p(t)$ is usually either the rate of change of q (velocity) or a multiple of this (momentum).

For the sake of concreteness, in this section q will represent the position of a particle moving along a straight line and acted on by a force F directed also along this line. Suppose the particle has mass m and let $q(t)$ be its position at time t. We start with Newton's equations and derive the corresponding Hamiltonian equations. You should realize that the resulting Hamiltonian is not the most general type of Hamiltonian on $\mathbb{R}^2$ but rather a special type associated with conservative mechanical systems.

In the most general case for Newtonian motion, the force on the particle depends on its position, its velocity, and the time: $F = F(q, \dot{q}, t)$, and the motion is governed by:

$$m\ddot{q} = F(q, \dot{q}, t).$$

In order to formulate this as a Hamiltonian system, we must restrict to the special case where the force F is *conservative*. This, by definition, means that it is independent of the time and velocities, and arises from a potential:

$$F(q) = -V'(q).$$

The function V is called a *potential function* for F, and the minus sign in the equation is inserted there for physical reasons. Note that the assumption that $F = -V'$ for some V, is equivalent to the assumption that F is continuous, since then we can always construct a potential function by integration:

$$V(q) \equiv - \int_{q_0}^{q} F(u)du,$$

where q_0 is some point in the domain of F. This assertion is for 1-dimensional systems only. In higher dimensions it is not true. In the 1-dimensional case we get a Hamiltonian system for describing the motion as follows. Newton's equation is

$$m\ddot{q} = -V'(q),$$

and is second order, but if we introduce, or define, the momentum p of the particle as being: $p = m\dot{q}$, then we can rewrite Newton's equation as a first-order system:

$$\dot{q} = p/m \qquad (9.5)$$
$$\dot{p} = -V'(q) \qquad (9.6)$$

These are *Hamilton's equations* for the motion of the particle. Note that going from Newton's equation to Hamilton's equations is similar to the reduction-to-1st-order technique. This need not always be the case in general. We see by inspection that (9.5)-(9.6) is a Hamiltonian system arising from the Hamiltonian:

$$H(q,p) = p^2/2m + V(q),$$

since $\partial H/\partial p = p/m$ and $-\partial H/\partial q = -V'$. The corresponding Hamiltonian vector field on $\mathbb{R}^2$ is:

$$X_H(q,p) = \left(p/m, \ -V'(q) \right).$$

and the analysis of the phase portrait using the techniques from the previous chapters is as follows.

The fixed points (or equilibrium points) come from solving $X_H(q, p) = 0$, i.e.,

$$p = 0 \qquad \text{zero momentum}$$
$$-V'(q) = 0 \qquad \text{zero force}$$

This is clear physically; if we start the particle with zero momentum (velocity) from a point q where the force vanishes, then the particle will be in equilibrium, i.e., remain there forever. Next we look at the linearization technique. The Jacobian matrix of X_H at any point (q, p) is:

$$A = X_H'(q, p) = \begin{bmatrix} 0 & 1/m \\ -V''(q) & 0 \end{bmatrix}.$$

At a fixed point $(q_0, 0)$ there is in general no particular reduction in the form of this matrix. The eigenvalues of A are determined from

$$\det(A - \lambda I) = \lambda^2 + V''(q_0)/m = 0.$$

Note that $(q_0, 0)$ is a nonsimple fixed point (i.e., $\det(A) = 0$) if and only if $V''(q_0) = 0$. This case gives us no information. Otherwise, there are only two possible type of *simple*, equilibrium points:

Fixed Points:

(1) $(V''(q_0) < 0)$. In this case the eigenvalues are $\lambda = \pm a$ (where $a = [-V''(q_0)/m]^{1/2}$), and the fixed point is a saddle point and is unstable.

(2) $(V''(q_0) > 0)$. In this case the eigenvalues are $\lambda = \pm ai$ (where $a = [V''(q_0)/m]^{1/2}$), and the fixed point is a possible center.

The possible center in Case (2) is an actual center because the Hamiltonian H is a Liapunov function for the fixed point. To see this note that

$$\nabla H(q, p) = \left(V'(q),\, p/m \right),$$

and Thus, $\nabla H \cdot X_H = 0$, everywhere on the domain of X_H. Furthermore, in Case (2), H has a local minimum at $(q_0, 0)$ since the Hessian of H:

$$\mathcal{H}_H = \begin{bmatrix} V''(q_0) & 0 \\ 0 & 1/m \end{bmatrix},$$

is clearly positive definite. This establishes that the Hamiltonian is a Liapunov function for the fixed points where the potential V has local minima.

Based on this analysis, we can immediately determine the fixed points and their type by looking at the graph of the potential function V. This is illustrated in Figure 9.1.

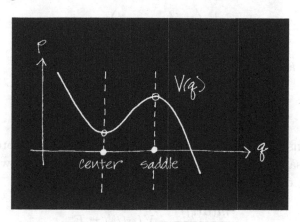

Figure 9.1: *Graph of the potential function V and determination of the fixed points.*

The fixed points occur along the q-axis (since $p = 0$ at a fixed point) at positions q where $V'(q) = 0$, i.e. where the tangent line to V is horizontal. The type of fixed point, determined by the two cases above, depends on whether $V''(q)$ is negative (i.e. the graph of V is concave down at q) or positive (i.e. the graph is concave up).

More complete information about the system and, in fact, an easy way to sketch the phase portrait by hand, can be derived from the conservation of energy principle.

9.1.1   Conservation of Energy

One way to derive the law for the conservation of energy in the 1-dimensional Hamiltonian system is to go back to Newton's version of the equation of motion:

$$m\ddot{q} = -V'(q).$$

Supposing that $q = q(t)$ is a solution of this and multiplying both sides of the equation by $\dot{q}$ gives

$$m\ddot{q}\dot{q} = -V'(q)\dot{q}.$$

By the chain rule, this is the same as:

$$\frac{d}{dt}\left(\frac{m\dot{q}^2}{2}\right) = \frac{d}{dt}\left(-V(q)\right),$$

or

$$\frac{d}{dt}\left(\frac{m\dot{q}^2}{2} + V(q)\right) = 0.$$

Hence there is a constant E, such that:

$$\frac{m\dot{q}^2}{2} + V(q) = E.$$

This is the conservation of energy principle. To be more precise about it, this principle says that if $\alpha : I \to \mathcal{O} \subseteq \mathbb{R}$ is a solution of Newton's equation then there is a constant E such that:

$$\frac{m(\dot{\alpha}(t))^2}{2} + V(\alpha(t)) = E,$$

for every $t \in I$. The quantity $m(\dot{\alpha}(t))^2/2$ is called the *kinetic energy* of the particle at time t and is clearly always positive. The quantity $V(\alpha(t))$ is the *potential energy* of the particle at time t. The constant E is called the *total energy* of the particle. Thus, as the particle moves about influenced by the force $F = -V'$, the kinetic and potential energies increase and decrease in such a way as to have their sum always constant.

In addition to the physical significance of the conservation of energy law, this law also gives us all the information we need to sketch the phase portrait of the 1-dimensional Hamiltonian system. To see this we go over to Hamilton's point of view and rewrite the conservation of energy law with $\dot{q} = p/m$ to get:

$$p^2/2m + V(q) = E,$$

which you will recognize as being simply expressed in terms of the Hamiltonian by

$$H(q,p) = E.$$

To be more precise, if the curve: $\gamma(t) = (\alpha(t), \beta(t))$, for $t \in I$, is a solution of Hamilton's equations (i.e., an integral curve of X_H), then there is a constant E such that:

$$H(\alpha(t), \beta(t)) = E,$$

for all $t \in I$. Of course since $\beta(t) = m\ddot{\alpha}(t)$, this last relationship is the same as before, but now we have the important interpretation that each integral curve of X_H lies on one of the level curves of the Hamiltonian function H. Thus, to get the phase portrait, it suffices to graph the level curves of H in the q-p plane. Because of the special form of the Hamiltonian we are using here, graphing these curves is actually quite easy. Thus, the level curves are the graphs of the equation

$$p^2/2m + V(q) = E,$$

for various values of E. For a given value of E, we can rearrange this equation and express p as a function of q. (Actually there are two functions of q involved, depending on the choice of the $\pm$ sign.):

$$p = f_\pm(q) \equiv \pm\sqrt{2m}\sqrt{E - V(q)}.$$

The domain of each of these functions is the set $\{q \,|\, V(q) \le E\}$, and the first derivative is

$$\frac{dp}{dq} = \frac{\pm\sqrt{2m}}{2} \frac{-V'(q)}{\sqrt{E - V(q)}}.$$

This shows that these functions have horizontal tangent lines at the same positions q that $-V$ does, and that they have vertical tangent lines for those q such that $V(q) = E$ and $V'(q) \ne 0$. Using this, we obtain the graph of the level curves as follows.

Suppose the potential V has graph as shown in Figure 9.1. The graph of $p = E - V(q)$ is obtained by flipping the graph of $p = V(q)$ over and translating it by the amount E.

The desired graphs of $f_+(q) = \sqrt{2m}\sqrt{E - V(q)}$ is qualitatively very similar to that of $p = E - V(q)$, except you must take into account the restrictions on the domain of f_+ and the introduction of vertical tangent lines. The graphs of $f_\pm$ are shown in Figure 9.2 for one choice of E.

By repeating this process for several values of E we obtain the plots of the level curves of H as shown in Figure 9.3.

The figure also shows the direction of flow of each integral curve on its corresponding level curve. This direction can be determined from looking at the direction of the vector:

$$X_H(q, p) = (p/m, -V'(q)),$$

at various points along the level curves. This vector is tangent to the level curve through (q, p) and, according to the above formula, has its horizontal

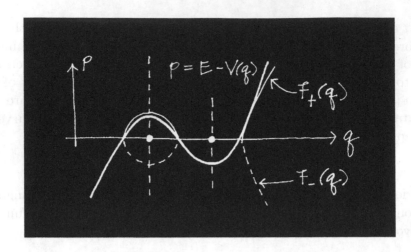

Figure 9.2: *Graphs of the two branches $f_\pm$ of a level curve for H.*

component pointing in the positive or negative direction depending whether p is positive or negative.

It is important to interpret the integral curves shown above in terms of the actual motion of a particle in 1-dimension, with $q(t)$ giving its position on the q-axis (the horizontal axis) at time t. For convenience, we take the mass $m = 1$, so that $p(t) = \dot{q}(t)$ is the velocity of the particle. Figure 9.4 shows some selected points for the analysis.

Initial point A represents a particle that starts in a position just to the left of the stable equilibrium point (the center point) with initial velocity in the positive direction. The particle will oscillate periodically about the equilibrium point, achieving its maximum separation from it when its velocity is zero and achieving its maximum velocity each time it passes through the stable equilibrium. Initial point B gives a similar particle motion except that it initially moves away from the equilibrium point and separates further from it than the one for point A.

Point C is on a separatrix and represents a particle starting to the left of the stable equilibrium and given just the right amount of velocity in the positive direction so that it passes only once through that equilibrium and thereafter approaches the unstable equilibrium with decreasing speed, taking forever to reach it. By contrast, point D represents a particle that starts further to the left of the stable equilibrium, but has sufficient initial velocity so that it passes through both equilibrium points (once) and escapes to infinity.

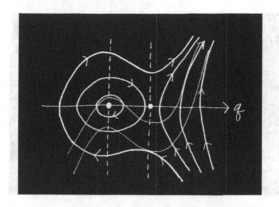

Figure 9.3: *Graphs of the level curves of H*

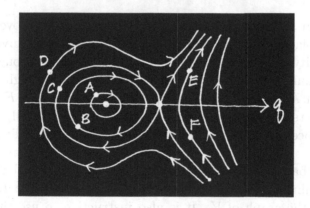

Figure 9.4: *Selected initial points for integral curves of the system.*

Point E represents a particle that starts to the right of the unstable equilibrium and has initial velocity in the positive direction. It moves directly off to infinity. Point F is similar except that the corresponding particle has initial velocity in the negative direction. It moves toward the unstable equilibrium point, slows down, turns around (when $\dot{q} = 0$), and then moves off to infinity.

In describing the motion as we did above it is often helpful to think in terms of a "ball rolling on the potential curve." This means that we view the graph of the potential function V as a mountain slope on which a ball is constrained to roll (with no friction). See Figure 9.5.

Placing the ball anywhere on the slope and giving it some (perhaps none) initial velocity (uphill or downhill) will produce a rolling motion that is qual-

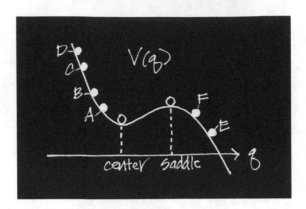

Figure 9.5: *A ball constrained to roll on the graph of* V.

itatively similar to the particle motions described in the above paragraphs.
For example, placing the ball at position A on the slope and giving it a small
downhill velocity will make the ball roll back and forth through the valley,
with its maximum separation from the valley floor a little higher up than its
initial position A. In contrast, placing the ball in position F on the slope
and giving it a relatively small uphill velocity will make it roll up toward the
peak, losing speed as it goes before turning around and rolling all the way
back downhill.

With this "ball on a curve" analogy, it is easy to distinguish the stable and
unstable equilibrium points—they correspond, respectively, to the valleys
and peaks on the graph of V. It is also instructive to use the analogy to
explain the motion corresponding to integral curves that are separatrices
(remembering that there is no friction). This is left as an exercise.

All the results developed above for the special 1-dimensional Hamiltonian
system (9.5)-(9.6), also apply to the general 1-dimensional Hamiltonian system:

$$\dot{q} = \frac{\partial H}{\partial p}(q,p)$$

$$\dot{p} = -\frac{\partial H}{\partial q}(q,p).$$

(Of course, the graphical techniques that rely on the potential function V do
not apply in the general case). The results for 1-degree of freedom also extend
to the general Hamiltonian system with n-degrees of freedom (except that
the visualization techniques are severely limited). We turn to a discussion
of this now.

Exercises 9.1

1. **(1-Dimensional Systems)** Consider the Hamiltonian systems with Hamiltonian $H(q, p) = p^2/(2m) + V(q)$, where $V : \mathcal{U} \to \mathbb{R}$ is one of the potential functions given below. For each of the ones assigned to you do the following:

   (i) Graph $V(q)$ and $-V(q)$ versus q.

   (ii) Sketch the phase portrait (by hand). Label and identify all fixed points and mark the directions on the integral curves. Be sure to sketch the separatrices (if any occur).

   (iii) Describe the motions associated with the various types of integral curves in the phase portrait. The motion is for a particle moving horizontally on the q axis with $q(t)$ representing its position at time t. You may take $m = 1$. For the description, it helps to think in terms of the "ball on the curve" analogy. Be sure to describe all the different sorts of motion, especially those corresponding to separatrices (if any occur).

   The potential functions are:

   (a) $V(q) = -2q + q^4$, on $\mathcal{U} = \mathbb{R}$.

   (b) $V(q) = 2q - q^4$, on $\mathcal{U} = \mathbb{R}$.

   (c) $V(q) = q^{-1}$, on $\mathcal{U} = (0, \infty)$.

   (d) $V(q) = q^3$, on $\mathcal{U} = \mathbb{R}$.

   (e) $V(q) = -q^3$, on $\mathcal{U} = \mathbb{R}$.

   (f) $V(q) = \dfrac{1}{q(q-1)}$, on $\mathcal{U} = (0, 1)$.

   (g) $V(q) = \dfrac{1}{q(1-q)}$, on $\mathcal{U} = (0, 1)$.

   (h) V as shown on the left in Figure 9.6.

   (i) V as shown on the right in Figure 9.6. Note the horizontal and vertical axes are suppose to be asymptotes for V.

2. Use a computer to graph the potential function V and the level curves of the Hamiltonian $H(q, p) = p^2/(2m) + V(q)$. (Take $m = 1$.)

   (a) $V(q) = q + \sin q$, on $\mathcal{U} = \mathbb{R}$.

   (b) $V(q) = q + 4\sin q$, on $\mathcal{U} = \mathbb{R}$.

   (c) $V(q) = 3q^2 - 2q^3$, on $\mathcal{U} = \mathbb{R}$.

   (d) $V(q) = 6q + 3q^2 - 2q^3$, on $\mathcal{U} = \mathbb{R}$.

   (e) $V(q) = \sin q$, on $\mathcal{U} = \mathbb{R}$.

   (f) $e^{-q}\sin(6q)$, on $\mathcal{U} = \mathbb{R}$.

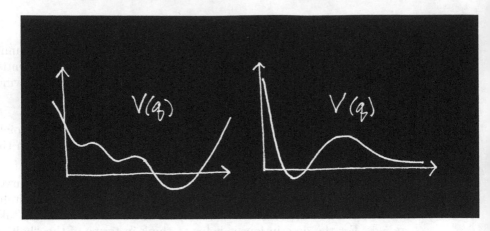

Figure 9.6: *Graphs of two potential functions V*

3. **(Conservative N-body Systems)** Consider a system of N bodies (or particles) with masses $m_1, \ldots, m_N$ and positions $\mathbf{r}_1(t), \ldots, \mathbf{r}_N(t)$ at time t. Here each $\mathbf{r}_i(t) = (x_i(t), y_i(t), z_i(t)) \in \mathbb{R}^3$, and for convenience we use the notation

$$\mathbf{r} \equiv (\mathbf{r}_1, \ldots, \mathbf{r}_N) \in \mathbb{R}^3 \times \cdots \times \mathbb{R}^3 = \mathbb{R}^{3N}.$$

The general form of the equations of motion for the system, governed by Newton's 2nd law, was given and discussed in detail in Chapter 8. A special case of this is when the forces acting on each body come from a *potential function* $V : \mathcal{U} :\to \mathbb{R}$, where $\mathcal{U}$ is an open subset of $\mathbb{R}^{3N}$. Then the equations of motion have the form

$$m_i \ddot{\mathbf{r}}_i = -\frac{\partial V}{\partial \mathbf{r}_i}(\mathbf{r}),$$

for $i = 1, \ldots, N$. *Note*: The notation with the partial derivative above means

$$\frac{\partial V}{\partial \mathbf{r}_i}(\mathbf{r}) \equiv -\left(\frac{\partial V}{\partial x_i}(\mathbf{r}), \frac{\partial V}{\partial y_i}(\mathbf{r}), \frac{\partial V}{\partial z_i}(\mathbf{r}), \right).$$

This system of differential equations is known as a *conservative N-body system*. Show that the corresponding 1st-order system of DEs is equivalent to a Hamiltonian system. Suggestion: Introduce the *momenta* $\mathbf{p}_i \equiv m_i \dot{\mathbf{r}}_i$, $i = 1, \ldots, N$, and let

$$q = \mathbf{r} = (\mathbf{r}_1, \ldots, \mathbf{r}_N), \qquad p = (\mathbf{p}_1, \ldots, \mathbf{p}_N).$$

Then determine an appropriate Hamiltonian $H = H(q, p)$ on $\mathcal{O} = \mathcal{U} \times \mathbb{R}^{3N}$.

9.2 Conservation Laws and Poisson Brackets

As we have seen the law for the conservation of energy was instrumental for our understanding of 1-D Hamiltonian systems. For n-dimensional Hamiltonian systems this conservation law is also helpful, but because of the added dimensions, additional conservation laws, as well as other information, are needed to fully understand the system.

Recall that if $H : \mathcal{O} \to \mathbb{R}$ is a function (Hamiltonian) on an open subset $\mathcal{O} \subseteq \mathbb{R}^{2n}$, then the corresponding Hamiltonian vector field on $\mathcal{O}$ is:

$$
\begin{aligned}
X_H(q,p) &= \left(\frac{\partial H}{\partial p}(q,p), -\frac{\partial H}{\partial q}(q,p) \right) \\
&= \left(\frac{\partial H}{\partial p_1}(q,p), \cdots, \frac{\partial H}{\partial p_n}(q,p), -\frac{\partial H}{\partial q_1}(q,p), \cdots - \frac{\partial H}{\partial q_n}(q,p) \right)
\end{aligned}
$$

We will almost always use the abbreviated form of X_H given in the first equation above. It's easy to see that X_H can also be expressed in the following form, which will be convenient to use from time to time:

$$
X_H = J\nabla H.
$$

Here $\nabla H = (\partial H/\partial q, \partial H/\partial p)$ is the gradient of H (written in the previous shorthand notation), and J is the $2n \times 2n$ matrix:

$$
J = \begin{bmatrix} 0 & I \\ -I & 0 \end{bmatrix}. \tag{9.7}
$$

This gives J in terms of four blocks of sub matrices, with 0 being the $n \times n$ zero matrix and I being the $n \times n$ identity matrix. J is called a *symplectic structure* on $\mathbb{R}^{2n}$.

Important Note: We have often identified the gradient ∇H with the Jacobian matrix H'. The latter, as defined in Appendix A, is a $1 \times 2n$ matrix. However, to conform to the usage of ∇H in specific equations involving matrix multiplication, we must use

$$
\nabla H \equiv (H')^T,
$$

as the definition of the gradient. Thus, $X_H = J\nabla H$ is a product of a $2n \times 2n$ matrix and a $2n \times 1$ matrix.

There is a standard bilinear form $\Omega : \mathbb{R}^{2n} \times \mathbb{R}^{2n} \to \mathbb{R}$ associated with J. It is called the *canonical symplectic form* on $\mathbb{R}^{2n}$ and is defined by

$$\Omega(v, w) = v \cdot Jw,$$

for $v, w \in \mathbb{R}^{2n}$. This makes $\mathbb{R}^{2n}$ into what is known as a *symplectic vector space* (and also a *symplectic manifold*). The ideas involved here have been abstracted and generalized in many directions and have become an important field of study in mathematics (cf. [LM 87], [ABK 92], [AM 78]). The material below will give an indication of the main concepts, however for simplicity we will, for the most part, not use Ω in the notation, but rather just write the expressions with J and the standard dot product on $\mathbb{R}^{2n}$

The first result we present is the basic conservation law associated with any Hamiltonian system:

Proposition 9.1 (Conservation of Energy) *Suppose* $t \mapsto (q(t), p(t))$, *for* $t \in I$ *is an integral curve of the Hamiltonian system* (9.1)-(9.2). *Then there exists a constant E such that:*

$$H(q(t), p(t)) = E,$$

for all $t \in I$.

Proof: The proof is elementary and just amounts to using the assumption that $t \mapsto (q(t), p(t))$ is an integral curve. In terms of components this means that for each $j = 1, \ldots, n$, and $t \in I$:

$$\dot{q}_j(t) \quad = \quad \frac{\partial H}{\partial p_j}(q(t), p(t))$$

$$\dot{p}_j(t) \quad = \quad -\frac{\partial H}{\partial q_j}(q(t), p(t)).$$

Now suppose $F : \mathcal{O} \to \mathbb{R}$ is any differentiable function, and suppose we calculate the rate of change of F along the given integral curve. Using the above and the chain rule we find:

$$\frac{d}{dt} F(q(t), p(t)) \tag{9.8}$$

$$= \sum_{j=1}^{n} \frac{\partial F}{\partial q_j}(q(t), p(t))\dot{q}_j(t) + \frac{\partial F}{\partial p_j}(q(t), p(t))\dot{p}_j(t)$$

$$= \sum_{j=1}^{n} \frac{\partial F}{\partial q_j}(q(t), p(t))\frac{\partial H}{\partial p_j}(q(t), p(t)) - \frac{\partial F}{\partial p_j}(q(t), p(t))\frac{\partial H}{\partial q_j}(q(t), p(t)).$$

This is a general result that we will need later. For the case at hand, we see that taking $F = H$ in this identity gives:

$$\frac{d}{dt} H\Big(q(t), p(t)\Big) = 0.$$

Thus, the composite function $H(q(t), p(t)), t \in I$ is a constant function, say $= E$, on I. $\square$

A geometrical interpretation of the proposition is this: each integral curve of a Hamiltonian system lies on one of the level hypersurfaces of the Hamiltonian, i.e., the trajectory traced out in phase space by an integral curve $t \mapsto (q(t), p(t))$ lies in the set:

$$S_H^E = \{(q, p) \in \mathcal{O} \,|\, H(q, p) = E\}.$$

This hypersurface is called an *energy hypersurface* for the system, and generally has dimension $2n - 1$ as a submanifold of $\mathbb{R}^{2n}$. Since the image of each integral curve is a 1-dimensional submanifold of $\mathbb{R}^{2n}$, the conservation of energy principle does not place enough restrictions on an integral curve to fully determine it. The exception to this is of course for $n = 1$, where the fact that each integral curve lies on a level curve of H fully determines all the integral curves. This suggests that for $n > 1$ we should look for additional conservation laws.

Because of the form of the conservation law: $H(q(t), p(t)) = E$ for all $t \in I$, the Hamiltonian H is called a constant of the motion (it has a constant value along any integral curve). So the search for other conservation laws, or constants of the motion, amounts to looking for such special functions.

Definition 9.2 (The Poisson Bracket) For a given Hamiltonian system on $\mathcal{O} \subseteq \mathbb{R}^{2n}$:

(1) A *constant of the motion* is a differentiable function $F : U \subseteq \mathcal{O} \to \mathbb{R}$, defined on an open dense set U, with the property: for each integral curve $(q(t), p(t)), t \in I$, that lies in U, there exists a constant K such that

$$F(q(t), p(t)) = K,$$

for all $t \in I$. The above equation is called the *conservation law* corresponding to the constant of the motion F. The function F is also known as a *first integral* for the Hamiltonian system. This concept is discussed in general in Chapter 7.

(2) For two functions $F, G : U \subseteq \mathcal{O} \to \mathbb{R}$. the *Poisson bracket* of F and G is the function $\{F, G\}$ on U defined by:

$$\{F, G\} = \sum_{j=1}^{n} \left(\frac{\partial F}{\partial q_j} \frac{\partial G}{\partial p_j} - \frac{\partial F}{\partial p_j} \frac{\partial G}{\partial q_j} \right) \tag{9.9}$$

Based on the computation given in equation (9.8), we easily obtain the results in the following corollary:

Corollary 9.1 *Suppose $H : \mathcal{O} \to \mathbb{R}$ is a Hamiltonian and $(q(t), p(t)), t \in I$ is an integral curve of the corresponding Hamiltonian system. Then for any differentiable function $F : \mathcal{O} \to \mathbb{R}$, one has*

$$\frac{d}{dt} F(q(t), p(t)) = \{F, H\}(q(t), p(t)),$$

for every $t \in I$. From this it follows that F is a constant of the motion if and only if

$$\{F, H\} = 0,$$

identically on $\mathcal{O}$.

Proof: Exercise.

The corollary is the first indication of how important the Poisson bracket is in the theory of Hamiltonian systems. Other results below will add more weight to this assertion. From an algebraic point of view, the Poisson bracket is interesting as well. It provides an exotic way of combining two functions F, G, to produce a new function $\{F, G\}$ and thus gives a product structure on the set of functions.

Specifically, let $C^\infty(\mathcal{O})$ be the set of all infinitely differentiable, real-valued functions on $\mathcal{O}$, endowed with the usual addition: $F + G$ of functions, and scalar multiplication: λF. Then the Poisson bracket

$$\{\cdot, \cdot\} : C^\infty(\mathcal{O}) \times C^\infty(\mathcal{O}) \to C^\infty(\mathcal{O}),$$

gives a binary operation on this set and has the following properties:

For E, F, G differentiable functions on $\mathcal{O}$, and any scalar λ one has:

(1) **Antisymmetry:** $\{F, G\} = -\{G, F\}$

(2) **Bilinearity:**

(a) $\{E+F,G\} = \{E,G\}+\{F,G\}$ and $\{E,F+G\} = \{E,F\}+\{E,G\}$.

(b) $\{\lambda F,G\} = \lambda\{F,G\}$ and $\{F,\lambda G\} = \lambda\{F,G\}$.

(3) **Jacobi Identity:**

$$\{E,\{F,G\}\} + \{F,\{G,E\}\} + \{G,\{E,F\}\} = 0.$$

These properties make $C^\infty(\mathcal{O})$ into a standard example of what is known as a *Lie algebra*. (See the exercises in the next section for an abstract definition of a Lie algebra.) More importantly, the subset

$$\mathcal{C}_H = \{F \mid \{F,H\} = 0\},$$

of all constants of the motion determined by the Hamiltonian H is a Lie subalgebra of $C^\infty(\mathcal{O})$. This is so because if F and G are constants of the motion then so are $F + G, \lambda F$, and $\{F,G\}$ (exercise).

The criterion $\{F,H\} = 0$ in Corollary 9.1 for determining constants of the motion can be interpreted as a partial differential equation:

$$\sum_{j=1}^{n}\left(\frac{\partial F}{\partial q_j}\frac{\partial H}{\partial p_j} - \frac{\partial F}{\partial p_j}\frac{\partial H}{\partial q_j}\right) = 0, \qquad (9.10)$$

that the constants of the motion F must satisfy. In the equation H is a given Hamiltonian and the equation specifies a relation that the partial derivatives of F must satisfy. Generally, constants of the motion are hard to determine, primarily because of the difficulty of describing the general solution of (9.10). However, the following formula, relating the Poisson bracket to the symplectic structure, is often useful in this task.

Proposition 9.2 *The Poisson bracket of $H, F : \mathcal{O} \to \mathbb{R}$ is given by*

$$\{F,H\} = \nabla F \cdot J\nabla H = \nabla F \cdot X_H. \qquad (9.11)$$

Using the proposition, it is easy to see that finding solutions F of $\{F,H\} = 0$ is equivalent to finding functions F whose gradients are perpendicular to the Hamiltonian vector field: $\nabla F \cdot X_H = 0$. This is often a convenient way to view constants of the motion.

Example 9.2 (Two Coupled Masses) Consider the system of two masses from Example 4.7, where the masses are equal, $m_1 = 1 = m_2$, and the three springs all have the same spring constant, $k_i = 1$, $i = 1, 2, 3$. Newton's equations for the motion are

$$\ddot{q}_1 = -2q_1 + q_2 \tag{9.12}$$
$$\ddot{q}_2 = q_1 - 2q_2. \tag{9.13}$$

With $q = (q_1, q_2)$ and $p = (p_1, p_2)$, we let $H : \mathbb{R}^4 \to \mathbb{R}$ be the Hamiltonian function defined by

$$H(q, p) = \tfrac{1}{2}(p_1^2 + p_2^2) + q_1^2 + q_2^2 - q_1 q_2.$$

Then it is easy to calculate that the corresponding Hamiltonian vector field is

$$X_H = \left(p_1, \ p_2, \ -2q_1 + q_2, \ q_1 - 2q_2 \right).$$

Thus, the Hamiltonian system for X_H is the equivalent 1st-order system for Newton's 2nd-order system. Define $F_1, F_2 : \mathbb{R}^4 \to \mathbb{R}$, by

$$F_1(q, p) = \tfrac{1}{4}(p_1 + p_2)^2 + \tfrac{1}{4}(q_1 + q_2)^2 \tag{9.14}$$
$$F_2(q, p) = \tfrac{1}{4}(p_1 - p_2)^2 + \tfrac{3}{4}(q_1 - q_2)^2. \tag{9.15}$$

Then it is easy to check that $\{F_i, H\} = \nabla F_i \cdot X_H = 0$, for $i = 1, 2$, and so F_1, F_2 are constants of the motion (exercise). It is also easy to verify that $\{F_1, F_2\} = \nabla F_1 \cdot X_{F_2} = 0$, and that $\nabla F_1, \nabla F_2$ are linearly independent (in fact, perpendicular) at each point (exercise).

   A more important interpretation of Corollary 9.1 (and Proposition 9.2) involves the constraint that the conservation law imposes on the possible motions of the system. From a geometrical standpoint, each constant of the motion constrains the motion of the Hamiltonian system to lie on a level set (also called a level hypersurface) in phase space. Thus, if F is a constant of the motion and c is a number in the range of F, then the *level set*:

$$S_F^c = \{ (q, p) \in \mathcal{O} \mid F(q, p) = c \},$$

is a $(2n - 1)$-dimensional submanifold (with suitable conditions on F) and for each point $(q, p) \in S_F^c$, the integral curve through (q, p) lies entirely on this submanifold: $\phi_t(q, p) \in S_F^c$, for every $t \in I_{(q,p)}$. The assertion about S_F^c

being a submanifold requires the proviso that the gradient of F be nonzero at each point of S_F^c and follows from the Submanifold Theorem in Appendix A.

If, in addition to the Hamiltonian $F_1 = H$, the system has other constants of the motion: $F_2, \ldots, F_k$, then each integral curve of the system is constrained to lie on level sets $F_i = c_i$, for $i = 1, \ldots k$, and thus must lie on the intersection of these sets:

$$S_{F_1 \cdots F_k}^{c_1 \cdots c_k} \equiv S_{F_1}^{c_1} \cap \cdots \cap S_{F_k}^{c_k}.$$

If we assume that the functions $F_1, \ldots, F_k$ are *functionally independent*, then, the Submanifold Theorem says that the set $S_{F_1 \cdots F_k}^{c_1 \cdots c_k}$ is a $(2n - k)$-dimensional manifold. Thus, functionally independent constants of the motion constrain the motion of the system to a particular lower dimensional manifold in phase space. The definition of independence for functions is as follows.

Definition 9.3 (Functional Independence)

(1) Functions $F_1, \ldots, F_k : \mathcal{O} \to \mathbb{R}$ on phase space are called *functionally independent* on an open set $U \subseteq \mathcal{O}$ if at each point $(q, p) \in U$, the vectors:

$$\nabla F_1(q, p), \ldots, \nabla F_k(q, p),$$

are linearly independent.

(2) The functions: $F_1, \ldots, F_k$ are said to be *in involution* on U if

$$\{F_i, F_j\} = 0,$$

for every $i, j = 1. \ldots, k$.

(3) If $F : \mathcal{O} \to \mathbb{R}$ is a function on phase space, with corresponding Hamiltonian vector field:

$$X_F = J \nabla F = (\frac{\partial F}{\partial p}, -\frac{\partial F}{\partial q}),$$

then, for convenience in the sequel, we will let the flow generated by X_F be denoted by ϕ^F

While the notion of independence for functions applies quite generally, in the Hamiltonian setting it is easy to see that $F_1, \ldots, F_k$ are functionally independent if and only if the vector fields $X_{F_1}, \ldots, X_{F_k}$ are linearly independent at each point (exercise). Also observe that for each $i = 1, \ldots, k$, the gradient vector ∇F_i is perpendicular to the level surface $S_{F_i}^{c_i}$ at each point (we assume that the gradient is nonzero along this surface). But by formula (9.11) for the Poisson bracket, we have

$$\nabla F_i \cdot X_{F_i} = \{F_i, F_i\} = 0,$$

and so it follows that X_{F_i} is tangent to the surface $S_{F_i}^{c_i}$. Furthermore if $F_1, \ldots, F_k$ are in involution, we have in addition that

$$\nabla F_j \cdot X_{F_i} = \{F_j, F_i\} = 0,$$

for each i, j, and at each point. Thus, not only is each X_{F_i} tangent to $S_{F_i}^{c_i}$, but also is tangent to each $S_{F_j}^{c_j}$, for each $j \neq i$. Consequently, each X_{F_i} is tangent to the submanifold $S_{F_1 \cdots F_k}^{c_1 \cdots c_k}$. Summarizing this discussion, we get:

Theorem 9.1 *Suppose $F_1, \ldots, F_k$ are constants of the motion for the Hamiltonian system: $(\dot{q}, \dot{p}) = X_H(q, p)$, and suppose that on an open set $U \subseteq \mathcal{O}$ these functions are*

(1) *functionally independent, and*

(2) *in involution: $\{F_i, F_j\} = 0, \ \forall i, j = 1, \ldots, k$.*

Then for values: $c_1, \ldots, c_k$ of these functions respectively, the set:

$$S_{F_1 \cdots F_k}^{c_1 \cdots c_k} = \{\, (q, p) \in U \mid F_i(q, p) = c_i, i = 1, \ldots, k \,\},$$

is a $(2n - k)$-dimensional submanifold of $\mathbb{R}^{2n}$, which contains all the integral curves: $\phi_t^{F_i}(q, p), \ t \in I_{(q,p)}$, that start at points $(q, p) \in S_{F_1 \cdots F_k}^{c_1 \cdots c_k}$. In particular, the Hamiltonian flow lies on this submanifold.

Furthermore, the vector fields $X_{F_1}, \cdots, X_{F_k}$ are linearly independent and tangent to $S_{F_1 \cdots F_k}^{c_1 \cdots c_k}$ at each of its points. Consequently, $k \leq n$ (since $k \leq 2n - k$).

The theorem indicates that we can hope to obtain at most n functionally independent constants of the motion that are in involution. This is the optimum case and is exceptional enough to warrant a special name.

Definition 9.4 (Completely Integrable Systems) For a given Hamiltonian $H : \mathcal{O} \subseteq \mathbb{R}^{2n} \to \mathbb{R}$, suppose $F_1 = H, F_2, \ldots, F_n$ are constants of the motion which, on $U \subseteq \mathcal{O}$, are functionally independent and in involution. Then the corresponding Hamiltonian system is called *completely integrable*, or more briefly: *integrable*.

Note that for $n = 1$, any Hamiltonian system is completely integrable. The coupled springs model in Example 11.2 is a standard example of a completely integrable Hamiltonian system. Each integral curve of the system is constrained to move on a two dimensional submanifold of $\mathbb{R}^4$ defined by the equations

$$\tfrac{1}{4}(p_1 + p_2)^2 + \tfrac{1}{4}(q_1 + q_2)^2 = c_1 \tag{9.16}$$

$$\tfrac{1}{4}(p_1 - p_2)^2 + \tfrac{3}{4}(q_1 - q_2)^2 = c_2. \tag{9.17}$$

The name *completely integrable system* comes from a classical result of Liouville that such systems can be "solved completely by means of quadratures, i.e., integrations." The term *integrable* systems is often used in the literature for completely integrable systems. However, as noted in Chapter 7, the term integrable system is used in another (but related) way for general systems of 1st-order equations. The result of Liouville (See Theorem 9.5 below) shows the extent to which a completely integrable Hamiltonian system is integrable in the sense of Chapter 7.

In addition to the assertions in the above theorem and Liouville's result, we discuss two other topics on completely integrable systems: invariance of complete integrability under canonical transformations and Arnold's theorem, which says that for a completely integrable system, the manifold $S_{F_1 \cdots F_n}^{c_1 \cdots c_n}$, if it is compact and connected, is diffeomorphic to the n torus T^n.

In general a diffeomorphism will *not* transform a Hamiltonian system into another Hamiltonian system. However, there is a special class of diffeomorphism which do, and because of this they deserve a special name.

Definition 9.5 (Canonical Transformations) Suppose $f : \mathcal{O} \to \overline{\mathcal{O}}$ is a diffeomorphism between open sets $\mathcal{O}, \overline{\mathcal{O}}$ in $\mathbb{R}^{2n}$. Then f is called a *canonical transformation* (or a *symplectic transformation*) if

$$f'(x) \, J \, f'(x)^T = J, \tag{9.18}$$

for every $x \in \mathcal{O}$. Here A^T denotes the transpose of the matrix A and J is the canonical symplectic matrix in (9.7).

The more modern term for such transformations is symplectic transformation, or *symplectomorphism*, because they preserve the symplectic structure (see the Exercises).

Example 9.3 (Symplectic Matrices) Suppose $f : \mathbb{R}^{2n} \to \mathbb{R}^{2n}$ is linear transformation, $f(x) = Ax$, for some $2n \times 2n$ matrix. Then since $f'(x) = A$ for every $x \in \mathbb{R}^{2n}$, it follows that the condition (9.18) reduces to the condition

$$AJA^T = J,$$

for linear transformations. A matrix A that satisfies this is called a *symplectic matrix*. It is easy to show that a symplectic matrix is invertible (see the exercises) and thus a linear transformation is a canonical transformation if and only if it is represented by a symplectic matrix.

A commonly occuring type of symplectic matrix is a block diagonal matrix of the form

$$A = \begin{bmatrix} B & 0 \\ 0 & E \end{bmatrix}, \tag{9.19}$$

where B, E are $n \times n$ matrices and 0 is the $n \times n$ zero matrix. It is easy to verify that a matrix of this form is symplectic if and only if $BE^T = I$ (exercise). Thus, in particular if $B = Q$ and $E = Q$, where Q is an orthogonal matrix, then A is a symplectic matrix.

The following theorem shows that symplectic transformations preserve Poisson brackets as well as the Hamiltonian nature of Hamiltonian systems (the latter being the motive for their definition).

Theorem 9.2 *Suppose $f : \mathcal{O} \to \overline{\mathcal{O}}$ is a canonical transformation between open sets of $\mathbb{R}^{2n}$. Then f transforms Hamiltonian vector fields into Hamiltonian vector fields. Specifically,*

$$f_*(X_H) = X_{H \circ f^{-1}}, \tag{9.20}$$

for any differentiable function $H : \mathcal{O} \to \mathbb{R}$. Furthermore f preserves the Poisson bracket, i.e.:

$$\{F \circ f^{-1}, H \circ f^{-1}\} = \{F, H\} \circ f^{-1}, \tag{9.21}$$

for any differentiable functions $F, H : \mathcal{O} \to \mathbb{R}$.

Proof: The proof will be an easy consequence of the chain rule and the definitions. Note also that we use the definition of the gradient mentioned above: $\nabla H \equiv (H')^T$.

First consider the push-forward of X_H by f. A direct calculation gives

$$
\begin{aligned}
f_*(X_H) &= (f' \circ f^{-1})(X_H \circ f^{-1}) \\
&= (f' \circ f^{-1}) J((H')^T \circ f^{-1}) \\
&= (f' \circ f^{-1}) J(f' \circ f^{-1})^T (f' \circ f^{-1})^{-T} (H' \circ f^{-1})^T \\
&= J(f' \circ f^{-1})^{-T} (H' \circ f^{-1})^T \\
&= J\left((H' \circ f^{-1})(f' \circ f^{-1})^{-1}\right)^T \\
&= J((H \circ f^{-1})')^T \\
&= X_{H \circ f^{-1}}
\end{aligned}
$$

To prove the second assertion about the Poisson bracket, it will be convenient to let $g = f^{-1}$. Note that g is also a canonical transformation (exercise) and so $g' J(g')^T = J$. Then using formula (9.11), we get

$$
\begin{aligned}
\{F \circ g, H \circ g\} &= ((F \circ g)')^T \cdot J((H \circ g)')^T \\
&= ((F' \circ g) g')^T \cdot J((H' \circ g) g')^T \\
&= (g')^T (F' \circ g)^T \cdot J(g')^T (H' \circ g)^T \\
&= (F' \circ g)^T \cdot g' J(g')^T (H' \circ g)^T \\
&= (F' \circ g)^T \cdot J(H' \circ g)^T \\
&= \{F, H\} \circ g
\end{aligned}
$$

This proves the theorem. $\square$

Corollary 9.2 *If $f : \mathcal{O} \to \overline{\mathcal{O}}$ is a canonical transformation, then any Hamiltonian system $\dot{x} = X_H(x)$, is differentiably equivalent to the Hamiltonian system $\dot{x} = X_{H \circ f^{-1}}(x)$. Furthermore, for any F*

$$\{F, H\} = 0,$$

if and only if

$$\{F \circ f^{-1}, H \circ f^{-1}\} = 0,$$

and consequently the Hamiltonian system with vector field X_H is completely integrable if and only if the one with vector field $X_{H \circ f^{-1}}$ is.

As in the general theory for transforming differential equations, the motivation behind canonical transformations is to use them for transforming Hamiltonian systems into simpler Hamiltonian systems.

Example 9.4 Consider the coupled masses system in Example 9.2. Newton's equations of motion are $\ddot{\mathbf{r}} = L\mathbf{r}$, where

$$L = \begin{bmatrix} -2 & 1 \\ 1 & -2 \end{bmatrix},$$

and we can use L to write the Hamiltonian for the system as

$$\begin{aligned} H(q,p) &= \tfrac{1}{2}(p_1^2 + p_2^2) + q_1^2 + q_2^2 - q_1 q_2 \\ &= \tfrac{1}{2}(p \cdot p) - \tfrac{1}{2}(q \cdot Lq), \end{aligned}$$

This form of the Hamiltonian suggests transforming to principal axes (Appendix C), i.e., diagonalizing L. For this let Q be the 2×2 orthogonal matrix

$$Q = 2^{-1/2} \begin{bmatrix} 1 & 1 \\ 1 & -1 \end{bmatrix},$$

formed from the normal mode vectors (cf. Example 5.7 in Chapter 5). Then $Q^T L Q = \text{diag}(-1, -3)$. Note also that $Q^{-1} = Q^T = Q$ for this matrix. Thus, if we let $f : \mathbb{R}^4 \to \mathbb{R}^4$, be the linear transformation:

$$f(q,p) = (Qq, Qp) = \begin{bmatrix} Q & 0 \\ 0 & Q \end{bmatrix} \begin{bmatrix} q \\ p \end{bmatrix},$$

then f is a canonical transformation and $f^{-1} = f$. Hence we get a new Hamiltonian

$$\widetilde{H}(q,p) = H(Qq, Qp) = \tfrac{1}{2}(p_1^2 + p_2^2) + \tfrac{1}{2}q_1^2 + \tfrac{3}{2}q_2^2,$$

that has a much simpler form and is equivalent, in the sense of the above corollary, to the original one. The two constants of the motion F_1, F_2 for H used in Example 11.2, transform into constants of the motion for $\widetilde{H}$:

$$\begin{aligned} \widetilde{F}_1 &= \tfrac{1}{2}p_1^2 + \tfrac{1}{2}q_1^2 \\ \widetilde{F}_2 &= \tfrac{1}{2}p_2^2 + \tfrac{3}{2}q_2^2. \end{aligned}$$

Of course if we had used the simpler Hamiltonian $\widetilde{H}$ from the start, the determination of these constants of the motion would have been easy.

Exercises 9.2

1. Prove Corollary 9.1. Be precise. You will have to use the Existence and Uniqueness Theorem from Chapter 3 at some point in your proof.

2. Prove that $F_1, \ldots, F_k$ are functionally independent on $U \subseteq \mathcal{O}$ if and only if $X_{F_1}(q, p), \ldots, X_{F_k}(q, p)$ are linearly independent at each point $(q, p) \in U$.

3. Suppose $F, G : U \subseteq \mathcal{O} \to \mathbb{R}$ are constants of the motion and $\lambda \in \mathbb{R}$. Use the Poisson bracket and its properties to show that $F + G, \lambda F$, and $\{F, G\}$ are constants of the motion.

4. For the Hamiltonian H and functions F_1, F_2 in Example 9.2, show that $\{F_i, H\} = \nabla F_i \cdot X_H = 0$, for $i = 1, 2$. Also show that $\{F_1, F_2\} = \nabla F_1 \cdot X_{F_2} = 0$, and that $\nabla F_1, \nabla F_2$ are linearly independent.

5. **(The Toda Molecule)** Let $H : \mathbb{R}^6 \to \mathbb{R}$ be the Hamiltonian

$$H(q, p) = \tfrac{1}{2}(p_1^2 + p_2^2 + p_3^2) + e^{q_1 - q_2} + e^{q_2 - q_3} + e^{q_3 - q_1}.$$

Consider the following two functions on $\mathbb{R}^6$:

$$F_1(q, p) = p_1 + p_2 + p_3,$$

and

$$
\begin{aligned}
F_2(q, p) = {} & \tfrac{1}{9}(p_1 + p_2 - 2p_3)(p_2 + p_3 - 2p_1)(p_3 + p_1 - 2p_2) \\
& - (p_1 + p_2 - 2p_3)e^{q_1 - q_2} - (p_2 + p_3 - 2p_1)e^{q_2 - q_3} \\
& - (p_3 + p_1 - 2p_2)e^{q_3 - q_1},
\end{aligned}
$$

Show that the Hamiltonian system $\dot{x} = X_H(x)$ is completely integrable by verifying that $\{F_1, H\} = 0$, $\{F_2, H\} = 0$, $\{F_1, F_2\} = 0$, and H, F_1, F_2 are functionally independent. You can use Maple's symbolic manipulations capabilities for this if you wish.

6. **(N Coupled Masses)** Generalize the result discussed in Example 9.4, by considering a system of N masses, with identical mass: $m_i = 1, i = 1, \ldots, N$, coupled by $N+1$ springs, as shown in Figure 4.17, and $k_i = 1, i = 1, \ldots, N+1$. The corresponding matrix A is the tridiagonal matrix discussed in Exercise 11 of Section 4.4. Use the results of that exercise to show that the Hamiltonian system governing the motion of the masses is completely integrable. Give the expressions for both Hamiltonians $H, \widetilde{H}$, and the constants of the motion $\widetilde{F}_i, i = 1, \ldots, N$, that make $\widetilde{H}$ completely integrable.

7. **(Symplectic Matrices)** Denote the set of all symplectic matrices by

$$\mathrm{Sp}(n) = \{\, A \in \mathcal{M}_{2n} \mid AJA^T = J \,\}.$$

Prove the following results.

(a) $J^2 = -I$ and $\det(J) = -1$.

(b) If A is symplectic, then $\det A = \pm 1$. Thus, A is invertible and A^{-1} is symplectic.

(c) A is symplectic if and only if $A^T J A = J$. This is the customary definition. We used the one above since it was more convenient for proving Theorem 9.2. *Hint:* Start with the assumption $A J A^T = J$ and multiply each side on the left by J.

(d) A is symplectic if and only A^T is.

These results say, among other things, that $\mathrm{Sp}(n)$ is a group, called the *symplectic group*, under matrix multiplication.

8. Suppose A is a $2n \times 2n$ matrix of the form

$$A = \begin{bmatrix} B & C \\ D & E \end{bmatrix},$$

where B, C, D, and E are $n \times n$ matrices. Show that A is symplectic if and only if $B^T D$ and $C^T E$ are symmetric matrices and $B^T E - D^T C = I$. *Note:* This is easiest if you use $A^T J A = J$ as the defining relation for a symplectic matrix. See the previous exercise. It is important to note that in the case $n = 1$, the matrix A is 2×2 and the above conditions say that A is symplectic if and only if $\det(A) = 1$.

9. Suppose $f : \mathcal{O} \to \overline{\mathcal{O}}$ is a canonical transformation. Show that f^{-1} is also a canonical transformation.

10. Let $\Omega(v, w) = v \cdot J w$ be the canonical symplectic form on $\mathbb{R}^{2n}$. A linear transformation $A : \mathbb{R}^{2n} \to \mathbb{R}^{2n}$ is said to *preserve the symplectic structure* of the symplectic vector space $(\mathbb{R}^{2n}, \Omega)$, if $\Omega(Av, Aw) = \Omega(v, w)$, for every $v, w \in \mathbb{R}^{2n}$. Show that A preserves the symplectic structure if and only if A is a symplectic matrix. More generally, a diffeomorphism $f : \mathcal{O} \to \overline{\mathcal{O}}$, between two open subsets $\mathcal{O}, \overline{\mathcal{O}}$ of $\mathbb{R}^{2n}$ is said to *preserve the symplectic structure* if $f^*(\Omega) = \Omega$. This latter condition means, by definition, that

$$\Omega\left(f_*(X)(f^{-1}(u)), f_*(Y)(f^{-1}(u))\right) = \Omega(X(u), Y(u)),$$

for any two vector fields $X, Y : \mathcal{O} \to \mathbb{R}^{2n}$, and for all $u \in \overline{\mathcal{O}}$. Show that f preserves the symplectic structure if and only if $A = f'(x)$ is a symplectic matrix for all $x \in \mathcal{O}$.

9.3   Lie Brackets and Arnold's Theorem

The Lie bracket of two vector fields is an important type of operation (nonassociative product) and occurs in several fields of study. The definition of this bracket is as follows. Suppose $X, Y : \mathcal{O} \subseteq \mathbb{R}^n \to \mathbb{R}^n$ are two vector fields with components: $X = (X^1, \ldots, X^n)$ and $Y = (Y^1, \ldots, Y^n)$, respectively. Then the *Lie bracket* of X with Y is the vector field $[X, Y]$ on $\mathcal{O}$ whose ith component function is:

$$[X, Y]^i = \sum_{j=1}^{n} \left(X^j \frac{\partial Y^i}{\partial x_j} - Y^j \frac{\partial X^i}{\partial x_j} \right). \tag{9.22}$$

One can verify that the Lie bracket has the following properties:

(1) **Antisymmetry:** $[X, Y] = -[Y, X]$.

(2) **Bilinearity:**

    (a) $[X + Y, Z] = [X, Z] + [Y, Z]$ and $[X, Y + Z] = [X, Y] + [X, Z]$.

    (b) $[\lambda X, Y] = \lambda [X, Y] = [X, \lambda Y]$.

(3) **Jacobi Identity:**

$$[X, [Y, Z]] + [Y, [Z, X]] + [Z, [X, Y]] = 0.$$

With this bracket as a nonassociative product, the collection of all (smooth) vector fields on $\mathcal{O}$ becomes a Lie algebra. This is what the above three properties say. These properties are considerably easier to verify (especially the Jacobi identity) if we make the following observations. Each vector field $X = (X^1, \ldots, X^n)$ gives rise a differential operator, also denoted by X:

$$X = \sum_{i=1}^{n} X^i \frac{\partial}{\partial x_i}. \tag{9.23}$$

This operator acts on a differentiable function $f : \mathcal{O} \to \mathbb{R}$, to give a new function $X(f)$ defined by:

$$X(f) = \sum_{i=1}^{n} X^i \frac{\partial f}{\partial x_i}.$$

Using this association, or different way of viewing vector fields, it's not hard to show from the above definition that

$$[X, Y](f) = X(Y(f)) - Y(X(f)). \tag{9.24}$$

Alternatively, this could be taken as the definition of $[X, Y]$, from which the above component expression could be derived.

In terms of differential operators, we see that the Lie bracket $[X, Y] = XY - YX$ measures the extent to which these operators do not commute. By definition, the operators X, Y *commute* when their Lie bracket vanishes: $[X, Y] = 0$. This is related to the commutativity of the corresponding flows as the following theorem shows.

Theorem 9.3 *Suppose X and Y are complete vector fields on $\mathcal{O} \subseteq \mathbb{R}^n$, and let ϕ_t, ψ_t denote the corresponding flows generated by X, Y, respectively. Then the following are equivalent:*

(1) $\phi_t \circ \psi_s = \psi_s \circ \phi_t$, *for every t, s.*

(2) $(\psi_s)_*(X) = X$, *for every s.*

(3) $[X, Y] = 0$.

Proof: Recall the notation: $\phi_t(x) = \phi(t, x)$, $\psi_s(x) = \psi(s, x)$, for the flows and the fact that they satisfy:

$$\frac{d}{dt}\phi_t(x) = X(\phi_t(x))$$
$$\frac{d}{ds}\psi_s(x) = Y(\psi_s(x))$$

Using this and the chain rule gives the identity:

$$\frac{d}{dt}\psi_s(\phi_t(x)) = \psi_s'(\phi_t(x))X(\phi_t(x)) \tag{9.25}$$

From these identities, we get the proof of the implications as follows.
(1) $\Longrightarrow$ (2). If we differentiate both sides of the equation: $\phi_t(\psi_s(x)) = \psi_s(\phi_t(x))$ with respect to t, and use identity (9.25) we get

$$X(\phi_t(\psi_s(x))) = \psi_s'(\phi_t(x))X(\phi_t(x)).$$

Taking $t = 0$ in this gives

$$X(\psi_s(x)) = \psi_s'(x)X(x).$$

Substituting: $x = \psi_{-s}(y) = \psi_s^{-1}(y)$ shows that (2) holds.

$(2) \implies (3)$ An equivalent version of (2) is

$$\psi'_s(x)X(x) = X(\psi_s(x)),$$

for all x and s. In terms of components this is

$$\sum_{j=1}^{n} \frac{\partial \psi^i}{\partial x_j}(x, s)X^j(x) = X^i(\psi_s(x)),$$

for all x and s. Differentiating both sides with respect to s and then taking $s = 0$ gives

$$\sum_{j=1}^{n} \frac{\partial Y^i}{\partial x_j}(x)X^j(x) = \sum_{j=1}^{n} \frac{\partial X^i}{\partial x_j}(x)Y^j(x).$$

This says that $[X, Y] = 0$.

$(3) \implies (1)$ This part is left as an exercise. $\square$

9.3.1 Arnold's Theorem

Arnold's theorem gives conditions under which the submanifold $S^{c_1 \cdots c_n}_{F_1 \ldots F_n}$, for a completely integrable system, is diffeomorphic to an n dimensional torus. Thus, before discussing the theorem, we need a motivating example and then the definition of the standard n torus T^n.

Example 9.5 (The Tori T^1, T^2, T^3) The usual torus is the surface in $\mathbb{R}^3$ which looks like an inner tube or surface of a doughnut. The viewpoint in this text is that curves, surfaces, and generally submanifolds in $\mathbb{R}^n$ are maps $f : U \subseteq \mathbb{R}^d \to \mathbb{R}^n$, for $d = 1, 2, \ldots$ However, we also view the image $f(U)$ as the geometric equivalent of a curve, surface, or submanifold. With these dual viewpoints, a torus in $\mathbb{R}^3$ can, on the one hand, be defined as a map $f : \mathbb{R}^2 \to \mathbb{R}^3$, of the form:

$$f(\theta, \phi) = \Big((a + b\cos\phi)\cos\theta, \ (a + b\cos\phi)\sin\theta, \ b\sin\phi \Big),$$

where $0 < b < a$. This is the surface in $\mathbb{R}^3$ obtained by revolving the circle $\phi \mapsto ((a + b\cos\phi), 0, b\sin\phi)$ about the z-axis. On the other hand, we also consider the image $f(\mathbb{R}^2)$, which is a set of points in $\mathbb{R}^3$, as a torus. In particular, we define the *standard 2-torus* in $\mathbb{R}^3$ as

$$T^2 \equiv f(\mathbb{R}^2),$$

with $b = 1, a = 2$ as the choice of constants. In a similar fashion, the *standard 1-torus* in $\mathbb{R}^2$ is the circle

$$T^1 \equiv f(\mathbb{R}),$$

where $f : \mathbb{R} \to \mathbb{R}^2$ is the map

$$f(\theta) = (\cos \theta, \sin \theta).$$

It is important to note that the independent variables θ and ϕ in each of these maps are angles. Thus, the three dimensional torus T^3 should come from a map f involving three angles: θ, ϕ, ψ, and you would guess, by analogy, that T^3 is a set of points in $\mathbb{R}^4$. Some thought will convince you that a suitable map $f : \mathbb{R}^3 :\to \mathbb{R}^4$ to define T^3 is

$$f(\theta, \phi, \psi) = \Big(a_1 \cos \theta, \, a_1 \sin \theta, \, a_2 \sin \phi, \, a_3 \sin \psi \Big),$$

where

$$
\begin{aligned}
a_1 &= b_1 + b_2 \cos \phi + b_3 \cos \phi \cos \psi \\
a_2 &= b_2 + b_3 \cos \psi \\
a_3 &= b_3.
\end{aligned}
$$

Here $b_1 > b_2 > b_3$ are constants. It is easy to see that $\theta \mapsto f(\theta, 0, 0)$ is a circle in the x-y plane, that $\phi \mapsto f(0, \phi, 0)$ is a circle in the x-z plane, and $\psi \mapsto f(0, 0, \psi)$ is a circle in the x-w plane. With a fixed choice for the b_i's, say $b_1 = 3, b_2 = 2, b_3 = 1$, we define the *standard 3-torus* in $\mathbb{R}^4$ as

$$T^3 \equiv f(\mathbb{R}^3).$$

Definition 9.6 (The Standard n Torus) The *standard n torus* in $\mathbb{R}^{n+1}$ is the set $T^n \equiv f(\mathbb{R}^n)$, where $f : \mathbb{R}^n \to \mathbb{R}^{n+1}$ is the map defined as follows. Denote the points in $\mathbb{R}^n$, by $\theta = (\theta_1, \ldots, \theta_n)$ and suppose $b_1 > b_2 > \cdots > b_n$ are any numbers. Then

$$f(\theta) = \Big(a_1(\theta) \cos \theta_1, \, a_1(\theta) \sin \theta_1, \, a_2(\theta) \sin \theta_2, \, \ldots, a_n(\theta) \sin \theta_n \Big),$$

where

$$a_k(\theta) = b_k + \sum_{j=k+1}^{n} b_j \Big(\prod_{i=k+1}^{j} \cos \theta_i \Big),$$

for $k = 1, \ldots, n-1$, and $a_n(\theta) = b_n$, for all θ.

The *standard n torus* in $\mathbb{R}^{n+1}$ is $T^n \equiv f(\mathbb{R}^n)$, where the choice of the b_i's is $b_i = n + 1 - i$.

Having defined the standard torus in each dimension, we need to now adopt the topologist's practice of calling any topological space S a torus if S is homeomorphic to a standard torus. Indeed, topologists like to think of tori as the spaces that arise from the following construction.

Definition 9.7 (The Decomposition Space) Suppose $f : U \to T$ is any map. Define an equivalence relation on the set U by $u_1 \sim u_2$ if and only if $f(u_1) = f(u_2)$. Then the *decomposition space* of f is the set

$$K_f = U/\sim,$$

of all equivalence classes of U under this equivalence relation.

The basic result for decomposition spaces needed here is

Proposition 9.3 *Suppose that U and T are topological spaces, with T compact, and that $f : U \to T$ is a continuous, open map from U onto T. Then the decomposition space K_f, endowed with the quotient topology, is homeomorphic to T.*

Proof: See [Dug 66] for the proof as well as the definitions of an open map and the quotient topology.

Example 9.6 It is easy to show that the standard n-torus T^n is compact (exercise), and because the sine and cosine functions have period 2π, it is also easy to see that $T^n = f(U)$, where U is the Cartesian product of the interval $[0, 2\pi]$ with itself n times:

$$U = [0, 2\pi] \times [0, 2\pi] \times \cdots \times [0, 2\pi] = [0, 2\pi]^n.$$

With a little work, one can show that f is an open map when restricted to U. Hence by the proposition, the n-torus T^n is homeomorphic to $K_f = [0, 2\pi]^n/\sim$. For topologists, K_f is the standard model for the n-torus.

In particular, we can think of the circle T^1 as being the same as $[0, 2\pi]/\sim$. The latter space is what results when one takes the interval $[0, 2\pi]$ and "identifies" the two endpoints $0 \sim 2\pi$. These are equivalent since $f(0) = f(2\pi)$, and they are the only points in the interval $[0, 2\pi]$ that get identified. Pictorially, one views the circle as arising from the interval by curling the interval around and joining its two endpoints.

In a similar fashion $T^2 \cong U/\sim$, where $U = [0, 2\pi] \times [0, 2\pi]$ is a square. Note that in this case the map $f = f(\theta, \phi)$ defining the torus, only identifies points on the boundary of the square. That is, $f(\theta, \phi) = f(\theta_*, \phi_*)$ if and only if either $\theta = \theta_*$ and $\phi, \phi_* \in \{0, 2\pi\}$ or $\phi = \phi_*$ and $\theta, \theta_* \in \{0, 2\pi\}$ (exercise). Geometrically, one conceives of the torus as arising from curling the square around so that the top and bottom sides are joined, thus forming a cylinder, and then bending this cylinder around so that its two boundary circles are joined, thus giving a torus.

Example 9.7 Consider the Hamiltonian system with the Hamiltonian $\widetilde{H}$ from Example 9.4. Hamilton's equations are

$$\begin{aligned} \dot{q}_i &= p_i \\ \dot{p}_i &= -\omega_i^2 q_i, \end{aligned}$$

for $i = 1, 2$ and $\omega_1 = 1$, $\omega_2 = \sqrt{3}$. For given initial displacements a, b of the two masses and no initial velocities, the corresponding integral curve is

$$\gamma(t) = \left(a \cos \omega_1 t, \ b \cos \omega_2 t, \ -\omega_1 a \sin \omega_1 t, \ -\omega_2 b \sin \omega_2 t, \right),$$

for $t \in \mathbb{R}$. Thus, this integral curve lies on the 2-dimensional submanifold (surface) of $\mathbb{R}^4$ defined by the map

$$f(t_1, t_2) = \left(a \cos \omega_1 t_1, \ b \cos \omega_2 t_2, \ -\omega_1 a \sin \omega_1 t_1, \ -\omega_2 b \sin \omega_2 t_2, \right),$$

for $(t_1, t_2) \in \mathbb{R}^2$. It is easy to check that this surface is also the intersection of the level sets of the two constants of the motion

$$\widetilde{F}_1 = \tfrac{1}{2}(p_1^2 + \omega_1^2 q_1^2), \qquad \widetilde{F}_2 = \tfrac{1}{2}(p_2^2 + \omega_2^2 q_2^2)$$

found in Example 9.4. That is,

$$f(\mathbb{R}^2) = S_{\widetilde{F}_1}^{a^2/2} \cap S_{\widetilde{F}_2}^{b^2/2}.$$

In addition, by the above discussion and proposition, this surface is homeomorphic to the decomposition space $K_f = ([0, 2\pi/\omega_1] \times [0, 2\pi/\omega_2])/ \sim$, for the map f. But K_f is homeomorphic to the standard torus T^2. With these identifications, we get that the integral curve γ lies on the torus $f(\mathbb{R}^2) \subset \mathbb{R}^4$.

We need one more fact, involving a relation between Lie and Poisson brackets, before stating and proving Arnold's theorem. For Hamiltonian vector fields X_F, X_G, we have the following important relation between the Lie bracket of the vector fields and the Poisson bracket of the underlying functions:

$$[X_F, X_G] = X_{\{G,F\}}.$$

This follows from a straight-forward computation and is left as an exercise. An easy consequence of this is that if $\{F, G\} = 0$, then $[X_F, X_G] = 0$, and therefore by the last theorem, the corresponding flows commute: $\phi_t^F \circ \phi_t^G = \phi_t^G \circ \phi_t^F$. This is a key ingredient in the following theorem.

Theorem 9.4 (Arnold's Theorem) *Suppose* $(\dot{q}, \dot{p}) = X_H(q, p)$ *is an integrable Hamiltonian system on* $\mathcal{O} \subseteq \mathbb{R}^{2n}$, *with* $F_1 = H, F_2, \ldots, F_n$ *being constants of the motion that are functionally independent and in involution. Suppose* $c_i, i = 1, \ldots, n$ *are values of the* F_i's *for which* $S_{F_1 \ldots F_n}^{c_1 \ldots c_n}$ *is compact and connected. Then* $S_{F_1 \ldots F_n}^{c_1 \ldots c_n}$ *is diffeomorphic to the* n *torus* T^n.

Proof: For convenience of notation let $M = S_{F_1 \ldots F_n}^{c_1 \ldots c_n}$. Since each X_{F_i} is tangent to M and by assumption M is compact, a standard result from manifold theory says that X_{F_i}, restricted to M is complete, and so the flow $\phi_t^{F_i}$ on M is defined for all $t \in \mathbb{R}$. Thus, each $\phi_t^{F_i}$ is a 1 parameter group of transformations of M. Putting all these flows together gives a n parameter group of transformations defined by:

$$\Psi_{\vec{t}} = \phi_{t_1}^{F_1} \circ \cdots \circ \phi_{t_n}^{F_n},$$

where we are using the notation:

$$\vec{t} = (t_1, \ldots, t_n) \in \mathbb{R}^n,$$

for the points in $\mathbb{R}^n$. The coordinates $t_1, \ldots, t_n$ of these points $\vec{t}$ are the n parameters that parametrize the group. We consider $\mathbb{R}^n$ as a group with vector addition being the group operation. To see that the group property holds, note that because $F_1, \ldots, F_n$ are in involution, the vector fields $X_{F_1}, \ldots, X_{F_n}$ commute with each other, and thus their flows $\phi_{t_1}^{F_1}, \ldots, \phi_{t_n}^{F_n}$ commute as well. From this, the verification of the group property is easy:

$$
\begin{aligned}
\Psi_{\vec{s}} \circ \Psi_{\vec{t}} &= \phi_{s_1}^{F_1} \circ \cdots \circ \phi_{s_n}^{F_n} \circ \phi_{t_1}^{F_1} \circ \cdots \circ \phi_{t_n}^{F_n} \\
&= \phi_{s_1}^{F_1} \circ \phi_{t_1}^{F_1} \circ \cdots \circ \phi_{s_n}^{F_n} \circ \phi_{t_n}^{F_n} \\
&= \phi_{s_1+t_1}^{F_1} \circ \cdots \circ \phi_{s_n+t_n}^{F_n} \\
&= \Psi_{\vec{s}+\vec{t}}
\end{aligned}
$$

Note also that $\Psi_{\vec{0}} = I$, the identity transformation, and because of this and the group property, inverses are given by: $\Psi_{\vec{t}}^{-1} = \Psi_{-\vec{t}}$.

Next we pick some point $(q, p) \in M$ and define a map: $f : \mathbb{R}^n \to M$ by:

$$f(\vec{t}) = \Psi_{\vec{t}}(q, p).$$

We will argue that f is onto, is locally a diffeomorphism, and that its decomposition space: $K_f = \mathbb{R}^n / \sim$, is diffeomorphic to the n torus T^n.

For this first note that for any $\vec{t} \in \mathbb{R}^n$, we have:

$$\frac{\partial f}{\partial t_j}(\vec{t}) = X_{F_j}(\Psi_{\vec{t}}(q, p)),$$

for $j = 1, \ldots, n$. This follows by taking the partial derivative of the identity:

$$f(\vec{t} + \vec{s}) = \Psi_{\vec{s}}\left(\Psi_{\vec{t}}(q, p) \right),$$

with respect to s_j and then setting $s = 0$. Consequently, the Jacobian matrix: $f'(\vec{t})$ has maximal rank, and so by the Inverse Function Theorem (Appendix A), f is a diffeomorphism on a neighborhood of $\vec{t}$. Using this and the fact that M is assumed to be both compact and connected, one can show that f is onto.

Now consider the decomposition space K_f for f that arises from the equivalence relation: $\vec{s} \sim \vec{t}$ if and only if $f(\vec{s}) \sim f(\vec{t})$. If we let:

$$G = \{ \vec{t} \in \mathbb{R}^n \mid \Psi_{\vec{t}}(q, p) = (q, p) \},$$

then, using the group property of $\Psi_{\vec{t}}$, it's easy to see that G is a subgroup of $\mathbb{R}^n$ and that $\vec{s} \sim \vec{t}$ if and only if $\vec{s} - \vec{t} \in G$. Thus, the decomposition space for f coincides with the factor group (cf. [Her 75, pp. 51-52]) of $\mathbb{R}^n$ by the subgroup G:

$$K_f = \mathbb{R}^n / G.$$

Now because f is locally a diffeomorphism about each point in $\mathbb{R}^n$, one can show that G must be a discrete subgroup of $\mathbb{R}^n$. A result from algebra (Cf. [Ar 78b, p. 276]) says that G must then be of the form:

$$G = \{ p_1 v_1 + \cdots + p_k v_k \mid \text{the } p_i\text{'s are integers} \},$$

where $v_1, \ldots, v_k$ are linearly independent vectors in $\mathbb{R}^n$. Using this, one can show that $\mathbb{R}^n / G$ is diffeomorphic to $T^k \times \mathbb{R}^{n-k}$. However, since M, and

therefore $\mathbb{R}^n/G$, is compact, we must have $k = n$. This completes the proof.
$\square$

The proof of Arnold's theorem (which was adapted from that in [Ar 78b, p. 271]) shows how the involution condition on the constants of the motion $F_1, \ldots, F_n$ is instrumental in determining the toroidal character of $S^{c_1 \ldots c_n}_{F_1 \ldots F_n}$ (assuming the later is compact and connected). The involution condition is also essential in showing that a completely integrable Hamiltonian system is locally integrable in the sense of Chapter 7, i.e., that there exist $n-1$ additional, independent constants of the motion $A_1, \ldots, A_{n-1}$. Before stating and proving this result, it will be convenient to introduce the following notation.

Suppose $W : U \subseteq \mathbb{R}^{2n} \to \mathbb{R}$ is a differentiable function. We have already used the notation

$$\frac{\partial W}{\partial p} \equiv \left(\frac{\partial W}{\partial p_1}, \ldots, \frac{\partial W}{\partial p_n} \right),$$

for the map $\partial W/\partial p : U \to \mathbb{R}^n$. With p fixed, the Jacobian of this map, with respect to q, is denoted by

$$\frac{\partial^2 W}{\partial q \partial p} = \begin{bmatrix} \frac{\partial^2 W}{\partial q_1 \partial p_1} & \cdots & \frac{\partial^2 W}{\partial q_n \partial p_1} \\ \vdots & & \vdots \\ \frac{\partial^2 W}{\partial q_1 \partial p_n} & \cdots & \frac{\partial^2 W}{\partial q_n \partial p_n} \end{bmatrix}.$$

In addition we will use the notation

$$\frac{\partial^2 W}{\partial p^2}, \quad \frac{\partial^2 W}{\partial q^2}, \quad \frac{\partial^2 W}{\partial p \partial q}.$$

It is easy to see that the first and second of these matrices are symmetric, while the third is not:

$$\left(\frac{\partial^2 W}{\partial p \partial q} \right)^T = \frac{\partial^2 W}{\partial q \partial p}.$$

Theorem 9.5 (Liouville) *Suppose* $(\dot{q}, \dot{p}) = X_H(q, p)$ *is a completely integrable Hamiltonian system on* $\mathcal{O} \subseteq \mathbb{R}^{2n}$, *with* $F_1, F_2, \ldots, F_n = H$ *being constants of the motion that are functionally independent and in involution. Let* $F = (F_1, \ldots, F_n)$ *and suppose* $(q_0, p_0) \in \mathcal{O}$ *is a point at which the matrix*

$$\frac{\partial F}{\partial p}(q_0, p_0) \equiv \left\{ \frac{\partial F_i}{\partial p_j}(q_0, p_0) \right\}_{i,j=1}^n,$$

is invertible. Then there exists a neighborhood $U \subseteq \mathcal{O}$ of (q_0, p_0) and a differentiable function $W : U \to \mathbb{R}$, which has the following properties.

(1) *For each integral curve $t \mapsto (q(t), p(t))$, $t \in I$ of the system that lies in U, there are constants $c_1, \ldots, c_n, b_1, \ldots, b_n$, such that*

$$F_i(q(t), p(t)) \;=\; c_i, \qquad\qquad (i = 1, \ldots, n) \qquad\qquad (9.26)$$

$$\frac{\partial W}{\partial p_i}(q(t), c) \;=\; b_i \qquad\qquad (i = 1, \ldots, n-1) \qquad\qquad (9.27)$$

$$\frac{\partial W}{\partial p_n}(q(t), c) \;=\; t + b_n, \qquad\qquad\qquad (9.28)$$

for all $t \in I$. Here $c = (c_1, \ldots, c_n)$.

(2) *Let $A_1, \ldots, A_n : U \to \mathbb{R}$, be the functions defined by*

$$A_i(q, p) = \frac{\partial W}{\partial p_i}(q, F(q, p)). \qquad\qquad (9.29)$$

Then $A_1, \ldots, A_{n-1}$ are constants of the motion. Furthermore, the $2n$ functions: $F_1, \ldots, F_n, A_1, \ldots, A_n$, are functionally independent.

(3) *The map $h : U \to \mathbb{R}^{2n}$ defined by*

$$h(q, p) = \Big(F(q, p), \; -A(q, p) \Big), \qquad\qquad (9.30)$$

where $A \equiv (A_1, \ldots, A_n)$, is a canonical transformation such that

$$h_*(X_H)(q, p) = (0, 0, \ldots, 0, -1),$$

for all $(q, p) \in h(U)$. Otherwise said, the transformed Hamiltonian $\widetilde{H} \equiv H \circ h^{-1}$, is

$$\widetilde{H}(q, p) = q_n,$$

for all $(q, p) \in h(U)$.

Proof: Define a map $f : \mathcal{O} \to \mathbb{R}^{2n}$ by

$$f(q, p) = (q, \; F(q, p)).$$

Then

$$f'(q, p) = \left[\begin{array}{cc} I & 0 \\ \frac{\partial F}{\partial q}(q, p) & \frac{\partial F}{\partial p}(q, p) \end{array} \right].$$

From the form of this and the hypotheses, this matrix is invertible at the point (q_0, p_0), and so by the Inverse Function Theorem, there is a neighborhood U of (q_0, p_0) on which f is a diffeomorphism. Let $f^{-1} : V = g(U) \to U$ denote the inverse of f. Then f^{-1} has the form

$$f^{-1}(q, p) = (q, \, G(q, p)),$$

and consequently we have the following identities for F and G:

$$F(q, G(q, p)) \quad = \quad p, \qquad \text{(for } (q, p) \in V) \qquad (9.31)$$
$$G(q, F(q, p)) \quad = \quad p, \qquad \text{(for } (q, p) \in U) \qquad (9.32)$$

We let $G = (G_1, \ldots, G_n)$ be the component expression for G. The involution condition on the F_i's enables us to establish the following claim.

Claim: On V, we have

$$\frac{\partial G_i}{\partial q_j} = \frac{\partial G_j}{\partial q_i},$$

for all $i, j = 1, \ldots, n$. To show this, suppose $(\bar{q}, \bar{p}) \in V$. With this fixed, choose $r > 0$ so that the ball $B = B((\bar{q}, \bar{p}), r)$ is contained in V. Define maps $g_i : B \to \mathbb{R}$ by

$$g_i(q, p) = p_i - G_i(q, \bar{p}),$$

for $i = 1, \ldots, n$. Then note that if $(q, p) \in B \cap S^{\bar{p}_1 \cdots \bar{p}_n}_{F_1 \cdots F_n}$, i.e., if $F(q, p) = \bar{p}$, then

$$g_i(q, p) = p_i - G_i(q, \bar{p}) = p_i - G_i(q, F(q, p)) = p_i - p_i = 0,$$

for each i. Thus, the g_i's are identically zero on $B \cap S^{\bar{p}_1 \cdots \bar{p}_n}_{F_1 \cdots F_n}$. By Theorem 11.1, the involution condition results in $S^{\bar{p}_1 \cdots \bar{p}_n}_{F_1 \cdots F_n}$ being invariant under each of the flows $\phi_t^{F_j}$: if $(q, p) \in S^{\bar{p}_1 \cdots \bar{p}_n}_{F_1 \cdots F_n}$, then $\phi_t^{F_j}(q, p) \in S^{\bar{p}_1 \cdots \bar{p}_n}_{F_1 \cdots F_n}$, for all t. Hence

$$g_i \left(\phi_{t_j}^{F_j}(q, p) \right) = 0,$$

for all t. Differentiating this with respect to t and then taking $t = 0$ gives

$$\{ g_i, \, F_j \}(q, p) = 0,$$

for all i, j and all $(q, p) \in B \cap S^{\bar{p}_1 \cdots \bar{p}_n}_{F_1 \cdots F_n}$. But recall that $\{g_i, F_j\} = X_{g_i} \cdot \nabla F_j$. Hence it follows that the vector field X_{g_i} is tangent to the submanifold

$B \cap S_{F_1 \cdots F_n}^{\overline{p}_1 \cdots \overline{p}_n}$. From this it is not difficult to show that flow of X_{g_i} leaves $B \cap S_{F_1 \cdots F_n}^{\overline{p}_1 \cdots \overline{p}_n}$ invariant: if $(q,p) \in B \cap S_{F_1 \cdots F_n}^{\overline{p}_1 \cdots \overline{p}_n}$, then $\phi_t^{g_i}(q,p) \in B \cap S_{F_1 \cdots F_n}^{\overline{p}_1 \cdots \overline{p}_n}$ for all t in an interval about 0 (exercise). Consequently, from the above observations

$$g_j\left(\phi_t^{g_i}(q,p)\right) = 0,$$

for all i, j, all t, and all $(q,p) \in B \cap S_{F_1 \cdots F_n}^{\overline{p}_1 \cdots \overline{p}_n}$. This in turn, exactly as above, leads to

$$\{\, g_j,\, g_i \,\}(q,p) = 0,$$

for all i, j and all $(q,p) \in B \cap S_{F_1 \cdots F_n}^{\overline{p}_1 \cdots \overline{p}_n}$. But an easy computation shows that

$$\{\, g_j,\, g_i \,\}(q,p) = \frac{\partial G_i}{\partial q_j}(q,\overline{p}) - \frac{\partial G_j}{\partial q_i}(q,\overline{p}),$$

for all $(q,p) \in B$. In particular, for $(q,p) = (\overline{q}, G(\overline{q},\overline{p}))$, we have by identity (9.31) that $F(q,p) = \overline{p}$ and hence

$$0 = \{\, g_j,\, g_i \,\}(\overline{q},\overline{p}) = \frac{\partial G_i}{\partial q_j}(\overline{q},\overline{p}) - \frac{\partial G_j}{\partial q_i}(\overline{q},\overline{p}).$$

This establishes the claim.

From the claim we get the existence of a differentiable map $W : V \to \mathbb{R}$, such that

$$\frac{\partial W}{\partial q_i}(q,p) = G_i(q,p),$$

for every i and all $(q,p) \in V$. Clearly the claim is a necessary set of conditions for the existence of the function W. The sufficiency of these conditions, however requires a topological condition on the set V (Cf. [Be 98, p. 232]). Rather than work at this level of generality, we can, for our purposes, always assume that V is an open ball in $\mathbb{R}^{2n}$, centered at $f(q_0, p_0)$. Then there is an integral formula for the construction of W (See the exercises).

Next, to see that the A_i's defined by (9.29) are constants of the motion, suppose that $t \mapsto (q(t), p(t))$, for $t \in I$, is an integral curve of the Hamiltonian system that lies in U. Then there are constants $c_1, \ldots, c_n$, such that $F(q(t), p(t)) = (c_1, \ldots, c_n) = c$, for all $t \in I$. But then by identity (9.32)

$$p(t) = G(q(t), F(q(t), p(t))) = G(q(t), c),$$

for all $t \in I$. Using this we find that

$$
\begin{aligned}
\frac{d}{dt}\Big[A_i(q(t), p(t))\Big] &= \frac{d}{dt}\left[\frac{\partial W}{\partial p_i}\Big(q(t), F(q(t), p(t))\Big)\right] \\
&= \frac{d}{dt}\left[\frac{\partial W}{\partial p_i}(q(t), c)\right] \\
&= \sum_{j=1}^{n} \frac{\partial^2 W}{\partial q_j \partial p_i}(q(t), c)\, q_j'(t) \\
&= \sum_{j=1}^{n} \frac{\partial G_j}{\partial p_i}(q(t), c)\, \frac{\partial H}{\partial p_j}(q(t), p(t)) \\
&= \sum_{j=1}^{n} \frac{\partial G_j}{\partial p_i}(q(t), c)\, \frac{\partial H}{\partial p_j}(q(t), G(q(t), c)) \\
&= \frac{\partial}{\partial p_i}\left[H\Big(q(t), G(q(t), p)\Big)\right]\Big|_{p=c} \\
&= \frac{\partial}{\partial p_i}\big[p_n\big] \\
&= \delta_{in}
\end{aligned}
$$

Thus, $A_1, \ldots, A_{n-1}$ are constants of the motion and $A_n(q(t), p(t)) = t + b_n$, for some constant b_n.

Next we show that the map h defined by (9.30) is a canonical transformation. For this, observe that because $\partial W/\partial q_i = G_i$ and G satisfies (9.32), we get the identity

$$
\frac{\partial W}{\partial q}(q, F(q, p)) = p, \tag{9.33}
$$

for all $(q, p) \in U$. Also observe the map h, in terms of W is

$$
h(q, p) = \left(F(q, p), -\frac{\partial W}{\partial p}(q, F(q, p))\right). \tag{9.34}
$$

Identity (9.33) and the form (9.34) of the definition for h are all that are needed to guarantee that h is canonical. We calculate the Jacobian matrix of h and show that this is a symplectic matrix at each point of V. The Jacobian matrix has the block form

$$
h' = \begin{bmatrix} B & C \\ D & E \end{bmatrix}, \tag{9.35}
$$

where, by the Chain Rule,

$$B = \frac{\partial F}{\partial q}, \quad C = \frac{\partial F}{\partial p},$$

and

$$D = -\frac{\partial^2 W}{\partial q \partial p} - \frac{\partial^2 W}{\partial p^2}\frac{\partial F}{\partial q}, \quad E = -\frac{\partial^2 W}{\partial p^2}\frac{\partial F}{\partial p}.$$

In these expressions we have suppressed the dependence on q and p. Now use Exercise 8 in Section 2, which says that a matrix of the form (9.35) will be symplectic if $B^T D, C^T E$ are symmetric matrices and if $B^T E - D^T C = I$. To verify this we are going to need the identity (9.33), or more specifically the following identities that arise by differentiating it:

$$\frac{\partial^2 W}{\partial q^2} + \frac{\partial^2 W}{\partial p \partial q}\frac{\partial F}{\partial q} = 0, \tag{9.36}$$

$$\frac{\partial^2 W}{\partial p \partial q}\frac{\partial F}{\partial p} = I \tag{9.37}$$

Then we see, using (9.36) and the remarks before the theorem, that

$$
\begin{aligned}
B^T D &= -\left(\frac{\partial F}{\partial q}\right)^T \frac{\partial^2 W}{\partial q \partial p} - \left(\frac{\partial F}{\partial q}\right)^T \frac{\partial^2 W}{\partial p^2}\frac{\partial F}{\partial q} \\
&= -\left(\frac{\partial^2 W}{\partial p \partial q}\frac{\partial F}{\partial q}\right)^T - \left(\frac{\partial F}{\partial q}\right)^T \frac{\partial^2 W}{\partial p^2}\frac{\partial F}{\partial q} \\
&= \frac{\partial^2 W}{\partial q^2} - \left(\frac{\partial F}{\partial q}\right)^T \frac{\partial^2 W}{\partial p^2}\frac{\partial F}{\partial q}.
\end{aligned}
$$

This is clearly a symmetric matrix. Also

$$C^T E = -\left(\frac{\partial F}{\partial p}\right)^T \frac{\partial^2 W}{\partial p^2}\frac{\partial F}{\partial p},$$

is a symmetric matrix. Checking the last condition, using (9.37) and the remarks before the theorem, we see that

$$
\begin{aligned}
B^T E - D^T C &= -\left(\frac{\partial F}{\partial q}\right)^T \frac{\partial^2 W}{\partial p^2}\frac{\partial F}{\partial p} + \left(\frac{\partial^2 W}{\partial q \partial p} + \frac{\partial^2 W}{\partial p^2}\frac{\partial F}{\partial q}\right)^T \frac{\partial F}{\partial p} \\
&= \frac{\partial^2 W}{\partial p \partial q}\frac{\partial F}{\partial p} \\
&= I
\end{aligned}
$$

This completes the verification that h' is symplectic at each point of U. But a symplectic matrix is invertible and thus, by the Inverse Function Theorem, h is a diffeomorphism on U. Hence h is a canonical transformation.

To see that $F_1, \ldots, F_n, A_1, \ldots, A_n$ are functionally independent on U, note that since $h = (F_1, \ldots, F_n, -A_1, \ldots, -A_n)$, the Jacobian matrix of h, expressed in terms of its rows, is

$$h' = [\nabla F_1, \ldots, \nabla F_n, -\nabla A_1, \ldots, -\nabla A_n].$$

As we have seen, this matrix is invertible at each point of U and so the rows are linearly independent at each point of U. The functional independence follows. It also follows from this and

$$\nabla F_i \cdot X_H = \{F_i, H\} = 0, \qquad \nabla A_i \cdot X_H = \{A_i, H\} = 0,$$

that

$$h_*(X_H)(q, p) = (0, 0, \ldots, 0, -1),$$

for every $(q, p) \in V$. This completes the proof of the theorem. $\square$

The above proof is modeled on the one given by Whittaker in [Wh 65, pp. 323-325]. However, the first edition of this text dates back to 1904 and so we have added modern notation and rigor to his arguments. The theorem is attributed to Liouville and is often paraphrased as saying that the solutions of a completely integrable system can be expressed by quadratures, or integrals. This can be explained roughly as follows.

Knowing n first integrals, $F_1, \ldots, F_n$ for the system, gives the equations

$$F_i(q(t), p(t)) = c_i, \qquad (i = 1, \ldots, n)$$

that an integral curve $t \mapsto (q(t), p(t))$ must satisfy. By the Implicit Function Theorem, we can solve these equations to explicitly get the momenta in terms of the positions and the constants

$$p_i(t) = G_i(q(t), c), \qquad (i = 1, \ldots, n)$$

Thus, if we can find a formula for $q(t)$, then the formula for $p(t)$ is given by the above. Finding a formula for $q(t)$ involves additional integrations (quadratures) because this is the process whereby the additional constants of the motion $A_1, \ldots, A_n$ are constructed. Indeed, these come from the function W that satisfies

$$\frac{\partial W}{\partial q_i}(q, p) = G_i(q, p), \qquad (i = 1, \ldots, n)$$

and heuristically W can be found by integrating:

$$W(q,p) = \int \sum_{i=1}^{n} G_i(q,p) \, dq_i.$$

Having constructed W by integration, then the other constants of the motion give the following equations that $q(t)$ must satisfy

$$\frac{\partial W}{\partial p_i}(q(t), c) = b_i \qquad (i = 1, \ldots, n-1)$$

$$\frac{\partial W}{\partial p_n}(q(t), c) = t + b_n.$$

Inverting these gives $q(t)$ explicitly in terms of the constants and the time

$$q(t) = \alpha(t, c, b),$$

and from this we get the explicit formula for the momenta

$$p(t) = G(\alpha(t, c, b), c).$$

This gives the "general solution" in that the formulas involve the $2n$ arbitrary constants $c_1, \ldots, c_n, b_1, \ldots, b_n$. Of course in practice, being able to calculate the integrals involved and then invert (solve) the resulting equations as indicated above can be an impossible task.

Also note that the theorem is local and is thus only a variation of Corollary 7.1, which guarantees that any system is integrable on a neighborhood of a nonfixed point (as (q_0, p_0) is here). And recall that Corollary 7.1 is a special case of the Flow Box Theorem, which says that near a nonfixed point, the flow is similar to a parallel flow in one direction. That is part of what Liouville's Theorem 9.5 says: X_H is differentiably equivalent to $X_{\widetilde{H}} = h_*(X_H)$, where $\widetilde{H}(q,p) = q_n$ and the flow for $X_{\widetilde{H}}$ is

$$\phi_t^{\widetilde{H}}(q,p) = (q_1, \ldots, q_n, p_1, \ldots, p_{n-1}, p_n - t).$$

However, Theorem 9.5 is stronger than Corollary 7.1 and the Flow Box Theorem since it gives a specific way to construct the diffeomorphism h from the constants of the motion $F_1, \ldots, F_n$. Namely

$$h(q,p) = \left(F(q,p), \, -\frac{\partial W}{\partial p}(q, F(q,p)) \right).$$

The examples below and the ensuing exercise problems will clarify what's involved in this construction.

Example 9.8 (1-d Mechanical Hamiltonians) For the case $n = 1$ and $H(q, p) = \frac{1}{2}p^2 + V(q)$, we have seen that the phase portrait and flow is easily determined from the graph of the potential function V. It is instructive to see how the construction in the proof of Theorem 9.5 works out in this special case. The only constant of the motion is $F = H$. In the proof of the theorem, the function G results from using the Implicit Function Theorem. A slightly different way of phrasing this is to say that G arises from solving the equation

$$F(q, x) = p$$

for x, giving

$$x = G(q, p).$$

For the special case in this example, the first equation is

$$\tfrac{1}{2}x^2 + V(q) = p,$$

and solving for x gives

$$x = \sqrt{2}\sqrt{p - V(q)}.$$

(We will use just the positive square root.) So $G(q, p) = \sqrt{2}\sqrt{p - V(q)}$. In the proof of the theorem the function W is then determined from $\frac{\partial W}{\partial q} = G(q, p)$. In the 1-d case, this amounts to letting W be an indefinite integral:

$$W(q, p) = \int G(q, p)dq = \int \sqrt{2}\sqrt{p - V(q)}\, dq. \tag{9.38}$$

This integral is explicitly computable for a few choices of potential function V (but in most other cases not). Doing this then gives the canonical transformation h:

$$h(q, p) = \left(\tfrac{1}{2}p^2 + V(q), \; -\frac{\partial W}{\partial p}(q, \tfrac{1}{2}p^2 + V(q)) \right). \tag{9.39}$$

which "straightens out" the flow.

Example 9.9 (A Harmonic Oscillator) Suppose $V(q) = \frac{1}{2}q^2$, so that $H(q, p) = \frac{1}{2}p^2 + \frac{1}{2}q^2$. Of course the level curves of H are all circles, but Figure 9.7 shows how this also arises using the standard technique of sketching the level curves from the graph of $-V$. For the sake of instruction, we will explicitly compute the expression of the canonical transformation h in Eq.(9.39).

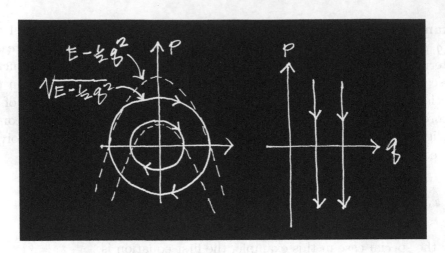

Figure 9.7: *Integral curves for $H = \frac{1}{2}p^2 + \frac{1}{2}q^2$ (on the left) and $H = q$ (on the right).*

First we need to calculate the integral that gives W:

$$W(q,p) = \int \sqrt{2}\sqrt{p - \tfrac{1}{2}q^2}\, dq = \int \sqrt{2p - q^2}\, dq.$$

For this use the trig substitution $q = \sqrt{2p}\sin\theta$ to get

$$
\begin{aligned}
W(q,p) &= \int \sqrt{2p - q^2}\, dq = \int 2p\cos^2\theta\, d\theta = 2p\int(\tfrac{1}{2} + \tfrac{1}{2}\cos 2\theta)d\theta \\
&= p(\theta + \tfrac{1}{2}\sin 2\theta) = p(\theta + \sin\theta\cos\theta) \\
&= p\left(\sin^{-1}\left(\frac{q}{\sqrt{2p}}\right) + \frac{q}{\sqrt{2p}}\sqrt{1 - \frac{q^2}{2p}}\right) \\
&= p\sin^{-1}\left(\frac{q}{\sqrt{2p}}\right) + \frac{q}{2}\sqrt{2p - q^2}
\end{aligned}
$$

From this we get

$$
\begin{aligned}
\frac{\partial W}{\partial p}(q,p) &= \sin^{-1}\left(\frac{q}{\sqrt{2p}}\right) + \frac{p \cdot \frac{q}{\sqrt{2}}(-\frac{1}{2}p^{-3/2})}{\sqrt{1 - \frac{q^2}{2p}}} + \frac{q}{2}\cdot\frac{1}{\sqrt{2p - q^2}} \\
&= \sin^{-1}\left(\frac{q}{\sqrt{2p}}\right)
\end{aligned}
$$

Using this in Formula (9.39) for the canonical transformation h gives

$$h(q,p) = \left(\tfrac{1}{2}p^2 + \tfrac{1}{2}q^2, \ -\sin^{-1}\left(\frac{q}{\sqrt{p^2+q^2}} \right) \right).$$

Next, let's check directly that this is a canonical transformation and that it does the required straightening. First we compute the Jacobian matrix of h and, using a little calculus and algebra, see that it reduces to

$$h'(q,p) = \left[\begin{array}{cc} q & p \\ \frac{-p}{p^2+q^2} & \frac{q}{p^2+q^2} \end{array} \right]$$

In 1-d, for h to be a canonical transformation the determinant of this matrix must be identically 1, i.e., $\det(h'(q,p)) = 1$ for all (q,p). (See Exercise 8 in 9.2.) But this is clearly the case. Finally, note that $X_H(q,p) = (p, -q)$ and so

$$X_{\widetilde{H}}(q,p) = h'(q,p)X_H(q,p) = \left[\begin{array}{cc} q & p \\ \frac{-p}{p^2+q^2} & \frac{q}{p^2+q^2} \end{array} \right] \left[\begin{array}{c} p \\ -q \end{array} \right] = \left[\begin{array}{c} 0 \\ -1 \end{array} \right],$$

which verifies the "straightening" of the flow for X_H. We can also verify this in another way. Namely, we compute h^{-1} (which is possible in this example) and then $\widetilde{H} = H \circ h^{-1}$. To compute h^{-1} we write h as

$$\bar{q} = \frac{1}{2}(p^2 + q^2)$$
$$\bar{p} = -\sin\left(\frac{q}{\sqrt{p^2 + q^2}} \right),$$

viewing it as a transformation $(q,p) \mapsto (\bar{q}, \bar{p})$. Then h^{-1} is the transformation $(\bar{q}, \bar{p}) \mapsto (q,p)$ which comes from solving the above system of equations for q and p in terms of $\bar{q}$ and $\bar{p}$. Using some standard algebra, we get

$$q = -\sqrt{2\bar{q}} \sin \bar{p}$$
$$p = \sqrt{2\bar{q}} \cos \bar{p}$$

Then clearly

$$\widetilde{H}(\bar{q}, \bar{p}) = H(h^{-1}(\bar{q}, \bar{p})) = \tfrac{1}{2}\left(-\sqrt{2\bar{q}} \sin \bar{p} \right)^2 + \tfrac{1}{2}\left(\sqrt{2\bar{q}} \cos \bar{p} \right)^2 = \bar{q}.$$

The proof Theorem 9.5 contains an important technique for constructing, or generating, canonical transformations. This technique is very useful even though it is often a local construction, so we record this as a separate result.

Corollary 9.3 (Generating Functions) *Suppose $W : V \to \mathbb{R}$ is a differentiable function on an open set V of $\mathbb{R}^{2n}$ and $F : U \to \mathbb{R}^n$ is a differentiable map on an open set U of $\mathbb{R}^{2n}$ such that $(q, F(q,p)) \in V$ and*

$$\frac{\partial W}{\partial q}(q, F(q,p)) = p, \tag{9.40}$$

for all $(q,p) \in U$. Then the map $h : U \to V$, defined by

$$h(q,p) = \left(F(q,p), -\frac{\partial W}{\partial p}(q, F(q,p)) \right), \tag{9.41}$$

is a canonical transformation.

The above corollary doesn't indicate where W and F come from. Of course in Theorem 9.5, $F = (F_1, \ldots, F_n)$ comes from knowing n constants of the motion: $F_1, \ldots, F_n$, and W is constructed as a solution of $\frac{\partial W}{\partial q} = G$, where G results from solving $F(q,x) = p$ for $x = G(q,p)$.

Another way to get W and F is to start with a fairly arbitrary function $W : V \subseteq \mathbb{R}^{2n} \to \mathbb{R}$ (conditions to follow) and then determine F from W, thereby resulting in a canonical transformation h via the corollary. This is why W is called a generating function for the canonical transformation h. Here's how this situation works: Start with any twice continuously differentiable function W for which you can

$$\text{solve } \frac{\partial W}{\partial q}(q,x) = p \quad \text{for } x = F(q,p). \tag{9.42}$$

Then condition (9.40) of the corollary is satisfied and so W generates a canonical transformation h via Formula (9.41). The solvability condition (9.42) requires that the map $f : V \subseteq \mathbb{R}^{2n} \to \mathbb{R}^{2n}$ given by $f(q,x) = (q, \frac{\partial W}{\partial q}(q,x))$ be invertible. Its inverse will have the form $f^{-1}(q,p) = (q, F(q,p))$ Of course, invertability is *locally* guaranteed by the Inverse Function Theorem on a neighborhood of any point (q_0, x_0) in V where

$$\det \left(\frac{\partial^2 W}{\partial q \partial x}(q_0, x_0) \right) \neq 0. \tag{9.43}$$

Thus, any twice continously differentiable function satisfying (9.43) at a point will generate a local canonical transformation.

Example 9.10 Consider the function $W(q, x) = \frac{1}{2}q^2x^3$, defined on all of $\mathbb{R}^2$. We check that this can generate a canonical transformation h for 1-d Hamiltonian systems and we explicitly compute h. From

$$\frac{\partial W}{\partial q}(q, x) = qx^3 \qquad \text{and} \qquad \frac{\partial^2 W}{\partial x \partial q}(q, x) = 3qx^2,$$

we see that $\frac{\partial^2 W}{\partial x \partial q}(q, x) \neq 0$ at all points except the origin and so W can locally serve as a generating function. To find F we solve

$$qx^3 = p,$$

for x, getting

$$x = \left(\frac{p}{q}\right)^{1/3} = q^{-1/3}p^{1/3}.$$

So $F(q, p) = q^{-1/3}p^{1/3}$. Next

$$\frac{\partial W}{\partial x}(q, x) = \frac{3}{2}q^2x^2$$

and

$$\frac{\partial W}{\partial x}(q, F(q, p)) = \frac{3}{2}q^2\left(q^{-1/3}p^{1/3}\right)^2 = \frac{3}{2}q^{4/3}p^{2/3}.$$

Consequently

$$h(q, p) = \left(F(q, p), \frac{\partial W}{\partial x}(q, F(q, p))\right) = \left(q^{-1/3}p^{1/3}, -\frac{3}{2}q^{4/3}p^{2/3}\right).$$

We check directly that this is a canonical transformation by computing the Jacobian matrix of h:

$$h'(q, p) = \begin{bmatrix} -\frac{1}{3}q^{-4/3}p^{1/3} & \frac{1}{3}q^{-1/3}p^{-2/3} \\ -2q^{1/3}p^{2/3} & -q^{4/3}p^{-1/3} \end{bmatrix}.$$

Then clearly $\det(h'(q, p)) = \frac{1}{3} + \frac{2}{3} = 1$, for all $(q, p) \neq (0, 0)$ and hence $h'(q, p)$ is a symplectic matrix.

Example 9.11 An example with 2-degrees of freedom will be instructive. So consider the function

$$W(q, x) = q_1x_1^2 + q_2x_2^2,$$

defined for all $(q, x) = (q_1, q_2, x_1, x_2) \in \mathbb{R}^4$. We get

$$\frac{\partial W}{\partial q}(q, x) = (x_1^2, x_2^2) \qquad \text{and} \qquad \frac{\partial^2 W}{\partial x \partial q}(q, x) = \begin{bmatrix} 2x_1 & 0 \\ 0 & 2x_2 \end{bmatrix},$$

and so $\det\left(\frac{\partial^2 W}{\partial x \partial q}(q, x)\right) = 4x_1 x_2$ is nonzero except where either x_1 or x_2 is zero. The equation $\frac{\partial W}{\partial q}(q, x) = p$ in this example is

$$(x_1^2, x_2^2) = (p_1, p_2),$$

which is a simple system of equations to solve for x_1, x_2. This gives $x_1 = \sqrt{p_1}$, $x_2 = \sqrt{p_2}$, and so

$$F(q, p) = (\sqrt{p_1}, \sqrt{p_2}).$$

Next

$$\frac{\partial W}{\partial x}(q, x) = (2q_1 x_1, 2q_2 x_2)$$

and

$$\frac{\partial W}{\partial x}(q, F(q, p)) = \left(2q_1 \sqrt{p_1}, 2q_2 \sqrt{p_2}\right).$$

Consequently

$$h(q, p) = \left(\sqrt{p_1}, \sqrt{p_2}, -2q_1 \sqrt{p_1}, -2q_2 \sqrt{p_2}\right).$$

To check directly that this is a canonical transformation, we compute the Jacobian matrix

$$h'(q, p) = \begin{bmatrix} 0 & 0 & \dfrac{1}{2\sqrt{p_1}} & 0 \\ 0 & 0 & 0 & \dfrac{1}{2\sqrt{p_2}} \\ -2\sqrt{p_1} & 0 & = \dfrac{q_1}{2\sqrt{p_1}} & 0 \\ 0 & -2\sqrt{p_1} & 0 & = \dfrac{q_2}{2\sqrt{p_2}} \end{bmatrix} = \begin{bmatrix} B & C \\ D & E \end{bmatrix}.$$

Here we have represented the 4×4 Jacobian matrix in terms of 2×2 matrices B, C, D, E. By the result in Exercise 8 in Section 9.2, $h'(q, p)$ is a symplectic matrix if and only if $B^T D$ and $C^T E$ are symmetric and $B^T E - D^T C = I$. But it's easy to see that $B^T D = 0$ and $C^T E$ is a diagonal matrix and $B^T E - D^T C = 0 - (-I) = I$.

It should also be noted that (completely) integrable systems are rather special. Wintner, in [W 47, pp. 143-144], gives a nice description of how perplexing this was to mathematicians and scientists in the 18th and 19th centuries when on the one hand the goal of solving differential equations was to find formulas for their general solutions, while on the other hand almost all of the important systems in physics failed to have sufficiently many first integrals. Wintner remarks that maybe the concept (or definition of) what integrability means is not right and concludes by saying: "All of this lies along the line of Poincaré's dictum, according to which a system is neither integrable, nor nonintegrable, but more or less integrable." (See also the brief historical discussion of integrability in the article [KGT 97, pp. 30-46]). Since Wintner's time, the topic of integrability has taken on renewed interest for a number of reasons. One of these is due to the KAM theorem (by Kolmogorov, Arnold, and Moser) which describes properties of Hamiltonian systems that are approximately integrable (see [Ar 78b, pp. 399-415], [HZ 94], [KLP 94] and the advanced text [Laz 93]).

Exercises 9.3

1. **(Lie Algebras)** A *Lie algebra* is a vector space $\mathcal{A}$, over the real or complex numbers, together with a binary operation

$$[\cdot, \cdot] : \mathcal{A} \times \mathcal{A} \to \mathcal{A},$$

called the *Lie bracket*, which satisfies

(1) **Antisymmetry:** $[X, Y] = -[Y, X]$, for all $X, Y \in \mathcal{A}$.

(2) **Bilinearity:**
  (a) $[X + Y, Z] = [X, Z] + [Y, Z]$ and $[X, Y + Z] = [X, Y] + [X, Z]$,
  (b) $[\lambda X, Y] = \lambda[X, Y] = [X, \lambda Y]$,

  for all $X, Y, Z \in \mathcal{A}$ and all scalars λ.

(3) **Jacobi Identity:**

$$[X, [Y, Z]] + [Y, [Z, X]] + [Z, [X, Y]] = 0,$$

  for all $X, Y, Z \in \mathcal{A}$.

Prove that each of the following is a Lie algebra.

(a) $\mathcal{A} = \mathcal{M}_n =$ the set of all $n \times n$ matrices (real or complex) with Lie bracket defined by
$$[A, B] \equiv AB - BA.$$

(b) $\mathcal{A}$ = the set of C^∞ vector fields $X : \mathcal{O} \to \mathbb{R}^n$ on an open set $\mathcal{O}$ in $\mathbb{R}^n$, with Lie bracket defined by equation (9.22) in the text. *Hint*: Make the identification of X with an operator on functions $F \in C^\infty(\mathcal{O})$ as suggested in the equation (9.23). With this identification, show that equation (9.24) holds. Now prove the required properties (1)-(3) for the Lie Bracket.

2. **(Poisson Brackets)** Consider the set $C^\infty(\mathcal{O})$ of C^∞ functions on an open set $\mathcal{O} \subseteq \mathbb{R}^{2n}$. This set, endowed with the Poisson bracket $\{\cdot, \cdot\}$ as its Lie bracket, is another prime example of a Lie algebra. This exercise is to prove this and some other related results.

(a) Show that the Poisson bracket is antisymmetric and bilinear.

(b) For $F \in C^\infty(\mathcal{O})$, let X_F denote the corresponding Hamiltonian vector field defined by

$$X_F = J\nabla F = \left(\frac{\partial F}{\partial p}, \frac{\partial F}{\partial p} \right).$$

Show by direct calculation that

$$X_{\{G,F\}} = [X_F, X_G],$$

where $[\cdot, \cdot]$ is the Lie bracket of vector fields, as defined in the text.

(c) Show that the Poisson bracket satisfies the Jacobi identity. You could try to do this directly, but here is an easier way.

As in the text, it is convenient to identify vector fields with differential operators. In the present setting, we identify the Hamiltonian vector field X_F with the differential operator

$$X_F = \sum_{i=1}^n \left(\frac{\partial F}{\partial p_i} \frac{\partial}{\partial q_i} - \frac{\partial F}{\partial q_i} \frac{\partial}{\partial p_i} \right).$$

With this identification, show that

$$\{F, G\} = X_G(F) = -X_F(G).$$

Then use this to prove the Jacobi identity for the Poisson bracket.

3. In the proof of Theorem 9.5, we had differentiable functions $G_1, \ldots, G_n : V \to \mathbb{R}$ on an open set V in $\mathbb{R}^{2n}$, such that

$$\frac{\partial G_i}{\partial q_j}(q, p) = \frac{\partial G_j}{\partial q_i}(q, p), \tag{9.44}$$

for every i, j and all $(q, p) \in V$. Assume that $V = B(0, r)$ is the open ball of radius r centered at the origin in $\mathbb{R}^{2n}$. Show that the function W defined on

V by

$$W(q,p) = \sum_{j=1}^{n} \int_0^1 G_j(sq,p)\, q_j ds,$$

makes sense for any $(q,p) \in V$ and that

$$\frac{\partial W}{\partial q_i}(q,p) = G_i(q,p),$$

for $i = 1, \ldots, n$. *Hint*: Differentiate under the integral sign, use the conditions (9.44), and then integrate by parts. Now generalize to the case where $V = B((q_0, p_0), r)$.

4. This exercise gives more examples of 1-d mechanical Hamiltonians: $H = \frac{1}{2}p^2 + V(q)$, for which we can directly verify the results of Liouville's Theorem 9.5. NOTE: $F = H$ is the constant of the motion in the 1-d case. For each of the potential functions V below (which you are assigned) do the following.

   (i) Compute (by hand): $W(q,p) = \int \sqrt{2}\sqrt{p - V(q)}\, dq$.

   (ii) Compute: $h(q,p) = \left(F(q,p), -\frac{\partial W}{\partial q}(q, F(q,p)) \right)$.

   (iii) Verify directly that h is a canonical transformation (for 1-d Hamiltonian systems) by computing $h'(q,p)$ and $\det(h'(q,p))$.

   (iv) Directly verify that $h'(q,p)X_H(q,p) = (0, -1)$ at all points (q,p) in the domain of h. If possible, compute h^{-1}. Then directly verify that $\tilde{H}(\bar{q}, \bar{p}) = \bar{q}$, where $\tilde{H} = H \circ h^{-1}$.

   (v) Use the graph of $-V$ to sketch the phase portrait for the Hamiltonian system corresponding to H. Compare this with that corresponding to $\tilde{H}$.

   Potential Functions:

   (a) $V(q) = q$.   (b) $V(q) = 0$.

   (c) $V(q) = -\frac{1}{2}q^2$. *Hint*: For the integral use a hyperbolic sine substitution $q = \sqrt{2p}\sinh x$. The work should directly compare with that in Example -, with x taking the place of the angle θ.

   (d) $V(q) = q^{-2}$. *Hint*: For the integral use the trig substitution $q = p^{-1/2}\sec\theta$.

5. **(Generating Functions)** For the functions W below (which you are assigned) do the following: (i) Calculate the corresponding canonical transformation h that W generates. (ii) Directly verify that h is a canonical transformation by calculating the Jacobian matrix $h'(q,p)$ and showing that it is a symplectic matrix.

(a) $W(x,q) = qx$ (b) $W(q,x) = q^2 x$

(c) $W(q,x) = \frac{1}{2} q^2 x^2$ (d) $W(q,x) = \frac{1}{k+1} q^{k+1} x^m$

(e) $W(q,x) = \frac{1}{2} \sin x$ (f) $W(q,x) = \ln(q+x)$

(g) $W(q,x) = q_1 x_1 + q_2 x_2$ (h) $W(q,x) = \frac{1}{2} q_1^2 x_1 + \frac{1}{2} q_2^2 x_2$

(i) $W(q,x) = \frac{1}{2} q_1^2 x_1^2 + q_2^2 x_2^2$ (j) $W(q,x) = \frac{1}{k+1} q_1^{k+1} x_1^m + \frac{1}{a+1} q_2^{a+1} x_2^b$

9.4 Liouville's Theorem

In this section we discuss another result of Liouville, which has become
known as "Liouville's theorem" even though Liouville is responsible for numerous other results in mechanics (for example, Theorem 9.5 above). Liouville's theorem says that volumes (or more precisely hypervolumes) in phase
space are preserved under the deformation induced by the flow of a Hamiltonian system. To be more specific, let ϕ^H denote the flow generated by the
Hamiltonian vector field X_H. Recall that the domain of ϕ^H is an open set in
$\mathcal{O} \times \mathbb{R}$ and so for each point $x_0 \in \mathcal{O}$, there is a product neighborhood $I \times U$
of $(0, x_0)$ contained in the domain of ϕ^H. Then the set

$$\phi_t^H(U)$$

represents the deformation of U under the flow at time $t \in I$. Liouville's
theorem says that for any such t this set has the same volume as that of U.

Liouville's theorem is essentially a special case of the transport theorem
from continuum mechanics. This theorem describes the time rate of change
of certain integral quantities associated with a continuum that is in motion.
Thus suppose that $X : \mathbb{R} \times \mathcal{O} \to \mathbb{R}^n$ is a time dependent vector field on an
open set $\mathcal{O}$ in $\mathbb{R}^n$, with corresponding flow: $\phi : \mathcal{D} \subseteq \mathbb{R} \times \mathbb{R}^n \to \mathbb{R}^n$. The set
$\mathcal{O}$ is thought of as a vessel which contains the continuum, like a fluid, and
$X(t,x)$ represents the velocity of the fluid flowing through the point $x \in \mathcal{O}$
at time t. While in continuum mechanics $n = 3$, the transport theorem holds
for any n.

To prove the general transport theorem, we will need the change of variables formula and the concept of Lebesgue measure (see, e.g., [Ru 74], [Jo
96]). The *Lebesgue measure*, denoted by λ, is a function defined on a large
class of subsets $B \subseteq \mathbb{R}^n$ and its value $\lambda(B)$ for any one of these subsets B is
interpreted as the "hyper-volume" of B. In particular, $\lambda(B)$ is the area of
B if $B \subseteq \mathbb{R}^2$, and $\lambda(B)$ is the volume of B if $B \subseteq \mathbb{R}^3$. In addition to giving
the hypervolume of sets in $\mathbb{R}^n$, the Lebesgue measure is also used to define

an integral, called the *Lebesgue integral*, which in some ways is an extension of, and more useful than, the Riemann integral. For suitable functions $f : \mathbb{R}^n \to \mathbb{R}$, the Lebesgue integral of f over B will be denoted by

$$\int_B f(x)\, dx,$$

where $dx = dx_1 dx_2 \cdots dx_n$ is heuristic notation standing for the "product" of the differentials. It follows from the Lebesgue theory (indeed the definitions) that the measure and the integral are related by

$$\lambda(B) = \int_B 1\, dx.$$

We can now state and prove the transport theorem.

Theorem 9.6 (Transport Theorem) *Suppose $f : \mathbb{R} \times \mathcal{O} \to \mathbb{R}$ is a C^1, time-dependent, scalar field on $\mathcal{O}$, and for $U \subseteq \mathcal{O}$, open, and I an interval, the product $I \times U$ is contained in the domain $\mathcal{D}$ of the flow ϕ generated by X. Then:*

$$\frac{d}{dt} \int_{\phi_t(U)} f\, dx = \int_{\phi_t(U)} \left[\frac{\partial f}{\partial t} + \nabla f \cdot X + \operatorname{div}(X) f \right] dx. \qquad (9.45)$$

In particular for $f = 1$, the constant 1 function, formula (9.45) gives the rate of change of the hypervolume of U as it is moved and deformed by the flow:

$$\frac{d}{dt} \lambda(\phi_t(U)) = \frac{d}{dt} \int_{\phi_t(U)} 1\, dx = \int_{\phi_t(U)} \operatorname{div}(X)\, dx. \qquad (9.46)$$

Hence, if X is divergence free, i.e., $\operatorname{div}(X) = 0$, then the hypervolume of U remains constant (is preserved) under the flow generated by X:

$$\lambda(\phi_t(U)) = \lambda(U),$$

for all $t \in I$.

Proof: The proof is rather elementary in that it is a direct application of the change of variables formula, differentiation under the integral sign, and several easy identities. Thus:

$$\frac{d}{dt} \int_{\phi_t(U)} f(t, x) dx = \frac{d}{dt} \int_U f(t, \phi_t(u)) \det(\phi_t'(u))\, du \qquad (9.47)$$

$$= \int_U \frac{\partial}{\partial t} \left[f(t, \phi_t(u)) \det(\phi_t'(u)) \right] du \qquad (9.48)$$

In the above we have used the fact that $\det(\phi_t'(u))$ must be positive for all u and t. To complete the proof, we need to compute the time derivative of the integrand in (9.48). For this note that:

$$
\begin{aligned}
\frac{\partial}{\partial t} f(t, \phi_t(u)) &= \left[\frac{\partial f}{\partial t}(t, x) + \sum_{j=1}^{n} \frac{\partial f}{\partial x_j}(t, x) \frac{\partial \phi^j}{\partial t}(t, u) \right]_{x = \phi_t(u)} \\
&= \left[\frac{\partial f}{\partial t}(t, x) + \sum_{j=1}^{n} \frac{\partial f}{\partial x_j}(t, x) X^j(t, x) \right]_{x = \phi_t(u)} \\
&= \left[\frac{\partial f}{\partial t}(t, x) + \nabla f(t, x) \cdot X(t, x) \right]_{x = \phi_t(u)}.
\end{aligned}
$$

We also need to compute the time derivative of the term involving the determinant, and so we look at a general identity for derivatives of determinantal quantities. Thus suppose $A = \{a_{ij}\}_{i,j=1,\ldots n}$ is an $n \times n$ matrix with entries a_{ij} being differentiable functions of t. Letting A be expressed in terms of its rows as $A = [R_1, \ldots, R_n]$, we have

$$
\frac{d}{dt} \det A = \frac{d}{dt} \det[R_1, \ldots, R_n] = \sum_{i=1}^{n} \det[R_1, \ldots, \frac{dR_i}{dt}, \ldots, R_n]. \qquad (9.49)
$$

This identity is easy to prove using the definition of the determinant:

$$
\det A = \sum_{\sigma \in \Pi_n} (-1)^{\sigma} a_{1\sigma 1} a_{2\sigma 2} \cdots a_{n\sigma n}.
$$

(exercise).

An additional calculation needed below is

$$
\begin{aligned}
&\nabla[X^i(\phi(t, u))] \\
&= \left(\frac{\partial}{\partial u_1}[X^i(t, \phi(t, u))], \ldots, \frac{\partial}{\partial u_n}[X^i(t, \phi(t, u))] \right) \\
&= \left(\sum_{j=1}^{n} \frac{\partial X^i}{\partial x_j}(t, \phi(t, u)) \frac{\partial \phi^j}{\partial u_1}(t, u), \ldots \sum_{j=1}^{n} \frac{\partial X^i}{\partial x_j}(t, \phi(t, u)) \frac{\partial \phi^j}{\partial u_n}(t, u) \right) \\
&= \sum_{j=1}^{n} \frac{\partial X^i}{\partial x_j}(t, \phi(t, u)) \nabla \phi^j(t, u). \qquad (9.50)
\end{aligned}
$$

Using this result and the identity (9.49) for differentiating a determinant gives

$$
\frac{\partial}{\partial t} \det(\phi_t'(u))
$$

$$= \frac{\partial}{\partial t} \det\left[\nabla \phi_t^1(u), \ldots, \nabla \phi_t^n(u)\right]$$

$$= \sum_{i=1}^{n} \det\left[\nabla \phi^1(t, u), \ldots, \frac{\partial}{\partial t}\nabla \phi^i(t, u), \ldots, \nabla \phi^n(t, u)\right]$$

$$= \sum_{i=1}^{n} \det\left[\nabla \phi^1(t, u), \ldots, \nabla \frac{\partial}{\partial t}\phi^i(t, u), \ldots, \nabla \phi^n(t, u)\right]$$

$$= \sum_{i=1}^{n} \det\left[\nabla \phi^1(t, u), \ldots, \nabla [X^i(t, \phi(t, u))], \ldots, \nabla \phi^n(t, u)\right]$$

$$= \sum_{i=1}^{n} \det\left[\nabla \phi^1(t, u), \ldots, \sum_{j=1}^{n}\frac{\partial X^i}{\partial x_j}(t, \phi(t, u))\nabla \phi^j(t, u), \ldots, \nabla \phi^n(t, u)\right]$$

$$= \sum_{i=1}^{n}\sum_{j=1}^{n}\frac{\partial X^i}{\partial x_j}(t, \phi(t, u)) \det\left[\nabla \phi^1(t, u), \ldots, \nabla \phi^j(t, u), \ldots, \nabla \phi^n(t, u)\right]$$

$$= \sum_{i=1}^{n}\frac{\partial X^i}{\partial x_i}(t, \phi(t, u)) \det(\phi_t'(u)) \tag{9.51}$$

$$= \operatorname{div}(X)(t, \phi(t, u)) \det(\phi_t'(u)). \tag{9.52}$$

The reduction to get (9.51) above comes from the property: the determinant is zero if two rows are the same. Thus if we use these calculations of the time derivatives, we see that the integral in (9.48) is

$$\int_U \left[\frac{\partial f}{\partial t}(t, x) + \nabla f(t, x) \cdot X(t, x) + \operatorname{div}(X)(t, x)f(t, x)\right]_{x=\phi_t(u)} \det(\phi_t'(u))du$$

$$= \int_{\phi_t(U)} \left[\frac{\partial f}{\partial t}(t, x) + \nabla f(t, x) \cdot X(t, x) + \operatorname{div}(X)(t, x)f(t, x)\right] dx,$$

where we have used the change of variables formula again. This completes the proof. $\square$

It is easy to see that any Hamiltonian vector field: $X_H = (\frac{\partial H}{\partial p}, -\frac{\partial H}{\partial q})$, is divergence free:

$$\operatorname{div}(X_H) = 0.$$

Thus by the transport theorem we get:

Corollary 9.4 (Liouville's Theorem) *For a Hamiltonian vector field X_H : $\mathcal{O} \to \mathbb{R}^n$, the flow ϕ^H preserves volumes in phase space. That is, if $U \subseteq \mathcal{O}$*

*is an open set and I is an interval of times for which $I \times U \subseteq \mathcal{D}$, where $\mathcal{D}$
is the domain of ϕ^H, then*

$$\lambda(\phi_t^H(U)) = \lambda(U).$$

for all $t \in I$.

Liouville's theorem leads to the general study of measure preserving maps
between two measure spaces and thus to ergodic theory (Cf. [Pa 81], [Pe
83]). All of these play a role in certain aspects of statistical mechanics.
We end the chapter with one application of the Transport Theorem which
pertains to these topics.

Theorem 9.7 (Poincaré Recurrence Theorem) *Suppose $X : \mathcal{O} \subseteq \mathbb{R}^n \to
\mathbb{R}^n$ is a complete vector field and is divergence free: $\operatorname{div}(X) = 0$. Suppose
$A \subseteq \mathcal{O}$ is a set of finite measure: $\lambda(A) < \infty$ and is also invariant under the
flow: $\phi_t(A) \subseteq A$, for every $t \in \mathbb{R}$. For a measurable subset B of A, let B_∞
be the following subset of B:*

$$B_\infty = \left\{ \; x \in B \; \left| \; \begin{array}{c} \exists \text{ a sequence } \{t_k\}_{k=1}^\infty \text{ with} \\ t_1 < t_2 < t_3 < \cdots \to \infty \\ \text{and } \phi_{t_k}(x) \in B, \; \forall k \end{array} \right. \right\}. \tag{9.53}$$

Then

$$\lambda(B_\infty) = \lambda(B). \tag{9.54}$$

*Thus, almost every point of B returns infinitely often to B under the flow
map.*

Remark: Equation (9.54) is equivalent to $\lambda(B \setminus B_\infty) = 0$, i.e., the set of
points in B that do not return to B infinitely often has measure zero.

Proof: Let $\tau > 0$ be any time increment and, relative to this fixed increment,
define sets Γ_k, $k = 0, 1, 2, \ldots$, by

$$\begin{aligned} \Gamma_k & \equiv \; \{ x \in A \, | \, \exists j \geq k \ni \phi_{j\tau}(x) \in B \} \\ & = \; \bigcup_{j=k}^\infty \phi_{j\tau}^{-1}(B). \end{aligned} \tag{9.55}$$

Then clearly from (9.55), the Γ_k's form a decreasing sequence of sets:

$$\Gamma_0 \supseteq \Gamma_1 \supseteq \Gamma_2 \supseteq \cdots$$

Also since $\phi_0(x) = x$, for every x, it is clear that $B \subseteq \Gamma_0$. Let Γ_∞ denote the intersection of the Γ_k's:

$$\Gamma_\infty \equiv \bigcap_{k=0}^{\infty} \Gamma_k$$
$$= \{x \in A \,|\, \forall k \,\exists j \geq k \ni \phi_{j\tau}(x) \in B\}$$
$$= \text{the set of points in } A \text{ that return to } B \text{ infinitely often.}$$

It is easy to see that

$$B \cap \Gamma_\infty \subseteq B_\infty \subseteq B, \tag{9.56}$$

and thus to prove the theorem all we have to show is

$$\lambda(B \cap \Gamma_\infty) = \lambda(B). \tag{9.57}$$

A key ingredient in proving this is to show that all the Γ_k's have the same measure (or hypervolume) and for this, the Transport Theorem is instrumental. In order to use this theorem, we first prove the following

Claim: $\qquad\qquad \phi_\tau(\Gamma_k) = \Gamma_{k-1} \qquad\qquad$ (for all k).

To prove the claim, suppose that $y \in \phi_\tau(\Gamma_k)$. Then there is an $x \in \Gamma_k$ such that $\phi_\tau(x) = y$. However, since $x \in \Gamma_k$, there is a $j \geq k$ such that $\phi_{j\tau}(x) \in B$. But then, using the semigroup property of the flow gives

$$\phi_{(j-1)\tau}(y) = \phi_{(j-1)\tau}(\phi_\tau(x)) = \phi_{j\tau}(x) \in B.$$

But this means that $y \in \Gamma_{k-1}$. Conversely, suppose that $y \in \Gamma_{k-1}$. Then there exists an $i \geq k - 1$ such that $\phi_{i\tau}(y) \in B$. Let $x \equiv \phi_{-\tau}(y)$, so that $x \in A$ (since A is invariant under the flow) and $\phi_\tau(x) = y$. Thus,

$$\phi_{(i+1)\tau}(x) = \phi_{(i+1)\tau}(\phi_{-\tau}(y)) = \phi_{i\tau}(x) \in B.$$

Hence $x \in \Gamma_k$. Since $y = \phi_\tau(x)$, this gives that $y \in \phi_\tau(\Gamma_k)$. This completes the proof of the Claim.

The Claim says that ϕ_τ maps Γ_k into Γ_{k-1}. However, since X is divergence free, the Transport Theorem guarantees that the hypervolume does not change under this mapping. Thus,

$$\lambda(\Gamma_{k-1}) = \lambda(\phi_\tau(\Gamma_k)) = \lambda(\Gamma_k),$$

for all k. This, together with a basic property of measures of hypervolumes, gives

$$\lambda(\Gamma_{k-1} \setminus \Gamma_k) = \lambda(\Gamma_{k-1}) - \lambda(\Gamma_k) = 0,$$

for all k. Thus, each of the annular rings $\Gamma_{k-1} \setminus \Gamma_k$ has measure zero. However, since

$$\Gamma_0 \setminus \Gamma_\infty = \bigcup_{k=1}^{\infty} (\Gamma_{k-1} \setminus \Gamma_k),$$

is a disjoint union, we have (by a basic property of measures) that

$$\lambda(\Gamma_0 \setminus \Gamma_\infty) = \sum_{k=1}^{\infty} \lambda(\Gamma_{k-1} \setminus \Gamma_k) = 0.$$

Now since $B \setminus \Gamma_\infty \subseteq \Gamma_0 \setminus \Gamma_\infty$, the above also gives that $\lambda(B \setminus \Gamma_\infty) = 0$. Using this gives

$$\begin{aligned}
\lambda(B) &= \lambda\Big((\Gamma_\infty \cap B) \cup (B \setminus \Gamma_\infty)\Big) \\
&= \lambda(\Gamma_\infty \cap B) + \lambda(B \setminus \Gamma_\infty) \\
&= \lambda(\Gamma_\infty \cap B).
\end{aligned}$$

Appendix A

Elementary Analysis

This appendix includes some material from analysis that will serve to augment several topics in the text.

A.1 Multivariable Calculus

We will need some elementary facts and notation from multivariable calculus. The space $\mathbb{R}^n$ is the set of all n-tuples:

$$x = (x_1, \ldots, x_n),$$

of real numbers: $x_i \in \mathbb{R}, i = 1, \ldots, n$. We will view $\mathbb{R}^n$ either as the canonical n-dimensional Eucludean space, whose elements $x \in \mathbb{R}^n$ are points in this space, or alternatively as an n-dimensional vector space, whose elements x are vectors (position vectors).

Multivariable calculus is the study of properties of maps (or functions):

$$f : \mathbb{R}^n \to \mathbb{R}^k,$$

between two Euclidean spaces. More generally, we will only assume that the domain of f is some (open) subset $\mathcal{O} \subseteq \mathbb{R}^n$ of $\mathbb{R}^n$, and denote this by:

$$f : \mathcal{O} \subseteq \mathbb{R}^n \to \mathbb{R}^k.$$

Applying f to a point $x = (x_1, \ldots, x_n) \in \mathcal{O}$ gives a point $f(x)$ in $\mathbb{R}^k$, and we denote the component form of this point by:

$$f(x) = (f^1(x), \ldots, f^k(x)).$$

Otherwise said, $f : \mathcal{O} \to \mathbb{R}^k$, is given via k real-valued functions: $f^i : \mathcal{O} \to \mathbb{R}, i = 1, \ldots, k$ of n variables. These functions are called the *component functions* of f. The map, or function, f is also called a *transformation* and is thought of as transforming points $x \in \mathcal{O}$ into points $y \in \mathbb{R}^k$. It's action is often denoted by:

$$y = f(x),$$

or in component form by:

$$
\begin{aligned}
y_1 &= f^1(x_1, \ldots, x_n) \\
y_2 &= f^2(x_1, \ldots, x_n) \\
&\vdots \\
y_k &= f^k(x_1, \ldots, x_n).
\end{aligned}
$$

A important special case of such a transformation is when $f : \mathbb{R}^n \to \mathbb{R}^k$ is a linear transformation:

$$f(x) = Ax,$$

where A is a $k \times n$ matrix, and x is written as a column vector in the above notation. Thus, this linear transformation written in vector form is:

$$
y = Ax = \begin{bmatrix} a_{11} & a_{12} & \cdots & a_{1n} \\ a_{21} & a_{22} & \cdots & a_{2n} \\ \vdots & \vdots & & \vdots \\ a_{k1} & a_{k2} & \cdots & a_{kn} \end{bmatrix} \begin{bmatrix} x_1 \\ x_2 \\ \vdots \\ x_n \end{bmatrix}.
$$

With the usual matrix operations, this transformation in component form is:

$$
\begin{aligned}
y_1 &= a_{11}x_1 + a_{12}x_2 + \cdots + a_{1n}x_n \\
y_2 &= a_{21}x_1 + a_{22}x_2 + \cdots + a_{2n}x_n \\
&\quad\cdot \\
&\quad\cdot \\
&\quad\cdot \\
y_k &= a_{k1}x_1 + a_{k2}x_2 + \cdots + a_{kn}x_n.
\end{aligned}
$$

$$(A.1)$$

When $k > 1$, some alternative terminology for a map $f : \mathcal{O} \to \mathbb{R}^k$ is to call f a *vector-valued function* on $\mathcal{O}$. Then $f(x)$, for $x \in \mathcal{O}$, is considered a vector in $\mathbb{R}^k$, rather than as a point in $\mathbb{R}^k$. A real-valued function $f : \mathcal{O} \to \mathbb{R}$ is often called a *scalar-valued function*, or a *scalar field* on $\mathcal{O}$. On the other hand, as was discussed in the text, a vector-valued function $f : \mathcal{O} \subseteq \mathbb{R}^n \to \mathbb{R}^n$ is called a *vector field* on $\mathcal{O}$.

Definition A.1 (Level sets) For a real-valued function $F : \mathcal{O} \to \mathbb{R}$ on an open set $\mathcal{O} \subseteq \mathbb{R}^n$ and a real number $k \in \mathbb{R}$, the set:

$$S_F^k = \{\, x \in \mathcal{O} \mid F(x) = k \,\} = F^{-1}(\{k\}),$$

is called a *level set* of F. Of course this set is empty when k is not in the range of F. For $n = 2$ the set S_F^k is called a *level curve* and for $n = 3$ the set S_F^k is called a *level surface*. For functions $F^1, \ldots, F^r$ on $\mathcal{O}$ and constants $k_1, \ldots, k_r$, the intersection of the respective level sets is denoted by

$$S_{F^1 \ldots F^r}^{k_1 \cdots k_r} \equiv S_{F^1}^{k_1} \cap S_{F^2}^{k_2} \cap \cdots \cap S_{F^r}^{k_r}.$$

Definition A.2 (Jacobian matrix) In the sequel unless specified otherwise $\mathcal{O}$ will denote an open subset of $\mathbb{R}^n$. A map $f : \mathcal{O} \to \mathbb{R}^k$ is is called *differentiable* on $\mathcal{O}$, if all the partial derivatives of the component functions $\partial f^i / \partial x_j$ exist on $\mathcal{O}$. When this is the case the *Jacobian matrix* of f at $x \in \mathcal{O}$ is the $k \times n$ mattrix:

$$f'(x) = \begin{bmatrix} \frac{\partial f^1}{\partial x_1}(x) & \cdots & \frac{\partial f^1}{\partial x_n}(x) \\[2mm] \frac{\partial f^2}{\partial x_1}(x) & \cdots & \frac{\partial f^2}{\partial x_n}(x) \\[2mm] \vdots & & \vdots \\[2mm] \frac{\partial f^k}{\partial x_1}(x) & \cdots & \frac{\partial f^k}{\partial x_n}(x) \end{bmatrix} \tag{A.2}$$

This matrix is also called the *derivative* of f at x. Thus, the derivative at x of a map from $\mathbb{R}^n$ to $\mathbb{R}^k$ is a $k \times n$ matrix.

Example A.1 For the sake of definiteness, we look at a few specific examples.

(1) For $n = 1$ and $k = 1$, the map f is just a real-valued function of a real variable, and its derivative, or Jacobian matrix, at x is just a 1×1 matrix, namely a number.

(2) Suppose $f : \mathbb{R} \to \mathbb{R}^2$ is given by:

$$f(t) = (t^3, t^2).$$

Then its Jacobian matrix at t is the 2×1 matrix:

$$f'(t) = \begin{bmatrix} 3t^2 \\ 2t \end{bmatrix} = (3t^2, 2t).$$

Note: Generally, when there is no risk of confusion, we identify row matrices and column matrices, thinking of them as vectors.

We can consider f as parametrizing a curve in $\mathbb{R}^2$, with t as the parameter. In this case $f'(t)$ is a tangent vector at the point $f(t)$ on the curve parametrized by f. See Figure A.1.

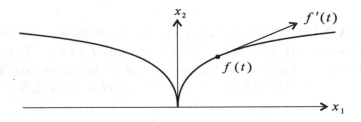

Figure A.1: *The curve $f(t) = (t^3, t^2)$.*

(3) For $n = 2$ and $k = 3$, suppose $f : \mathbb{R}^2 \to \mathbb{R}^3$ is the map defined by the formula:

$$f(x_1, x_2) = \left(x_1 + x_2^2, \ x_1^2 + x_2, \ 1 + \sin(x_1 x_2) \right).$$

Then f is differentiable on $\mathbb{R}^2$, and its derivative at x is the 3×2 matrix:

$$f'(x_1, x_2) = \begin{bmatrix} 1 & 2x_2 \\ 2x_1 & 1 \\ x_2 \cos(x_1 x_2) & x_1 \cos(x_1 x_2) \end{bmatrix} \tag{A.3}$$

In particular at the point $x = (0, 0)$, the Jacobian matrix is:

$$f'(0, 1) = \begin{bmatrix} 1 & 0 \\ 0 & 1 \\ 0 & 0 \end{bmatrix}.$$

As in the last example, we can think of f as parametrizing a *submanifold* in $\mathbb{R}^3$, where x_1, x_2 are the parameters. In this case the

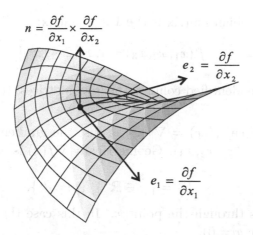

Figure A.2: *The surface* $f(x_1, x_2) = (x_1 + x_2^2, x_1^2 + x_2, 1 + \sin(x_1 x_2))$.

submanifold is a surface since there are two parameters. Figure A.2 shows a part of this surface corresponding to parameters $(x_1, x_2) \in [-0.5, 0.5] \times [0.5, 0.5]$.

For $x = (0, 0)$, the figure also shows the two vectors

$$e_1(x) = \frac{\partial f}{\partial x_1}(x) = \left(\frac{\partial f^1}{\partial x_1}(x), \frac{\partial f^2}{\partial x_1}(x), \frac{\partial f^3}{\partial x_1}(x) \right) = (1, 0, 0)$$

$$e_2(x) = \frac{\partial f}{\partial x_2}(x) = \left(\frac{\partial f^1}{\partial x_2}(x), \frac{\partial f^2}{\partial x_2}(x), \frac{\partial f^3}{\partial x_2}(x) \right) = (0, 1, 0),$$

as well as the cross product of these vectors

$$n(x) = \frac{\partial f}{\partial x_1}(x) \times \frac{\partial f}{\partial x_2}(x) = (0, 0, 1),$$

plotted at the point $f(x) = (0, 0, 1)$ on the surface. The vectors $e_1(x), e_2(x)$ are tangent to the surface at x and determine the tangent plane at this point. The vector $n(x)$ is normal to the surface at the point $f(x)$. Note also that $e_1(x)$ and $e_2(x)$ are the two columns of the Jacobian matrix

$$f'(x) = \left[\frac{\partial f}{\partial x_1}(x), \frac{\partial f}{\partial x_2}(x) \right] = \left[e_1(x), e_2(x) \right].$$

(4) Suppose $f : \mathbb{R}^3 \to \mathbb{R}$ is given by:

$$f(x_1, x_2, x_3) = x_1^2 + x_2^2 + x_3^2,$$

Then the Jacobian matrix is the 1×3 matrix:

$$f'(x_1, x_2, x_3) = [2x_1, 2x_2, 2x_3],$$

Here we have identified column and row matrices, thinking of them as vectors.

In this example, $f'(x) = \nabla f(x)$ is just the gradient (vector) of f at the point $x = (x_1, x_2, x_3)$. Geometrically $\nabla f(x)$ is normal to the level surface of f

$$S_f^k = \{\, y \in \mathbb{R}^3 \mid f(y) = k \,\},$$

which passes through the point x. In this case the level surface is a sphere (when $x \neq 0$).

(5) Suppose $f : \mathbb{R}^3 \to \mathbb{R}^2$ is given by:

$$f(x_1, x_2, x_3) = (x_1 - x_2 + 5x_3, 2x_1 + x_2 - 4x_3).$$

Then the derivative of f at any point is the 2×3 matrix:

$$f'(x_1, x_2, x_3) = \begin{bmatrix} 1 & -1 & 5 \\ 2 & 1 & -4 \end{bmatrix}.$$

In general, if $f : \mathbb{R}^n \to \mathbb{R}^k$ is given by matrix multiplication:

$$f(x) = Ax,$$

where A is an $n \times k$ matrix, then its derivative is a constant (matrix):

$$f'(x) = A,$$

for every $x \in \mathbb{R}^n$.

We will often need to require that the maps, or functions, f we deal with have some specific degree of differentiablity and continuity. For this there is some standard terminology:

Definition A.3 (Differentiability: C^k Maps) Suppose $f : \mathcal{O} \subseteq \mathbb{R}^n \to \mathbb{R}^k$ is a map. Then f is called *continuous*, or C^0, if each of its component functions f^i, $i = 1, \ldots, n$ is continuous. If each of these component functions is differentiable (i.e. has 1st-order derivatives existing on $\mathcal{O}$), then f is called *differentiable*. If in addition these 1st-order partial derivatives are continuous functions, then f is called a C^1 function. More generally, f is called C^r function (for $r = 1, 2, \ldots$) if each of its component functions has partial derivatives existing up to the rth-order and the rth-order derivatives are continuous. The map f is called C^∞ if each of its component functions has derivatives existing to all orders. Often we will not particularly care about the precise degree of differentiability of f and will call f a *smooth* function if it is at least a C^1 function.

A.2 The Chain Rule

We present here, for your convenience, the statement of the chain rule for vector-valued functions $f : \mathcal{O} \subseteq \mathbb{R}^n \to \mathbb{R}^k$ of several variables. It is a natural extension of the chain rule studied in undergraduate calculus. In fact, with the proper notation and with the concept that the dervative $f'(x)$ at $x \in \mathcal{O}$ is a matrix, the general case looks identical to the undergraduate case.

Suppose that $g : U \subseteq \mathbb{R}^p \to \mathbb{R}^n$ and $f : \mathcal{O} \subseteq \mathbb{R}^n \to \mathbb{R}^k$ are differentiable maps, with domains $U, \mathcal{O}$ being open sets in $\mathbb{R}^p, \mathbb{R}^n$ respectively. Assuming the range of g is contained in the domain of f, i.e., $g(U) \subseteq \mathcal{O}$, we can form the composite function: $f \circ g : U \to \mathbb{R}^k$. See Figure A.3.

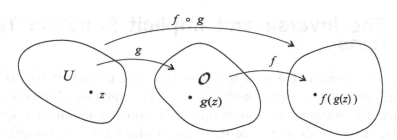

Figure A.3: *Composition of functions.*

The *chain rule* simply says that the composite function is differentiable and its derivative at $z \in \mathbb{R}^p$ is given by:

$$(f \circ g)'(z) \;=\; f'(g(z))g'(z) \tag{A.4}$$

$$= \begin{bmatrix} \frac{\partial f^1}{\partial x_1}(g(z)) & \cdots & \frac{\partial f^1}{\partial x_n}(g(z)) \\ \vdots & & \vdots \\ \frac{\partial f^k}{\partial x_1}(g(z)) & \cdots & \frac{\partial f^k}{\partial x_n}(g(z)) \end{bmatrix} \begin{bmatrix} \frac{\partial g^1}{\partial z_1}(z) & \cdots & \frac{\partial g^1}{\partial z_p}(z) \\ \vdots & & \vdots \\ \frac{\partial g^n}{\partial z_1}(z) & \cdots & \frac{\partial g^n}{\partial z_p}(z) \end{bmatrix}$$

This says that the Jacobian matrix $(f \circ g)'(z)$ (which is a $k \times p$ matrix) is the product of the Jacobian matrix $f'(g(z))$ (a $k \times n$ matrix) and the Jacobian matrix $g'(z)$ (an $n \times p$ matrix).

An important special case of this is when g is a curve in $\mathcal{O}$. So suppose $\alpha : I \to \mathcal{O}$ is a differentiable curve, then $f \circ \alpha : I \to \mathbb{R}^k$ is a differentiable curve, and

$$\frac{d}{dt}(f \circ \alpha)(t) = f'(\alpha(t))\alpha'(t) \tag{A.5}$$

$$= \begin{bmatrix} \frac{\partial f^1}{\partial x_1}(\alpha(t)) & \cdots & \frac{\partial f^1}{\partial x_n}(\alpha(t)) \\ \vdots & & \vdots \\ \frac{\partial f^k}{\partial x_1}(\alpha(t)) & \cdots & \frac{\partial f^k}{\partial x_n}(\alpha(t)) \end{bmatrix} \begin{bmatrix} \frac{d\alpha^1}{dt}(t) \\ \vdots \\ \frac{d\alpha^n}{dt}(t) \end{bmatrix} \tag{A.6}$$

This says that the time derivative of the ith component of the curve $f \circ \alpha$ is given by:

$$\frac{d}{dt}(f^i \circ \alpha)(t) = \sum_{j=1}^{n} \frac{\partial f^i}{\partial x_j}(\alpha(t))\frac{d\alpha_j}{dt}(t) = \nabla f^i(\alpha(t)) \cdot \alpha'(t),$$

which is perhaps the more familar version of the chain rule in this case.

A.3 The Inverse and Implicit Function Theorems

The Inverse Function Theorem is one of the most important results in the multivariable calculus and is useful in proving a number of closely related theorems, for example, the Submanifold Theorem and Implicit Function Theorem. We provide proofs of these latter theorems here specifically for the reason of showing how the Inverse Function Theorem is central to these theorems. This is the real value of presenting the proofs and will perhaps give you an understanding of how to derive other results in the literature.

We do not prove the Inverse Function Theorem, but refer you to [Ru 76] for this.

Theorem A.1 (Inverse Function Theorem) *Suppose* $f : \mathcal{O} \to \mathbb{R}^n$ *is a* C^1 *function on an open set* $\mathcal{O}$ *in* $\mathbb{R}^n$. *If* $c \in \mathcal{O}$ *is a point at which the Jacobian matrix* $f'(c)$ *is invertible, then* f *is locally invertible on a neighborhood of* c. *Specifically, there is a neighborhood* $U \subseteq \mathcal{O}$ *of* c *and a neighborhood* V *of* $f(c)$ *such that*

$$f : U \to V,$$

is 1-1, onto, and its inverse $f^{-1} : V \to U$ *is* C^1. *Furthermore the derivative of the inverse function is given by*

$$(f^{-1})'(y) = [f'(f^{-1}(y))]^{-1}, \tag{A.7}$$

for all $y \in V$.

The proof of the derivative formula (A.7) follows easily form the chain rule (exercise). The form of this formula (and the proof of its validity) is *exactly* the same as the 1-variable case in calculus, except now the expressions involve matrices.

The next result is instrumental in developing the notion of what an abstract manifold is. We do not discuss general manifolds (topological, differentiable, or analytic) in this text, but rather work with concrete ones as described in the next to the last section. All these manifolds, like curves and surfaces, are, by definition, maps $h : U \to \mathbb{R}^n$, with suitable properties. An alternative description of such manifolds is as the intersection of level sets. Thus, while a curve in $\mathbb{R}^3$ is a map rather than a set of points, we can alternatively consider the intersection of two level surfaces: $F^1(x, y, z) = k_1, F^2(x, y, z) = k_2$, with appropriate conditions on F^1, F^2, as being a "curve" as well. The next theorem gives the extent to which this alternative view is the same as the original definition.

Theorem A.2 (Submanifold Theorem) *Suppose* $F^1, \ldots, F^r : \mathcal{O} \to \mathbb{R}$ *are smooth functions* (C^1 *functions*) *on an open set* $\mathcal{O}$ *in* $\mathbb{R}^n$ (*with* $r \leq n$) *and* $k_1, \ldots, k_r$ *are constants. Let*

$$S^{k_1 \cdots k_r}_{F^1 \cdots F^r} = \{ x \in \mathcal{O} \mid F^1(x) = k_1, \ldots, F^r(x) = k_r \},$$

be the intersection of the corresponding level sets. If $c \in S^{k_1 \cdots k_r}_{F^1 \cdots F^r}$ *is a point such that*

$$\nabla F^1(c), \ldots, \nabla F^r(c),$$

are linearly independent, then there is a neighborhood $V \subseteq \mathcal{O}$ *of* c, *an open set* $U \subseteq \mathbb{R}^{n-r}$, *and a differentiable map*

$$h : U \to S^{k_1 \cdots k_r}_{F^1 \cdots F^r} \cap V,$$

that is 1-1 and onto. Furthermore h *has the form*

$$h(u) = \left(h^1(u), \ldots, u_1, \ldots, u_{n-r}, \ldots, h^n(u) \right), \tag{A.8}$$

i.e., there are indices $\ell_1 < \cdots < \ell_{n-r} \in \{1, \ldots, n\}$, *such that* $h^{\ell_i}(u) = u_i$, *for* $i = 1, \ldots, n - r$. *See Figure A.4. Hence the Jacobian matrix* $h'(u)$ *has rank* $n - r$ *for all* $u \in U$, *and the inverse of* h *is given by projection:*

$$h^{-1}(x) = (x_{\ell_1}, \ldots, x_{\ell_{n-r}}), \tag{A.9}$$

for all $x = (x_1, \ldots, x_n) \in S_{F^1 \ldots F^r}^{k_1 \cdots k_r} \cap V$.

Proof: For convenience of notation let $F : \mathcal{O} \to \mathbb{R}^r$ be the function

$$F(x) = (F^1(x), \ldots, , F^r(x)).$$

We can express the $r \times n$, Jacobian matrix of F at c either in terms of its rows or in terms of its columns as:

$$F'(c) = \begin{bmatrix} \nabla F^1(c) \\ \vdots \\ \nabla F^r(c) \end{bmatrix} = \begin{bmatrix} \dfrac{\partial F}{\partial x_1}(c), \ldots, \dfrac{\partial F}{\partial x_n}(c) \end{bmatrix}$$

This matrix has rank r, since it has r linearly independent rows (the gradient vectors), and since the rank of a matrix is also the same as the number of linearly independent columns, there are r linearly independent columns as well. The proof is most clear in the case where the linearly independent columns come last. So we do this case first and then show how to reduce the general case to this special case.

Case (1): Assume the last r columns of $F'(c)$, i.e., the vectors:

$$\frac{\partial F}{\partial x_{n-r+1}}(c), \ldots, \frac{\partial F}{\partial x_n}(c),$$

are linearly independent. Define $\psi : \mathcal{O} \to \mathbb{R}^n$ by

$$\psi(x) = (x_1, \ldots, x_{n-r}, F(x)),$$

for $x = (x_1, \ldots, x_n) \in \mathcal{O}$. Then the Jacobian matrix of ψ at c is

$$\psi'(c) = \begin{bmatrix} 1 & & 0 & 0 & & 0 \\ & \ddots & & & \ddots & \\ 0 & & 1 & 0 & & 0 \\ \dfrac{\partial F}{\partial x_1}(c) & \cdots & \dfrac{\partial F}{\partial x_{n-r}}(c) & \dfrac{\partial F}{\partial x_{n-r+1}}(c) & \cdots & \dfrac{\partial F}{\partial x_n}(c) \end{bmatrix}.$$

This is a block diagonal matrix with the $(n-r) \times (n-r)$ identity matrix as the upper lefthand block and the $r \times r$ matrix $\{\partial F^i(c)/\partial x_j\}_{j=n-r+1\cdots n}^{i=1\cdots r}$ as the lower right-hand block. Thus, $\psi'(c)$ is an invertible matrix. By the Inverse Function Theorem, there is a neighborhood B of c and a neighborhood W of $\psi(c)$, such that $\psi : B \to W$ has a C^1 inverse $\phi : W \to B$. See Figure A.4.

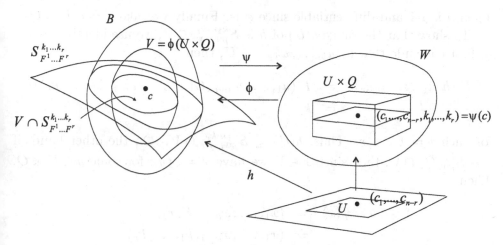

Figure A.4: *The construction of the map h.*

It is easy to show (exercise) that ϕ must have the form

$$\phi(y) = (y_1, \ldots, y_{n-r}, G(y)),$$

for each $y = (y_1, \ldots, y_n) \in W$, where

$$G(y) = (G^1(y), \ldots, G^r(y)),$$

and the functions $G^i : W \to \mathbb{R}$ are differentiable functions. Furthermore because $\phi = \psi^{-1}$, the function G satisfies the identities:

$$F^i\left(y_1, \ldots, y_{n-r}, G(y)\right) = y_{n-r+i}, \qquad (i = 1, \ldots, r) \qquad \text{(A.10)}$$

for each $y \in W$. Now since $\psi(c) \in W$ and

$$\psi(c) = (c_1, \ldots, c_{n-r}, F^1(c), \ldots, F^r(c)) = (c_1, \ldots, c_{n-r}, k_1, \ldots, k_r),$$

there exists a neighborhood $U \subseteq \mathbb{R}^{n-r}$ of $(c_1, \ldots, c_{n-r})$ and a neighborhood $Q \subseteq \mathbb{R}^r$ of $(k_1, \ldots, k_r)$ such that $U \times Q \subseteq W$. See Figure A.4. Define

$h : U \to B$ by

$$
\begin{aligned}
h(u_1, \ldots, u_{n-r}) &= \phi(u_1, \ldots, u_{n-r}, k_1, \ldots, k_r) \\
&= \left(u_1, \ldots, u_{n-r}, G(u_1, \ldots, u_{n-r}, k_1, \ldots, k_r) \right)
\end{aligned}
$$

Then h is 1-1 and differentiable since ϕ is. Finally we take $V = \phi(U \times Q)$.

To show that the range $h(U)$ of h is $S_{F^1 \ldots F^r}^{k_1 \ldots k_r} \cap V$, we use identity (A.10) to first conclude that if $(u_1, \ldots, u_{n-r}) \in U$, then

$$
\begin{aligned}
F^i(h(u_1, \ldots, u_{n-r})) &= F^i(u_1, \ldots, u_{n-r}, G(u_1, \ldots, u_{n-r}, k_1, \ldots, k_r)) \\
&= k_i
\end{aligned}
$$

for each $i = 1, \ldots, r$. Thus, $h(U) \subseteq S_{F^1 \ldots F^r}^{k_1 \ldots k_r} \cap V$. On the other hand, if $x \in S_{F^1 \ldots F^r}^{k_1 \ldots k_r} \cap V$, then since $x \in V$ we have $x = \phi(y)$, for some $y \in U \times Q$. Then

$$
\begin{aligned}
y = \psi(x) &= (x_1, \ldots, x_{n-r}, F(x)) \\
&= (x_1, \ldots, x_{n-1}, k_1, \ldots, k_r)
\end{aligned}
$$

Consequently,

$$
\begin{aligned}
x = \phi(y) &= \phi(y_1, \ldots, y_{n-1}, k_1, \ldots, k_r) \\
&= h(y_1, \ldots, y_{n-1}).
\end{aligned}
$$

Thus, $h(U) = S_{F^1 \ldots F^r}^{k_1 \ldots k_r} \cap V$. It is clear that h has the form stated in equation (A.8), and thus the proof of the theorem is complete in this case.

Case (2): In general we can only assume that there are indices $j_1 < \ldots < j_r \in \{1, \ldots, n\}$ such that

$$
\frac{\partial F}{\partial x_{j_1}}(c), \ldots, \frac{\partial F}{\partial x_{j_r}}(c),
$$

are linearly independent. Let $\ell_1 < \cdots < \ell_{n-r}$ be the other $n - r$ indices and let P be the $n \times n$ permutation matrix such that

$$
P(x_1, \ldots, x_n) = (x_{\ell_1}, \ldots, x_{\ell_{n-r}}, x_{j_1}, \ldots, x_{j_r}),
$$

for all $(x_1, \ldots, x_n) \in \mathbb{R}^n$. Note that P is obtained from the identity matrix I by permuting its rows (or columns) in the same way as the x_i's. Also P

is its own inverse. Let $\tilde{\mathcal{O}}$ be the open set $\tilde{\mathcal{O}} = P\mathcal{O}$ and so also $P\tilde{\mathcal{O}} = \mathcal{O}$. Define $\tilde{F} : \tilde{\mathcal{O}} \to \mathbb{R}^r$ by

$$\tilde{F}(\tilde{x}) = F(P\tilde{x}),$$

for $\tilde{x} \in \tilde{\mathcal{O}}$. If we let $\tilde{c} = Pc$, then $P\tilde{c} = c$, so $\tilde{F}(\tilde{c}) = F(c)$ and, by the chain rule,

$$
\begin{aligned}
\tilde{F}'(\tilde{c}) &= F'(P\tilde{c})P = F'(c)P \\
&= \left[\frac{\partial F}{\partial x_{\ell_1}}(c), \ldots, \frac{\partial F}{\partial x_{\ell_{n-r}}}(c), \frac{\partial F}{\partial x_{j_1}}(c), \ldots, \frac{\partial F}{\partial x_{j_r}}(c) \right]
\end{aligned}
$$

Note that the last r columns of the matrix $\tilde{F}'(\tilde{c})$ are linearly independent, and thus we can apply the theorem to $\tilde{F}$ since it satisfies the assumptions in Case (1) above. Thus, there is a neighborhood $\tilde{V}$ of $\tilde{c}$, an open set $\tilde{U}$ in $\mathbb{R}^{n-r}$, and differentiable map

$$\tilde{h} : \tilde{U} \to S^{k_1 \cdots k_r}_{\tilde{F}^1 \cdots \tilde{F}^r} \cap \tilde{V},$$

that is 1-1, onto and has the form

$$\tilde{h}(\tilde{u}) = (\tilde{u}_1, \ldots, \tilde{u}_{n-r}, \tilde{h}^{n-r+1}(\tilde{u}), \ldots, \tilde{h}^n(\tilde{u})),$$

for all $\tilde{u} = (\tilde{u}_1, \ldots, \tilde{u}_{n-r}) \in \tilde{U}$. Now let $V = \tilde{V}$ and define $h : U \to \mathbb{R}^n$ by

$$h = P \circ \tilde{h}.$$

Then it is easy to prove that

$$P\left(S^{k_1 \cdots k_r}_{\tilde{F}^1 \cdots \tilde{F}^r} \cap \tilde{V} \right) = S^{k_1 \cdots k_r}_{F^1 \cdots F^r} \cap V,$$

and that h has the properties stated in the theorem (exercise). With this the proof is complete. $\square$

The Implicit Function Theorem is a direct corollary of the Submanifold Theorem. Indeed they are essentially the same result, phrased in two different ways.

Corollary A.1 (Implicit Function Theorem) *Suppose* $F^1, \ldots, F^r : \mathcal{O} \to \mathbb{R}$ *are smooth functions* (C^1 *functions) on an open set* $\mathcal{O}$ *in* $\mathbb{R}^n$ *(with* $r \leq n$*) and* $k_1, \ldots, k_r$ *are constants. Denote by* F *the function* $F : \mathcal{O} \to \mathbb{R}^r$

defined by $F(z) = (F^1(z), \ldots, F^r(z))$ *and let* $k = (k_1, \ldots, k_r)$. *Suppose* $c = (a, b) \in \mathcal{O}$, *with* $a \in \mathbb{R}^{n-r}$ *and* $b \in \mathbb{R}^r$, *satisfies the system of* r *equations in* n *unknowns:*

$$F(x, y) = k,$$

which is written in vector form, with with $x \in \mathbb{R}^{n-r}$ *and* $y \in \mathbb{R}^r$. *Suppose in addition that the vectors*

$$\frac{\partial F}{\partial z_{n-r+1}}(c), \ldots, \frac{\partial F}{\partial z_n}(c), \tag{A.11}$$

are linearly independent. Then there is a neighborhood $U \subseteq \mathbb{R}^{n-r}$ *of the point* $a = (c_1, \ldots, c_{n-r})$ *and differentiable maps* $G^1, \ldots, G^r : U \to \mathcal{O}$ *such that*

$$F\Big(x, G(x)\Big) = k,$$

for all $x = (x_1, \ldots, x_{n-r}) \in U$. *Here* $G(x) \equiv (G^1(x), \ldots, G^r(x))$.

Proof: This is the result in Case (1) from the last proof. $\square$

In essence the Implicit Function Theorem gives conditions under which we can solve the system of equations $F(x, y) = k$, for y as a function of x, giving an explicitly defined function $y = G(x)$ that satisfies the system of equations. Note the theorem says nothing about how many explicit solutions G there are.

A.4 Taylor's Theorem and The Hessian

The multivariable version of Taylor's Theorem can be expressed in two distinct ways, depending on the notation one prefers. Here we state the theorem using the notation involving multilinear forms on $\mathbb{R}^n$ (See Appendix C for the definition of multilinear forms).

Definition A.4 (Higher-Order Derivatives) Suppose $f : \mathcal{O} \subseteq \mathbb{R}^n \to \mathbb{R}^k$ is a vector-valued function on an open set $\mathcal{O}$ in $\mathbb{R}^n$. Assuming that f has partial derivatives to the rth-order, the rth *derivative of* f at $x \in \mathcal{O}$, denoted by $f^{(r)}(x)$, is the multilinear map $f^{(r)}(x) : (\mathbb{R}^n)^r \to \mathbb{R}^n$ defined by

$$f^{(r)}(x)(v^1, v^2, \ldots, v^r) = \sum_{i_1, i_2, \ldots, i_r} \frac{\partial^r f}{\partial x_{i_1} \partial x_{i_2} \cdots \partial x_{i_r}}(x) \, v_{i_1}^1 v_{i_2}^2 \cdots v_{i_r}^r.$$

Here $v^j = (v_1^j, \ldots, v_n^j) \in \mathbb{R}^n$, for $j = 1, \ldots, r$, and the sum is over all choices of indices $i_1, \ldots, i_r \in \{1, \ldots, n\}$. Furthermore

$$\frac{\partial^r f}{\partial x_{i_1} \cdots \partial x_{i_r}}(x) = \left(\frac{\partial^r f^1}{\partial x_{i_1} \cdots \partial x_{i_r}}(x), \ldots, \frac{\partial^r f^n}{\partial x_{i_1} \cdots \partial x_{i_r}}(x) \right) \in \mathbb{R}^n,$$

for $i_1, \ldots, i_r \in \{1, \ldots, n\}$.

The definition includes the case of the 1st derivative, since we identify matrices with linear maps. Thus, $f'(x)$ is the linear map $f'(x) : \mathbb{R}^n \to \mathbb{R}^k$ given by

$$f'(x)v = \sum_{i=1}^n \frac{\partial f}{\partial x_i}(x)\, v_i.$$

The expression on the right is a linear combination of the columns of the Jacobian matrix of f.

For for the 2nd derivative, it is traditional to use the notation $f^{(2)}(x) = f''(x)$. This is the bilinear map on $\mathbb{R}^n$ given by

$$f''(x)(v, w) = \sum_{i,j=1}^n \frac{\partial^2 f}{\partial x_i \partial x_j}(x)\, v_i w_j,$$

for $v, w \in \mathbb{R}^n$.

Based on this notation, we can state Taylor's Theorem in the following form.

Theorem A.3 (Taylor's Theorem) *Suppose* $f : \mathcal{O} \subseteq \mathbb{R}^n \to \mathbb{R}^k$ *is a* C^{r+1}, *vector-valued function on an open set in* $\mathbb{R}^n$ *and that* $x \in \mathcal{O}$ *and* $h \in \mathbb{R}^n$, *with* h *small enough so that the line segment joining* x *and* $x + h$ *lies in* $\mathcal{O}$. *Then there exists a point* c *on the line segment joining* x *and* $x + h$ *such that*

$$\begin{aligned} f(x + h) &= f(x) + f'(x)h + \tfrac{1}{2}f''(x)(h, h) + \cdots &\text{(A.12)} \\ &+ \tfrac{1}{r!} f^{(r)}(x)(h, \ldots, h) + \tfrac{1}{(r+1)!} f^{(r+1)}(c)(h, \ldots, h). \end{aligned}$$

Furthermore, with the appropriate understanding that c *depends on* h, *we have*

$$\lim_{h \to 0} \frac{|f^{(r+1)}(c)(h, \ldots, h)|}{|h|^{r+1}} = 0,$$

with $|\cdot|$ *indicating the respective Euclidean norms on* $\mathbb{R}^k$ *and* $\mathbb{R}^n$.

Proof: See [MH 93] for the case when $k = 1$. Then the $k > 1$ is an easy corollary (exercise). $\square$

When f is C^∞, then the following power series in h

$$\sum_{j=0}^{\infty} \frac{1}{j!} f^{(j)}(x)(h, \dots, h)_j = f(x) + f'(x)h + \tfrac{1}{2} f''(x)(h, h) + \cdots,$$

is known as the *Taylor series* for f at x. With suitable bounds on the partial derivatives of f, one can use Taylor's Theorem to prove that convergence of the Taylor series to the value of the function:

$$f(x + h) = \sum_{j=0}^{\infty} \frac{1}{j!} f^{(j)}(x)(h, \dots, h)_j.$$

If this happens for all $x + h$ in a neighborhood of x, then f is called *analytic* at x. If f is analytic at each point $x \in \mathcal{O}$, then it is called an *analytic* function on $\mathcal{O}$.

Taylor's Theorem is useful in many ways and in particular forms the basis for the classification of the local extrema of a real-valued function.

Definition A.5 (Critical Points and the Hessian) Suppose $f : \mathcal{O} \rightarrow \mathbb{R}$ is a real-valued function on an open set $\mathcal{O}$ in $\mathbb{R}^n$.

(1) f has a *local maximum value* at a point $c \in \mathcal{O}$ if there is a neighborhood U of c ($c \in U \subseteq \mathcal{O}$) such that $f(x) \leq f(c)$ for every $x \in U$. Similarly, f is said to have a *local minimum value* at c if there is a neighborhood U of c such that $f(c) \leq f(x)$ for all $x \in U$. In either case, whether $f(c)$ is a local maximum or minimum value, the number $f(c)$ is called a *local extreme value* of f.

(2) Assume that the first order partials of f exist at each point in $\mathcal{O}$. A point $c \in \mathcal{O}$ is called a *critical point* of f if $f'(c) = 0$.

(3) If the 2nd-order partials of f exist at $x \in \mathcal{O}$, then the *Hessian* of f at x is the $n \times n$ matrix

$$\mathcal{H}_f(x) = \begin{bmatrix} \frac{\partial^2 f}{\partial x_1^2}(x) & \cdots & \frac{\partial^2 f}{\partial x_1 \partial x_n}(x) \\ \vdots & & \vdots \\ \frac{\partial^2 f}{\partial x_n \partial x_1}(x) & \cdots & \frac{\partial^2 f}{\partial x_n^2}(x) \end{bmatrix}.$$

As in the one variable case, the local extreme values occur at critical points and the Hessian can be used to determine whether the values are local maximum or minimum values. Note that in the one variable case, the Hessian is the 1×1 matrix $\mathcal{H}_f(x) = [f''(x)]$ and the graph of f is concave up at x if $f''(x)$ is positive and is concave down at x if $f''(x)$ is negative. Thus, the test for a local minimum or maximum value at x is to check whether $f''(x)$ is positive or negative. Of course the test yields no information if $f''(x) = 0$. For example, the function $f(x) = x^4$ has a minimum value of 0 at $x = 0$, but the second derivative test fails to tell us this.

In the multivariable case, the Hessian is a symmetric matrix, and as such determines a bilinear form on $\mathbb{R}^n$ (see Appendix C for a brief discussion of bilinear forms). Many texts define the Hessian as a bilinear form. In any event, it is important to note that this bilinear form is just the 2nd derivative of f at x as defined above. That is

$$f''(x)(v, w) = \mathcal{H}_f(x)v \cdot w.$$

Also see Appendix C for the definition of what it means for a bilinear form, or symmetric matrix, to be positive (or negative) definite. This is the criterion used in the 2nd derivative test in the following theorem.

Theorem A.4 (Local Maximum/Minimum Values) *Let $f : \mathcal{O} \to \mathbb{R}$ be a real-valued function on an open set $\mathcal{O}$ in $\mathbb{R}^n$. Suppose f has a local extreme value at c. If the 1st-order partials of f exist at c, then $f'(c) = 0$. Thus, the extreme values of f occur at critical points.*

Suppose the 2nd-order partials of f exist at c and that $f'(c) = 0$. If the Hessian $\mathcal{H}_f(c)$ is positive definite, then $f(c)$ is a local minimum value of f. On the other hand if $\mathcal{H}_f(c)$ is negative definite, then $f(c)$ is a local maximum value of f.

Proof: Again see [MH 93] for the proof. $\Box$

Intuitively, the use of the Hessian in determining whether an extreme value is a local minimum or maximum is based on Taylor's Theorem as follows. If f has a local extrema at c (and is suitably differentiable), then $f'(c) = 0$ and so near c, the Taylor series expansion is

$$f(c + h) = f(c) + \tfrac{1}{2}\mathcal{H}_f(c)h \cdot h + \tfrac{1}{3!}f'''(c)(h, h, h) + \cdots,$$

for h sufficiently small. If the Hessian is positive definite, then $\mathcal{H}_f(c)h \cdot h > 0$ for all h and so if all but the first two terms in the Taylor series are neglected, we get that $f(c + h) > f(c)$. Similarly, we get an intuitive argument for showing that $f(c + h) < f(c)$, when the Hessian is negative definite, i.e., when $\mathcal{H}_f(c)h \cdot h < 0$ for all h.

In any particular example it is usually difficult to determine if the Hessian matrix is positive definite, negative definite, or neither. Note that the last option yields no information about the type of extreme value. From the theory in Appendix C, we can always look at the eigenvalues of the matrix (which are usually difficult to compute as well) to determine positive or negative definiteness. Thus, if all the eigenvalues of $\mathcal{H}_f(c)$ are positive, then $\mathcal{H}_f(c)$ is positive definite, and if all the eigenvalues are negative then $\mathcal{H}_f(c)$ is negative definite. Otherwise $\mathcal{H}_f(c)$ is neither positive nor negative definite. Thus, when $\mathcal{H}_f(c)$ is an invertible matrix, there is a famous third possiblity for the type of critical point that can occur.

Definition A.6 (Saddle Points) Suppose $f : \mathcal{O} \subseteq \mathbb{R}^n \to \mathbb{R}$ has 2nd-order partial derivatives existing at $c \in \mathcal{O}$ and that c is a critical point ($f'(c) = 0$). Then c is called a *saddle point* of f if $f(c)$ is neither a maximum or minimum value of f and the Hessian $\mathcal{H}_f(c)$ is invertible.

By the theory in Appendix C, the point c is a saddle point whenever the Hessian matrix $\mathcal{H}_f(c)$ has some positive eigenvalues and some negative eigenvalues, but zero is *not* an eigenvalue (assuming f is twice differentiable).

A.5 The Change of Variables Formula

Theorem A.5 (Change of Variables) *Suppose that U and V are open subsets of $\mathbb{R}^n$, and $\phi : U \to V$ is a continuously differentiable function that is 1-1, onto, and $\det(\phi'(u)) \neq 0$, for every $u \in U$. Then for any continuous function $f : V \to \mathbb{R}$, we have:*

$$\int_{\phi(U)} f(x)dx = \int_U f(\phi(u)) \, |\det \phi'(u))| \, du. \tag{A.13}$$

Here: $dx = dx_1 dx_2 \cdots dx_n$ and $du = du_1 du_2 \cdots du_n$ represent the Lebesgue measure λ on $\mathbb{R}^n$, and the integrals in formula (A.13) are Lebesgue integrals. In particular for $f = 1$, the constant 1 function, formula (A.13) gives:

$$\lambda(\phi(U)) = \int_{\phi(U)} 1 \, dx = \int_U |\det(\phi'(u))| \, du, \tag{A.14}$$

where λ denotes the Lebesgue measure. (See [Ru 76] for a proof.)

Appendix B

Lipschitz Maps and Linearization

For the sake of reference we collect together here a more detailed discussion of some of the topics in the text. These topics, Lipschitz functions, the contraction mapping principle, and the Linearization Theorem, were all mentioned briefly in the main body of the text, but in the interest of pursuing other subjects, many instructors will not wish to cover the additional details discussed here. This is particularly true for the proof of the Linearization Theorem, which is perhaps best left to independent study by the students.

As in the proof of the Existence and Uniqueness Theorem, and elsewhere in the text, we find it convenient to use the ℓ_1 norm

$$\|x\| = \sum_{i=1}^{n} |x_i|,$$

on elements $x = (x_1, \ldots, x_n)$ of $\mathbb{R}^n$, because we think the proofs are simpler with this choice. Thus, on elements x in $\mathbb{R}^n$, the notation $\|x\|$ is used exclusively for the ℓ_1 norm of x. Other norms on $\mathbb{R}^n$ will be denoted differently. For example $|x| = (\sum_{i=1}^{n} x_i^2)^{1/2}$ will denote the Euclidean (or ℓ_2) norm of x and below Theorem B.3 refers to a special norm on $\mathbb{R}^n$ denoted by $\|\cdot\|_0$.

Similar comments apply to the norm

$$\|A\| = \max\{ |a_{ij}| \mid i, j \in \{1, \ldots, n\}\},$$

on the space $\mathcal{M}_n$ of $n \times n$ matrices $A = \{a_{ij}\}_{i,j=1}^{n}$. While this norm is easy to compute and work with, it has the disadvantage that the number n appears in many of the inequalities we need. However, in the context of matrices, the notation $\|\cdot\|$ will refer exclusively to this norm and other norms on $\mathcal{M}_n$ will be denoted differently.

B.1 Norms

In the context of abstract vector spaces V, we use the notation $\|\cdot\|$ for an abstract norm on V. This concept is defined as follows.

Definition B.1 (Norm) A norm on a vector space V is map $\|\cdot\| : V \to \mathbb{R}$, that has the following properties. For all $v, w \in V$ and scalars λ:

(1) $\|v\| \geq 0$ and $\|v\| = 0$ iff $v = 0$.

(2) $\|\lambda v\| = |\lambda| \|v\|$.

(3) $\|v + w\| \leq \|v\| + \|w\|$.

It is an important fact that *all* norms on a finite dimensional vector space (like $\mathbb{R}^n$ and $\mathcal{M}_n$) are equivalent in the following sense.

Definition B.2 (Equivalent Norms) Two norms $\|\cdot\|_1$ and $\|\cdot\|_2$ on a vector space V are called *equivalent* if there are positive constants α, β such that

$$\alpha\|v\|_1 \leq \|v\|_2 \leq \beta\|v\|_1, \tag{B.1}$$

for all $v \in V$.

It is well-known that equivalent norms determine the same topology on V and thus certain analytical results, like convergence of sequences, are the same with respect to either norm. In addition, it is easy to see that if an inequality like

$$\|v\|_2 \leq K_2\|w\|_2,$$

holds for one norm, and $\|\cdot\|_1$ is an equivalent norm with relations as shown in (B.1), then

$$\|v\|_1 \leq K_1\|w\|_1,$$

where $K_1 = \beta K_2/\alpha$. Therefore, the definitions and results below that involve norm inequalities with certain constants, which are stated for the ℓ_1 norm on $\mathbb{R}^n$, are independent of the norm, *provided* that the magnitude of the constants does not matter. For example, the definition of a Lipschitz function is independent of the norm, but the definition of a contraction map involves a constant that must be less than 1 and so this concept does depend on which norm is used. In addition, the result stated in Theorem B.3 involves a special norm $\|\cdot\|_0$ on $\mathbb{R}^n$ and is not true for all norms. The proof of this result requires a special norm on matrices different from the one discussed above. This norm is an operator norm.

Definition B.3 (Operator Norm) If V is a finite dimensional vector space with norm $\|\cdot\|$ and $A : V \to V$ is a linear operator, then the *operator norm* of A is defined by

$$\|A\|_* \equiv \sup\{ \frac{\|Av\|}{\|v\|} \mid v \in V, \, v \neq 0 \}.$$

With regard to this norm, which depends on the given norm on V, one has

$$\|Av\| \leq \|A\|_* \|v\|,$$

for all $v \in V$.

B.2 Lipschitz Functions

In the Existence and Uniqueness Theorem, we required X to be a C^1 function in the spatial variables $x \in \mathbb{R}^n$. An alternative theorem can be obtained by requiring that X be a Lipschitz function (relative to the spatial variables) instead.

Definition B.4 (Lipschitz Functions) A function $f : B \to \mathbb{R}^p$ on a subset $B \subseteq \mathbb{R}^n$ is called *Lipschitz* on B if there is a constant $k > 0$ such that

$$\|f(x) - f(y)\| \leq k\|x - y\| \qquad\qquad \forall x, y \in B. \qquad\qquad (\text{B.2})$$

The inequality (B.2) is called a *Lipschitz condition* on f. Here $\|\cdot\|$ denotes the ℓ_1 norm on $\mathbb{R}^p$ and $\mathbb{R}^n$ respectively. It is easy to show that if inequality (B.2) holds for the ℓ_1 norms, then it holds, with possibly a different choice of k, for any other norms on $\mathbb{R}^p$ and $\mathbb{R}^n$ respectively.

The least of all the constants k for which inequality (B.2) holds is called the *Lipschitz constant* for f and is denoted by $\mathrm{Lip}(f)$. Thus,

$$\|f(x) - f(y)\| \leq \mathrm{Lip}(f)\|x - y\| \qquad\qquad \forall x, y \in B,$$

and $\mathrm{Lip}(f)$ is the least constant for which this is true. It is easy to see the value of $\mathrm{Lip}(f)$ *does* depend on the choice of norms.

A function $f : U \to \mathbb{R}^n$ is called *locally Lipschitz* on U if each point $c \in U$ has an open neighborhood $B \subseteq U$ such that f is Lipschitz on B.

The following proposition is phrased in terms of the ℓ_1 norm on $\mathbb{R}^n$, but is clearly true for any norm on $\mathbb{R}^n$.

Proposition B.1 *Suppose $f : \mathcal{O} \to \mathbb{R}^p$ is a C^1 function on an open set $\mathcal{O} \subseteq \mathbb{R}^n$. Then f is locally Lipshitz on $\mathcal{O}$. More specifically, for each $c \in \mathcal{O}$ and $r > 0$ such that $\overline{B}(c, r) \subset \mathcal{O}$, there is a constant $k > 0$ such that*

$$\|f(x) - f(y)\| \le k\|x - y\|,$$

for all $x, y \in \overline{B}(c, r)$.

Proof: The argument for the proof is almost exactly the same as the one we gave in deriving inequality (3.8) in Theorem 3.1 for the vector field X (except now $X = f$ does not depend on the time and it has values in $\mathbb{R}^p$). Thus, the essential ingredient in getting the locally Lipschitz condition is the Mean Value Theorem, which follows from the assumption that f is C^1. $\square$

There is an easy extension of the above result if $f : S \times \mathcal{O} \to \mathbb{R}^p$ also depends continuously on a parameter $s \in S$, with S a compact subset of $\mathbb{R}^\ell$. Then one has

$$\|f(s, x) - f(s, y)\| \le k\|x - y\|,$$

for all $s \in S$ and $x, y \in \overline{B}(c, r)$, whenever $\overline{B}(c, r) \subset \mathcal{O}$.

Note that a Lipschitz function (or a locally Lipschitz function) need not be differentiable, and so not C^1 either. The standard example of this is the absolute value function $f : \mathbb{R} \to \mathbb{R}$, $f(x) = |x|$. This function is Lipschitz on $\mathbb{R}$, with Lipschitz constant $\text{Lip}(f) = 1$, but f is not differentiable at $x = 0$.

On the other hand the condition of being locally Lipschitz is all that is required for obtaining and existence and uniqueness result exactly like that in Theorem 3.1. The formulation of the result for the time depependent case is left as an exercise, but the result for the autonomous case is easily stated as follows.

Corollary B.1 *Suppose $X : \mathcal{O} \to \mathbb{R}^n$ is a locally Lipschitz vector field on an open set $\mathcal{O} \subseteq \mathbb{R}^n$. Then for each point $c \in \mathcal{O}$, there exists a curve $\alpha : I \to \mathcal{O}$, with $0 \in I$, that satisfies the initial value problem:*

$$\begin{aligned} x' &= X(x) \\ x(0) &= c. \end{aligned}$$

Furthermore, if $\gamma : J \to \mathcal{O}$ is any other solution for the initial value problem, then

$$\alpha(t) = \gamma(t) \qquad\qquad \text{for every } t \in I \cap J. \tag{B.3}$$

Proof: It is easy to see that a locally Lipschitz function is continuous and this is all that is needed to get inequality (3.7) in the proof of Theorem 3.1 (but now with X autonomous). We have already mentioned that the other inequality (3.8) there is just the local Lipschitz condition. Thus, with appropriate modifications of the proof there, we get the proof of the corollary (exercise). $\square$

A rather nice result concerning Lipschitz vector fields is that they are complete, as the next theorem below shows. To prove the theorem we need an inequality known as Gronwall's inequality. This inequality is quite remarkable in that it is elementary to prove, but has some ingenious uses. Besides being useful in proving the theorem below, we will find that Gronwall's inequality is instrumental in proving several aspects of the Linearization Theorem as well.

Lemma B.1 (Gronwall's Inequality) *Suppose that $m \geq 0, b > 0$, and that $u, h : [0, b] \to [0, \infty)$ are continuous, nonnegative functions that satisfy*

$$u(t) \leq m + \int_0^t h(s)u(s)\,ds, \qquad \text{for all } t \in [0, b] \qquad \text{(B.4)}$$

Then u satisfies the inequality

$$u(t) \leq m e^{\int_0^t h(s)\,ds}, \qquad \text{for all } t \in [0, b]. \qquad \text{(B.5)}$$

Proof: The proof is actually quite simple if we let F be the function defined by the right-hand side of inequality (B.4), i.e.,

$$F(t) \equiv m + \int_0^t h(s)u(s)\,ds,$$

for $t \in [0, b]$. Then F is a nonnegative function and, by the Fundamental Theorem of Calculus, is differentiable, with

$$F'(t) = h(t)u(t) \leq h(t)F(t),$$

for all $t \in [0, b]$. To proceed further, we assume that $m > 0$. Then $F(t) > 0$ for all $t \in [0, b]$, and so we divide the last inequality above by $F(t)$ and rewrite it as

$$\frac{d}{dt} \ln F(t) = \frac{F'(t)}{F(t)} \leq h(t),$$

for all $t \in [0, b]$. Using the second part of the Fundamental Theorem of Calculus and the last inequality gives

$$\ln F(t) - \ln m = \int_0^t \frac{d}{dt} \ln F(s)\, ds \le \int_0^t h(s)\, ds,$$

for all $t \in$. Thus, since the natural exponential function is an increasing function, we have

$$e^{\ln F(t) - \ln m} \le e^{\int_0^t h(s)\, ds},$$

Rearranging this gives Gronwall's inequality (B.5).

Thus, we have finished the proof for the case that $m > 0$. In the case that $a = 0$, let $\{m_j\}_{j=1}^\infty$ be sequence of positive numbers that converges to 0. Then clearly

$$u(t) \le m_j + \int_0^t h(s) u(s)\, ds,$$

for all $t \in [0, b]$, and all j. Consequently,

$$u(t) \le m_j e^{\int_0^t h(s)\, ds},$$

for all $t \in [0, b]$ and all j. Letting $t \to \infty$ in the last inequality gives $u(t) \le 0$, for all $t \in [0, b]$. This shows that Gronwall's inequality (B.5) also holds when $m = 0$. $\square$

Theorem B.1 *If $X : \mathcal{O} \to \mathbb{R}^n$ is Lipschitz on $\mathcal{O}$, then X is complete, i.e., for each $c \in \mathcal{O}$, the maximal interval of existence I_c for the integral curve passing through c at time 0 is $I_c = \mathbb{R}$.*

Proof: By assumption there is a positive constant k such that

$$\|X(x) - X(y)\| \le k\|x - y\| \qquad \text{for all } x, y \in \mathcal{O}.$$

Supose $c \in \mathcal{O}$ and let $\alpha_c : I_c \to \mathcal{O}$ be the maximal integral curve of X with $\alpha_c(0) = c$. Suppose the maximal interval $I_c = (a_c, b_c)$. We have to show that $b_c = \infty$ and $a_c = -\infty$.

Suppose, if possible, that $b_c < \infty$. Then for $t \in [0, b_c)$, we get from the Lipschits condition that

$$
\begin{aligned}
\|\alpha_c(t) - c\| &= \left\| \int_0^t X(\alpha_c(s))\, ds \right\| = \left\| \int_0^t \left[X(c) + (X(\alpha_c(s)) - X(c)) \right] ds \right\| \\
&\le \|X(c)\| b_c + \int_0^t \|X(\alpha_c(s)) - X(c)\|\, ds \\
&\le m + \int_0^t k\|\alpha_c(s) - c\|\, ds.
\end{aligned}
$$

Here we have let $m \equiv \|X(c)\| \, b_c$. We now apply Gronwall's inequality with $h(s) = k$ and $u(s) = \|\alpha_c(s) - c\|$ for $s \in [0, b]$. The result is that

$$\|\alpha_c(t) - c\| \le m e^{\int_0^t k \, ds} = m e^{kt} \le m e^{kb_c},$$

for all $t \in [0, b_c)$. If we let $r = m e^{kb_c}$, then the above inequality says that $\alpha_c(t) \in \bar{B}(c, r)$, for all $t \in [0, b_c)$. Since $\bar{B}(c, r)$ is compact and convex, Corollary 3.2 applies to gives us that $b_c = \infty$. This contradicts our assumption that $b_c < \infty$.

In a similar fashion, if we assume that $a_c > -\infty$, then minor modifactions of the above argument lead us to the contradiction that $a_c = -\infty$. It is an exercise to fill in the details. This completes the proof. $\square$

B.3   The Contraction Mapping Principle

The proof of Theorem 3.1, on the existence and uniqueness of solutions to initial value problems, exhibits one motivation for a principle that has become known as the *contraction mapping principle*. While this principle was initially identified with the method of successive approximations, or Picard's iterative scheme, it soon elevated to the status of a principle because of the large number of uses found for it in many other areas of mathematics. In particular, Theorem B.4 below, which is a central element in proving the Linearization Theorem, uses the contraction mapping principle in two different, and clever, ways (due to Hartman [Ha 82]).

The contraction mapping principle was stated and discussed from a very general point of view in Exercise 2 from Section 3.2 (which you should read and work through if you have the time). However, here we only need the version of it for maps from a Banach space V to itself. A Banach space V is a real vector space with a norm, $\| \cdot \| : V \to [0, \infty)$, which is complete with respect to this norm, i.e., every Cauchy sequence in V converges to an element of V.

Theorem B.2 (Contraction Mapping Principle) *Suppose $T : V \to V$ is a map from a Banach space V to itself and there is a constant $0 < q < 1$ such that*

$$\|T(x) - T(y)\| \le q\|x - y\|, \tag{B.6}$$

for all $x, y \in V$. Then T has a unique fixed point. That is, there is one, and only one, point $c \in V$, such that $T(c) = c$.

Proof: Imitate the latter part of the proof of Theorem 3.1.

If T is a contraction map as in the above theorem, with $T(c) = c$ the unique fixed point $c \in V$, then since

$$\|T(x) - c\| \leq q\|x - c\|, \tag{B.7}$$

for any $x \in V$, it is easy to see that by iteration, we have

$$\|T^k(x) - c\| \leq q^k\|x - c\|, \tag{B.8}$$

for all $k = 1, 2, 3, \ldots$ (exercise). Hence the sequence $\{T^k(x)\}_{k=1}^\infty$ of iterates of T converges to c for any $x \in V$.

Condition (B.7) (or equivalently Condition (B.8)) is weaker than Condition (B.6) for a T to be a contraction map. The former merely says that T maps each point x closer to c than it was. The latter says that T maps each pair of points closer together than they originally were. In the theory of discrete dynamical systems (cf. Exercise 2 in Section 3.5), one often has a map T that is not a contraction map but has known fixed points c. A property like Condition (B.8) allows one to conclude that c is an asymptotically stable fixed point for the discrete system. Corollary (B.2) below shows that this condition follows from a simple assumption on the eigenvalues of the Jacobian matrix $A = T'(c)$, namely that they all have modulus less than 1. In the case where T is a linear map, the next theorem gives an even stronger result.

To motivate this theorem and its proof, consider the 3×3 matrix

$$J = J_3(\lambda) = \begin{bmatrix} \lambda & 1 & 0 \\ 0 & \lambda & 1 \\ 0 & 0 & \lambda \end{bmatrix},$$

with $0 < \lambda < 1$. Because of the ones above the diagonal, J is not a contraction relative to the Euclidean norm $|\cdot|$ in $\mathbb{R}^3$. However, as noted in Chapter 5, we can write $J = \lambda I_3 + N_3$, where N_3 is the nilpotent matrix:

$$N_3 = \begin{bmatrix} 0 & 1 & 0 \\ 0 & 0 & 1 \\ 0 & 0 & 0 \end{bmatrix}.$$

If we let $H(t) = \lambda I_3 + tN_3$, for $t \in [0, 1]$, then $H(0) = \lambda I_3$ is a contraction map (for any norm on $\mathbb{R}^3$) and $H(1) = J$ is the matrix we started with.

Further note that $H(t)$ and $H(1) = J$ are similar matrices for all $t > 0$. Indeed, it is easy to verify that $Q_t^{-1}H(1)Q_t = H(t)$, where

$$Q_t = \begin{bmatrix} 1 & 0 & 0 \\ 0 & t & 0 \\ 0 & 0 & t^2 \end{bmatrix}.$$

It seems reasonable that by taking t close to 0, yet positive, we can guarantee that $H(t)$ is a contraction map. The next theorem makes this assertion precise.

Theorem B.3 *Suppose A is an $n \times n$ real matrix and that all the eigenvalues of A have modulus ≤ 1. Let $\rho \leq 1$ be the largest of all the moduli of the eigenvalues. If q is any number with $\rho \leq q \leq 1$, then there exists a norm $\|\cdot\|_0$ on $\mathbb{R}^n$ such that*

$$\|Ax\|_0 \leq q\|x\|_0, \tag{B.9}$$

for every $x \in \mathbb{R}^n$. In the special case when all the eigenvalues have modulus strictly less that one, then we may choose $q < 1$.

Proof: Let P be an invertible matrix that conjugates A into its Jordan form:

$$P^{-1}AP = J = \begin{bmatrix} J_{k_1}(\lambda_1) \\ & \ddots \\ & & J_{k_r}(\lambda_r) \\ & & & C_{2m_1}(a_1, b_1) \\ & & & & \ddots \\ & & & & & C_{2m_s}(a_s, b_s) \end{bmatrix}$$

Then λ_j, $j = 1, \ldots, r$ and $\mu_j^{\pm} = a_j \pm b_j i$, $j = 1, \ldots, s$, are the real and complex eigenvalues of A and by assumption

$$|\lambda_j| \leq \rho, \qquad |\mu_j^{\pm}| = (a_j^2 + b_j^2)^{1/2} \leq \rho,$$

for all j. Let D and M be the following block-diagonal matrices:

$$D = \begin{bmatrix} \lambda_1 I_{k_1} \\ & \ddots \\ & & \lambda_r I_{k_r} \\ & & & D_{2m_1}(a_1, b_1) \\ & & & & \ddots \\ & & & & & D_{2m_s}(a_s, b_s) \end{bmatrix} \tag{B.10}$$

$$M = \begin{bmatrix} N_{k_1} & & & & & \\ & \ddots & & & & \\ & & N_{k_r} & & & \\ & & & M_{2m_1} & & \\ & & & & \ddots & \\ & & & & & M_{2m_s} \end{bmatrix} \tag{B.11}$$

Here I_k denotes the $k \times k$ identity matrix, $D_{2m}(a,b)$ is the $2m \times 2m$ matrix given in equation (6.11), N_k is the $k \times k$ matrix given in equation (6.8), and M_{2m} is the $2m \times 2m$ matrix given in equation (6.12). Using these, we define

$$H(t) = D + tM,$$

for $t \in \mathbb{R}$. Then it is easy to see that $H(0) = D$ and $H(1) = J$.

Now D is essentially a diagonal matrix. If we decompose a vector $v \in \mathbb{R}^n$ according to the block structure of D, say $v = (v^1, \ldots, v^r, w^1, \ldots, w^s)$, then

$$\begin{aligned} Dv &= D(v^1, \ldots, v^r, w^1, \ldots, w^s) \\ &= (\lambda_1 v^1, \ldots, \lambda_r v^r, D_{2m_1}(a_1, b_1)w^1, \ldots, D_{2m_s}(a_s, b_s)w^s). \end{aligned}$$

Thus, if we let $|\cdot|$ denote the Euclidean norm on $\mathbb{R}^p$, for any value of p, then one can verify that $|D_{2m}(a,b)w| = (a^2 + b^2)^{1/2}|w|$, for all $w \in \mathbb{R}^{2m}$. Consequently,

$$\begin{aligned} &|Dv| \\ &= [\,|\lambda_1|^2|v^1|^2 + \cdots + |\lambda_r|^2|v^r|^2 + (a_1^2 + b_1^2)|w^1|^2 + \cdots + (a_s^2 + b_s^2)|w^s|^2\,]^{1/2} \\ &\leq \rho\,[\,|v^1|^2 + \cdots + |v^r|^2 + |w^1|^2 + \cdots + |w^s|^2\,]^{1/2} \\ &= \rho|v|, \end{aligned}$$

for all $v \in \mathbb{R}^n$. This gives, with respect to the operator norm $\|\cdot\|_*$ associated to the Euclidean norm,

$$\|D\|_* \leq \rho.$$

Now choose t such that $0 < t < (q - \rho)/\|M\|_*$. For this value of t, which is fixed for the remainder of the proof, we get

$$\begin{aligned} \|H(t)\|_* &= \|D + tM\|_* \leq \|D\|_* + t\|M\|_* \\ &< \rho + (q - \rho) = q. \end{aligned}$$

Next note that there is an invertible, diagonal matrix Q_t that will conjugate $J = H(1)$ into $H(t)$:

$$Q_t^{-1} J Q_t = H(t).$$

To see this observe that $B_t^{-1} N_k B_t = t N_k$, where $B_t = \text{diag}(1, t, t^2, \ldots, t^{-k+1})$. Similarly $C_t^{-1} M_{2m} C_t = t M_{2m}$, for an appropriate diagonal matrix C_t and from the B_t's and C_t's one can construct the matrix Q_t (exercise).

Now let $R_t = P Q_t$, so that we have

$$R_t^{-1} A R_t = H(t).$$

We use R_t to define a new norm on $\mathbb{R}^n$ by

$$\|x\|_0 \equiv |R_t^{-1} x|,$$

where $|x|$ is the usual Euclidean norm on $\mathbb{R}^n$. The we easily get the following:

$$
\begin{aligned}
\|Ax\|_0 &= |R_t^{-1} A x| \\
&= |R_t^{-1} A R_t R_t^{-1} x| \\
&= |H(t) R_t^{-1} x| \\
&\leq \|H(t)\|_* |R_t^{-1} x| \\
&\leq q \|x\|_0,
\end{aligned}
$$

for all $x \in \mathbb{R}^n$. This completes the proof. $\square$

Corollary B.2 *Suppose $f : \mathcal{O} \to \mathbb{R}^n$ is a C^1 map with fixed point $f(c) = c$ at $c \in \mathcal{O}$. Let $A = f'(c)$ and suppose all the eigenvalues of A have modulus less than one. If ρ is the largest of these moduli, then for any q such that $\rho < q < 1$, there is a norm $\| \cdot \|_0$ on $\mathbb{R}^n$ and a $\delta > 0$ such that*

$$\|f^k(x) - c\|_0 < q^k \|x - c\|_0, \tag{B.12}$$

for all $x \in B(c, \delta) \equiv \{ x \mid \|x - c\|_0 < \delta \}$ and all positive integers k. Hence for every point $x \in B(x, \delta)$, the sequence $\{f^k(x)\}_{k=1}^{\infty}$ of iterates converges to the fixed point c.

Proof: Choose q_0 so that $\rho < q_0 < q$ and use the theorem above to get a norm $\| \cdot \|_0$ on $\mathbb{R}^n$ such that

$$\|Ax\|_0 < q_0 \|x\|_0,$$

for every $x \in \mathbb{R}^n$. Now since $f'(c) = A$ and $f(c) = c$, it follows from the definition of the derivative that

$$\lim_{x \to c} \frac{\|f(x) - c - A(x - c)\|_0}{\|x - c\|_0} = 0.$$

(For some, this *is* the definition of the derivative at c). Hence if we let $\varepsilon = q - q_0$, the above limits says that we can choose $\delta > 0$ so that

$$\|f(x) - c - A(x - c)\|_0 < \varepsilon \|x - c\|_0,$$

for all $x \in B(c, \delta)$. Now a general property for any norm is that

$$\|v\| - \|w\| \leq \|v - w\|.$$

Applying this with the above gives

$$
\begin{aligned}
\|f(x) - c\|_0 \quad &< \quad \|A(x - c)\|_0 + \varepsilon \|x - c\|_0 \\
&< \quad q_0 \|x - c\|_0 + \varepsilon \|x - c\|_0 \\
&= \quad q \|x - c\|_0,
\end{aligned}
$$

for all $x \in B(c, \delta)$. Since $q < 1$, the above inequality gives $\|f(x) - c\|_0 < q\delta < \delta$, for $x \in B(c, \delta)$, and so $f(x)$ lies in $B(c, \delta)$ as well. Then we can apply the inequality to $f(x)$ and get

$$\|f^2(x) - c\|_0 < q\|f(x) - c\|_0 < q^2 \|x - c\|_0.$$

Continuing like this, we inductively get the inequality (B.12). $\square$

Corollary B.3 *Suppose A is an $n \times n$ real matrix and that all the eigenvalues of A have modulus greater than one. Let $\mu > 1$ be the smallest of all the moduli of the eigenvalues. If q is any number with $0 < q < \mu$, then there exists a norm $\| \cdot \|_1$ on $\mathbb{R}^n$ such that*

$$\|A^{-1}x\|_1 < q\|x\|_1, \tag{B.13}$$

for every $x \in \mathbb{R}^n$. Equivalently

$$q^{-1}\|x\|_1 < \|Ax\|_1, \tag{B.14}$$

for every $x \in \mathbb{R}^n$. Hence A is an expansion and the iterates $\{A^k x\}_{k=1}^{\infty}$, for $x \neq 0$, diverge to infinity.

Proof: Note that A is invertible and for any $\lambda \neq 0$

$$Av = \lambda v \iff A^{-1}v = \lambda^{-1}v.$$

Thus, the eigenvalues of A^{-1} are the reciprocals of the eigenvalues of A. Applying the above theorem to A^{-1} yields the result. $\square$

B.4 The Linearization Theorem

The Linearization Theorem states that on a neighborhood of a hyperbolic fixed point c of a vector field X, the nonlinear system $x' = X(x)$ is topologically equivalent to the linear system $y' = Ay$, where $A = X'(c)$. By translating, we can assume, without loss of generality, that the fixed point is the origin $c = 0$.

To prove this theorem, we first extend X to a vector field $\widetilde{X}$ on all of $\mathbb{R}^n$ whose flow differs from the linear flow by a family of Lipschitz maps (Lemma B.2 and Theorem B.4). Then after proving a basic, and useful, result about perturbations of a linear map by a Lipshitz map (Theorem B.5), we prove the theorem itself. The techniques and approach here are derived from those in Hartman's book [Ha 82] and are variations of others in the more recent literature (cf. [Per 91], [Rob 95]).

We should also mention that there is an analogous linearization theorem for maps associated with discrete dynamical systems whose proof is entirely similar. See the material on the electronic component for discrete dynamical systems.

Lemma B.2 *Suppose $X : \mathcal{O} \to \mathbb{R}^n$ is a C^1 vector field on $\mathcal{O} \subseteq \mathbb{R}^n$ with the origin as a fixed point: $X(0) = 0$. Let $A = X'(0)$ and let $R : \mathcal{O} \to \mathbb{R}^n$ be defined by*

$$R(x) = X(x) - Ax,$$

for $x \in \mathcal{O}$. If $\delta > 0$ is given, then we can choose $b > 0$, with $B(0, b) \subseteq \mathcal{O}$, and a C^1 vector field $\widetilde{R} : \mathbb{R}^n \to \mathbb{R}^n$ on $\mathbb{R}^n$ such that:

(1) *$\widetilde{R}(x) = R(x)$ for $\|x\| \leq b/2$ and $\widetilde{R}(x) = 0$ for $\|x\| \geq b$.*

(2) *$\|\widetilde{R}(x) - \widetilde{R}(y)\| \leq \delta \|x - y\|$ for all $x, y \in \mathbb{R}^n$.*

Thus, $\widetilde{R}$ is Lipschitz on $\mathbb{R}^n$, agrees with R on $\overline{B}(0, b/2)$, and vanishes outside $\overline{B}(0, b)$.

Proof: Note that $R(0) = X(0) - A0 = 0$. Also $R'(x) = X'(x) - A$ for every $x \in \mathcal{O}$ and so in particular $R'(0) = 0$. From this and the fact that R' is continuous at 0, we are assured the existence of a $b > 0$ such that $B(0, b) \subseteq \mathcal{O}$ and

$$\left| \frac{\partial R^i}{\partial x_j}(x) \right| \leq \frac{\delta}{8n}, \tag{B.15}$$

for all $x \in B(0, b)$ and all $i, j = 1, \ldots, n$. One can use this and the Mean Value Theorem, exactly as in the proof of the Existence and Uniqueness Theorem 3.1 (see inequality (3.8)), to show that

$$|R^i(x) - R^i(y)| \le \frac{\delta}{8n}\|x - y\|, \tag{B.16}$$

for all $x, y \in \overline{B}(0, b)$ and all $i = 1, \ldots, n$. In particular, for $y = 0$ this gives (since $R(0) = 0$) that

$$|R^i(x)| \le \frac{\delta}{8n}\|x\|, \tag{B.17}$$

for all $x \in \overline{B}(0, b)$ and all $i = 1, \ldots, n$. We construct $\widetilde{R}$ by multiplying R by a suitable function that is zero outside of $B(0, b)$.

First we define a function $\mu : \mathbb{R} \to \mathbb{R}$ by

$$\mu(r) = \begin{cases} 1 & \text{if } r \in (-\infty, b^2/4] \\ 1 - \frac{1}{27b^6}[12(4r - b^2)^2(b^2 - r) + (4r - b^2)^3] & \text{if } r \in (b^2/4, b^2) \\ 0 & \text{if } r \in [b^2, \infty) \end{cases} \tag{B.18}$$

Then it is easy to see that $0 \le \mu(r) \le 1$ for all r and μ is differentiable on $\mathbb{R}$, with derivative $\mu'(r) = -24(4r - b^2)(b^2 - r)/27b^6$ for $r \in (b^2/4, b^2)$ and $\mu'(r) = 0$ elsewhere. Thus, μ' is continuous on $\mathbb{R}$. It is also easy to calculate that μ' has a minimum value of $-2/b^2$ at $r = 5b^2/8$ and a maximum value of 0. Thus,

$$|\mu'(r)| \le \frac{2}{b^2}, \tag{B.19}$$

for all $r \in \mathbb{R}$. Using this function, we now define $\widetilde{R} : \mathbb{R}^n \to \mathbb{R}^n$ by

$$\widetilde{R}(x) = \begin{cases} \mu(\|x\|^2)\, R(x) & \text{if } \|x\| \le b \\ 0 & \text{if } \|x\| > b \end{cases}. \tag{B.20}$$

Then clearly $\widetilde{R}$ satisfies Property (1) of the theorem. Also $\widetilde{R}$ is differentiable with continuous first order partial derivatives:

$$\frac{\partial \widetilde{R}^i}{\partial x_j}(x) = \mu'(\|x\|^2)\, 2\|x\| \frac{x_j}{|x_j|}\, R^i(x) + \mu(\|x\|^2) \frac{\partial R^i}{\partial x_j}(x), \tag{B.21}$$

for $x \in B(0, b)$. Thus, for $\|x\| < b$

$$\left|\frac{\partial \widetilde{R}^i}{\partial x_j}(x)\right| \le |\mu'(\|x\|^2)|\, 2\|x\|\, |R^i(x)| + \mu(\|x\|^2)\left|\frac{\partial R^i}{\partial x_j}(x)\right|$$

$$\le \frac{2}{b^2} \cdot 2b \cdot \frac{\delta}{8n} \cdot b + \frac{\delta}{8n} = \frac{5\delta}{8n} < \frac{\delta}{n}$$

Since

$$\frac{\partial \widetilde{R}^i}{\partial x_j}(x) = 0,$$

for $\|x\| \geq b$, the last inequality holds in this case too, and we get that

$$\left| \frac{\partial \widetilde{R}^i}{\partial x_j}(x) \right| < \frac{\delta}{n},$$

for all $x \in \mathbb{R}^n$ and all i, j. From this and the Mean Value Theorem it follows (as in the proof of the Existence and Uniqueness Theorem) that

$$\|\widetilde{R}(x) - \widetilde{R}(y)\| \leq \delta \|x - y\|,$$

for all $x, y \in \mathbb{R}^n$. This completes the proof of the lemma. $\square$

Theorem B.4 *Suppose $X : \mathcal{O} \to \mathbb{R}^n$ is a C^1 vector field on $\mathcal{O} \subseteq \mathbb{R}^n$ with the origin as a fixed point: $X(0) = 0$. Let $A = X'(0)$ and suppose that $\tau > 0, \varepsilon > 0$ are given. Then we can choose $b > 0$, with $B(0, b) \subseteq \mathcal{O}$, and a C^1 vector field $\widetilde{X} : \mathbb{R}^n \to \mathbb{R}^n$, which is Lipschitz on $\mathbb{R}^n$, such that*

$$\widetilde{X}(x) \quad = \quad X(x) \qquad \text{for } \|x\| \leq b/2 \tag{B.22}$$
$$\widetilde{X}(x) \quad = \quad Ax \qquad \text{for } \|x\| \geq b. \tag{B.23}$$

and such that the flow $\widetilde{\phi}$ for $\widetilde{X}$ has the form:

$$\widetilde{\phi}_t(x) = e^{At}x + g_t(x), \tag{B.24}$$

for $x \in \mathbb{R}^n$, $t \in \mathbb{R}$. Furthermore, the function $g(t, x) = g_t(x)$ has the following properties: (1) g is continuous on $\mathbb{R} \times \mathbb{R}^n$, (2) $g_t(x) = 0$ for $\|x\| \geq b$ and for all $t \in \mathbb{R}$ (So in particular each g_t is bounded), and (3) for all $x, y \in \mathbb{R}^n$ and $t \in [-\tau, \tau]$

$$\|g_t(x) - g_t(y)\| \leq \varepsilon \|x - y\|. \tag{B.25}$$

Proof: Let $R : \mathcal{O} \to \mathbb{R}^n$ be the vector field defined by $R(x) = X(x) - Ax$, for $x \in \mathcal{O}$. Then R is the remainder term in the 1st-order Taylor polynomial approximation of X at $x = 0$:

$$X(x) = Ax + R(x),$$

for $x \in \mathcal{O}$. Choose $\delta > 0$ such that

$$\delta e^{\tau \delta} < \varepsilon e^{-2n\tau \|A\|}.$$

By the last Lemma, we can choose $b > 0$ and a vector field $\widetilde{R} : \mathbb{R}^n \to \mathbb{R}^n$ which is an extension of of R from the neighborhood $B(0, b/2)$ to a vector field on the whole plane having Lipschitz constant $\mathrm{Lip}(\widetilde{R}) \leq \delta$. Further $\widetilde{R}$ vanishes outside $B(0, b)$. Then the vector field $\widetilde{X} : \mathbb{R}^n \to \mathbb{R}^n$ defined by

$$\widetilde{X}(x) = Ax + \widetilde{R}(x),$$

for $x \in \mathbb{R}^n$, clearly satisfies the two conditions (B.22)-(B.23) in this theorem. For convenience let $K = n\|A\| + \delta$. Then since $\widetilde{X}(x) - \widetilde{X}(y) = A(x - y) + \widetilde{R}(x) - \widetilde{R}(y)$, we get that

$$\|\widetilde{X}(x) - \widetilde{X}(y)\| \leq K\|x - y\|, \tag{B.26}$$

for all $x, y \in \mathbb{R}^n$. Thus, $\widetilde{X}$ is Lipschitz on $\mathbb{R}^n$ and so by Theorem B.1, its flow $\widetilde{\phi}$ is defined on all of $\mathbb{R} \times \mathbb{R}^n$. Let $\phi : \mathcal{D} \to \mathbb{R}^n$ be the flow generated by X. Then because of conditions (B.22)-(B.23), we have

$$\widetilde{\phi}_t(x) = \phi_t(x) \quad \text{for } \|x\| \leq b/2 \text{ and } t \in I_x \tag{B.27}$$
$$\widetilde{\phi}_t(x) = e^{At}x \quad \text{for } \|x\| \geq b \text{ and } t \in \mathbb{R} \tag{B.28}$$

Thus, defining $g : \mathbb{R} \times \mathbb{R}^n \to \mathbb{R}^n$ by

$$g_t(x) = \widetilde{\phi}_t(x) - e^{At}x,$$

gives us a function for which $g_t = 0$ outside the ball $B(0, b)$, for all $t \in \mathbb{R}$.

We have left to prove the Lipschitz estimate on the g_t's. This requires several uses of Gronwall's inequality. The details are as follows.

First note that the integral version of the system $x' = \widetilde{X}(x) = Ax + \widetilde{R}(x)$ is precisely the statement that the flow $\widetilde{\phi}$ satisfies

$$\widetilde{\phi}_t(x) = x + \int_0^t \widetilde{X}(\widetilde{\phi}_s(x))\, ds = x + \int_0^t [A\widetilde{\phi}_s(x) + \widetilde{R}(\widetilde{\phi}_s(t))]\, ds. \tag{B.29}$$

Thus, from the inequality (B.26), we get that

$$\|\widetilde{\phi}_t(x) - \widetilde{\phi}_t(y)\| \leq \|x - y\| + \int_0^t K\|\widetilde{\phi}_s(x) - \widetilde{\phi}_s(y)\|\, ds,$$

for all $x, y \in \mathbb{R}^n$ and all $t \geq 0$. Applying Gronwall's inequality to this gives

$$\|\widetilde{\phi}_t(x) - \widetilde{\phi}_t(y)\| \leq e^{K|t|}\|x - y\|, \tag{B.30}$$

for all $x, y \in \mathbb{R}^n$ and all $t \geq 0$. *Note*: Since t was assumed to be nonnegative in deriving the above inequality, we do not need the absolute value $|t|$ in the exponent on the right-hand side. However, if we repeat the above argument for $t < 0$, use $\int_t^0$ and $-t = |t|$ we will get that inequality (B.30) holds for all $t \in \mathbb{R}$. Next we want to do a similar thing for g_t, i.e., derive an estimate and then use Gronwall's inequality. For this note that

$$e^{At}(x - y) = x - y + \int_0^t Ae^{As}(x - y)\, ds,$$

for all $x, y \in \mathbb{R}^n$ and all $t \in \mathbb{R}$. This is just the integral version of $x' = Ax$ applied to its flow. Using this and the definition of g, we get, for $t \geq 0$,

$$
\begin{aligned}
g_t(x) - g_t(y) &= \tilde{\phi}_t(x) - \tilde{\phi}_t(y) - e^{At}(x - y) \\
&= x - y + \int_0^t \left(A[\tilde{\phi}_s(x) - \tilde{\phi}_s(y)] + \tilde{R}(\tilde{\phi}_s(x)) - \tilde{R}(\tilde{\phi}_s(y)) \right) ds \\
&\quad - e^{At}(x - y) \\
&= \int_0^t [\tilde{R}(\tilde{\phi}_s(x)) - \tilde{R}(\tilde{\phi}_s(y))]\, ds \\
&\quad + \int_0^t A[\tilde{\phi}_s(x)) - \tilde{\phi}_s(y)) - e^{As}(x - y)]\, ds \\
&= \int_0^t [\tilde{R}(\tilde{\phi}_s(x)) - \tilde{R}(\tilde{\phi}_s(y))]\, ds + \int_0^t A[g_s(x) - g_s(y)]\, ds
\end{aligned}
$$

Hence for $t \in [0, \tau]$, we have

$$\|g_t(x) - g_t(y)\| \leq \int_0^t \|\tilde{R}(\tilde{\phi}_s(x)) - \tilde{R}(\tilde{\phi}_s(y))\|\, ds + \int_0^t n\|A\|\, \|g_s(x) - g_s(y)\|\, ds.$$
(B.31)

In order to apply Gronwall's inequality to this, we replace the first integral by the following estimate

$$
\begin{aligned}
\int_0^t \|\tilde{R}(\tilde{\phi}_s(x)) - \tilde{R}(\tilde{\phi}_s(y))\|\, ds &\leq \delta \int_0^t \|\tilde{\phi}_s(x) - \tilde{\phi}_s(y)\|\, ds \\
&\leq \delta \int_0^t e^{K|s|}\|x - y\|\, ds \leq \delta \tau e^{K\tau}\|x - y\|.
\end{aligned}
$$

Using this in inequality (B.31), we get

$$\|g_t(x) - g_t(y)\| \leq \delta \tau e^{K\tau}\|x - y\| + \int_0^t n\|A\|\, \|g_s(x) - g_s(y)\|\, ds,$$

for all $x, y \in \mathbb{R}^n$ and all $t \in [0, \tau]$. Hence applying Gronwall's inequality, and recalling that $K = n\|A\| + \delta$, gives

$$
\begin{aligned}
\|g_t(x) - g_t(y)\| &\leq \delta \tau e^{K\tau} \|x - y\| e^{n\|A\| |t|} \\
&\leq \delta \tau e^{K\tau} \|x - y\| e^{n\|A\|\tau} \\
&= \delta e^{\tau\delta} e^{2n\tau\|A\|} \|x - y\| \\
&< \varepsilon \|x - y\|,
\end{aligned}
$$

In a similar way one can show that this inequality also holds for $t \in [-\tau, 0]$ and thus it holds for all $t \in [-\tau, \tau]$. This completes the proof. $\square$

Theorem B.5 *Suppose L is an invertible $n \times n$ block-diagonal matrix of the form*

$$
L = \begin{bmatrix} B & 0 \\ 0 & C \end{bmatrix},
$$

with B an $m \times m$ matrix and C a $p \times p$ matrix. Let

$$
a = \max\{ m\|B\|, \, p\|C^{-1}\| \},
$$

and assume that $a < 1$. Let c be a number such that $a < c < 1$ and let

$$
\varepsilon = \tfrac{1}{2} \cdot \min\{ c - a, \, \frac{c}{n\|L^{-1}\|} \}.
$$

Denote by $C_b^0(\mathbb{R}^n)$ the set of all bounded continuous functions $g : \mathbb{R}^n \to \mathbb{R}^n$ on $\mathbb{R}^n$ and let

$$
\mathcal{L}_\varepsilon \equiv \{ g \in C_b^0(\mathbb{R}^n) \mid \mathrm{Lip}(g) < \varepsilon \}.
$$

Then the following results hold:

(1) *If $g \in \mathcal{L}_\varepsilon$ then $L + g$ is a homeomorphism of $\mathbb{R}^n$.*

(2) *If $g, h \in \mathcal{L}_\varepsilon$, then there is a unique $v_{gh} \in C_b^0(\mathbb{R}^n)$ such that:*

$$
(L + g) \circ (I + v_{gh}) = (I + v_{gh}) \circ (L + h). \tag{B.32}
$$

     Here $I : \mathbb{R}^n \to \mathbb{R}^n$ is the identity map.

(3) *The map $I + v_{gh}$ in (2) is in fact a homeomorphism of $\mathbb{R}^n$.*

Proof:

(1) We use the contraction mapping principle to show that that $L + g$ is a 1-1 and onto. Thus, suppose $y \in \mathbb{R}^n$ and let $T : \mathbb{R}^n \to \mathbb{R}^n$ be defined by $T(x) = L^{-1}(y - g(x))$. Then clearly, for any $x \in \mathbb{R}^n$,

$$T(x) = x \qquad \text{if and only if} \qquad Lx + g(x) = y.$$

Thus, if T has a fixed point then $L + g$ is onto and if this fixed point is unique then $L + g$ must be 1-1. But the following easy estimate shows that T is a contraction:

$$\begin{aligned}
\|T(u) - T(z)\| &= \|L^{-1}(g(u) - g(z))\| \\
&\leq n\|L^{-1}\| \cdot \mathrm{Lip}(g) \cdot \|u - z\| \\
&= < n\|L^{-1}\| \cdot \varepsilon \cdot \|u - z\| \leq c\|u - z\|.
\end{aligned}$$

Hence, since $\mathbb{R}^n$, with the norm $\|\cdot\|$, is complete, T has a unique fixed point. Thus, $L + g$ is a bijection. Clearly $L + g$ is continuous. It is a well-known result from topology that a continuous bijection on $\mathbb{R}^n$ is a homeomorphism of $\mathbb{R}^n$.

(2) To prove the second assertion, we again use the contraction mapping principle, but now the contraction will be a map $T : C_b^0(\mathbb{R}^n) \to C_b^0(\mathbb{R}^n)$. For the construction of T, we need to split, or decompose, the elements $v \in C_b^0(\mathbb{R}^n)$ into parts: $v = (v^1, v^2)$, with $v^1 : \mathbb{R}^n \to \mathbb{R}^m$ and $v^2 : \mathbb{R}^n \to \mathbb{R}^p$. This decomposition is based on the block-diagonal decomposition of L. Since we are using the ℓ_1 norm $\|\cdot\|$ on $\mathbb{R}^n$, it is easy to see that for $i = 1, 2$,

$$\|v^i(x)\| \leq \|v(x)\| = \|v^1(x)\| + \|v^2(x)\|,$$

for all $x \in \mathbb{R}^n$. Here $\|\cdot\|$ also is used to denote the ℓ_1 norm on $\mathbb{R}^m$ and $\mathbb{R}^p$. We use a norm on $C_b^0(\mathbb{R}^n)$ that is the maximum of the two sup norms on the elements in this decomposition: for $v = (v^1, v^2) \in C_b^0(\mathbb{R}^n)$ define

$$\|v\| \equiv \max\{\, \sup_x \|v^1(x)\|, \ \sup_x \|v^2(x)\| \,\}.$$

With this norm, $C_b^0(\mathbb{R}^n)$ is a Banach space. The decomposition of $C_b^0(\mathbb{R}^n)$ also carries over to Lipschitz maps g in $\mathcal{L}_\varepsilon$. Note that with the splitting $g = (g^1, g^2)$, each map g^1, g^2 is Lipschitz and since, for $i = 1, 2$,

$$\|g^i(x) - g^i(y)\| \leq \|g(x) - g(y)\| \leq \mathrm{Lip}(g)\|x - y\|,$$

it follows that $\text{Lip}(g^i) \leq \text{Lip}(g)$. Using these observations and definitions about g, v, and the norms, it is easy to verify the following inequality, which will be used in the proof below:

$$
\begin{aligned}
&\|g^i(x + v(x)) - g^i(x + w(x))\| \\
\leq\ &\text{Lip}(g^i)\|v(x) - w(x)\| \\
=\ &\text{Lip}(g^i)\Big[\|v^1(x) - w^1(x)\| + \|v^2(x) - w^2(x)\|\Big] \\
\leq\ &2\,\text{Lip}(g^i)\|v - w\| \\
\leq\ &2\,\text{Lip}(g)\|v - w\|,
\end{aligned}
\tag{B.33}
$$

for $i = 1, 2$, $x \in \mathbb{R}^n$, and $v, w \in C_b^0(\mathbb{R}^n)$. With these preliminaries out of the way, we can proceed to the contstruction of the map $T : C_b^0(\mathbb{R}^n) \to C_b^0(\mathbb{R}^n)$.

Suppose $g, h \in \mathcal{L}_\varepsilon$. To prove the existence of v_{gh} satisfying equation (B.32), we expand that equation to give the equivalent equation:

$$
L + L \circ v_{gh} + g \circ (I + v_{gh}) = L + h + v_{gh} \circ (L + h).
$$

or

$$
L \circ v_{gh} + g \circ (I + v_{gh}) - h = v_{gh} \circ (L + h).
$$

Now split this equation into two parts based on the decomposition of $C_b^0(\mathbb{R}^n)$ discussed above:

$$
\begin{aligned}
B \circ v_{gh}^1 + g^1 \circ (I + v_{gh}) - h^1 &= v_{gh}^1 \circ (L + h) \\
C \circ v_{gh}^2 + g^2 \circ (I + v_{gh}) - h^2 &= v_{gh}^2 \circ (L + h)
\end{aligned}
$$

For convenience let $f \equiv L + h$, which by Part (1) is a homeomorphism. Now rewrite the above two equations as

$$
\begin{aligned}
B \circ v_{gh}^1 \circ f^{-1} + g^1 \circ (I + v_{gh}) \circ f^{-1} - h^1 \circ f^{-1} &= v_{gh}^1 \\
C^{-1}\Big[v_{gh}^2 \circ f - g^2 \circ (I + v_{gh}) + h^2\Big] &= v_{gh}^2
\end{aligned}
$$

View this as the fixed point equation $T(v_{gh}) = v_{gh}$, where $T : C_b^0(\mathbb{R}^n) \to C_b^0(\mathbb{R}^n)$ is the map defined by

$$
T(v) = \begin{bmatrix} B \circ v^1 \circ f^{-1} + g^1 \circ (I + v) \circ f^{-1} - h^1 \circ f^{-1} \\ C^{-1}\big[v^2 \circ f - g^2 \circ (I + v) + h^2\big] \end{bmatrix}.
$$

Here in the notation, we have split the $T(v) = (T(v)^1, T(v)^2)$ into parts and written it as a column vector. It is clear that if $v \in C_b^0(\mathbb{R}^n)$, then $T(v)$ is a continuous function. $T(v)$ is also bounded because, using the remarks above about the norms, we easily see that

$$\|T(v)(x)\| \leq \left(m\|B\| + p\|C^{-1}\|\right)\|v\| + \left(1 + p\|C^{-1}\|\right)(\|g\| + \|h\|),$$

for all $x \in \mathbb{R}^n$. Thus, $T(v) \in C_b^0(\mathbb{R}^n)$, for every $v \in C_b^0(\mathbb{R}^n)$. The major calculation is to show that T is a contraction.

For this, suppose $v, w \in C_b^0(\mathbb{R}^n)$ and $x \in \mathbb{R}^n$. For convenience let $y = f^{-1}(x)$, $z = f(x)$. Then, using inequality (B.33), we get the following inequalities for the first component in the decomposition of $T(v)(x) - T(w)(x)$:

$$
\begin{aligned}
& \|T(v)^1(x) - T(w)^1(x)\| \\
\leq\ & \|B[v^1(y) - w^1(y)]\| + \|g^1(y + v(y)) - g^1(y + w(y))\| \\
\leq\ & m\|B\|\|v^1(y) - w^1(y)\| + 2\operatorname{Lip}(g)\|v - w\| \\
\leq\ & m\|B\|\|v - w\| + 2\operatorname{Lip}(g)\|v - w\| \\
\leq\ & (a + 2\varepsilon)\|v - w\| \\
\leq\ & c\|v - w\|. \hspace{5cm} \text{(B.34)}
\end{aligned}
$$

In a similar fashion, for the second component we get

$$
\begin{aligned}
& \|T(v)^2(x) - T(w)^2(x)\| \\
\leq\ & p\|C^{-1}\|\left[\|v^2(z) - w^2(z)\| + \|g^2(x + v(x)) - g^2(x + w(x))\|\right] \\
\leq\ & +p\|C^{-1}\|\left[\|v - w\| + 2\operatorname{Lip}(g)\|v - w\|\right] \\
\leq\ & a(1 + 2\varepsilon)\|v - w\| = (a + 2a\varepsilon)\|v - w\| \\
\leq\ & c\|v - w\|. \hspace{5cm} \text{(B.35)}
\end{aligned}
$$

Now take the sup over all $x \in \mathbb{R}^n$ in Inequalities (B.34)-(B.35) and use the definition of the norm on $C_b^0(\mathbb{R}^n)$, to get

$$\|T(v) - T(w)\| \leq c\|v - w\|.$$

This holds for all $v, w \in C_b^0(\mathbb{R}^n)$, and since $c < 1$, we have that T is a contraction. Thus, there exists a unique $v_{gh} \in C_b^0(\mathbb{R}^n)$ such that $T(v_{gh}) = v_{gh}$.

(3) Suppose $g, h, k \in \mathcal{L}_\varepsilon$. The uniqueness assertion in Part (2) says that if $H : \mathbb{R}^n \to \mathbb{R}^n$ is a continuous map that satisfies $(L + g) \circ H = H \circ (L + h)$

and has $H - I$ bounded, then $H - I = v_{gh}$. We use this to derive the following results. For convenience we let

$$P_{gh} = I + v_{gh}.$$

Then it is easy to see that

$$(L + g) \circ P_{gh} \circ P_{hk} = P_{gh} \circ (L + h) \circ P_{hk} = P_{gh} \circ P_{hk} \circ (L + k).$$

Further

$$P_{gh} \circ P_{hk} - I = (I + v_{gh}) \circ (I + v_{hk}) - I = v_{hk} + v_{gh} \circ (I + v_{hk}),$$

is clearly bounded (and continuous). Hence, by the initial comment above, we get

$$P_{gh} \circ P_{hk} - I = v_{gk},$$

and this holds for all $g, h, k \in \mathcal{L}_\varepsilon$. Now it is clear that $v_{gg} = 0$ for all $g \in \mathcal{L}_\varepsilon$, and thus from the last equation above we get

$$P_{gh} \circ P_{hg} = I \qquad \text{and} \qquad P_{hg} \circ P_{gh} = I.$$

These show that P_{gh} is a bijection and has inverse $P_{gh}^{-1} = P_{hg}$. Hence $I + v_{gh}$ is a homeomorphism. $\square$

Theorem B.6 (Hartman-Grobman Linearization Theorem) *Suppose $X : \mathcal{O} \to \mathbb{R}^n$ is a C^1 vector field and that the origin $0 \in \mathcal{O}$ is a hyperbolic fixed point of X. Let $A = X'(0)$. Then there is a neighborhood $U \subseteq \mathcal{O}$ of 0, a C^1 vector field $\widetilde{X} : \mathbb{R}^n \to \mathbb{R}^n$, which is Lipschitz on $\mathbb{R}^n$, and a homeomorphism $f : \mathbb{R}^n \to \mathbb{R}^n$ such that*

(1) $\widetilde{X} = X$ on U,

(2) The flow $\widetilde{\phi} : \mathbb{R} \times \mathbb{R}^n \to \mathbb{R}^n$ for $\widetilde{X}$ satisfies

$$\widetilde{\phi}_t(x) = f^{-1}(e^{At} f(x)), \tag{B.36}$$

for all $x \in \mathbb{R}^n$ and all $t \in \mathbb{R}$.

Hence the system $x' = X(x)$, restricted to the neighborhood U is topologically equivalent to the linear system $y' = Ay$ restricted to $f^{-1}(U)$.

Proof: We first prove the theorem for the case when A is a Jordan form with its real and complex Jordan blocks grouped according to the real parts of their eigenvalues. Then we show how the general case can be reduced to this. Thus, assume A has the form

$$A = \begin{bmatrix} J_1 & 0 \\ 0 & J_2 \end{bmatrix},$$

where J_1 is an $m \times m$ Jordan canonical form with all its eigenvalues having negative real parts and J_2 is a $p \times p$ Jordan canonical form with all of its eigenvalues having positive real parts. Apply the Linear Stability Theorem to J_1 and $-J_2$ to get a time $\tau > 0$ such that

$$n\|e^{\tau J_1}\| < 1, \qquad n\|e^{-\tau J_2}\| < 1.$$

Then let

$$L = e^{\tau A} = \begin{bmatrix} e^{\tau J_1} & 0 \\ 0 & e^{\tau J_2} \end{bmatrix} = \begin{bmatrix} B & 0 \\ 0 & C \end{bmatrix}.$$

Consequently, $a \equiv \max\{\, m\|B\|,\ p\|C^{-1}\|\,\} < 1$, and we can choose a c such that $a < c < 1$. Further let

$$\varepsilon = \tfrac{1}{2} \cdot \min \Big\{\, a - c,\ \frac{c}{n\|L^{-1}\|} \,\Big\}.$$

Now apply Theorem B.4 with these values for τ and ε to get the vector field $\tilde{X}$ and neighborhood $B(0, b)$ with the properties stated in that theorem. The flow for $\tilde{X}$ has the form $\tilde{\phi}_t(x) = e^{At}x + g_t(x)$, for $x \in \mathbb{R}^n$ and $t \in \mathbb{R}$. In particular

$$\tilde{\phi}_\tau(x) = e^{A\tau}x + g_\tau(x),$$

and by the choice of constants, $g_\tau \in \mathcal{L}_\varepsilon$, with $L \equiv e^{A\tau}$. Thus, Theorem B.5 applies (taking $g = g_\tau$ and $h = 0$ in Part (2)) and gives a unique $v \in C_b^0(\mathbb{R}^n)$, such that

$$\tilde{\phi}_\tau \circ (I + v) = (I + v) \circ e^{A\tau}. \tag{B.37}$$

That theorem further guarantees that $H \equiv I + v$ is a homeomorhism of $\mathbb{R}^n$. To finish the proof of this theorem, we take $r = b/2$ and prove that $f = H^{-1}$ is the desired homeomorphism.

The crux of the matter now is to show that equation (B.37) holds for any time $t \in \mathbb{R}$, not just for time τ. For this suppose $t \in \mathbb{R}$ and let

$$w \equiv \tilde{\phi}_t \circ (I + v) \circ e^{-tA} - I$$

$$
\begin{aligned}
&= \ \widetilde{\phi}_t \circ (e^{-tA} + v \circ e^{-tA}) - I \\
&= \ (g_t + e^{tA}) \circ (e^{-tA} + v \circ e^{-tA}) - I \\
&= \ g_t \circ (e^{-tA} + v \circ e^{-tA}) + e^{tA} \circ v \circ e^{-tA} \quad\quad \text{(B.38)}
\end{aligned}
$$

From the definition of w in the first line above, we see that $w = v$ if and only if

$$
\widetilde{\phi}_t \circ (I + v) \circ e^{-tA} = I + v.
$$

This last equation is what we need, so it suffices to show that $w = v$. Now since g_t and v are bounded continuous functions, the last line in equation (B.38), shows that w is also bounded and continuous (exercise). Furthermore, we have

$$
\begin{aligned}
\widetilde{\phi}_\tau \circ (I + w) \circ e^{-\tau A} &= \ \widetilde{\phi}_\tau \circ \widetilde{\phi}_t \circ (I + v) \circ e^{-tA} \circ e^{-\tau A} \\
&= \ \widetilde{\phi}_t \circ \widetilde{\phi}_\tau \circ (I + v) \circ e^{-\tau A} \circ e^{-tA} \\
&= \ \widetilde{\phi}_t \circ (I + v) \circ e^{-tA} \\
&= \ I + w.
\end{aligned}
$$

But v is the unique, bounded, continuous map such that $\widetilde{\phi}_\tau \circ (I+v) \circ e^{-\tau A} = I + v$. Hence $w = v$.

Finally, with $H = I + v$, we have shown that $\widetilde{\phi}_t \circ H = H \circ e^{tA}$, on $\mathbb{R}^n$ for all $t \in \mathbb{R}$. Taking $f = H^{-1}$ and rearranging gives the desired result:

$$
\widetilde{\phi}_t(x) = f^{-1}(e^{At} f(x)), \quad\quad \text{(B.39)}
$$

for all $x \in \mathbb{R}^n$ and all $t \in \mathbb{R}$. Thus, we have proved Parts (1) and (2) of the theorem hold for the case when A has the special form stated above.

To prove the theorem for the general case, note that there is an invertible matrix P such that

$$
P^{-1}AP = J = \begin{bmatrix} J_1 & 0 \\ 0 & J_2 \end{bmatrix},
$$

where J is a permutation of the Jordan form for A so that all the Jordan blocks with eigenvalues having negative real parts come first. Use P to define a vector field Y on $\overline{\mathcal{O}} \equiv P^{-1}\mathcal{O}$, by

$$
Y(y) = P^{-1}X(Py).
$$

This is just the push-forward of X by the map P^{-1}. Now since $Y(0) = 0$ and $Y'(0) = J$, we can use the first part of the proof on Y. Thus, there is a vector field $\widetilde{Y} : \mathbb{R}^n \to \mathbb{R}^n$, a homeomorphism $F : \mathbb{R}^n \to \mathbb{R}^n$, and an $r > 0$, such that

(1) $\tilde{Y} = Y$ on $B(0, r)$,

(2) The flow $\tilde{\psi} : \mathbb{R} \times \mathbb{R}^n \to \mathbb{R}^n$ for $\tilde{X}$ satisfies

$$\tilde{\psi}_t(y) = F^{-1}(e^{Jt} F(y)), \tag{B.40}$$

for all $y \in \mathbb{R}^n$ and all $t \in \mathbb{R}$.

If we let

$$\tilde{X} = P \circ \tilde{Y} \circ P^{-1} \quad \text{and} \quad f = P \circ F \circ P^{-1},$$

then, by Exercise 1, Section 6.3, and by equation (B.40), the flow $\tilde{\phi}$ for $\tilde{X}$ satisfies

$$
\begin{aligned}
\tilde{\phi}_t(x) &= P\tilde{\psi}_t(P^{-1}x) \\
&= PF^{-1}\left(e^{tJ} F(P^{-1}x)\right) \\
&= PF^{-1}\left(P^{-1} e^{tA} PF(P^{-1}x)\right) \\
&= f^{-1}(e^{tA} f(x)),
\end{aligned}
$$

for all $x \in \mathbb{R}^n$ and all $t \in \mathbb{R}$. It is also clear that $\tilde{X}(x) = X(x)$ for all x in $U \equiv P(B(0, r))$. This proves that Parts (1) and (2) of the theorem hold in the general case.

Finally, we must show topological equivalence of the restricted systems. Suppose first that $\alpha : J \to U$ satisfies $\alpha'(t) = X(\alpha(t))$, for all $t \in J$. We can assume $0 \in J$. Let $x = \alpha(0)$. Then since α lies in U and $\tilde{X} = X$ there, we also have $\alpha'(t) = \tilde{X}(\alpha(t))$, for all $t \in J$. Hence $\alpha(t) = \tilde{\phi}_t(x)$, for all $t \in J$. Hence by equation (B.37), we have

$$f(\alpha(t)) = f(\tilde{\phi}_t(x)) = e^{tA} f(x).$$

Thus, $f \circ \alpha$ is an integral curve of the system $y' = Ay$, restricted to $f(U)$. Conversely, suppose that $\beta : J \to f(U)$ satisfies $\beta'(t) = A\beta(t)$ for every $t \in J$. We can assume $0 \in J$. Let $y = \beta(0)$. Then $\beta(t) = e^{tA} y$ for all $t \in J$, and consequently by equation (B.37), we have

$$f^{-1}(\beta(t)) = f^{-1}(e^{tA} y) = \tilde{\phi}_t(f^{-1}(y)),$$

for all $t \in J$. Hence $f^{-1} \circ \beta$ is an integral curve of the system $x' = \tilde{X}(x)$ restricted to U. Hence it is also an integral curve of the system $x' = X(x)$ restricted to U. This completes the proof of the theorem. $\square$

Appendix C

Linear Algebra

In this appendix we collect together a number of definitions and results from linear algebra that supplement the material in the text and provide background material for readers not familar with these topics.

C.1 Vector Spaces and Direct Sums

We discuss here an important result connected with an $n \times n$ matrix (with real entries), considered as a linear operator $A : \mathbb{R}^n \to \mathbb{R}^n$. The result relies on the concepts of direct sums of subspaces and invariance of subspaces, which we now define.

Definition C.1

(1) Suppose $V_1, \ldots, V_k$ are subspaces of $\mathbb{R}^n$. We say that $\mathbb{R}^n$ is a *direct sum* of these subspaces if each vector $v \in \mathbb{R}^n$ has a *unique* representation as a sum of vectors from $V_1, \ldots, V_k$. That is, each $v \in \mathbb{R}^n$ can be expressed as

$$v = v_1 + \cdots + v_k,$$

with $v_j \in V_j$, for $j = 1, \ldots, k$, and this representation of v is unique in the sense that, if also

$$v = v_1' + \cdots + v_k',$$

with $v_j' \in V_j$, for $j = 1, \ldots, k$, then

$$v_j = v_j', \qquad \text{for } j = 1, \ldots, k.$$

It is not hard to see that this uniqueness criterion is equivalent to the following condition: If

$$0 = v_1 + \cdots + v_k,$$

579

with $v_j \in V_j$, for $j = 1, \ldots, k$, then $v_j = 0$, for all j, i.e., there is a unique representation of 0 as a sum of vectors from the V_j's. Namely as $0 = 0 + \cdots + 0$. To symbolize that $\mathbb{R}^n$ is a direct sum of these subspaces we write

$$\mathbb{R}^n = V_1 \oplus \cdots \oplus V_k.$$

(2) If A is an $n \times n$ matrix and V is a subspace of $\mathbb{R}^n$, then, viewing $A : \mathbb{R}^n \to \mathbb{R}^n$ as a linear transformation, we let

$$AV = \{ Av \,|\, v \in V \},$$

denote the *image*, or *range space* of A. This is always a subspace of $\mathbb{R}^n$. The subspace V is said to be *invariant under* A if

$$AV \subseteq V.$$

The following proposition is very elementary, yet is of basic importance for much of the theory in for linear operators in linear algebra. In particular when applied to the decomposition of $\mathbb{R}^n$ in terms of the generalized eigenspaces of A, the proposition leads to the Jordan canonical form for A.

Proposition C.1 *Suppose that*

$$\mathbb{R}^n = V_1 \oplus \cdots \oplus V_p$$

is a direct sum of subspaces V_j, each of which is invariant under A. For each j suppose $\{v_i^j\}_{j=1,\ldots,n_j}$ is a basis for V_j, and let P be the $n \times n$ matrix formed by using these vectors as columns:

$$P = [v_1^1 \cdots v_{n_1}^1 \cdots v_1^p \cdots v_{n_p}^p].$$

Then $P^{-1}AP$ is a block diagonal matrix:

$$P^{-1}AP = \begin{bmatrix} B_1 & & \\ & \ddots & \\ & & B_p \end{bmatrix}, \tag{C.1}$$

where the blocks B_j are $n_j \times n_j$ matrices.

Proof: Since $AV_j \subseteq V_j$, for each j, it follows that for each basis vector v_i^j, the vector Av_i^j is in V_j and thus can be represented in terms of the basis, say

$$Av_i^j = \sum_{k=1}^{n_j} b_{ki}^j v_k^j. \tag{C.2}$$

The numbers b_{ki}^j form the entries of a matrix: $B_j \equiv \{b_{ki}^j\}$. Now all we have to do is verify that $P^{-1}AP$ is a block diagonal matrix with these blocks. As is customary, we let

$$[v_1^j \cdots v_{n_j}^j]$$

denote the matrix with $v_1^j, \ldots, v_{n_j}^j$ as columns. Then by the way matrix mutiplication works, we have

$$A[v_1^j \cdots v_{n_j}^j] = [Av_1^j \cdots Av_{n_j}^j],$$

and

$$[v_1^j \cdots v_{n_j}^j]B_j = [\sum_{k=1}^{n_j} b_{k1}^j v_k^j \cdots \sum_{i=1}^{n_j} b_{kn_j}^j v_k^j].$$

Hence by (C.2), we get

$$A[v_1^j \cdots v_{n_j}^j] = [v_1^j \cdots v_{n_j}^j]B_j.$$

From this and the definition of the matrix P, it is easy to see that:

$$AP = P \begin{bmatrix} B_1 & & \\ & \ddots & \\ & & B_p \end{bmatrix},$$

and this gives the result of the theorem.

The transpose operation on matrices is a simple but important concept. Thus, if $A = \{a_{ij}\}$ is a matrix, then its *transpose* is the matrix $A^T = \{a_{ij}^T\}$, where $a_{ij}^T = a_{ji}$, for each i and j. It is easy to verify following fundamental relation between the dot product on $\mathbb{R}^n$ and the transpose operation:

$$Av \cdot w = v \cdot A^T w,$$

for all $v, w \in \mathbb{R}^n$. In dealing with eigenvalues of A, since they can be complex numbers, we will need the above identity for the dot product in $\mathbb{C}^n$. Specifically, we can think of the real matrix A as a complex $n \times n$ matrix and get an operator $A : \mathbb{C}^n \rightarrow \mathbb{C}^n$, that operates on vectors $v \in \mathbb{C}^n$ in the natural way: $Av = A(u + iw) = Au + iAw$, where $v = u + iw$ is the representation of v in terms of its real and imaginary parts $u, w \in \mathbb{R}^n$. Then it is not hard to see that $Av \cdot \tilde{v} = v \cdot A^T \tilde{v}$, for all $v, \tilde{v} \in \mathbb{C}^n$.

There are several important types of matrices, whose definitions involve the transpose operation.

Definition C.2 An $n \times n$ matrix A is called *symmetric* if $A^T = A$ and is called *skew symmetric* if $A^T = -A$. An $n \times n$ matrix Q is called an *orthogonal* if $Q^T Q = I$. (Note that this implies that $QQ^T = I$ as well.)

C.2 Bilinear Forms

There are several results concerning symmetric bilinear forms which are indispensable in a number of areas of mathematics and which we have used at various places in the text. A brief discussion of these is given here. We begin with the more general notion of a multilinear map.

Definition C.3 (Multilinear Maps) Suppose V is a vector space and let V^r denote the Cartesian product of V with itself r times:

$$V^r \equiv V \times V \times \cdots \times V.$$

A map $\beta : V^r \rightarrow U$, with range in another vector space U is called *multilinear* if for all $v_1, \ldots, v_r, w \in V$ and all scalars a:

$$\beta(v_1, \ldots, v_i + aw, \ldots, v_r) = \beta(v_1, \ldots, v_i, \ldots, v_r) + a\beta(v_1, \ldots, w, \ldots, v_r),$$

for each $i \in \{1, \ldots, r\}$. Thus, β is a linear map with respect to any one of its arguments, when all the other arguments are held fixed. In the special case when $r = 2$, the map β is called *bilinear*.

When U is the field of scalars (for us, either $\mathbb{R}$ or $\mathbb{C}$), then β is called a *multilinear form*.

A bilinear map β is called *symmetric* if $\beta(v, w) = \beta(w, v)$ for all $v, w \in V$.

There are many important examples of multilinear maps. Here are a few that are important for the material in the text.

Example C.1

(1) For a function $f : \mathcal{O} \subseteq \mathbb{R}^n \to \mathbb{R}^k$, with partial derivatives existing to the rth order at $x \in \mathcal{O}$, the rth *derivative of f at x* is a multilinear map $f^{(r)}(x) : (\mathbb{R}^n)^r \to \mathbb{R}^k$. See the definition in the section on Taylor's theorem in Appendix A.

(2) In Chapter 7, we defined a map $N : (\mathbb{R}^n)^{n-1} \to \mathbb{R}^n$, called the *normal operator*, which generalizes the cross product operator. Namely, $N(v_1, \ldots, v_{n-1})$ is a vector in $\mathbb{R}^n$ that is orthogonal to each v_i, for $i = 1, \ldots, n-1$. The map N is a good example of a multilinear map.

(3) The dot product (or inner product) on $\mathbb{R}^n$ is a bilinear form

$$\beta(v, w) = v \cdot w = \sum_{j=1}^{n} v_j w_j,$$

which is clearly a symmetric bilinear form.

(4) If $A = \{a_{ij}\}$ is any $n \times n$ matrix, then

$$\beta_A(v, w) \equiv Av \cdot w = \sum_{i,j=1}^{n} a_{ij} v_i w_j,$$

is the standard bilinear form associated with A. If A is a symmetric matrix, then β_A is a symmetric bilinear form. When A is the identity matrix $A = I$, then β_I is the dot product on $\mathbb{R}^n$ in Part (3).

(5) The inner product on $\mathbb{C}^n$ is the bilinear form

$$\beta(v, w) = v \cdot w \equiv \sum_{j=1}^{n} v_j \overline{w}_j,$$

where $\overline{w}_j$ denotes the complex conjugate of the complex number w_j. For convenience we always consider $\mathbb{R}^n \subset \mathbb{C}^n$ and then the dot product of vectors in $\mathbb{C}^n$ reduces to the usual real dot product when restricted to $\mathbb{R}^n$.

Part (4) of the above example exhibits a relationship between $n \times n$ matrices A and corresponding bilinear forms β_A on $\mathbb{R}^n$, which allows us to identify any bilinear form β with a matrix A that represents it with respect to the standard unit vector basis $\{\varepsilon_1, \ldots, \varepsilon_n\}$ for $\mathbb{R}^n$. That is, if β is a given bilinear form, then $\beta = \beta_A$, where $A = \{a_{ij}\}$ is the $n \times n$ matrix with entries

$$a_{ij} = \beta(\varepsilon_j, \varepsilon_i),$$

for $i, j = 1, \ldots, n$. This representation allows us to identify properties of a bilinear form with corresponding properties of the matrix that represents it.

Definition C.4 (Positive/Negative Definiteness) Suppose $\beta : V \times V \to \mathbb{R}$ is a symmetric, bilinear form. Then β is called

(1) *nondegenrate* if $v \in V$ and $\beta(v, w) = 0$, for every $w \in V$, imply that $v = 0$.

(2) *positive definite* if $\beta(v, v) > 0$ for all $v \neq 0$ in V.

(3) *negative definite* if $\beta(v, v) < 0$ for all $v \neq 0$ in V.

(4) *positive semi-definite* if $\beta(v, v) \geq 0$ for all $v \in V$.

(5) *negative semi-definite* if $\beta(v, v) \leq 0$ for all $v \in V$.

A symmetric matrix A is called *positive definite* if the corresponding bilinear form $\beta_A(v, w) = Av \cdot w$, is positive definite, i.e., if $Av \cdot v > 0$ for all $v \neq 0$. Similarly one calls A *negative definite, positve semi-definite*, or *negative semi-definite*, if β_A is. Note also that β_A is nondegenerate if and only if A is invertible (exercise). Also note that if β is positive or negative definite, then it is nondegenerate.

C.3   Inner Product Spaces

The standard dot product (or inner product) $v \cdot w$ of vectors $v, w \in \mathbb{R}^n$ determines the fundamental geometry of $\mathbb{R}^n$ and is essential in defining many additional concepts connected with "orthogonality" in $\mathbb{R}^n$. The abstract notion of an inner product on a vector space is defined as follows

Definition C.5 (Inner Product Spaces) Suppose V is a vector space over the reals. An *inner product* on V is a symmetric, positive definite bilinear form $\beta : V \times V \to \mathbb{R}$, on V. The pair (V, β) is called an *inner product space*. Relative to the inner product β we can defined the following concepts.

(1) Vectors $v, w \in V$ are called *orthogonal* (or *perpendicular*) if $\beta(v, w) = 0$. If U is any subset of V, then the set

$$U^\perp = \{\, v \in V \mid \beta(v, u) = 0, \ \forall u \in U \,\},$$

is the set of vectors which are perpendicular to U.

(2) The *norm* associated with the inner product is $|v| = \beta(v, v)^{1/2}$. The number $|v|$ is called the *norm*, or *length*, of the vector v.

(3) If V is finite dimensional, then an *orthonormal basis* for V is a basis $\{e_1, \ldots, e_n\}$, consisting of mutually orthogonal vectors of unit length, i.e., $\beta(e_i, e_j) = \delta_{ij}$, for every i, j.

(4) A linear map $T : V \to V$ is called an *isometry* if $\beta(Tv, Tv) = \beta(v, v)$, for every $v \in V$.

Comment: Throughout the rest of this section β will be a given inner product on V and for notational convenience we will use a dot for the inner product:

$$v \cdot w \equiv \beta(v, w).$$

Also we assume V is finite dimensional.

A fundamental construction in inner product spaces is the following.

Proposition C.2 (Gram-Schmidt Orthogonalization Process)
Suppose $v_1, \ldots v_k$ are linearly independent vectors in an inner product space. Then there exist orthonormal vectors $e_1, \ldots, e_k$, with the same span, i.e.,

$$e_i \cdot e_j = \delta_{ij} \qquad \text{(for } i, j = 1, \ldots k),$$

and

$$\text{span}\{e_1, \ldots, e_k\} = \text{span}\{v_1, \ldots, v_k\}.$$

Proof: The Gram-Schmidt process is simply described as follows. First normalize: $e_1 \equiv v_1/|v_1|$. Then orthogonalize:

$$\tilde{e}_2 \equiv v_2 - (v_2 \cdot e_1)e_1.$$

Normalize again: $e_2 \equiv \tilde{e}_2/|\tilde{e}_2|$. Then orthogonalize

$$\tilde{e}_3 \equiv v_3 - (v_3 \cdot e_1)e_1 - (v_3 \cdot e_2)e_2,$$

and continue in this fashion to construct all the e_i's. At the ith step we have

$$\tilde{e}_i = v_i - \sum_{j=1}^{i-1} (v_i \cdot e_j)\, e_j.$$

and this is easily seen to be orthogonal to each $e_p, p = 1, \ldots, i-1$. By induction we can assume that

$$\text{span}\{e_1, \ldots, e_{i-1}\} = \text{span}\{v_1, \ldots, v_{i-1}\}.$$

From this we get that $|\tilde{e}_i| \neq 0$. Otherwise we would have

$$v_i = \sum_{j=1}^{i-1} (v_i \cdot e_j)\, e_j \in \text{span}\{e_1, \ldots, e_{i-1}\} = \text{span}\{v_1, \ldots, v_{i-1}\},$$

which contradicts the assumption that $v_1, \ldots, v_{i-1}, v_i$ are linearly independent. Thus, at any stage in the process, we can always normalize: $e_i \equiv \tilde{e}_i/|\tilde{e}_i|$. It is not hard to now show that $\text{span}\{e_1, \ldots, e_i\} = \text{span}\{v_1, \ldots, v_i\}$.

$\square$

We can always use the Gram-Schmidt process to construct an orthonormal basis for V and, as the following proposition shows, we can also use it to demonstrate that each subspace of V has an orthogonal complement.

Proposition C.3 (Orthogonal Complements) *If U is any subspace of a finite dimensional inner product space, then $U^\perp$ is also a subspace and*

$$V = U \oplus U^\perp.$$

Proof: It is not hard to show that $U^\perp$ is a subspace of $\mathbb{R}^n$. Furthermore, if U itself is a subspace, then $U^\perp$, called the *orthogonal complement* of U, allows us to express V as a direct sum $V = U \oplus U^\perp$. To see this, note that we can always choose an orthonormal basis $\{e_1, \ldots, e_p\}$ for U. Then for any $v \in V$, we can write

$$v = \sum_{j=1}^{p} (v \cdot e_j)e_j + \left(v - \sum_{j=1}^{p} (v \cdot e_j)e_j\right).$$

The first summand is clearly in U, and it is easy to check that the second summand is orthogonal to each e_k, and hence to each $u \in U$. This gives that $U + U^\perp = V$. To see that the above expression for v is unique, suppose that $v = u + w$ with $u \in U$ and $w \in U^\perp$. The $u = \sum_{j=1}^{p} c_j e_j$, for scalars $c_1, \ldots, c_p$. However, by the orthogonality property of the basis for U, it is easy to see that $v \cdot e_k = (u + w) \cdot e_k = u \cdot e_k = c_k$. From this it follows that $u = \sum_{j=1}^{p} (v \cdot e_j) e_j$, and hence also that $w = v - \sum_{j=1}^{p} (v \cdot e_j) e_j$. $\square$

The theorem below shows that in a finite dimensional inner product space, any symmetric bilinear form can be represented by a diagonal matrix with respect to some choice of an orthonormal basis. This is viewed by some as the Principal Axes Theorem, but we reserve this name for the theorem presented in the next section, which is actually the matrix version of the theorem here (together with some extra details).

Theorem C.1 *Suppose* $\alpha : V \times V :\to \mathbb{R}$ *is a symmetric, bilinear form on a finite dimensional inner product space* (V, β). *Then there is an orthonormal basis* $\{e_1, e_2, \ldots, e_n\}$ *for* V *and real numbers* $\lambda_1, \lambda_2, \ldots, \lambda_n$ *such that*

$$\alpha(e_i, e_j) = \lambda_i \delta_{ij}, \tag{C.3}$$

for all $i, j = 1, \ldots, n$.

Proof: If $\alpha(v, v) = 0$, for all $v \in V$, then it follows that

$$0 = \alpha(v + w, v + w) = \alpha(v, v) + 2\alpha(v, w) + \alpha(w, w) = 2\alpha(v, w),$$

for all $v, w \in V$. Hence α is identically zero and so any orthonormal basis $\{e_1, \ldots, e_n\}$ will do, if we choose $\lambda_i = 0$, for $i = 1, \ldots, n$. Thus, we are done in this case.

On the other hand, suppose there is a $v_1 \in V$, such that $\alpha(v_1, v_1) \neq 0$. Let $e_1 = v_1/|v_1|$ and $\lambda_1 \equiv \alpha(e_1, e_1)$. Letting $E_1 = \text{span}\{e_1\}$, we get $V = E_1 \oplus E_1^\perp$.

Now restrict α to $E_1^\perp \times E_1^\perp$. Then we can repeat the argument above but now with $\alpha|_{E_1^\perp \times E_1^\perp}$. That is, if $\alpha(v, v) = 0$ for every $v \in E_1^\perp$, then $\alpha|_{E_1^\perp \times E_1^\perp}$ is identically zero and so we can choose any orthonormal basis $\{e_2, \ldots, e_n\}$ for $E_1^\perp$, take $\lambda_i = 0$, for $i = 2, \ldots, n$, and be done with the proof. On the other hand, if there exists a $v_2 \in E_1^\perp$ such that $\alpha(v_2, v_2) \neq 0$, then we can let $e_2 = v_2/|v_2|$ and $\lambda_2 = \alpha(e_2, e_2)$, to get $\alpha(e_i, e_j) = \lambda_i \delta_{ij}$, for $i, j = 1, 2$. Then let $E_2 = \text{span}\{e_1, e_2\}$, so that $V = E_2 \oplus E_2^\perp$.

Restricting α to $E_2^\perp \times E_2^\perp$, we continue on with the same process, which after a finite number of steps terminates: either α resticted to $E_k^\perp \times E_k^\perp$ is identically zero for some $k < n$ or after n steps, we get $E_n = V$. $\square$

C.4   The Principal Axes Theorem

The Principal Axes Theorem is a basic result that we have used in several places in the text. We phrase this theorem in the following way, which is slightly different than some statements of it in the literature, but is more general and convenient. *Note:* The dot product $v \cdot w$ is the usual inner product on $\mathbb{R}^n$ and orthogonality is with respect to this inner product unless otherwise specified.

Theorem C.2 (Principal Axes Theorem) *Suppose H and G are symmetric $n \times n$ matrices, with G positive definite. Define a real-valued function $f : \mathbb{R}^n \setminus \{0\} \to \mathbb{R}$ by*

$$f(x) = \frac{Hx \cdot x}{Gx \cdot x},$$

for $x \neq 0$ in $\mathbb{R}^n$. Then

(1) *All the eigenvalues of $G^{-1}H$ are real.*

(2) *Eigenvectors of $G^{-1}H$ corresponding to distinct eigenvalues are orthogonal with respect to the inner product β_G, i.e., if $G^{-1}Hv = \lambda v$ and $G^{-1}Hw = \mu w$, with $\lambda \neq \mu$, then $Gv \cdot w = 0$.*

(3) *c is a critical point of f if and only if $f(c)$ is an eigenvalue of $G^{-1}H$ with c as a corresponding eigenvector.*

(4) *f has a maximum and a minimum value on $\mathbb{R}^n \setminus \{0\}$, and these values are the largest and smallest of the eigenvalues of $G^{-1}H$.*

(5) *$\mathbb{R}^n$ has a basis $\{e_1, \ldots, e_n\}$ of eigenvectors of $G^{-1}H$:*

$$G^{-1}He_i = \lambda_i e_i \qquad (\text{for } i = 1, \ldots, n),$$

that are orthonormal with respect to the inner product β_G:

$$Ge_i \cdot e_j = \delta_{ij} \qquad (\text{for } i, j = 1, \ldots, n).$$

Proof:

(1) Suppose λ is an eigenvalue of $G^{-1}H$ and let $v \in \mathbb{C}^n$ be a corresponding eigenvector: $G^{-1}Hv = \lambda v$ and $v \neq 0$. Then $Hv = \lambda Gv$, and so

$$\lambda(Gv \cdot v) = (\lambda Gv) \cdot v = Hv \cdot v = v \cdot Hv = v \cdot (\lambda Gv) = \overline{\lambda}(Gv \cdot v).$$

However, since $v \neq 0$ and G is positive definite, we have $Gv \cdot v \neq 0$. Thus, the above equation forces $\lambda = \overline{\lambda}$. Hence λ is real.

(2) If $\lambda \neq \mu$ are eigenvalues of $G^{-1}H$ and v, w are corresponding eigenvectors, then $Hv = \lambda Gv$ and $Hw = \mu Gw$. From (1), λ and μ are real and $v, w \in \mathbb{R}^n$. As in the proof of (1) we get

$$\lambda(Gv \cdot w) = Hv \cdot w = v \cdot Hw = v \cdot (\mu Gw) = \mu(v \cdot Gw) = \mu(Gv \cdot w).$$

Since $\lambda \neq \mu$, this forces $Gv \cdot w = 0$.

(3) From the definition of f we get the identity:

$$Hx \cdot x = f(x)\, Gx \cdot x,$$

for all $x \in \mathbb{R}^n$. Otherwise said

$$\sum_{i,j=1}^{n} h_{ij}x_i x_j = f(x) \sum_{i,j=1}^{n} g_{ij}x_i x_j.$$

Taking the partial derivative of both sides of this equation with respect to x_k gives

$$2\sum_{j=1}^{n} h_{kj}x_j = \frac{\partial f}{\partial x_k}(x)\,(Gx \cdot x) + 2f(x)\sum_{j=1}^{n} g_{kj}x_j.$$

Writing this in vector form gives

$$2Hx = (Gx \cdot x)\nabla f(x) + 2f(x)Gx,$$

or

$$G^{-1}Hx = f(x)x + \tfrac{1}{2}(Gx \cdot x)G^{-1}\nabla f(x).$$

Hence if c is a critical point of f, i.e., $\nabla f(c) = 0$, the above gives $G^{-1}Hc = f(c)c$. Conversely if $G^{-1}Hc = f(c)c$, then the above gives that $\tfrac{1}{2}(Gc \cdot c)G^{-1}\nabla f(c) = 0$ and hence $\nabla f(c) = 0$.

(4) It is easy to see from the definition of f that $f(rx) = f(x)$, for every real number $r \neq 0$ and every $x \neq 0$ in $\mathbb{R}^n$. Hence the values of f on $\mathbb{R}^n \setminus \{0\}$ are the same as it values on the unit sphere $S = \{\, x \in \mathbb{R}^n \,|\, |x| = 1 \,\}$. But since S is compact, there are points $c_1, c_n \in S$, where f assumes it minimum and maximum values. Thus, $\lambda_1 \equiv f(c_1) \leq \lambda_n = f(c_n)$, are the minimum and maximum values of f on all of $\mathbb{R}^n \setminus \{0\}$:

$$\lambda_1 \leq f(x) \leq \lambda_n,$$

for all $x \in \mathbb{R}^n - \{0\}$. By Part (3), λ_1 and λ_n are eigenvalues of $G^{-1}H$. Further, if λ is any eigenvalue of $G^{-1}H$, say $G^{-1}Hv = \lambda v$, then $Hv = \lambda Gv$ and so

$$\lambda = \frac{Hv \cdot v}{Gv \cdot v} = f(v).$$

Hence $\lambda_1 \leq \lambda \leq \lambda_n$. So λ_1 and λ_n are the minimum and maximum eigenvalues of $G^{-1}H$.

(5) If we let $\beta_H(v, w) = H \cdot w$ and $\beta_G(v, w) = Gv \cdot w$ be the standard bilinear forms associated with matrices H and G, then β_H is symmetric and β_G is an inner product on $\mathbb{R}^n$. Thus, we can apply Theorem – to β_H, with $(\mathbb{R}^n, \beta_G)$ as the inner product space. Hence there is a basis $\{e_1, \ldots, e_n\}$ for $\mathbb{R}^n$ that is orthonormal with respect to β_G, i.e.,

$$Ge_i \cdot e_j = \delta_{ij}, \tag{C.4}$$

for all i, j, and real numbers $\lambda_1, \ldots, \lambda_n$ such that

$$He_i \cdot e_j = \lambda_i \delta_{ij}, \tag{C.5}$$

for all i, j. Now let $P = [e_1 \cdots e_n]$ be the $n \times n$ matrix with $e_1, \ldots, e_n$ as its columns and $D = \operatorname{diag}(\lambda_1, \ldots, \lambda_n)$, be the diagonal matrix with $\lambda_1, \ldots, \lambda_n$ on the diagonal. Then P is invertible and equations (–)-(–) say that

$$\begin{aligned} P^T G P &= I \\ P^T H P &= D \end{aligned}$$

From the first equation we get $G^{-1} = PP^T$. Using this in the second equation gives, with some rearrangement,

$$G^{-1}HP = G^{-1}P^{-T}D = PP^T P^{-T}D = PD.$$

But this says that $G^{-1}He_i = \lambda_i e_i$, for $i = 1, \ldots, n$. $\square$

Corollary C.1 (Spectral Theorem) *Suppose A is an $n \times n$ symmetric matrix (with real entries). Then all the eigenvalues $\lambda_1, \ldots, \lambda_n$ of A are real, and and eigenvectors corresponding to distinct eigenvalues are orthogonal. Furthermore, $\mathbb{R}^n$ has an orthonormal basis $\{e_1, \ldots, e_n\}$ consisting of eigenvectors of A:*

$$Ae_i = \lambda_i e_i,$$

$i = 1, \ldots, n$. Thus, A is diagonalizable by an orthogonal transformation:

$$P^{-1}AP = \begin{bmatrix} \lambda_1 & & \\ & \ddots & \\ & & \lambda_n \end{bmatrix},$$

where $P = [e_1 \cdots e_n]$ is the orthogonal matrix with $e_1, \ldots, e_n$ as its columns.

In addition, if the labelling is done so that $\lambda_1 \leq \lambda_2 \leq \cdots \leq \lambda_n$, then λ_1 and λ_n are the absolute maximum and minimum values of the function

$$f(x) = \frac{Ax \cdot x}{x \cdot x},$$

on $\mathbb{R}^n \setminus \{0\}$, and $e_1, e_2, \ldots, e_n$ are critical points of f, such that $f(e_i) = \lambda_i, i = 1, \ldots, n$.

Note: In the Principal Axes Theorem and its corollary, we can assume the vectors $e_1, \ldots, e_n$ in the basis form a "right-handed system", i.e., are such that the matrix $P = [e_1 \cdots e_n]$, formed with the e_i's as columns, has positive determinant: $\det(P) > 0$. If this is not the case then we can always relabel the e_i's (and the λ_i's too) so that this is the case. Then the *principal axes* are the lines through the origin determined by the vectors $e_1, \ldots, e_n$. In the case of the corollary, these axes are mutually orthogonal in the Euclidean sense. In general they are only orthogonal with respect to the inner product β_G.

There are numerous applications of the Principal Axes Theorem, two of which in this text deal with moments of inertia in Chapter 9 and principal curvatures in Chapter 10.

C.5 Generalized Eigenspaces

In this section we prove the Jordan Canonical Form Theorem. We prove what is known as the *real form* of this theorem, which takes a little more effort than the case where one works over the complex field $\mathbb{C}$. In either case the proof is long, but worthy of study since it involves many important tools and ideas from algebra and linear algebra. Most of this material directly extends to operators on finite dimensional vectors spaces of other fields besides $\mathbb{R}$ or $\mathbb{C}$.

Definition C.6 (Minimal and Characteristic Polynomials) Suppose A is an $n \times n$ matrix.

(1) A *monic* polynomial is one whose coefficient of its highest power is 1. The *minimal polynomial* for A is the monic polynomial:

$$p(x) = a_0 + a_1 x + \cdots + a_{k-1}x^{k-1} + x^k,$$

of least degree such that:

$$p(A) = a_0 I + a_1 A + \cdots + a_{k-1}A^{k-1} + A^k = 0.$$

Note that polynomials $f(x)$ for which $f(A) = 0$ always exist. To see this, observe that the set $\mathcal{M}_n$ of all $n \times n$ matrices is a finite dimensional vector space with dimension n^2. Because of this, the $n^2 + 1$ vectors: $I, A, \ldots, A^{n^2}$ cannot be linearly independent. Thus, there is a non-trivial linear combination of these that gives the zero matrix. From this we get a polynomial f such that $f(A) = 0$. *Note:* If f is a polynomial such that $f(A) = 0$, then the minimal polynomial p for A divides f. To verify this note that by the *Euclidean division algorithm* there are polynomials q, r such that $f(x) = p(x)q(x) + r(x)$, and either $r = 0$ or the degree of r is strictly less than p. This forces $r = 0$, because otherwise we would have $r(A) = f(A) - p(A)q(A) = 0$ and so p would not be the minimal polynomial.

(2) The *characteristic polynomial* is the polynomial p_A of degree n defined by

$$p_A(x) = \det(A - xI).$$

This polynomial determines the eigenvalues of A and has many important uses, however it is the minimal polynomial that plays the central role in the decomposition theorem below.

To relate the characteristic and minimal polynomials, we must appeal to the *Cayley-Hamilton theorem* which says that $p_A(A) = 0$. Thus, from the above remarks it follows that $p(x)$ divides $p_A(x)$. Furthermore, it can be shown that the characteristic and minimal poynomials have the same roots. Consequently, their respective factorizations into linear and irreducible quadratic factors are the same except for multiplicities. Thus, if

$$p(x) = (x - \lambda_1)^{k'_1} \cdots (x - \lambda_r)^{k'_r} q_1(x)^{m'_1} \cdots q_s(x)^{m'_s}, \qquad (C.6)$$

then

$$p_A(x) = (-1)^n (x - \lambda_1)^{k_1} \cdots (x - \lambda_r)^{k_r} q_1(x)^{m_1} \cdots q_s(x)^{m_s}, \qquad (C.7)$$

where $k'_1 \leq k_1, \ldots, k'_r \leq k_r, m'_1 \leq m_1, \ldots, m'_s \leq m_s$. In the above, the irreducible quadratic factors q_j have the form

$$q_j(x) = (x - a_j)^2 + b_j^2,$$

and thus the roots of $p(x)$ and $p_A(x)$ (which are the eigenvalues of A) are

$$\lambda_1, \ldots, \lambda_r, a_1 \pm b_1 i, \ldots, a_s \pm b_s i.$$

It is implied that the numbers listed above are distinct and, since the characteristic polynomial $p_A(x)$ has degree n, it follows that

$$n = k_1 + \cdots + k_r + 2m_1 + \cdots + 2m_s.$$

We will use the notation and factorizations shown in equations (C.6)-(C.7), for $p(x)$ and $p_A(x)$ throughout the rest of this section.

Definition C.7 (Generalized Eigenspaces) Suppose that a complete list of the distinct eigenvalues of A is $\lambda_1, \ldots, \lambda_r, a_1 \pm b_1 i, \ldots, a_s \pm b_s i$. The *generalized eigenspaces* for A are the subspaces:

$$GE_{\lambda_j} = \{ v \in \mathbb{R}^n \,|\, (A - \lambda_j I)^k v = 0 \,, \text{for some } k \}$$

and

$$GE_{a_j \pm b_j i} = \{ v \in \mathbb{R}^n \,|\, [(A - a_j I)^2 + b_j^2 I]^m v = 0 \,, \text{for some } m \}$$

It is easy to show that the generalized eigenspaces are actually subspaces and that they are invariant under A since A commutes with $A - \lambda_j I$ and $(A - a_j I)^2 + b_j^2 I$ (exercise). It is also clear that $E_{\lambda_j} \subseteq GE_{\lambda_j}$ for the real eigenvalues λ_j. As we have seen, for some matrices, the generalized eigenspaces coincide with the eigenspaces. In general, however, this will not be the case.

The following theorem is the fundamental tool for obtaining the Jordan canonical form for the matrix A:

Theorem C.3 (Primary Decomposition Theorem) *Suppose that:*

$$\lambda_1, \ldots, \lambda_r, a_1 \pm b_1 i, \ldots, a_s \pm b_s i,$$

is a complete list of the distinct eigenvalues of A. Then $\mathbb{R}^n$ is the direct sum of the generalized eigenspaces of A:

$$\mathbb{R}^n = GE_{\lambda_1} \oplus \cdots \oplus GE_{\lambda_r} \oplus GE_{a_1 \pm b_1 i} \oplus \cdots \oplus GE_{a_s \pm b_s i}, \qquad (C.8)$$

and each generalized eigenspace is invariant under A. Furthermore, the minimal polynomials for A restricted to GE_{λ_j} and $GE_{a_j \pm b_j i}$ are

$$F_j(x) \equiv (x - \lambda_j)^{k'_j}, \qquad G_j(x) \equiv ((x - a_j)^2 + b_j^2)^{m'_j}, \qquad (C.9)$$

respectively. Here k'_j and m'_j are the multiplicities of λ_j and $a_j + b_j i$ as roots of the minimal polynomial $p(x)$. On the other hand if $k_j \geq k'_j$ and $m_j \geq m'_j$ denote the multiplicities of these eigenvalues as roots of the characteristic polynomial $p_A(x)$, then

$$\dim(GE_{\lambda_j}) = k_j$$
$$\dim(GE_{a_j \pm b_j i}) = 2m_j.$$

Proof: The proof uses some interesting algebraic properties of the ring $\mathbb{R}[x]$ of polynomials. Most of these have been abstracted to the more general setting of ring theory in modern algebra.

Consider the minimal polynomial for A in factored form

$$p(x) = (x - \lambda_1)^{k'_1} \cdots (x - \lambda_r)^{k'_r} q_1(x)^{m'_1} \cdots q_s(x)^{m'_s},$$

with irreducible quadratic factors $q_j(x) = (x - a_j)^2 + b_j^2$. Define some associated polynomials $g_1, \ldots, g_r, h_1, \ldots, h_s$, by

$$g_j(x) = \frac{p(x)}{(x - \lambda_j)^{k'_j}}$$

$$h_j(x) = \frac{p(x)}{q_j(x)^{m'_j}}.$$

Then it should be evident that $g_1, \ldots, g_r, h_1, \ldots, h_s$ are *relatively prime*. This means that no polynomial of positive degree divides each of these polynomials. Now consider the following set of polynomials

$$\mathcal{I} = \{\, \alpha_1 g_1 + \cdots + \alpha_r g_r + \beta_1 h_1 + \cdots + \beta_s h_s \,|\, \alpha_j, \beta_j \in \mathbb{R}[x] \,\},$$

i.e., $\mathcal{I}$ consists of all polynomials of the form $f = \alpha_1 g_1 + \cdots + \alpha_r g_r + \beta_1 h_1 + \cdots + \beta_s h_s$, where the α_j's and β_j are *any* polynomials. This is an *ideal* of $\mathbb{R}[x]$, which means it is a subset with the properties: (1) If $f, g \in \mathcal{I}$, then $f + g \in \mathcal{I}$, and (2) If $f \in \mathcal{I}$ and γ is any polynomial then $\gamma f \in \mathcal{I}$. We would like to show that $\mathcal{I}$ has an alternative expression. For this let f_0 be a monic polynomial in $\mathcal{I}$ with least degree. Then for any $f \in \mathcal{I}$, we can use the Euclidean division algorithm to write $f = f_0 q + r$, where either $r = 0$ or r has degree less than f_0. But since $r = q f_0 - f \in \mathcal{I}$, it must be the case that $r = 0$. This proves that each element f in $\mathcal{I}$ is divisible by f_0. Otherwise said

$$\mathcal{I} = \{ \gamma f_0 \mid \gamma \in \mathbb{R}[x] \}.$$

Now each g_j and each h_j is in $\mathcal{I}$, so they are divisible by f_0. Hence f_0 is a common factor of $g_1, \ldots, g_r, h_1, \ldots, h_s$. But since these are relatively prime, f_0 must be a constant, and indeed $f_0 = 1$, since it was assumed to be monic. We have thus shown that $1 \in \mathcal{I}$, and so by the original definition of $\mathcal{I}$, there exist polynomials $\alpha_1, \ldots, \alpha_r, \beta_1, \ldots, \beta_s$, such that

$$1 = \alpha_1(x)g_1(x) + \cdots + \alpha_r(x)g_r(x) + \beta_1(x)h_1(x) + \cdots + \beta_s(x)h_s(x), \quad \text{(C.10)}$$

for all x. If we replace x in this identity by the matrix A, we get the following matrix identity

$$I = \alpha_1(A)g_1(A) + \cdots + \alpha_r(A)g_r(A) + \beta_1(A)h_1(A) + \cdots + \beta_s(A)h_s(A). \quad \text{(C.11)}$$

This is the fundamental identity that makes the decomposition theorem work. Indeed, if $v \in \mathbb{R}^n$, the identity gives

$$\begin{aligned}
v &= \alpha_1(A)g_1(A)v + \cdots + \alpha_r(A)g_r(A)v + \beta_1(A)h_1(A)v + \cdots + \beta_s(A)h_s(A)v \\
&= v_1 + \cdots + v_r + w_1 + \cdots + w_s,
\end{aligned}$$

where $v_j = \alpha_j(A)g_j(A)v$ and $w_j = \beta_j(A)h_j(A)v$. Then $v_j \in GE_{\lambda_j}$, since

$$(A - \lambda_j I)^{m'_j} v_j = (A - \lambda_j I)^{m'_j} \alpha_j(A)g_j(A)v = \alpha_j(A)p(A)v = 0.$$

Similarly $w_j \in GE_{a_j \pm b_j i}$. This shows that

$$\mathbb{R}^n = GE_{\lambda_1} + \cdots + GE_{\lambda_r} + GE_{a_1 \pm b_1 i} + \cdots + GE_{a_s \pm b_s i}$$

To prove that this sum is direct, we introduce the following subspaces

$$\begin{aligned}
V_j &= \{ v \in \mathbb{R}^n \mid (A - \lambda_j I)^{k'_j} v = 0 \} \\
W_j &= \{ v \in \mathbb{R}^n \mid q_j(A)^{m'_j} v = 0 \}.
\end{aligned}$$

Then clearly $V_j \subseteq GE_{\lambda_j}$ and $W_j \subseteq GE_{a_j \pm b_j i}$. We want to argue that equality holds in both cases. So suppose $v \neq 0$ is in GE_{λ_j}. Then there is a k such that $(A - \lambda_j I)^k v = 0$. Assume that k is the least positive integer such that this holds. If $k \leq k'_j$, then clearly

$$(A - \lambda_j I)^{k'_j} v = (A - \lambda_j I)^{k'_j - k} (A - \lambda_j I)^k v = 0,$$

and so $v \in V_j$. On the other hand suppose $k > k'_j$. Now by the minimality of k, it follows that $w = (A - \lambda_j I)^{k-1}$ is not zero. However, it is easy to see that

$$w = \alpha_1(A)g_1(A)w + \cdots + \alpha_r(A)g_r(A)w + \beta_1(A)h_1(A)w + \cdots + \beta_s(A)h_s(A)w = 0,$$

since $p(A)$ will be a factor in each of the above summands. This contradiction shows that we must have $v \in V_j$. So $V_j = GE_{\lambda_j}$. Similarly $W_j = GE_{a_j \pm b_j i}$.

With this established, we can easily prove the directness of the sum. Thus, if

$$0 = v_1 + \cdots + v_r + w_1 + \cdots + w_s,$$

with $v_j \in GE_{\lambda_j}$ and $w_j \in GE_{a_j \pm b_j i}$, then applying $g_j(A)$ to the above equation gives

$$0 = g_j(A)[v_1 + \cdots + v_r + w_1 + \cdots + w_s] = g_j(A)v_j.$$

However, $g_j(x)$ and $(x - \lambda_j)^{k'_j}$ are relatively prime polynomials. So (as above) there are polynomials α, β, such that

$$1 = \alpha(x)g_j(x) + \beta(x)(x - \lambda_j)^{k'_j},$$

for all x. Replacing x by A gives the matrix identity

$$I = \alpha(A)g_j(A) + \beta(A)(A - \lambda_j I)^{k'_j}.$$

Hence
$$v_j = \alpha(A)g_j(A)v_j + \beta(A)(A - \lambda_j I)^{k'_j} v_j = 0.$$

Similarly each $w_j = 0$. This proves that the reperesentation of $\mathbb{R}^n$ as a sum of the generalized eigenspaces is a direct sum.

We next prove that the minimal polynomials for A restricted to each of the summands are given by the F_j's and G_j's in equation (C.9). For this let $\widetilde{F}_j$ and $\widetilde{G}_j$ be the minimal polynomials for A restricted to GE_{λ_j} and $GE_{a_j \pm b_j i}$ respectively. Then since these subspaces are equal to V_j and W_j, it is clear that $F_j(A)v = 0$, for every $v \in V_j$ and $G_j(A)w = 0$, for every $w \in W_j$. Hence $\tilde{F}_j$ divides F_j and $\tilde{G}_j$ divides G_j. Thus,

$$\tilde{F}_j(x) = (x - \lambda_j)^{\tilde{k}_j}, \quad \tilde{G}_j(x) = ((x - a_j)^2 + b_j^2)^{\tilde{m}_j},$$

for integers $\tilde{k}_j \leq k'_j$, $\tilde{m}_j \leq m'_j$. On the other hand, let $\tilde{p} = \tilde{F}_1 \cdots \tilde{F}_r \tilde{G}_1 \cdots \tilde{G}_s$. Then because of the direct sum decomposition, it is easy to see that $\tilde{p}(A)v = 0$, for every $v \in \mathbb{R}^n$. Hence p divides $\tilde{p}$, and so $k'_j \leq \tilde{k}_j$, $m'_j \leq \tilde{m}_j$. Thus, equality must hold.

We finally prove the assertions about the dimensions of the generalized eigenspaces. Thus, let $\dim(GE_{\lambda_j}) = \ell_j$ and $\dim(GE_{a_j \pm b_j i}) = n_j$. We need to show that $\ell_j = k_j$ and $n_j = 2m_j$. Because of the direct sum decomposition of $\mathbb{R}^n$ into the generalized eigenspaces, Proposition C.1 and its proof say that we can choose *any* bases for these subspaces, say $\{v_i^j\}_{i=1\cdots\ell_j}$ and $\{w_i^j\}_{i=1\cdots n_j}$ and get

$$P^{-1}AP = \begin{bmatrix} B_1 & & & & & \\ & \ddots & & & & \\ & & B_r & & & \\ & & & C_1 & & \\ & & & & \ddots & \\ & & & & & C_s \end{bmatrix}. \tag{C.12}$$

Here $P = [v_1^1, \ldots, v_{\ell_1}^1, \ldots, w_1^s, \ldots, w_{n_s}^s]$ is the matrix formed with all the basis elements as columns and the matrix on the left is block diagonal with the B_j's and C_j's on the diagonal. These latter matrices have dimensions $\ell_j \times \ell_j$ and $n_j \times n_j$ respectively. We claim that the characteristic polymomial for B_j is $p_{B_j}(x) = (-1)^{\ell_j}(x - \lambda_j)^{\ell_j}$. To see this let $P_j = [v_1^j, \ldots, v_{\ell_j}^j]$ be the $n \times \ell_j$ matrix formed from the chosen basis for GE_{λ_j}. Then since $AP_j = P_j B_j$, and $(A - \lambda_j I)^{k_j}$ is identically zero on GE_{λ_j}, we get that

$$0 = (A - \lambda_j I)^{k'_j} P_j u = P_j (B_j - \lambda_j I)^{k'_j} u,$$

for all $u \in \mathbb{R}^{\ell_j}$. But since the columns of P_j are linearly independent, this forces $(B_j - \lambda_j I)^{k'_j} u = 0$ for all $u \in \mathbb{R}^{\ell_j}$. Hence the characteristic polynomial for B_j has the claimed form.

Similarly let $P_j = [w_1^j, \ldots, w_{n_j}^j]$ be the $n \times n_j$ matrix formed from the chosen basis for $GE_{a_j \pm b_j i}$. Then since $AP_j = P_j C_j$, and $[(A - a_j I)^2 + b_j^2 I]^{m'_j}$

is identically zero on $GE_{a_j \pm b_j i}$, we get that

$$0 = \left[(A - a_j I)^2 + b_j^2 I\right]^{m'_j} P_j u = P_j \left[(C_j - a_j I)^2 + b_j^2 I\right]^{m'_j} u,$$

for every $u \in \mathbb{R}^{n_j}$. But this gives that $F_j(C_j) = 0$, where $F_j(x) = [(x - a_j)^2 + b_j^2]^{m'_j}$. So the minimal polynomial for C_j divides $F_j(x)$. Thus, the only eigenvalues of C_j are $a_j \pm b_j i$, from which it follows that the dimension of C_j is even, say $n_j = 2t_j$. Then the characteristic polynomial for C_j must be $p_{C_j}(x) = [(x - a_j)^2 + b_j^2]^{t_j}$.

Thus, from equation (C.12) and properties of determinants, we get

$$
\begin{aligned}
p_A(x) &= \det(A - xI) = \det(P^{-1}(A - xI)P) = \det(P^{-1}AP - xI) \\
&= \det(B_1 - xI) \cdots \det(B_r - xI) \det(C_1 - xI) \cdots \det(C_s - xI) \\
&= p_{B_1}(x) \cdots p_{B_r}(x) p_{C_1}(x) \cdots p_{C_s}(x).
\end{aligned}
$$

Comparing exponents involved in the polynomials on each side of this equation gives $k_j = \ell_j$ and $m_j = t_j$ for each j. This completes the proof. $\square$

The theorem shows that the block structure in equation (C.12) results from any choice of bases for the generalized eigenspaces. In fact this structure only requires a decomposition as a direct sum of subspaces that are invariant under A. If we now exploit the fact that the subspaces are generalized eigenspaces and pick the basis for each in a special way, we will get the Jordan canonical form for A. The following discussion shows how these bases are selected.

First consider the real eigenvalues $\lambda \equiv \lambda_j$ and for convenience let $V = GE_\lambda$. Then $N \equiv A - \lambda I$ is nilpotent when restricted to V and has $k' \equiv k'_j$ as its index of nilpotency. This means that $N^{k'} v = 0$ for every $v \in V$ and no lower power $k < k'$ has this property.

Lemma C.1 *Suppose M is an $n \times n$ matrix and $V \subseteq \mathbb{R}^n$ is a subspace that is invariant under M. If M is nilpotent on V, say $M^k v = 0$, for all $v \in V$, and if $a \neq 0$ is any real number, then $aI - M$, when restricted to V is 1-1 and onto. In particular, if $(aI - M)v = 0$, for $v \in V$, then $v = 0$.*

Proof: Let $Q = a^{-1}M$. Then clearly Q is nilpotent on V as well: $Q^k v = 0$, for all $v \in V$. Now let B be the following $n \times n$ matrix

$$B = I + Q + Q^2 + \cdots + Q^{k-1}.$$

Then B commutes with Q (and also M) and

$$BQ = Q + Q^2 + \cdots + Q^{k-1} + Q^k.$$

Thus, $B(I - Q) = B - BQ = I - Q^k$, and so $B(I - Q)v = v$ for all $v \in V$. Consequently,

$$a^{-1}B(aI - M)v = v,$$

for all $v \in V$. This shows that $aI - M$ is 1-1 and, since B and $aI - M$ commute, this equation also shows that $aI - A$ is onto. $\Box$

The lemma is instrumental in choosing the basis we need for $V = GE_\lambda$. Since $N = A - \lambda I$ has index k', the matrix $N^{k'-1}$ is not identically zero on V. So let $v \in V$ be such that $N^{k'-1}v \neq 0$. Then the vectors

$$v, Nv, N^2v, \ldots, N^{k'-1}v,$$

are linearly independent vectors in V. If not, there are constants $a_0, a_1, \ldots a_{k'-1}$, such that

$$0 = a_0v + a_1Nv + \cdots + a_{k'-1}N^{k'-1}v = \left(a_0 + a_1N + \cdots + a_{k'-1}N^{k'-1}\right)v.$$

Let a_p be the first nonzero a_i. Then the above expression reduces to

$$0 = \left(a_pN^p + \cdots + a_{k'-1}N^{k'-1}\right)v = (a_p + M)N^pv,$$

where $M \equiv a_{p+1}N + \cdots + a_{k'-1}N^{k'-p-1}$. Now M is nilpotent on V since we can write it as $M = NB$, where b commutes with N. Thus, by Lemma C.1, $a_p + M$ is 1-1 on V, and so from $(a_p + M)N^pv = 0$, we conclude that $N^pv = 0$. But since $p \leq k' - 1$, we conclude that $N^{k'-1}v = 0$. This contradicts our initial selection of v.

Now we label the linearly independent vectors $v, Nv, \ldots, N^{k'-1}$ in reverse order

$$v_1 = N^{k'-1}v, \ v_2 = N^{k'-2}v, \ldots, v_{k'} = v.$$

This gives us $Nv_1 = 0, Nv_2 = v_1, \ldots, Nv_{k'} = v_{k'-1}$. But since $N = A - \lambda I$, we can write these equations as

$$
\begin{aligned}
Av_1 &= \lambda v_1 \\
Av_2 &= \lambda v_2 + v_1 \\
&\ \vdots \\
Av_{k'} &= \lambda v_{k'} + v_{k'-1}
\end{aligned}
\tag{C.13}
$$

In matrix form these equations are

$$A[v_1, \ldots, v_{k'}] = [v_1, \ldots, v_{k'}] \begin{bmatrix} \lambda & 1 & 0 & 0 & \cdots & 0 \\ 0 & \lambda & 1 & 0 & \cdots & 0 \\ & & & \ddots & & \\ 0 & 0 & \cdots & 0 & \lambda & 1 \\ 0 & 0 & 0 & \cdots & 0 & \lambda \end{bmatrix}. \tag{C.14}$$

Otherwise said, $AP_1 = P_1 J_{k'}(\lambda)$, where $P_1 = [v_1, \ldots, v_{k'}]$ is the $n \times k'$ matrix formed from the v_j's as columns. Thus, we are done with the construction of the desired basis if the span: $V_1 \equiv \text{span}\{v_1, \ldots, v_{k'}\}$, is all of GE_λ. Note that equations (C.13) give an algorithm that can be used to successively compute the elements in the basis. Namely, first compute an eigenvector v_1 coresponding to λ. With this known, then compute a solution v_2 of the equation $Av_2 = \lambda v_2 + v_1$. Continue in this fashion, finally determining a solution $v_{k'}$ of the equation $Av_{k'} = \lambda v_{k'} + v_{k'-1}$.

If $V_1 \neq GE_\lambda$, then one can show that $GE_\lambda = V_1 \oplus W$ for some subspace W that is invariant under N (Cf. [Her 75, Lemma 6.5.4, p. 295]). Let k'' be the index of nilpotency of N when restricted to W (so that necessarily $k'' \leq k'$). Then as above, there are linearly independent vectors $w_1, \ldots, w_{k''}$ in W, such that $Nw_1 = 0$ and $Nw_j = w_{j-1}$, for $j = 2, \ldots, k''$. Letting P_2 be the $n \times k''$ matrix formed with these vectors as colunms, we get

$$A[P_1, P_2] = [P_1, P_2] \begin{bmatrix} J_{k'}(\lambda) & 0 \\ 0 & J_{k''}(\lambda) \end{bmatrix}.$$

Let $V_2 = \text{sp}\{w_1, \ldots, w_{k''}\}$. If $V_2 = W$ then we are done, otherwise we can continue the process. Since GE_λ is finite dimensional, this process must terminate after finitely many steps and we end up with sequences $\{v_1, \ldots, v_{k'}\}, \{w_1, \ldots, w_{k''}\}, \ldots, \{u_1, \ldots, u_{k^{(p)}}\}$, of cyclic vectors for N that span subspaces $V_1, V_2, \ldots, V_p$, with dimensions $k' \geq k'' \geq \cdots \geq k^{(p)}$ and such that $GE_\lambda = V_1 \oplus \cdots \oplus V_p$. Then $k' + k'' + \cdots + k^{(p)} = \dim(GE_\lambda) = k$, and we will have $AP = P \text{diag}(J_{k'}(\lambda), J_{k''}(\lambda), \ldots, J_{k^{(p)}}(\lambda))$, where P is the $n \times k$ matrix formed from all the cyclic vectors.

Now consider one of the subspaces $G_{a_j \pm b_j i}$ of $\mathbb{R}^n$ corresponding to a pair of complex conjugate eigenvalues of A. To select the desired basis for $G_{a_j \pm b_j i}$, it will be convenient to make an excursion into the complex domain.

We can consider A as a complex matrix, $A : \mathbb{C}^n \to \mathbb{C}^n$, and note that the Primary Decomposition Theorem applies to complex matrices. The proof is the same. However, the key assumption in the proof is that the minimal polynomial is factored completely into irreducible factors. But over $\mathbb{C}$, the quadratic factors split into linear factors

$$\left[(x - a_j)^2 + b_j^2\right]^{m_j'} = (x - \mu_j)^{m_j'}(x - \overline{\mu}_j)^{m_j'},$$

where $\mu_j = a_j + b_j i$. If we let $GE_\mu \equiv \{\, v \in \mathbb{C}^n \,|\, (A - \mu)^k v = 0,\ \text{for some } k\,\}$, then the primary decomposition of $\mathbb{C}^n$ is

$$\mathbb{C}^n = GE_{\lambda_1} \oplus \cdots \oplus GE_{\lambda_r} \oplus GE_{\mu_1} \oplus GE_{\overline{\mu}_1} \oplus \cdots \oplus GE_{\mu_s} \oplus GE_{\overline{\mu}_s}.$$

Now consider one of the complex eigenvalues $\mu = a + bi \equiv a_j + b_j i$, with $m' \equiv m_j'$ being the multiplicity of this eigenvalue as a root of the minimal polynomial and $m = m_j$ its multiplicity as a root of the characteristic polynomial. Then by the Primary Decomposition Theorem, the dimension of GE_μ over the complex numbers is m and $A - \mu I$ is nilpotent of index m' on GE_μ. Exactly as in the discussion for the real case above, we get vectors $v_1, \ldots, v_{m'}$ in GE_μ, which are linearly independent over $\mathbb{C}$ and which satisfy

$$
\begin{aligned}
Av_1 &= \mu v_1 \\
Av_2 &= \mu v_2 + v_1 \\
&\ \ \vdots \\
Av_{m'} &= \mu v_{m'} + v_{m'-1}.
\end{aligned}
$$

If we split each of the complex vectors v_j into its real and imaginary parts $v_j = u_j + i w_j$, then it is not hard to see that $u_1, w_1, \ldots, u_{m'}, w_{m'}$ are $2m'$, linearly independent vectors in $\mathbb{R}^n$ and the above equations can be written as

$$
\begin{aligned}
A[u_1, w_1] &= [u_1, w_1]C \\
A[u_2, w_2] &= [u_2, w_2]C + [u_1, w_1] \\
&\ \ \vdots \\
A[u_{m'}, w_{m'}] &= [u_{m'}, w_{m'}]C + [u_{m'-1}, w_{m'-1}],
\end{aligned}
$$

where $[u_j, w_j]$ denotes the $n \times 2$ matrix with u_j, w_j as columns and

$$C = \begin{bmatrix} a & b \\ -b & a \end{bmatrix}.$$

Combining all these into one matrix equation gives

$$A[u_1, w_1, \ldots, u_{m'}, w_{m'}] = [u_1, w_1, \ldots, u_{m'}, w_{m'}] \begin{bmatrix} C & I & 0 & 0 & \cdots & 0 \\ 0 & C & I & 0 & \cdots & 0 \\ & & & \ddots & & \\ 0 & 0 & \cdots & 0 & C & I \\ 0 & 0 & 0 & \cdots & 0 & C \end{bmatrix}.$$

Now each $v_j \in GE_\mu$ and so $(A - \mu I)^k v_j = 0$, for some k. But then

$$0 = (A - \overline{\mu} I)^k (A - \mu I)^k v_j = \left[(A - aI)^2 + b^2 I \right]^k v_j.$$

This shows that $u_j, w_j \in G_{a \pm bi}$ for each j.

Continuing in this fashion, working in the complex domain and then transferring the result to the real domain we get sequences

$$\{v_1^1, \ldots, v_{m'}^1\}, \{v_1^2, \ldots, v_{m''}^2\}, \ldots, \{v_1^p, \ldots, v_{m^{(p)}}^p\},$$

of complex, cyclic vectors for $A - \mu I$, which span subspaces $V_1, V_2, \ldots, V_p$ of $\mathbb{C}^n$, with dimensions $m' \geq m'' \geq \cdots \geq m^{(p)}$ and such that $GE_\mu = V_1 \oplus \cdots \oplus V_p$. Then $m' + m'' + \cdots + m^{(p)} = \dim_{\mathbb{C}}(GE_\lambda) = m$. Decomposing into real and imaginary parts: $v_j^k = u_j^k + i w_j^k$, we get $2m$ linearly independent, real vectors $\{u_j^k, w_j^k\}$ in $GE_{a \pm bi}$, which therefore must be a basis for this subspace. Using these vectors to form the columns of a $n \times 2m$ matrix P gives

$$AP = P \operatorname{diag}(C_{2m'}(a, b), C_{2m''}(a, b), \ldots, C_{m^{(p)}}(a, b)).$$

This is part of the Jordan canonical form that corresponds to $GE_{a \pm bi}$. All together the discussion shows how to choose bases for each of the generalized eigenspaces so that if P denotes the $n \times n$ matrix formed from these bases, appropriately ordered, then $AP = PJ$, where J is the Jordan canonical form for A.

C.6 Matrix Analysis

We present here some background material on matrix analysis that is used to define the matrix exponential e^A in terms of a series. This analysis is also useful in other respects and is a special case of the analysis on Banach spaces and Banach algebras (or more generally complete, topologiacal vector spaces and topologiical algebras).

We let $\mathcal{M}_n$ denote the collection of all real $n \times n$ matrices. Recall that $\mathcal{M}_n$ is a vector space under the usual addition and scalar multiplication operations on matrices (It's also an algebra as well, using matrix multiplication). In order to apply the familar notions of limits, sequences, series, etc. to matrices we need to have a norm defined on $\mathcal{M}_n$. There are many matrix norms to choose from, but one that is convenient for us here is the following.

Definition C.8 For $A \in \mathcal{M}_n$, the *norm* $\|A\|$ of A is defined to be the largest of all the absolute values of the entries of A. That is, if $A = \{a_{ij}\}$ then:

$$\|A\| = \max\{\,|a_{ij}| \mid i, j \in \{1, \cdots, n\}\,\} \tag{C.15}$$

It's easy to verify, directly from the definition, that this function $\|\cdot\| : \mathcal{M}_n \to \mathbb{R}$, has the following properties:

(1) $\|A\| \geq 0$ and $\|A\| = 0$ iff $A = 0$

(2) $\|\lambda A\| = |\lambda|\|A\|$.

(3) $\|A + B\| \leq \|A\| + \|B\|$.

(4) $\|AB\| \leq n\|A\|\|B\|$.

The first three properties (1)-(3) are the general axioms for what is required for any abstract function $\|\cdot\|$ to be an norm on a vector space. Property (3) is known as the *triangle inequality* for a norm.

Using a matrix norm like this allows us to extend all the results on sequences and series of real numbers to sequences and series of matrices. Furthermore, we will see that these results *look* entirely like their real number counterparts.

Suppose $\{A_k\}_{k=0}^{\infty} = \{A_0, A_1, A_2, \cdots\}$ is a sequence of matrices. In terms of the entries, the kth matrix of this sequence looks like:

$$A_k = \{a_{ij}^k\} = \begin{bmatrix} a_{11}^k & a_{12}^k & \cdots & a_{1n}^k \\ a_{21}^k & a_{22}^k & \cdots & a_{2n}^k \\ \vdots & \vdots & & \vdots \\ a_{n1}^k & a_{n2}^k & \cdots & a_{nn}^k \end{bmatrix} \tag{C.16}$$

The norm $\|\cdot\|$ gives us a way of measuring distances between the elements in our space $\mathcal{M}_n$ of matrices, and this in turn leads to the natural notion of the limit of a sequence of matrices:

Definition C.9

(1) The *distance* between two matrices $A, B \in \mathcal{M}_n$ is $\|A - B\|$.

(2) A sequence $\{A_k\}_{k=0}^{\infty}$ of matrices, in $\mathcal{M}_n$, is said to *converge* to the matrix B (or have limit $= B$), if $\forall \epsilon > 0, \exists$ an N such that:

$$\|A_k - B\| < \epsilon, \qquad\qquad (\text{C.17})$$

for every $k \geq N$. When this is the case, we use the customary notation:

$$\lim_{k \to \infty} A_k = B.$$

(3) A sequence $\{A_k\}_{k=0}^{\infty}$ of matrices is called a *Cauchy sequence* if $\forall \epsilon > 0, \exists$ an N such that:

$$\|A_k - A_p\| < \epsilon, \qquad\qquad (\text{C.18})$$

for every $k, p \geq N$.

The above definition gives us the abstract notion of when two matrices A and B are close together, namely when the distance between them $\|A - B\|$ is small. The notion of a Cauchy sequence of matrices is the same as for sequences of real numbers: it is a sequence whose terms are as close together as we wish if we go far enough out in the sequence. The following proposition is easy to prove from the definitions, and is left as an exercise.

Proposition C.4 *Suppose $\{A_k\}_{k=0}^{\infty}$ is a sequence of matrices. Then*

(1) *The sequence converges to a matrix B:*

$$\lim_{k \to \infty} A_k = B,$$

if and only if each of the sequences of entries converge to the corresponding entries of B, i.e., $\forall i, j$:

$$\lim_{k \to \infty} a_{ij}^k = b_{ij}.$$

(2) *The sequence of matrices is Cauchy if and only if each of the sequences of its entries is Cauchy.*

(3) *If the sequence of matrices is Cauchy then the sequence converges.*

Parts (1) and (2) of the proposition say, in essence, that we really do not need a matrix norm to deal with limits of matrices. However, this is true only in theory. In practice we will see that there are numerous occassions, like the ratio and root tests discussed below, where the norm is most useful.

Part (3) is the completeness property: the vector space $\mathcal{M}_n$ is called a complete normed vector space because of this property. Complete spaces are convenient, since it is often easy to check to see if a sequence is a Cauchy sequence. We will only need this concept briefly below in the discussion of series.

Definition C.10 Suppose $\{C_k\}_{k=0}^{\infty}$ is a sequence of $n \times n$ matrices. Then the series of matrices:

$$\sum_{k=0}^{\infty} C_k, \tag{C.19}$$

is said to *converge* to the matrix C, (or *have sum* equal to C), if the corresponding sequence of partial sums converges to C. When this is the case we write:

$$\sum_{k=0}^{\infty} C_k = C.$$

Recall that, just as for series of real numbers, the corresponding sequence of partial sums for (C.19) is: $\{S_k\}_{k=0}^{\infty}$, where:

$$
\begin{aligned}
S_0 &= C_0 \\
S_1 &= C_0 + C_1 \\
S_2 &= C_0 + C_1 + C_2 \\
&\vdots \\
S_k &= C_0 + C_1 + C_2 + \cdots + C_k \\
&\vdots
\end{aligned}
\tag{C.20}
$$

Note that any series $\sum_{k=0}^{\infty} C_k$ of $n \times n$ matrices is comprised of n^2 series of real numbers: $\sum_{k=0}^{\infty} c_{ij}^k$, formed from the entries of matrix series. Furthermore, by the last proposition, it is easy to see that the series of matrices $\sum_{k=0}^{\infty} C_k$ converges to a matrix C if and only if for each i, j, the corresponding series of i-jth entries $\sum_{k=0}^{\infty} c_{ij}^k$ converges to the i-$j$the entry c_{ij} of C.

Symbollically:

$$
\sum_{k=0}^{\infty}
\begin{bmatrix}
c_{11}^k & \cdots & c_{1n}^k \\
c_{21}^k & \cdots & c_{2n}^k \\
\vdots & \vdots & \vdots \\
c_{n1}^k & \cdots & c_{nn}^k
\end{bmatrix}
=
\begin{bmatrix}
\sum_{k=0}^{\infty} c_{11}^k & \cdots & \sum_{k=0}^{\infty} c_{1n}^k \\
\sum_{k=0}^{\infty} c_{21}^k & \cdots & \sum_{k=0}^{\infty} c_{2n}^k \\
\vdots & \vdots & \vdots \\
\sum_{k=0}^{\infty} c_{n1}^k & \cdots & \sum_{k=0}^{\infty} c_{nn}^k
\end{bmatrix}.
\tag{C.21}
$$

Despite the fact, just stated, that covergence of series of matrices is equivalent to the convergence of all the series of i-jth entries, it is often quite impossible to examine the series of entries effectively. In such circumstances it is easier to apply various convergence tests, like the ratio and root tests, directly to the matrix series, using the matrix norm. For example, given a matrix series: $\sum_{k=0}^{\infty} C_k$, the ratio test goes as follows. Suppose

$$
\lim_{k \to \infty} \|C_{k+1}\|/\|C_k\| = \rho,
$$

exists. Then the matrix series converges absolutely if $\rho < 1$, and diverges if $\rho > 1$. The test gives no information if $\rho = 1$. By absolute convergence we mean:

Definition C.11 The matrix series $\sum_{k=0}^{\infty} C_k$ is said to *converge absolutely* if the series: $\sum_{k=0}^{\infty} \|C_k\|$, of real numbers, converges.

The following proposition shows that absolute convergence is a stronger notion than just ordinary convergence.

Proposition C.5 *If the series $\sum_{k=0}^{\infty} C_k$ converges absolutely then it converges.*

Proof: By assumption the series $\sum_{k=0}^{\infty} \|C_k\|$ of real numbers converges. This means the corresponding sequence of partial sums: $\{\sigma_k\}$, with $\sigma_k = \sum_{j=0}^{k} \|C_j\|$ is a convergent sequence, and therefore is also a Cauchy sequence. Thus, given $\epsilon > 0$, we can choose K so that $\forall k > p \geq K$:

$$
|\sigma_k - \sigma_p| < \epsilon.
$$

But then, using the triangle inequality for the norm $\| \cdot \|$, it follows that the sequence of matrix partial sums: $\{S_k\}$, with $S_k = \sum_{j=0}^{k} C_j$, satisfies the following, $\forall k > p \geq K$:

$$\|S_k - S_p\| = \|\sum_{j=p+1}^{k} C_j\| \tag{C.22}$$

$$\leq \sum_{j=p+1}^{k} \|C_j\| \tag{C.23}$$

$$= \sum_{j=0}^{k} \|C_j\| - \sum_{j=0}^{p} \|C_j\| \tag{C.24}$$

$$= |\sigma_k - \sigma_p| < \epsilon \tag{C.25}$$

This says that the sequence $\{S_k\}_{k=0}^{\infty}$ of matrix partial sums is Cauchy, and thus by part (3) of the last proposition, this sequence converges. Hence the series $\sum_{k=0}^{\infty} C_k$ converges. $\square$

Example C.2 An important application of the last proposition deals with the series:

$$\sum_{k=0}^{\infty} \frac{1}{k!} A^k = I + A + \frac{1}{2!} A^2 + \frac{1}{3!} A^3 + \cdots \tag{C.26}$$

where A is an $n \times n$ matrix, and I is the identity matrix. To see that this matrix series converges for any matrix A, we use the ratio test, looking at the ratio:

$$\frac{\|\frac{1}{(k+1)!} A^{k+1}\|}{\|\frac{1}{k!} A^k\|} = \frac{1}{k+1} \frac{\|A A^k\|}{\|A^k\|}$$

$$\leq \frac{1}{k+1} \frac{n\|A\|\|A^k\|}{\|A^k\|} = \frac{n}{k+1} \|A\|. \tag{C.27}$$

Since the last expression on the right-hand side in (C.27) tends to zero as $k \to \infty$, we see that $\rho = 0$ in the ratio test. Hence the series (C.26) converges absolutely, and therefore converges. The matrix that this series converges to is denoted suggestively by:

Definition C.12

$$e^A = \sum_{k=0}^{\infty} \frac{1}{k!} A^k. \tag{C.28}$$

The matrix e^A is called the *matrix exponential*, and $A \to e^A$, actually defines a function with domain and codomain $\mathcal{M}_n$. This function has many properties of the ordinary exponential function, as the following theorem shows.

Theorem C.4 *The matrix exponential defined by the series* (C.28) *has the properties:*

(1) *If A and B commute: $AB = BA$, then $e^{A+B} = e^A e^B$.*

(2) *For any A, the matrix e^A is invertible, and has inverse given by: $(e^A)^{-1} = e^{-A}$.*

(3) *If P is any invertible matrix, then $e^{P^{-1}AP} = P^{-1}e^A P$.*

(4) *If $Av = \lambda v$, then $e^A v = e^\lambda v$. In particular this gives an association between the eigenvalues/vectors of A and e^A.*

Proof: We prove property (1), and leave the others as exercises. We use a result (without proof) about *Cauchy multiplication* of two series: If the series: $\sum_{k=0}^{\infty} C_k$ and $\sum_{k=0}^{\infty} D_k$ are absolutely convergent series, converging to C and D respectively: $\sum_{k=0}^{\infty} C_k = C$ and $\sum_{k=0}^{\infty} D_k = D$, then the following series converges to the product of C and D:

$$\sum_{k=0}^{\infty} \left(\sum_{p=0}^{k} C_p D_{k-p} \right) = CD. \tag{C.29}$$

Intuitively, this result says, under the stated conditions, that the infinite distributive law holds:

$$
\begin{aligned}
CD &= (C_0 + C_1 + C_2 + \cdots)(D_0 + D_1 + D_2 + \cdots) \\
&= \left\{
\begin{aligned}
&C_0 D_0 + C_0 D_1 + C_0 D_2 + \cdots \\
&\quad + C_1 D_0 + C_1 D_1 + \cdots \\
&\quad\quad + C_2 D_0 + \cdots \\
&\qquad\qquad\qquad \vdots
\end{aligned}
\right.
\end{aligned}
\tag{C.30}
$$

If you add up the columns displayed in (C.30) you will arrive at the terms in the parentheses in equation (C.29).

If we now apply this result to e^A and e^B, each of which is represented by an absolutely convergent series (take $C_k = A^k/k!$ and $D_k = B^k/k!$), we get

$$e^A e^B = \sum_{k=0}^{\infty} \left(\sum_{p=0}^{k} \frac{1}{p!(k-p)!} A^p B^{k-p} \right). \tag{C.31}$$

On the other hand, since A and B commute, we can use the binomial theorem on expressions like $(A + B)^k$, to arrive at the following:

$$
\begin{aligned}
e^{A+B} &= \sum_{k=0}^{\infty} \frac{1}{k!}(A + B)^k \\
&= \sum_{k=0}^{\infty} \frac{1}{k!}\left(\sum_{p=0}^{k} \frac{k!}{p!(k-p)!} A^p B^{k-p} \right) \qquad\text{(C.32)} \\
&= e^A e^B. \qquad\text{(C.33)}
\end{aligned}
$$

This completes the proof of property (1). Parts (2),(3), and (4) are left for the exercises. $\square$

C.6.1   Power Series with Matrix Coefficients

The last bit of matrix analysis we will need deals with power series of the form:

$$
\sum_{k=0}^{\infty} (t - t_0)^k C_k = C_0 + (t - t_0)C_1 + (t - t_0)^2 C_2 + (t - t_0)^3 C_3 + \cdots, \quad\text{(C.34)}
$$

where the C_k's are given $n \times n$ matrices. These matrices are the coefficients of the power series, and the way we have written the power series (C.34) may look a little strange, but that's because it is traditional to write the scalars $(t - t_0)^k$ to the left of the matrices C_k.

It turns out that most all the techniques and results you studied in connection with power series with real coefficients carry over to the matrix coefficient case here. For example, given the coefficents: $\{C_k\}_{k=0}^{\infty}$, and the time t_0, one can ask for what values t in (C.34) does the series converge ? Of course the series converges for $t = t_0$, but to determine if it converges for other values of t, one usually applies some version of the root test. Specifically (just as in the real coefficient case), let $0 \leq R \leq \infty$ be the number determined by:

$$
R = \left(\limsup_{k\to\infty} \|C_k\|^{1/k} \right)^{-1}. \qquad\text{(C.35)}
$$

The lim sup in (C.35) is rather technical, but coincides with the ordinary limit when this latter limit exists. The number R (which can be 0 or ∞ depending on the series involved) is called the *radius of convergence* for the power series (C.34). It is often difficult to compute R, sometimes because

(C.35) is difficult to work with. It is a theorem that R can also be computed from:

$$R = \left(\lim_{k \to \infty} \frac{\|C_{k+1}\|}{\|C_k\|} \right)^{-1},$$

provided the limit in the parentheses exists.

The importance of the radius of convergence is that it helps describe the largest set of times t for which the power series converges. The next theorem makes this more specific, and also makes precise the assertion that a power series may be differentiated term by term. Before stating the theorem, we need a definition of what is meant by the derivative of a matrix-valued function.

Definition C.13 Suppose C is a matrix-valued function: $C : I \to \mathcal{M}_n$, defined on an open interval I, given in terms of its entries by:

$$C(t) = \{c_{ij}(t)\} = \begin{bmatrix} c_{11}(t) & \cdots & c_{1n}(t) \\ \vdots & & \vdots \\ c_{n1}(t) & \cdots & c_{nn}(t) \end{bmatrix}. \tag{C.36}$$

Then C is said to be *differentiable on I*, if each of the real-valued functions c_{ij} is differentable on I, and when this is the case, the *derivative* of C at $t \in I$ is defined to be the matrix:

$$C'(t) = \{c'_{ij}(t)\} = \begin{bmatrix} c'_{11}(t) & \cdots & c'_{1n}(t) \\ \vdots & & \vdots \\ c'_{n1}(t) & \cdots & c'_{nn}(t) \end{bmatrix}. \tag{C.37}$$

Theorem C.5 *Let R be the radius of convergence of the matrix coefficent power series: $\sum_{k=0}^{\infty}(t - t_0)^k C_k$. If $R = 0$, then the series converges only for $t = t_0$. On the other hand, if $R > 0$, then for each t in the interval $I = (t_0 - R, t_0 + R)$, the series converges absolutely, and therefore converges to a matrix $C(t)$:*

$$C(t) = \sum_{k=0}^{\infty}(t - t_0)^k C_k, \tag{C.38}$$

thus defining a matrix-valued function $C : I \to \mathcal{M}_n$. The convergence the series (C.38) to C is uniform on compact subsets of I. If t is outside the interval $[t_0 - R, t_0 + R]$, the series $\sum_{k=0}^{\infty}(t - t_0)^k C_k$ diverges. Nothing can be said in general about whether the series converges or not when t is one of the end points of I.

Furthermore, the matrix-valued function C defined by formula (C.38) *is differentiable on I, with derivative given by:*

$$C'(t) = \sum_{k=1}^{\infty} k(t - t_0)^{k-1} C_k. \tag{C.39}$$

The power series in the formula (C.39) *has the same radius of convergence R as the original power series.*

The following proposition shows that the product rule holds for derivatives when considering the following types of products.

Proposition C.6 *Suppose that $A, B : I \to \mathcal{M}_n$ are differentiable matrix-valued functions on I, and that $\gamma \to \mathbb{R}^n$ is a differentiable curve in $\mathbb{R}^n$. Then the matrix-valued function: $(AB)(t) \equiv A(t)B(t)$ and the curve: $(A\gamma)(t) \equiv A(t)\gamma(t)$, are differentiable on I, and the product rule holds:*

$$(AB)'(t) = A'(t)B(t) + A(t)B'(t) \tag{C.40}$$
$$(A\gamma)'(t) = A'(t)\gamma(t) + A(t)\gamma'(t), \tag{C.41}$$

Appendix D

Electronic Contents

The electronic component consists of folders which contain Maple worksheets (*.mws files) that supplement and complement the text material. There is one folder for each release of Maple. For example, the DE-MapleV folder is for users of Maple V Release 5 and the DE-Maple6 is for users of Maple 6. Otherwise the files in each folder are identical. To conveniently use the files, start Maple and open the table of contents file, CDtoc.mws, which has hyperlinks to all the other files.

Electronic Component Table of Contents

Bibliography

[AM 78]   R. Abraham and J. Marsden, *Foundations of Mechanics*, 2nd Ed., Addison-Wesley, Reading, MA, 1978.

[AvM 88]   M. Adler and P. Van Moerbecke, *Algebraic Completely Integrable Systems: A Systematic Approach*, Academic Press, New York, 1988.

[ABK 92]   B. Aebischer, M. Borer, M. Kälin, Ch. Leuenberger, and H. Reimann, *Symplectic Geometry*, Birkhäuser, Boston, 1992.

[Am 90]   H. Amann, *Ordinary Differential Equations: An Introduction to Nonlinear Analysis*, Walter de Gruyter, Berlin, 1990.

[Ar 78a]   V. I. Arnold, *Ordinary Differential Equations*, MIT Press, Boston, 1978.

[Ar 78b]   V. I. Arnold, *Mathematical Methods of Classical Mechanics*, Springer-Verlag, New York, 1978.

[BBT 03]   Olivier Babelon, Denis Bernard, and Michel Talon, *Introduction to Classical Integrable Systems*, Cambridge Univ. Press, Cambridge, 2003.

[Bel 53]   R. Bellman, *Stability Theory of Differential Equations*, McGraw-Hill, New York, 1953.

[Be 98]   D. Betounes, *Partial Differential Equations for Computational Science with Maple and Vector Analysis*, Springer-Verlag, New York, 1998.

[BCG 91]   R. L. Bryant, S. S. Chern, R. B. Gardner, H. L. Goldschmidt, and P. A. Griffiths, *Exterior Differential Systems*, Springer-Verlag, New York, 1991.

[Bu 85]   T. A. Burton, *Stability and Periodic Solutions of Ordinary and Functional Differential Equations*, Academic Press, New York, 1985.

[Ca 77]   J. L. Casti, *Dynamical Systems and Their Applications: Linear Theory*, Academic Press, New York, 1977.

[CE 83]   P. Collet and J-P. Eckmann, *Iterated Maps on the Interval as Dynamical Systems*, Birkhäuser, Boston, 1983.

[Co 65]   W. A. Coppel. *Stability Theory of Differential Equations*, D. C. Heath, Boston, 1965.

[Cr 94]   Jane Cronin, *Differential Equations: Introduction and Qualitative Theory*, 2nd Ed., Dekker, New York, 1994.

[Cr 08]   Jane Cronin, *Differential Equations: Introduction and Qualitative Theory*, 3nd Ed., Chapman & Hall, Boca Raton, 2008.

[Di 74]   J. Dieudonné, *Treatise on Analysis*, Vol. IV, Academic Press, New York, 1974.

[Dev 86]   R. L. Devaney, *An Introduction to Chaotic Dynamical Systems*, Benjamin-Cummings , Menlo Park, 1986.

[Du 70]   James Dugundji, *Topology*, Allyn and Bacon, Boston, 1970.

[Fi 86]   M. Fiedler, *Special Matrices and their Applications in Numerical Mathematics*, Kluwer, Dordrecht, 1986.

[Go 59]   H. Goldstein, *Classical Mechanics*, Addison-Wesley, Reading, 1959.

[GS 85]   M. Golubitsky and D. G. Schaeffer, *Singularities and Groups in Bifurcation Theory*, Springer-Verlag, New York, 1985.

[Ha 82]   P. Hartman, *Ordinary Differential Equations*, Birkhäuser, Boston, 1982.

[Her 75]   I. N. Herstein, *Topics in Algebra*, Wiley, 2nd Ed., New York, 1975.

[HZ 94]   Helmut Hofer and Eduard Zehnder, *Symplectic Invariants and Hamiltonian Dynamics*, Birkhäuser, Boston, 1994.

[HS 99]   Po-Fang Hsieh and Yasutaka Sibuya, *Basic Theory of Ordinary Differential Equations*, Springer-Verlag, New York, 1999.

[Irw 80]   M. C. Irwin, *Smooth Dynamical Systems*, Academic Press, New York, 1980.

[Jo 93]   Frank Jones, *Lebesgue Integration on Euclidean Space*, Jones and Bartlett, Boston, 1993.

[Ko 89]   H. Koçak, *Differential and Difference Equations through Computer Experiments*, 2nd Ed., Springer-Verlag, New York, 1989.

[KGT 97]   Y. Kosmann-Schwarzbach, B. Grammaticos, and K. Tamizhmani, Eds., *Integrability of Nonlinear Systems*, Springer-Verlag, New York, 1997.

[KLP 94]   S. Kuksin, V. Lazutkin, and J. Pöschel, Eds., *Seminar on Dynamical Systems*, Birkhäuser, Boston, 1994.

[Ku 95]   Y. A. Kuznetsnov, *Elements of Applied Bifurcation Theory*, Springer-Verlag, New York, 1995.

[Laz 93]   V. Lazutkin, *KAM Theory and Semiclassical Approximations to Eigenfunctions*, Springer-Verlag, New York, 1993.

[LM 87]   P. Libermann and C. Marle, *Symplectic Geometry and Analytical Mechanics*, D. Reidel, Boston, 1987.

[MT 95]   J. Marion and S. Thorton, *Classical Dynamics of Particles and Systems*, 4th Ed., Harcourt Brace, Orlando, 1995.

[Ma 04]   Nelson Markley, *Principles of Differential Equations*, John Wiley & Sons, New Jersey, 2004.

[MH 93]   J. E. Marsden and M. J. Hoffman, *Elementary Classical Analysis*, 2nd Ed., W. H. Freeman, New York, 1993.

[Mar 92]   Mario Martelli, *Discrete Dynamical Systems and Chaos*, Longman House, Essex, 1992.

[Mer 97]   D. R. Merkin, *Introduction to the Theory of Stability*, Springer-Verlag, New York, 1997.

[Mo 73] J. Moser, *Stable and Random Motion in Dynamical Systems*, Princeton University Press, Princeton, 1973.

[PM 82] J. Palis and W. de Melo, *Geometric Theory of Dynamical Systems: An Introduction*, Springer-Verlag, New York, 1982.

[Par 80] B. N. Parlett, *The Symmetric Eigenvalue Problem*, Prentice-Hall, Englewood Cliffs, NJ, 1980.

[Pa 81] William Parry, *Topics in Ergodic Theory*, Cambridge University Press, Cambridge, 1981.

[Per 91] L. Perko, *Differential Equations and Dynamical Systems*, Springer-Verlag, New York, 1991.

[Per 03] L. Perko, *Differential Equations and Dynamical Systems*, 3rd Ed., Springer-Verlag, New York, 2001.

[Pe 83] K. E. Petersen, *Ergodic Theory*, Cambridge University Press, Cambridge, 1983.

[Po 90] H. Poincaré, Sur le probléme des trois corps et les équations de la dynamique, *Acta Math.* **13** (1890) p. 1-271.

[Po 57] H. Poincaré, *Les Méthodes Nouvelles de la Méchanique Céleste*, Vol. I, II, III, Dover Publications, New York, 1957 (unabridged republication of the volumes originally published in 1892, 1893, and 1899).

[Pon 62] L. S. Pontryagin, *Ordinary Differential Equations*, Addison-Wesley, Reading, MA, 1962.

[Rob 95] Clark Robinson, *Dynamical Systems: Stability, Symbolic Dynamics, and Chaos*, CRC Press, Boca Raton, 1995.

[RM 80] N. Rouche and J. Mawhin, *Ordinary Differential Equations: Stability and Periodic Solutions*, Pitman, Boston, 1980.

[Ru 76] Walter Rudin, *Principles of Mathematical Analysis*, 3rd Ed., McGraw-Hill, New York, 1976.

[Sl 70] W. Slebodzinski, *Exterior Forms and their Applications*, Polish Scientific Publishers, Warszawa, 1970.

[Th 78]   W. Thirring, *A Course in Mathematical Physics 1: Classical Dynamical Systems*, Springer-Verlag, New York, 1978.

[Ver 90]   F. Verhulst, *Nonlinear Differential Equations and Dynamical Systems*, Springer-Verlag, New York, 1990.

[Wa 98]   Wolfgang Walter, *Ordinary Differential Equations*, Springer-Verlag, New York, 1998.

[Wh 65]   E. T. Whittaker, *A Treatise on the Analytical Dynamics of Particles and Rigid Bodies*, Cambridge University Press, Cambridge, 1965.

[W 47]   A. Wintner, *The Analytical Foundations of Celestial Mechanics*, Princeton, Princeton University Press, 1947.

Index